Environmental Assessment of Products

Environmental Assessment of Products

Volume 1
Methodology, tools and case studies in product development

Henrik Wenzel
Michael Hauschild
and
Leo Alting

Methodology and results from the EDIP programme (Environmental Design of Industrial Products). A cooperation programme between the Institute for Product Development, Technical University of Denmark, five Danish industrial companies, the confederation of Danish Industries and the Danish Environmental Protection Agency.

CHAPMAN & HALL
London · Weinheim · New York · Tokyo · Melbourne · Madras

Published by Chapman & Hall, 2–6 Boundary Row, London SE1 8HN, UK

Chapman & Hall, 2–6 Boundary Row, London SE1 8HN, UK

Chapman & Hall GmbH, Pappelallee 3, 69469 Weinheim, Germany

Chapman & Hall USA, 115 Fifth Avenue, New York, NY 10003, USA

Chapman & Hall Japan, ITP-Japan, Kyowa Building, 3F, 2-2-1 Hirakawacho, Chiyoda-ku, Tokyo 102, Japan

Chapman & Hall Australia, 102 Dodds Street, South Melbourne, Victoria 3205, Australia

Chapman & Hall India, R. Seshadri, 32 Second Main Road, CIT East, Madras 600 035, India

First edition 1997

Layout and figures: Ipsen & Parmo, Copenhagen, Denmark
Vignettes: Flemming Vestergaard, Copenhagen, Denmark
Translation: Ian Bock, Language & Data, Ringkøbing, Denmark
Printed in Great Britain by the University Press, Cambridge

ISBN 0 412 80800 5

A catalogue record for this book is available from the British Library

∞ Printed on permanent acid-free text paper, manufactured in accordance with ANSI/NISO Z39.48-1992 and ANSI/NISO Z39.48-1984 (Permanence of Paper)

Contents

Preface

The aim of this book is to support industry in their effort to design environmentally friendly products. The book comprises a method and a manual for life cycle assessment of products and it includes examples of how industrial companies have used the method succesfully in the design of more environmentally friendly products.

The method has been developed over a period of four years under the Danish EDIP programme (**E**nvironmental **D**esign of **I**ndustrial **P**roducts) by a team representing the Technical University of Denmark, five Danish industrial companies, the Confederation of Danish Industries and the Danish Environmental Protection Agency. The method is coheherent and operational and it is well documented by a large variety of examples including five different complex electromechanical products. It guides the user through the inventory and assessment of environmental impacts of products and shows how various products and design solutions during product development can be compared. The method is supported by a base of data for the assessments of environmental impacts and is thus designed as a tool which will make it possible for the user to start on life cycle assessment at once. The book also guides the user through the identification of environmental improvement potentials in the product and the setting of environmental specifications within the general concept of overall commercial optimization.

The partnership between industry, authorities and university has been highly fruitful. Throughout the project the companies have tested the method and incorporated it into their product development, and industry's requirements for the method have thus been incorporated from its inception. International developments in the field of life cycle assessment have been followed closely and have also been incorporated in the methodology. The method has been reviewed by external experts within the field of life cycle assessment of products and by about 30 experts within each of the types of environmental effects entering into the method's impact assessment phase.

Review comments during the early stage of the method's development were offered by Tomas Ekvall (Chalmers Industriteknik, Sweden), Kim Christiansen (Krüger Consult A/S, Denmark) and Anders Schmidt (dk-Teknik, Energy and Environment, Denmark), and later by Tomas Ekvall, Sven-Olof Ryding (Confederation of Swedish Industries), Gitte Pedersen (Novo Nordisk A/S, Denmark) and Bo Pedersen Weidema (Institute for Product Development, Denmark). We thank the reviewers for their work and their comments, which have been of great assistance in our work with the method.

Our special thanks are due to our secretary Christine Molin for language revision, proofreading and indefatigable and devoted assistance during the entire project.

Most of the work was financed as part of the EDIP programme by the Danish Ministry of the Environment and Energy and by the participating industrial companies and, for the section on the working environment, also by the Danish Working Environment Foundation.

To work in such depth over four years on development of the method has been a privilege, and we are truly grateful for having been given this opportunity. We hope that the method will gain wide acceptance, and that it will be able to help cover the demand found in industry, the authorities and consumer organizations for uniform criteria for the environmental assessment of products.

Henrik Wenzel · Michael Hauschild · Leo Alting
Institute for Product Development, Technical University of Denmark

Introduction

Interest in the environmental qualities of products has been increasing over the last 10 years. From an earlier focus on only a few product types such as foodstuffs and packaging, interest is now being expressed in environmentally friendly products within almost all product types.

The authorities are working with environmental labelling regulations and plans for green purchasing in the public sector, and in several European countries the environmental authorities are working on product-oriented policy initiatives.

Companies are working on environmental management and cleaner technology, and many companies have begun to set requirements on their subcontractors in order to ensure sound environmental qualities in products throughout the life cycle. Environmental considerations are thus gaining importance throughout the chain of supply and demand.

Consumers have become more conscious of environmental problems in industrial products and are making increasing demands on products' environmental qualities.

But with the growing interest in products' environmental qualities, new questions have also arisen. What are the environmental impacts of a product? How are they measured? How can different types of environmental impacts be compared? How can different products be compared?

Answering these questions is by no means an unambiguous matter, and the answers which have been given have therefore often been different. Previous experience has shown that, depending on the method used, different conclusions can be arrived at when products' environmental qualities are compared. And a significant need has arisen for development and standardization of assessment methods. Development of the EDIP method presented in this book is one step in the attempts to achieve more standardized and well-founded methods.

Choices of the EDIP method

"Environmental Assessment of Industrial Products" is designed as a handbook: the life cycle assessment method is made operational through the recommendation of concrete choices and actions in all stages of the method and through the provision of factors for the requisite calculations. There are still many elements in a life cycle assessment for which several options are available, and where no unambiguous answer can be found. The choices recommended under the EDIP method are therefore not objective; in many cases, other choices are arguable and this will be of significance for the result of the life cycle assessment.

The significant choices in a life cycle assessment method must be evident, i.e. criteria of selection and reasons for the choices must be known and transparent. The EDIP method therefore provides concrete recommendations in all situations, partly to highlight the choices and partly to ensure a common methodological basis when different products and solutions are compared. The choices made via the EDIP method are accentuated in the review of the method, and alternatives are shown so that the user of the method can assess the choices and, if necessary, make a different decision.

The book's target group, the environmental specialist

The book is directed primarily at the environmental specialist who must carry out environmental assessment of products, and it shows the interplay with the product designer in the work of developing more environmentally friendly products.

The environmental specialist can be employed in the company as environmental manager or environmental employee. The specialist may also be an external consultant on which the company draws when concrete tasks have to be carried out.

The user of the life cycle assessment method described in this book must have a technical and chemical background knowledge corresponding to that of an engineering education or a relevant university education. He or she should be well oriented within the environmental field, but does not require special experience in environmental assessment of products or with product development. Certain parts of the life cycle assessment method require an environmental-chemical and ecotoxicological expertise in the user, but these parts can be carried out by external consultants.

The book provides a thorough introduction to the subject and is therefore also suitable for use in the teaching of students in environmental assessment of products in engineering colleges and the technological and natural science faculties of universities.

The book is not directed in the first instance to the designer. Nonetheless, Part III of the book, describing the environmental tasks in product development, and the collection of case histories in Part V will also be of interest to the designers responsible of developing new products.

In the experience of the authors, carrying out a product life cycle assessment requires a greater knowledge of the environment than that generally required from the product developer or designer. When environmental considerations are to be built into new products, collaboration is therefore required between the designer and the environmental specialist. The task of the environmental specialist is to assess the product's environmental characteristics. The task of the designer is to build the environmental requirements into the product.

The EDIP programme has focused on the use of life cycle assessment within product development. The life cycle assessment method is therefore suitable for product development, but as will be clear from the review, it is generally useful for environmental assessment of products and it can also be used in applications within areas other than product development.

Structure of the book

The book consists of five Parts.

Part I provides an introduction to environmental assessment of products under the following headings:

- A cross-disciplinary review of the status of the environment with focus on the various environmental effects and current efforts to combat them.
- A discussion of how society is regulating the impacts on the environment and what developments can be expected in the area.
- An introduction to environmental assessment of products and a brief historical review and international status report.
- An introduction to the most basic requirements placed on the life cycle assessment method by product development.

The object of this introduction to the subject is: firstly to show what environmental assessment of products is, and how it enters into society's strategy for the environmental area; and then to introduce environmental assessment into product development as a prelude to Part II.

The method is reviewed in Part II. Emphasis is placed on describing the method's various elements and on discussing the choices made in determination of the foundation of the method. The review of the method is supplemented by numerous examples which aid understanding of the procedure and the goal.

Part III of the book describes the life cycle assessment method's use in product development and reviews the types of questions which the environmental specialist can expect from the designer under the various parts of the product development procedure. The review is structured in accordance with the five significant environmental tasks in product development: environmental assessment of reference products, environmental diagnosis, environmental specification, environmental assessment of product concepts, and environmental assessment of detailed solutions.

Part IV of the book is the Toolbox, where concrete guidelines for action are given in a detailed review of the life cycle assessment method, step by step. Part IV is structured as a handbook of environmental assessment of products with a chapter for each of the life cycle assessment method's phases.

Part V presents the life cycle assessment method applied in practice to five different electromechanical products. The examples are for both the environmental specialist and the designer, and illustrate a paradigm for cooperation and for how environmental tasks can be solved during product development.

Finally, the book provides a glossary with an explanation of the units and variables that are introduced, and the abbreviations that are used in the book.

This book for the environmental specialist is supplemented by the book "Scientific Background for Environmental Assessment of Industrial Products" (Hauschild and Wenzel, 1997g), which documents the method's principles of assessment and which is therefore necessary for an understanding of the background of the method.

Guide to reading

To some degree the five parts of the book can be read independently of one another, but the introduction to the life cycle assessment method in Part II should be read before the Toolbox in Part IV.

The reader who possesses a solid and broad background knowledge within the area of the environment can happily skim Part I and then proceed to the subsequent parts. But for most environmental specialists this first part of the book will offer new knowledge, and its thorough introduction to the basis of life cycle assessment of products will serve future work well.

The description of the life cycle assessment method and its phases in Part II should be read by all with responsibility for carrying out an environmental assessment of a product. This section reviews the considerations behind the actual design of the method, and the choices made en route are presented here.

When Part II has been read, the Toolbox in Part IV of the book offers the concrete procedure for life cycle assessment. In future use of the method it is the Toolbox in particular which functions as handbook and as a work of reference with concrete factors for calculation and examples of forms for use in the collection and inventory of data.

Part III illustrates the tasks of the environmental specialist in collaboration with the designer to build environmental considerations into new products. This section presents the types of questions which the environmental specialist should anticipate from the designer, and a review is provided of how the method is to be used to generate the requisite knowledge.

Part V gives case histories for five different products illustrating the procedure of life cycle assessment and the types of results obtained when using the EDIP method. All case histories are focused on the implementation of the life cycle assessment at the various stages of the product development procedure and amply illustrate how environmental considerations can be built into the new generations of products.

Part I. Introduction to Environmental Assessment of Products

- *What is the condition of the environment?*
- *What are companies and the authorities doing about it?*
- *What is environmental assessment of products?*
- *How does life cycle assessment relate to product development?*

1. The state of the environment

Various international status reports on the condition of the environment have been prepared over the last few decades. One report which, more than any of its predecessors, has come to shape the discussion of our civilization's effect on our common environmental basis was prepared by a commission established by the UN in 1983. The object was to prepare a status report for the relationship between environment and development in the world today, and on this background to propose a global programme for changes. The commission was named the Brundtland Commission after its chairman Gro Harlem Brundtland, who subsequently became prime minister of Norway.

The Brundtland Commission's report was given the title "Our Common Future", and it described scarcity of resources, population growth, environmental impacts and unequal distribution of economic welfare and growth as interacting threats to our common future on earth. "Sustainable development" was introduced as the only possible and acceptable development if human civilization is to avoid collapse in the near future. A sustainable development was defined as "... development that meets the needs of the present without compromising the ability of future generations to meet their own needs" (p. 43 in the World Commission on Environment and Development, 1987). With this definition of the concept of sustainability the focus was on our environment's ability to fulfil our needs and those of future generations. Included were:

- The material resources, non-renewable and renewable.
- Biological diversity, genetic resources.
- The health of the environment in which our own and future generations must live.

The concept of sustainable development can also include environmental qualities which fulfil other, more aesthetic needs, for example experiencing a varied natural and undisturbed environment.

In contrast to the earlier environmental status reports of similar nature, the Brundtland Commission's report gained major significance for the international political discussion on the human impact on the environment. The report led in 1992 to a UN-sponsored World Conference on Environment and Development (UNCED) in Rio de Janeiro.

The Rio Conference enjoyed the participation of heads of governments from 118 countries and resulted in two international treaties - one on climatic changes and one on biodiversity. A programme for international cooperation on integration of development processes and environmental considerations, which is to ensure a responsible development for the earth into the 21st century, also resulted (Agenda 21; e.g. United Nations, 1993).

In recent years attempts have been made to define the concept of sustainable development in the form of "the environmental space", which is an attempt to quantify the resource consumption and the environmental impact which can be accommodated for the individual person without exceeding the level of sustainable development (Spangenberg, 1995).

Developments in human impact on the environment

In previous centuries human influence on the environment was of local extent, concentrated especially around urban communities, but today it is often major regions, or perhaps the entire earth, which is affected. While previous environmental impacts were of a predominantly physical character, a significant portion of today's environmental impact can be blamed on emissions of chemicals toxic to the environment.

The reasons underlying this development are:

- Constantly growing human activity.
  Over the last 40 years the industrialized countries have experienced an increase in their material standard of living which is without precedent in world history. Continued economic growth is expected for the industrialized countries, and growth in the developing countries will have to be even greater if the difference from the industrialized countries is to be reduced. At the same time, world population has grown exponentially. By mid-1995 the world's population was 5.7 billion. With an annual growth of 90-100 million, the population will double in the next 40-50 years. Population growth is stagnating in the industrialized countries but it is high in developing countries, and according to a 1992 UN prediction, world population in the year 2050 will be at least 10-11 billion (United Nations, 1992). As a result the total global production is increasing, as are the environmental impacts which the production is generating.

- Use of many new chemicals foreign to the environment.
  In the last half of the 20th century there has been an explosive growth in the number of chemicals in general use. Production has become industrialized to an increasing extent, and based on new chemicals which in most cases are foreign to the environment. Within the course of a few decades, the former chemical universe based on natural substances has been replaced by a much more extensive chemical universe based principally on petrochemistry (crude oil products). When the chemicals do not occur naturally, they can have unexpected and often unforeseeable effects on the environment, which has not been exposed to them before.

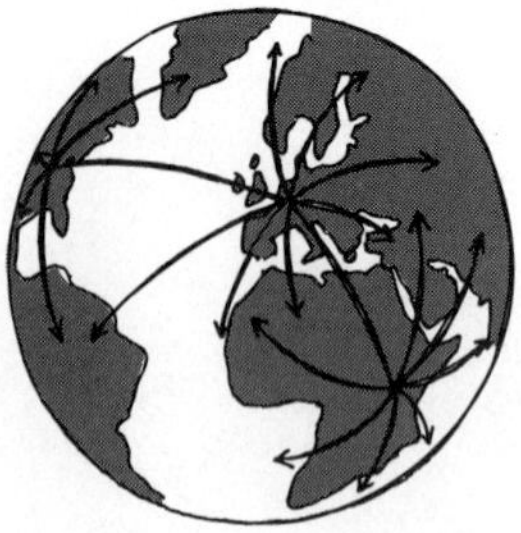

- Use of increasingly larger parts of the earth.
  Human activities are being extended to include areas which were previously pristine, and the environment over most of the earth is therefore affected. Deeper layers of the earth's surface are also being involved in human activities to an increasing extent, for example in exploration for oil, coal, gas and metal resources.

The following sections review the effects of man's current impacts on the environment. The review takes its point of departure in the damage which can be observed today or is expected in the future as a consequence of these impacts.

Firstly, a brief review of current knowledge on the effect in question is given, followed by some of the unfortunate consequences which can be expected to arise, followed finally by a summary of the initiatives which have been taken with respect to the impacts contributing to the effect.

Global impacts and environmental effects

As noted, some human impacts on nature have already reached such an extent that they are influencing all parts of the earth and may thus be regarded as global. This applies to the exploitation of non-renewable resources and it applies to chemical impacts on the environment from substances with the ability to spread to sensitive parts of the global ecosystems.

For global environmental impacts, the substances concerned include carbon dioxide, CFCs, mercury, DDT and PCBs, which share the following characteristics:

- long life in the environment, leading to wide dispersal before they are degraded or immobilized,
- high mobility in the environment, leading to transport to all parts of the global environment, including those parts which are sensitive to the substances' impacts.

The substances are often emitted in such large quantities that the effects can be seen globally, even if the substances are greatly diluted on dispersal.

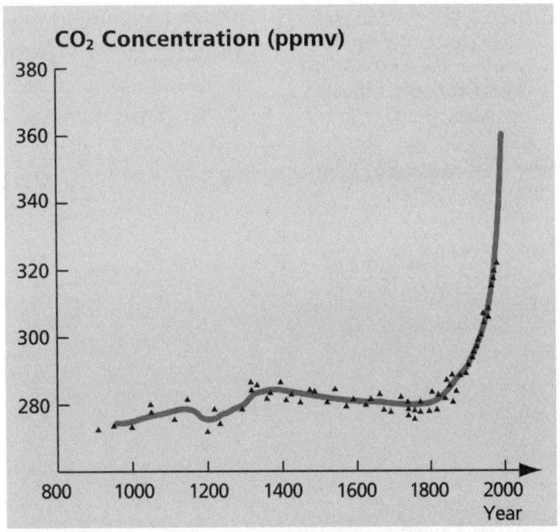

Figure 1.1. Atmospheric concentration of CO_2 during the last thousand years. The unit is parts per million by volume. After Schimel et al., 1995

Climatic changes as a consequence of global warming

The man-made global warming is attributed to the emission of gases which retain heat radiation which would otherwise be lost from the earth into space. The gases thereby contribute to the heating of the atmosphere.

What do we know?

The atmospheric content of all of the most significant greenhouse gases has increased strongly over the last 100 years, especially after the Second World War. This is shown for CO_2 in Figure 1.1. Most of the increase is attributable to combustion of the fossil fuels oil, coal and natural gas. In the same period, the mean global temperature has increased by a good half a degree, as shown in Figure 1.2. The climate may show large natural variations, and it can therefore be difficult to identify man-made changes before they have reached a certain extent. Nevertheless, it is estimated today that the man-made global warming impact has already caused measurable changes in climate (Houghton *et al.*, 1996).

Figure 1.2. Global annual average temperature expressed as the deviation from mean temperature 1951-1980. Reproduced after World Meteorological Organization, 1991

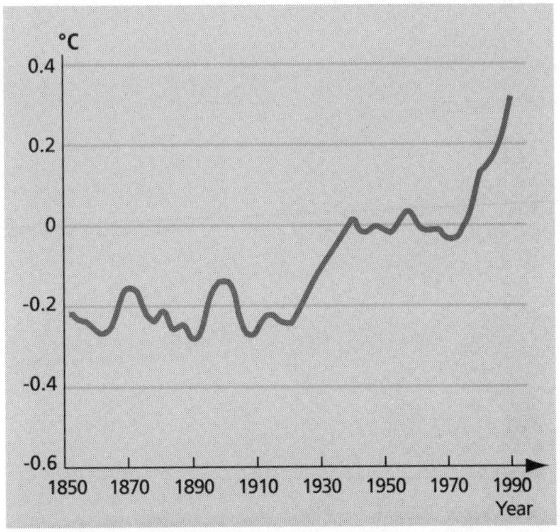

What do we fear?

Sudden regional changes in existing climatic systems, e.g. as a consequence of a changed course in major ocean currents or a melting of polar ice caps and glaciers in mountainous areas, are perspectives causing concern. Such changes will lead to alterations in the conditions of life for man and ecosystems at a speed which will render gradual adaptation impossible. Industrialization of large parts of the Third World could increase future emissions of CO_2, the most important greenhouse gas, dramatically.

What are we doing?

The UN's 1992 Climate Convention required the 154 signatory countries to stabilize emissions of CO_2 and other greenhouse gases at the 1990 level by the year 2000. Some countries have adopted more ambitious goals. In Denmark, the Ministry of Energy's plan of action in the area has set a target for reduction of CO_2 emissions in the year 2005 by 20% compared to the 1988 level. The most important halocarbons, which are also greenhouse gases, will be phased out quickly over the next few years following special agreements concerning this group of substances (see Stratospheric ozone depletion below).

Figure 1.3. Ozone content of the stratosphere 1957-1994, with 1979 as reference year
a) Decrease in spring concentration of ozone in the stratosphere over Antarctica (Albritton et al., 1995b)
b) Decrease in the annual mean concentration of ozone in the stratosphere over Denmark (Statistics Denmark, 1995)

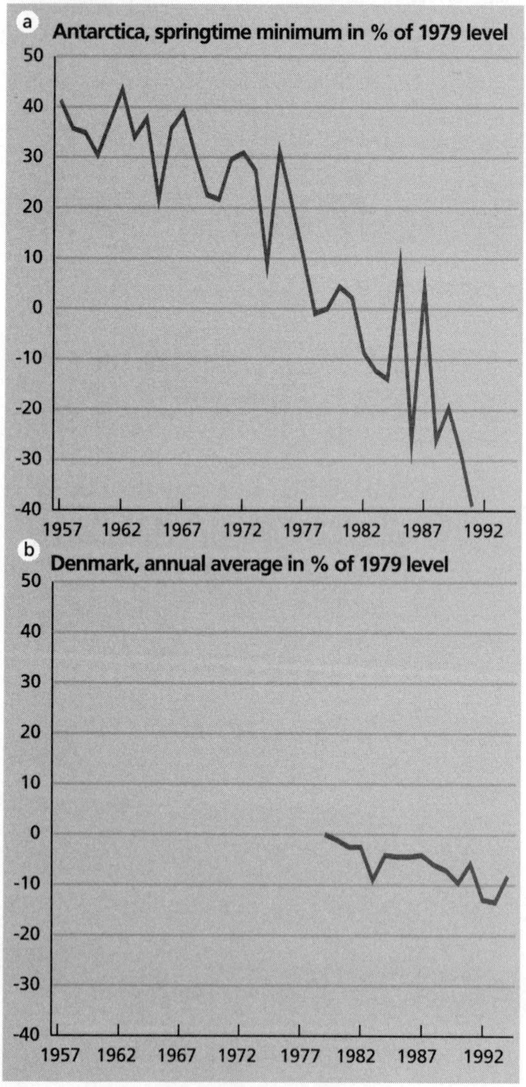

Stratospheric ozone depletion

Formation and breakdown of ozone were originally in balance in the earth's stratosphere at an altitude of 15-40 km, but the breakdown has been accelerated by man-made emissions of halocarbons, i.e. organic compounds containing chlorine or bromine, and which are sufficiently long-lived to reach the stratosphere. The reduced ozone level in the stratosphere means a higher intensity of dangerous UV radiation in the sunlight reaching the surface of the earth.

What do we know?

Man-made emissions of halocarbons (CFCs, tetrachloromethane, 1,1,1-trichloroethane, HCFCs, halons and methyl bromide) have led to a steady increase in the concentration of chlorine and bromine compounds in the stratosphere over the past 40 years. It has been possible over the last 20 years to confirm that the mean ozone concentration in the stratosphere has decreased steadily, apparently more rapidly in recent years. Since the middle of the 1980s, a radical depletion of the ozone layer over the South Pole (the "ozone hole") has been observed during early spring. In the last few years an accelerated depletion of the ozone layer over the North Pole has also been observed. Figures 1.3a and 1.3b illustrate the trend in stratospheric ozone content over Antarctica and over northern Europe.

Depletion of the ozone layer is accompanied by an increased intensity of the UV radiation reaching the earth's surface over large parts of the southern and northern hemispheres. Increased exposure to UV radiation can lead to increased frequency of skin cancer and cataracts and is responsible for a reduced efficiency of immune defences in animals and humans and for damage to the photosynthetic system in algae and higher plants.

Currently (1997) there is agreement that man-made halocarbons are an important reason for the depletion of stratospheric ozone. The magnitude of ozone depletion is expected to peak within the next 5-10 years and then to decrease as a result of the agreements already in force for the phasing out of halocarbons (see below). But irrespective of current efforts to slow ozone depletion, it is expected that the critical lower limit for formation of the ozone hole will be exceeded at least until the middle of the 21st century.

What do we fear?

If ozone depletion reaches the same extent over the North Pole as it has over the South Pole, large and densely populated areas in Europe and North America will be exposed to strongly increased UV radiation with associated extensive damage to health.

What are we doing?

With the Montreal Protocol and subsequent revisions (see e.g. UNEP, 1993), the international community has agreed on an unusually strong control of a group of environmentally hazardous substances. Under these agreements, the industrialized world's production and use of CFCs, tetrachloromethane and 1,1,1-trichloroethane was phased out by 1 January 1996. In developing countries the deadline for phasing out of these gases is 2010. In industrialized countries the production of halons ceased as of 1 January 1994. Production and use of HCFCs stabilized in 1996 and will be phased out by the year 2030. At the same time, the use of methyl bromide was stabilized in 1995. Some industrialized countries have further tightened these deadlines. In Denmark, HCFCs will be phased out gradually by 2002, and methyl bromide will be phased out by 1 January 1998. Even the strong CFCs will, however, continue for a long time to seep into the environment from insulating foam and refrigeration and air-conditioning systems in use and from the disposal of products such as refrigerators which are already in use.

Persistent toxicity

Over the last few decades, attention has been drawn to a number of different substances which, intentionally or unintentionally, have been produced as a consequence of human activities and which are persistent, that is, highly resistant to degradation in the environment. Some of these environmentally foreign substances also possess characteristics which enable them to accumulate in living organisms until they reach toxic concentrations. If they are also mobile in the environment and are emitted in sufficient quantities, the substances' toxicity to man or ecosystems may become a global environmental problem. The list of globally distributed environmentally hazardous substances is steadily growing, but among the best known are mercury, the insecticide DDT, the group of polychlorinated biphenyls (PCBs) and the group of dioxins and furans. In 1997 these substances amount to a global problem.

More information on persistent toxicity is provided in the section on regional environmental impacts as it is still exceptional for persistent toxicity to be of global extent.

Consumption of non-renewable resources

Many of the resources on which we base our material world are regenerated either not at all or so slowly that it is without practical significance for us. They can be regarded as capital which is not accruing any interest. We often use them in an irreversible manner, for example when we burn fossil resources such as oil and coal, thereby reducing the opportunity for future generations to use these resources to meet their material needs.

What do we know?

For several of the important metal resources, the current rate of consumption means that the exploitable reserves of ores known today will be used up in a few decades. Large parts of the Third World are currently experiencing a rapid industrialization, and a rising material standard of living can be expected for large populations in the Far East. Even if the industrialized world uses its non-renewable resources more effectively, increased pressure on them can still be expected in future years.

What do we fear?

Reduced availability of one or more important non-renewable resources can lead to international tensions, and in the worst case to armed conflicts. The industrialized world's dependence on access to large quantities of oil has thus been behind several instances of this in the last few decades, with the Gulf War in 1991 as the most dramatic to date.

What are we doing?

In large parts of the western world, attempts are underway to promote the use of such non-depletable energy sources as solar and wind energy. There is also focus in society on replacing the use of new materials with the recycling of old. But as a rule, recycling initiatives are driven by the desire to minimize the amount of waste produced in the community, rather than to save resources.

In production, there is focus in many places on minimizing consumption of materials, often for economic reasons.

Regional environmental impacts

In contrast to global impacts, regional impacts only affect the environment in the region where the source of the impact is found. 'Region' in this context means an area stretching from around 100 to 1000 kilometres, depending on the nature of the impact and the sensitivity of the environment. On a regional scale, more short-lived substances can also contribute to environmental impacts.

In contrast to the local impacts discussed at the end of the chapter, regional environmental impacts are said to be caused by "diffuse" sources, meaning that the impacts cannot be traced back to a single point source, either because they arise from sources which are far away, or because the pollution is arising as a consequence of a highly intricate network of small local sources.

Damage to forests and other vegetation

The forests in large parts of the industrialized world are suffering damage from air-borne pollution which either damages the leaves of plants or ruins the soil in which the plants are growing.

What do we know?
The damage is caused by a combination of different influences which have an impact on plants. Some of the influences are man-made:

- Emission of acidifying compounds, especially sulphur dioxide, oxides of nitrogen and ammonia, which attack leaves and acidify the soil. In large parts of Europe, critical loads for acidification have been exceeded in forest soils.
- Emission of organic compounds which are broken down in the atmosphere with photochemical formation of ozone and other reactive compounds which attack leaves.
- Emission of nitrogenous compounds, especially ammonia and oxides of nitrogen, which fertilize forest trees into a forced growth and create the basis for an altered species composition in nutrient-poor ecosystems, e.g. heathlands and raised bogs. In many parts of Europe the critical loads for nitrogen enrichment have been exceeded.
- Increased intensity of UV radiation caused by stratospheric ozone depletion damages the photosynthetic organelles in leaf cells.
- Extreme climatic conditions such as drought and high winds damage perennial plants in particular. The frequency of such climatic extremes may already be increased as a consequence of man-made global warming.

Figure 1.4. In extreme cases the damage from air pollution can kill the forest, as here in the Czech Republic (Photo: Biofoto)

What do we fear?
Increased extent of the damage, with consequent loss of major ecological and recreational values.

What are we doing?
There is an international convention under the UN's Economic Commission for Europe, UNECE, on reduction of long-range transboundary air pollution. A number of European countries have signed separate protocols on reduction in emissions of sulphur dioxide, oxides of nitrogen and volatile organic compounds (UNECE, 1991). The agreements have led to significant reductions in SO_2 emissions from many countries, while emissions of volatile organic compounds and oxides of nitrogen are largely unchanged for most countries due to a strong growth in road transport over the last decade, which more than offset the improvements achieved in control of car engine combustion processes and in the treatment of flue gases from coal-fired power plants.

Damage to lakes and coastal waters

Our coastal waters and many of our lakes are suffering the effects of deposition of excessive water-borne or air-borne compounds containing the nutrients nitrogen and phosphorus. In addition, many Scandinavian lakes have been acidified as a consequence of air-borne pollution.

What do we know?

The growth of algae in aquatic ecosystems is most often limited by the availability of either nitrogen or phosphorus. Addition of the nutrient which limits growth causes increased growth of algae, possibly an algal bloom. This effect, known as eutrophication, has been occurring locally about towns for a long time. In Denmark, individual episodes of oxygen depletion in fjords and the Baltic Sea are known back to the beginning of the twentieth century, but in our time the eutrophication has extended much more widely. Figure 1.5 shows the extent of oxygen depletion in inner Danish waters in the summer and autumn of 1995. Oxygen depletion has been an annually recurring problem attributed to the fact that the large populations of algae occurring because of eutrophication sink to the bottom and are broken down with consumption of the oxygen in the bottom layers.

Benthic animals are thus exterminated over large parts of the sea floor. Catches of cod and plaice have fallen substantially over recent years, probably at least partly because periods of oxygen depletion are exterminating the bottom-dwelling animals on which the fish feed. The nutrient enrichment is also a contributing cause to the increased growth of various poisonous species of algae in coastal waters and lakes.

Nitrogen is normally the limiting nutrient in coastal waters. The nitrogen load is attributed particularly to leaching out from extensive agricultural areas, but air-borne emissions of oxides of nitrogen from power plants and transport also contribute significant quantities. Emissions of acidifying gases which can travel large distances are affecting sensitive lakes in large parts of Scandinavia. The clear waters of acid lakes resulting from the direct emission of acid compounds into the lakes have been known in individual cases in Norway and Sweden for almost a hundred years. But the problem has increased in extent, and today several thousand lakes are affected in Norway, Sweden and Finland as well as in central European mountain areas.

Figure 1.5. Extent of oxygen depletion in Danish coastal waters in the summer and autumn of 1995 (based on the Danish Environmental Protection Agency, 1995a)

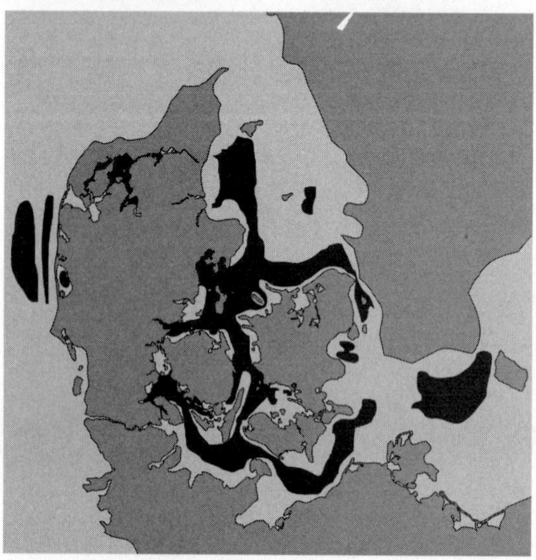

The acidification of lakes is caused by air pollution, especially by sulphur dioxide and oxides of nitrogen, which is ultimately deposited in the lakes. A large portion of the acidifying compounds which end in Swedish and southern Norwegian lakes originates in emissions from Danish and English power plants.

What do we fear?

Further nutrient enrichment of coastal waters, but also of larger marine areas like the Baltic Sea and the North Sea, can lead to dramatic and permanent changes in the species composition of the marine ecosystems and threaten the fishing industry in these areas. Large parts of the benthic vegetation of the Baltic Sea are thus already dead, and industrialization of eastern European agriculture can be expected to further increase the impact on the Baltic Sea.

What are we doing?
Agreements have been signed between affected countries and targets have been set to limit nutrient enrichment of the marine environment in coastal waters in various regions in Europe. For the North Sea and the Baltic Sea, a reduction in emissions of nutrients by 50% is thus being sought, but these agreements have not yet (1997) had any detectable effects.

Attempts are being made to reduce the emission of acidifying substances as described above under 'Damage to forests'.

Persistent toxicity

Some of the substances emitted are not easily degraded in the environment and can build up to concentrations which cause toxic effects to man or ecosystems.

The heavy metals formerly attracted considerable attention, but over the last decade it has become clear that many different organic substances can also contribute to persistent toxicity.

What do we know?
Of the order of 100,000 industrially produced chemicals are currently registered for marketing, and thus use, within the EU (EEC, 1990). Very little or nothing is known about the behaviour of most substances in the environment and their potentially toxic effects on different organisms. For those substances investigated to date, it is clear that:

- Some substances are not easily degraded in our sewage treatment plants or in the environment, and others have been found to have stable degradation products. These substances or their degradation products will remain in the environment long after they are emitted.

- Some substances accumulate in living organisms until they reach levels at which they are toxic to the organism. One of the most commonly cited examples is the insecticide DDT, which is concentrated via the food chains in birds of prey and which previously caused thin-shelled eggs which could not be hatched. Another, more recent example is the occurrence of polychlorinated biphenyls, PCBs, and dioxins in mother's milk in women, reaching concentrations which have generated a debate on whether breast feeding of babies should be discouraged.

- Some substances have been shown to find their way into parts of the environment where they have not been expected. Recent examples are the discovery of many agricultural pesticides in ground water which is used as drinking water and the spreading of plasticizers from PVC in water, soil and air.

- Some substances have been shown to have unexpected toxic effects. An example is the recent discovery that a range of different industrial chemicals or their degradation products in the environment mimic the female hormone oestrogen (see Table 1.1).

Substances with effect as environmental oestrogens (summer 1995)	
Alkyl phenols Nonyl phenol Octyl phenol	Degradation products from detergents widely used in e.g. cleaning products and paint
Phthalates BBP DBP	Plasticizers in certain types of soft PVC plastic. Phthalates are among the most commonly found man-made chemicals in the environment
Pesticides DDT, DDE	Now prohibited in large parts of the world
Methoxychlor Lindane	Prohibited in Denmark Prohibited in Denmark
Others Bisphenol A	Component in epoxy materials and polycarbonate
PCB, some	Now prohibited in large parts of the world

Table 1.1. Substances foreign to the environment for which an oestrogenic effect has been demonstrated (Toppari et al., 1995)

The presence of hazardous substances in the environment is often cited as a probable contributor to:

- the increasing frequency of allergy in children and adults,
- increased risk of development of several different forms of cancer,
- reduced sperm counts in men in most of the industrialized world,
- reduced reproductive capacity in many of the animal species at the top of the food chains.

Currently (1997) there is, however, still no definitive proof of these assumed relationships.

We do, however, know one thing: there is much we don't know about the persistent toxicity of thousands of different substances.

What do we fear?

The impact, and thus the possibilities of damage, of so many different chemicals is difficult both to monitor and to control. There is also the possibility of discovering new undesirable and hitherto unknown effects on health and environment from some of the many substances which we emit.

What are we doing?

Within the EU, a systematic risk assessment of the industrial chemicals which are produced and used in greatest quantities within Europe has been initiated. There are 2,000-3,000 chemicals used or imported by individual importers within the EU in quantities exceeding 1,000 tons per annum. An agreement has been reached within the EU on risk assessment of 79 chemicals (EC, 1994).

There are agreements for both the North Sea and the Baltic Sea on reduction of emissions of toxic substances by 50% from the end of the 1980s to 1995. There is consensus in the most recent agreements on the North Sea concerning the goal that emissions of environmentally hazardous substances into the North Sea must have ceased by the year 2020 (Danish Ministry of Environment and Energy, 1995).

Substances which are particularly problematic because of their persistent toxicity will be phased out gradually, as was the case for lead in petrol, or prohibited, as has been done for PCBs or the most problematic agricultural pesticides.

Local environmental impacts

Local environmental effects are those occurring about significant individual sources.

Local impacts are limited to the immediate vicinity of the source or influence, and the geographic extent of the immediate effect is therefore normally at most a few kilometres.

Physical impacts

Deforestation is one of the first serious physical impacts to which man has subjected the environment.

Today, physical impacts have reached a much higher level as a consequence of the number of people, and not least the use of machinery which increases man's capacity to do mechanical work many times.

What do we know?

Significant physical environmental impacts the world over are caused by construction works such as the building of roads, bridges and harbours. Large quantities of materials are moved and the surroundings can be affected by noise and vibrations both during the construction phase and after the finished installations are taken into use.

Figure 1.6. Clearing of South American rain forest (Photo: M. Hauschild)

Agricultural cultivation of the earth also has significant physical impact on the environment. Small ponds are drained and filled in so that the earth can be cultivated, thus removing the habitats of many species. To facilitate mechanized agriculture, live hedges are being removed between adjoining fields, and the remaining habitats appear as small enclaves isolated from one another, with very little opportunity for the dispersal and exchange of species. Marshes are being drained and rivers and watercourses straightened to devote the areas to the cultivation of crops, thus increasing the leaching out of nitrogen to our coastal areas and reducing the formation of groundwater.

The primeval forest in large parts of Europe has been cleared over the last 1,000 years. A similar clearing of forests is occurring today in many areas in the Far East and South America. Clearing of the primeval forest is occurring for sale of timber, cultivation of the cleared ground or mining for minerals, as was the case in Europe. But large parts of the earth's rain forest ecosystems exist on poor soil, and most of the nutrients are found in the terrestrial plant biomass. Forests hold the soil in place and are also of major significance for local climate. The clearing of rain forest therefore leads to loss of the nutrients and the topsoil, and leaves large areas barren and eroded and without scope for significant cultivation.

Table 1.2. Annual extent of rain forest clearing in various tropical countries (World Resources Institute, 1990)

Extent of rain forest clearing in tropical countries

Country	Clearing % per year	Year
Brazil	2.2	1987
Cameroon	0.6	1976-86
Costa Rica	7.6	1977-83
India	4.1	1975-82
Indonesia	0.8	1979-84
Myanmar	2.1	1975-81
Philippines	1.5	1981-88
Thailand	2.5	1978-85
Vietnam	2.0	1976-81

Among physical impacts on the environment are:

- loss of or splitting up of important habitats required by various forms of life,
- loss of aesthetic qualities and cultural assets,
- soil erosion and loss of arable land,
- lowering of the water table,
- flooding due to a reduced drainage capacity of rivers and watercourses, e.g. as a consequence of clearing of forest on mountain sides adjacent to the river, leading to major surface run-off, erosion of the slopes and increased sedimentation in the rivers.

What do we fear?
A gradual degradation of the natural environment as a consequence of a widespread occurrence of local physical impact on the environment almost everywhere on earth. One consequence is a loss of species which is particularly serious in the clearing of rain forests, which are among the most diverse ecosystems on earth.

What are we doing?
Regulation of physical impacts of the environment is being sought in many countries through various forms of regional and local planning, and in the submission of environmental impact assessments covering both chemical and physical impacts for major construction works.

Local toxicity

Local toxicity to ecosystems or humans can have the same causes as the persistent toxicity discussed above under regional environmental impacts. The probability of experiencing local effects of an emission is, however, generally greater, because dilution of the substances emitted is small in the immediate vicinity, so that the substances occur in greater concentrations on a local than on a regional scale. The higher concentrations also mean that acute toxicity can occur, i.e. toxic effects which appear shortly after the release.

What do we know?
Chronic effects as discussed under persistent toxicity can occur locally, in some cases also for substances which are not persistent. Acute ecotoxicological effects can also arise around effluent discharges from sewage treatment plants or from industry. Under unfavourable circumstances, acute toxicity to humans can occur via the breathing of air pollution, e.g. during periods of smog. In the more environmentally regulated industrialized countries, both acute toxicity and acute ecotoxicity generally occur only rarely, partly because of the extensive use of treatment plants both in industry and for municipal emissions. They occur in other parts of the world, most notably in connection with accidents, but also as acute toxicity following emission of untreated industrial effluent and municipal waste water.

What do we fear?

- New and undesirable effects from the substances which we emit to the environment.
- Technologies which expose the user to hazardous substances and which are so complicated that accidents are increasingly more difficult to avoid: wrecks involving oil tankers, accidents involving chemical transport, emissions from nuclear power plants.
- Local pollution of drinking water from leaching of chemicals or landfills.

What are we doing?

In most of the industrialized world, industry has very greatly reduced toxic emissions from production processes over the last two decades. Originally, this occurred mainly through establishment of treatment plants, later by integrating environmental considerations into production, i.e. by the gradual introduction of cleaner technologies. As a consequence of this development, industrial emissions as a whole have become much less toxic, and many industrialized countries today are experiencing no problems with acute ecotoxicity from industrial emissions. In return, the treatment plants have provided a residual product in the form of sludge from the treatment of sewage. In many cases the sludge has a content of substances toxic to the environment which makes it unsuitable as an agricultural fertilizer, and it must therefore be dumped or burned.

Monitoring of man-made pollutants in groundwater has increased. Future limit values for the content of hazardous chemicals in drinking water are under discussion within the EU.

Attempts are being made within the EU to reduce air-borne emissions from road transport by tightening requirements concerning car engine combustion processes. The effect of this initiative is, however, counteracted by a rise in road transport, especially in personal transport.

Renewable resources

Renewable resources are resources such as groundwater, populations of fish, forests and other forms of biological resources which are not depleted by man's use if they can be regenerated at the same rate. Renewable resources can therefore be used without infringing upon the resources of future generations.

As the rate of regeneration can vary greatly from area to area, use of renewable resources is regarded as a local to regional impact. It is primarily in the local areas that care can be taken that the resource is not used faster than it can be regenerated.

What do we know?
In many parts of the world, renewable resources are being exploited at a rate which will ensure that they will decrease, and perhaps disappear entirely, to the detriment of present and future generations:

- Fish populations are being depleted.
- Forests are being cleared without regard for their regeneration with the same biodiversity.
- Agricultural practices are leading to an impoverishment of the soil and loss of its humus content.
- Groundwater is being extracted faster than it can regenerate. It is being polluted by penetration of salt water in coastal areas or by the dispersion of local pollution.

What do we fear?
Permanent loss of the opportunity of using the renewable resources which are a significant basis for the functioning of society.

What are we doing?
Use of some of the renewable resources is subject to various forms of national and international regulation, e.g. international fishing agreements for protection of fish populations or national and international initiatives for the promotion of sustainable forestry.

2. Society's environmental focus

Since the beginning of environmental legislation in the industrialized world in the 1970s, considerable attention has been paid to industrial pollution. Over the last 20 years, both companies and authorities have therefore accumulated considerable experience in reducing and controlling the environmental impact of industry, and both parties have built up organizational structures for the environmental work.

Companies have directed their attention to the treatment of waste water and air pollution and the introduction of cleaner technology, and many companies have established environment departments and appointed employees with responsibility for environmental matters.

Official regulation of pollution has occurred primarily via legislation and arrangements concerning supervision and approval. An administrative system has been established to undertake this work. Official responsibility for environmental administration is divided among central and regional authorities, and the practical work is carried out by decentralized technical administrations and national environmental agencies.

Efforts have mainly been concentrated on emissions from material production, product manufacturing and disposal as illustrated in Figure 2.1.

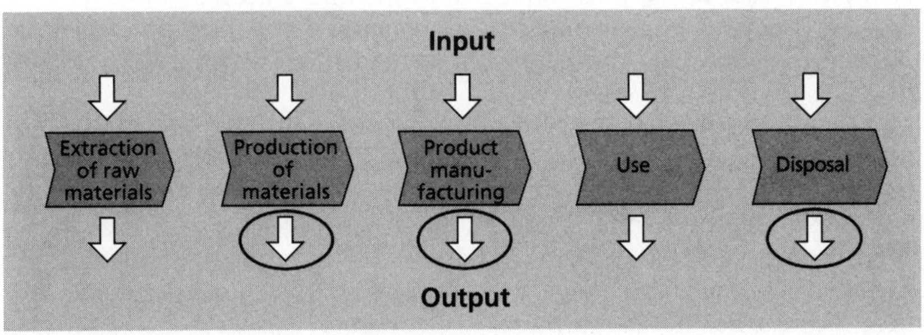

Figure 2.1. Environmental focus in the industrialized world

This focus has had a marked effect on emissions. An example for manufacturing companies in Denmark is shown in Figure 2.2a, which illustrates the development in various pollution parameters for emissions of waste water from industrial production over a 10-year period (Wenzel *et al.*, 1990). Virtually all pollution parameters in industrial waste water emissions in this period have been reduced by 80-90% for Denmark as a whole.

But at the same time, a number of other circumstances have been pulling in the opposite direction with respect to impacts on the environment. The material standard of living has been rising, and consumption of products, and therewith resources, has risen virtually in step with the falling impact of emissions from companies (Figure 2.2b).

Since environmental impacts are attributed in increasing degree to a rising supply and consumption of products, control solely on the basis of a concentrated effort directed towards individual emissions from production processes is becoming more difficult.

Even if the biggest individual emissions are reduced, the total environmental impact in some areas will still rise as a consequence of the

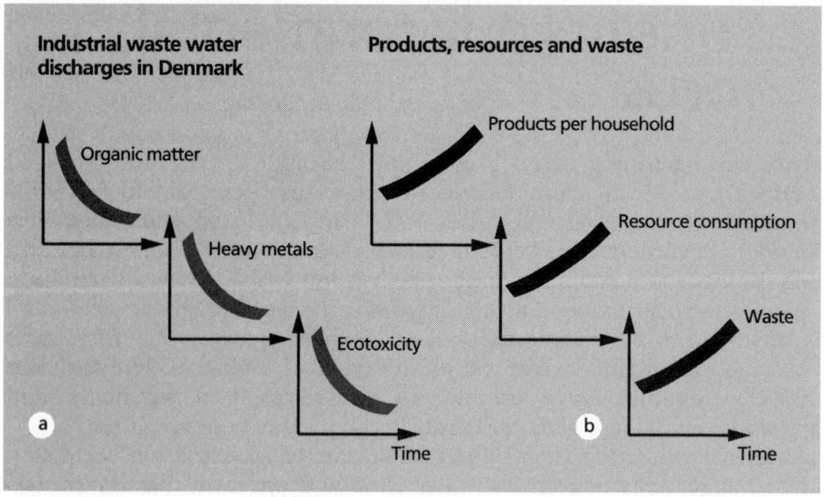

Figure 2.2a. Examples of the development in emissions of industrial waste water in Denmark over 10 years (Wenzel et al., 1990)

Figure 2.2b. Diagrammatic outline of the development in the quantity of products, resource consumption and quantities of waste over the same period

growing flow of products and the activities to which they are leading. It is the "many little streams which make the big river", and it is not possible to control environmental problems attributable to the flow of products by regulating the emissions from the manufacturing processes.

In recent years this recognition has influenced environmental administration, and the focus is changing from processes to products, so that attention is paid to the total impact from the entire product system. To a large extent, future environmental management will therefore occur at the interface between company and customer, as illustrated in Figure 2.3.

Through his or her choice of products, the consumer has a chance of influencing manufacturers toward reducing the environmental impacts of their products, by demanding environmentally friendly products, and the customer's attitude to the product's environmental characteristics therefore becomes a decisive factor. The role of the authorities in environmental regulation will change character, from direct management to a more indirect but no less important role, namely ensuring the right rules of play and parameters of competition in the market of the future,

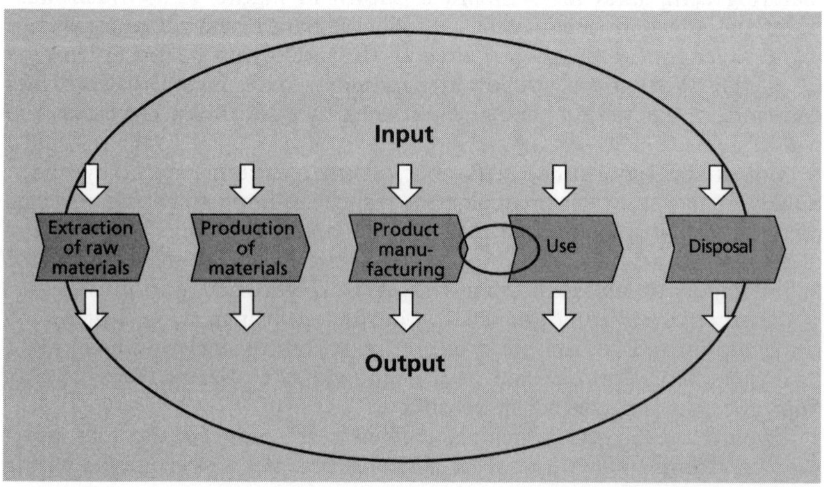

Figure 2.3. Future environmental focus in the industrialized world

including ensuring the reliability of the environmental information on the product.

Legislation and standardization

Environmental legislation in several countries has started to reflect the environmental focus on products and their product systems. Examples of actual and forthcoming initiatives in the area are shown in Figure 2.4. In 1995 the Dutch Ministry of the Environment published a memorandum on Dutch environmental policy in the product area, and Denmark has likewise published a proposal for a product-oriented environmental policy in 1996.

Among other initiatives in the area, Germany, Sweden and the Netherlands are preparing systems for the return of electronic waste, and the EU has adopted a packaging directive which prescribes environmental assessment of the packaging for its entire product system.

Standards for environmental management include formulations specifying that the environmental impacts of products should also be considered beyond the factory gates. The most important are BS7750, the EU system ECO-management and Audit Scheme (EMAS), and a forthcoming international standard in the ISO 14000 series. The reference list includes some of the standards noted.

Until now it has not been usual for companies to include the product's full product system in environmental management, but this is

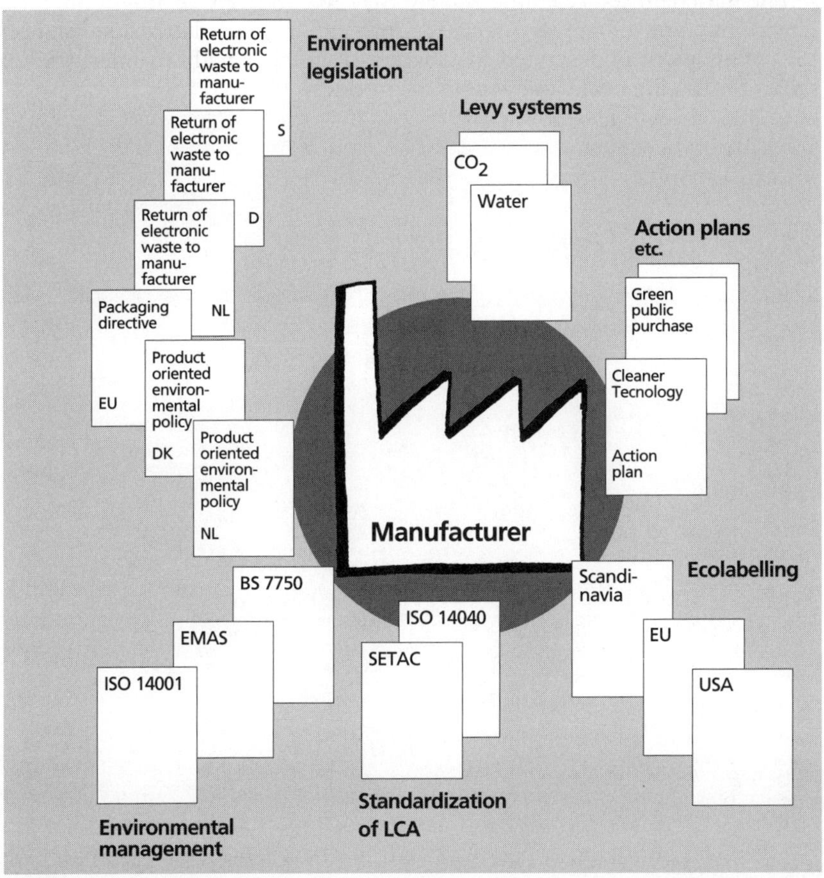

Figure 2.4. Examples of community initiatives of significance for a product-oriented environmental policy

expected to come. The Danish Ministry of the Environment and Energy and the Ministry of Commerce have established the Environmental Management Council, the duties of which include management of a subsidy system within the area of environmental management. Other Danish subsidy systems are already product-oriented, including the Ministry of the Environment and Energy's Action Plan for Cleaner Technology. Danish national authorities and institutional bodies must also consider the environment on an equal footing with other considerations in public purchases, and Danish counties and municipalities have been urged to do the same.

Within the EU, the Member States have adopted criteria for positive environmental labelling of products such as dishwashers and washing machines, toilet rolls and paper towelling, compost-based soil improvement agents, paint, laundry detergents and electric light bulbs. Work is proceeding on preparation of criteria for further product groups, including insulating materials, textiles, writing paper, refrigerators and hair spray. The criteria are based on an environmental assessment of typical product systems for the product type.

A number of environment-related taxes have been introduced in Denmark, but for the time being they are generally not directly related to the product. Examples include the tax on waste water discharge and the CO_2 tax.

Finally, work is proceeding internationally on standardization of the methods of environmental assessment of products, both within the environmental chemists' scientific society SETAC (Society of Environmental Toxicology and Chemistry) and the international standards organization ISO. This work is described in Chapter 4 together with an international status for the life cycle assessment of products.

3. Environmental assessment of products - what does it say?

The new environmental focus on the product is bringing environmental considerations closer to the source of the impact on the environment, namely the demand and the supply of products, and the competitive factors which drive them both. When the *product* is subjected to environmental assessment, environmental considerations can enter the chain of supply and demand and be a parameter of competition on an equal footing with the other parameters for the product such as price and quality. If both customers and manufacturers emphasize the product's environmental properties, environmental considerations can be included even before the product is made, and inappropriate solutions and technologies can be avoided from the start. The chain of suppliers and buyers often consists of many links, and if sound environmental properties are demanded, they will propagate along the whole chain and have a very great effect: see Figure 3.1.

Figure 3.1. Simplified outline of the links in the chain of supply and demand

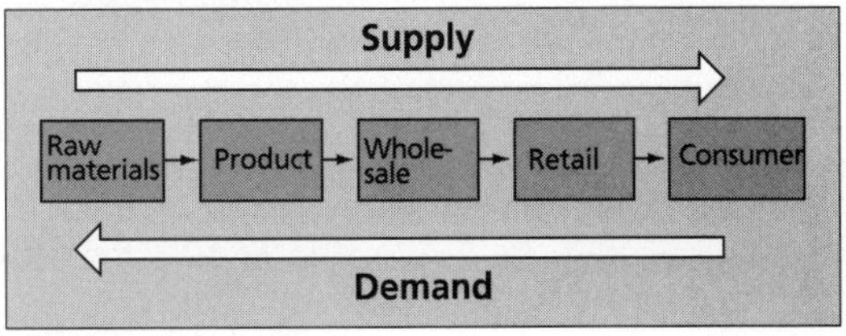

The product approach includes some major potentials for environmental improvements, which to date were used only to a limited extent and only for a few product types, e.g. organic foods, organic cotton and other products which have been assigned an environmental label.

The service provided by the product

A definition of a product is that it is of benefit to the user: it provides a service. If there is no definable user service, it is not a product. A consequence of focusing on the product is therefore making the product's *service* the focal point. The service, or the covering of the user's needs, is the reason for the product's existence, and it is therefore also the service which is ascribed all of the environmental impacts caused by the product.

But a service can be provided in many, often very different ways. The promise of environmental assessment of products is the possibility of comparing the environmental consequences of widely differing ways of providing a service, for it is thus possible to develop products and to select products on the basis of environmental considerations, and to achieve major environmental gains.

The functional unit

Figure 3.2 illustrates examples of various ways of providing a service. The object of the figure is to show how differently the same service can be provided. The tea can be stored in many different ways while it is drunk, the hair can be done in widely differing ways, and there are different ways of mowing the lawn.

To ensure that the different ways of providing a service are comparable, the service must be defined and precisely quantified. This is called determining a *functional unit* for the product. This functional unit is the fixed reference point for the environmental assessment. When alternative solutions are compared, everything but the functional unit can be permitted to vary, because the solutions are alternatives to one another only

Figure 3.2. Different ways of providing the same service

when they provide the same service. Figure 3.2 describes the functional unit for the three services, and the different solutions are all specified as providing the functional unit which has been determined. Because of different life spans for the different types of products, a different number of products is used to provide the same functional unit. This is an important point in the environmental assessment of products.

Consider which solutions in Figure 3.2 are the most attractive environmentally.

Assessment of environmental impacts

The environmental assessment of products is a new discipline, but it builds on the experiences gained within the area of the environment over the last few decades. The method of assessment in the environmental assessment of products is especially bound to the principles developed within the environmental assessment of chemicals and mixtures of chemicals, as found, for example, in waste water emissions and industrial airborne emissions. These principles are well established, and they are described as an introduction to make it easier to explain the principle of the environmental assessment of products.

Environmental assessment of chemicals and mixtures of chemicals

The impact of a substance on the environment depends on three things, viz.

1) How much is emitted?
2) How hazardous is it?
3) In what quantities and concentrations does it reach those parts of the environment where it can exert its effect?

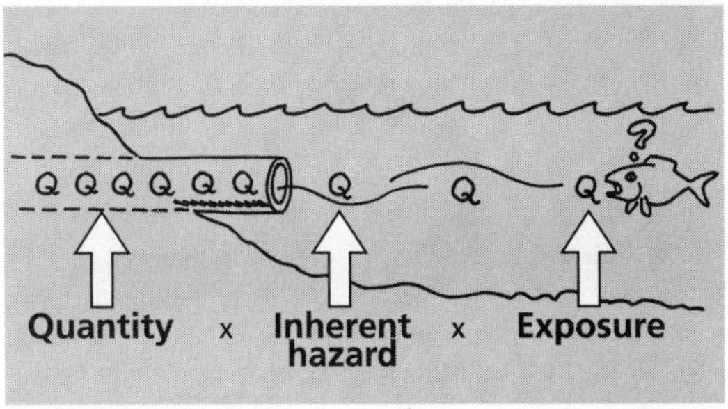

Figure 3.3. The impact of a chemical in the environment depends of three factors

This relationship is universal in the sense that it applies to emissions to air, water and ground, and that it applies to both individual chemicals and to mixtures of chemicals. Figure 3.3 illustrates the principle for emission of waste water. The relationship is expressed as:

Impact = quantity · impact potential · exposure

where the three factors in the expression represent the above three questions. The *impact potential* is the substance's inherent hazardousness, i.e. its ability to trigger a given impact. The exposure is the degree to which the substance reaches parts of the environment where the impact can be exerted, e.g. the degree to which a CFC emission reaches the ozone layer, or a toxic substance reaches an organism. This general relationship is well described in the literature; see, for example, Pedersen *et al.* (1994), who treat the assessment of waste water toxicity.

When the resulting environmental impacts of an emission are to be analysed, these three factors are often investigated separately, and the study of each factor has developed into independent professional environmental disciplines. A substantial knowledge is therefore available within each field, which can be used in the environmental assessment of products.

The question now is: what is involved in the environmental assessment of a product? How does this differ from the environmental assessment of an emission?

Environmental assessment of products

Figure 3.4 illustrates a product, an electric mixer. As it stands, it has no environmental impact, and in principle it can stand there for years without affecting the environment. The product in the figure only "has" an environmental impact by force of its past and its future.

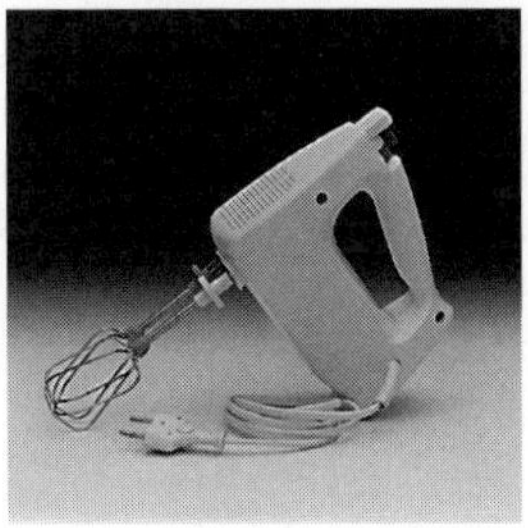

Figure 3.4. A product

The environmental impact of the product derives from the processes into which it enters. It is the processes which exchange substances or energy with the surroundings, and only if there is an exchange with the surroundings can there be an environmental impact.

The process's environmental exchanges

An *exchange* with the environment is defined as an *input* to a process, an *output* from the process, or an *internal interaction* with an operator (a worker) of the process. When, for example, the product is painted during manufacture, or when it uses energy when in operation, it enters into a process which has exchanges with the surroundings. Figure 3.5 illustrates the process's exchanges.

In general there is therefore only one way of making an environmental assessment of a product, and that is to assess the processes into which it enters. The task thus resembles the task already known, to carry out an

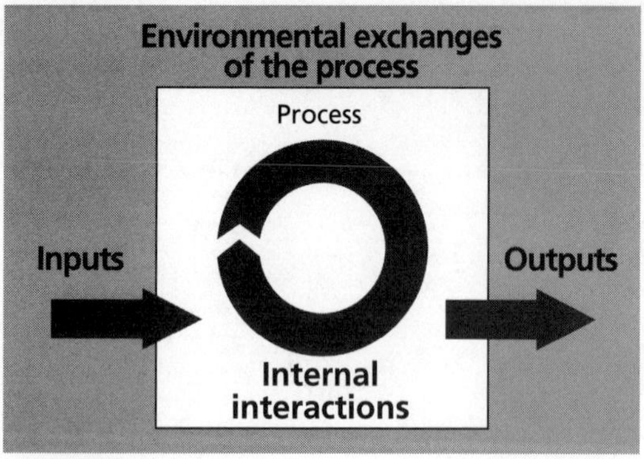

Figure 3.5. The process's environmental exchanges

environmental assessment of an emission. The product merely enters into numerous processes with many emissions which must be assessed together. In other words, numerous environmental assessments must be *put together*. The new element in the task is in this, and in assessing the many categories of impacts at the same time: various categories of resource consumption, environmental impacts and impacts on the working environment.

The task thus differs in the way that it is the whole and the overview which are important. Detailed knowledge of the individual emission must thus be dispensed with.

The product system

Even the simplest products enter into a large number of processes. The "life" of the product starts right from the first extraction of raw materials, and continues throughout manufacture, use and disposal of the product. The processes into which the product enters are collectively termed its *product system*.

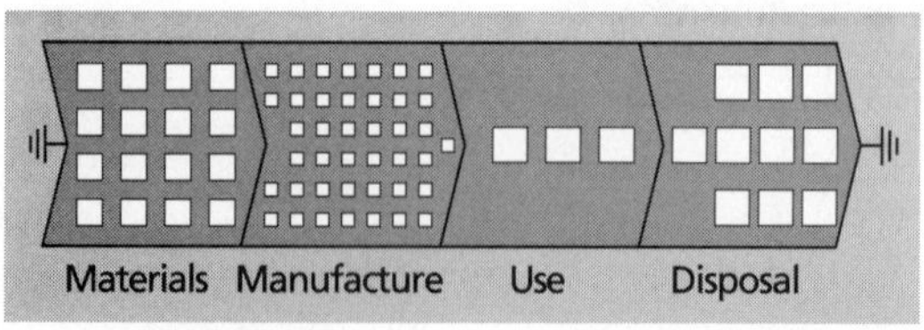

Figure 3.6. Conceptual outline of the product system

Figure 3.6 shows a conceptual outline of a product system. Four stages of the product's life are illustrated: 1) the material stage, 2) the manufacturing stage, 3) the use stage and 4) the disposal stage. A number of processes, which together comprise the stage, are shown in each stage.

The outline in Figure 3.6 shows what is understood by a product system, but it is not suitable for illustrating the extent and the variation within the product system.

The actual degree of detail is better given by the detailed model of the product system in Figure 3.7.

Figure 3.7 illustrates an additional dimension, namely geographic variation. In many cases, the same process will occur in several different places, and for products with a high proportion of exports this will apply in particular to the use and the disposal of the product. Processes in material production can also occur in several different places, because there are often different suppliers of the same product, and raw material extraction often occurs over large parts of the world.

The product system's main and lateral streams

Each of the small squares in Figure 3.7a symbolizes a process in the product's main stream, i.e. the stream of materials entering into the product. For almost all processes there are, however, a number of lateral streams associated with the process. The lateral streams are energy or ancillary substances, such as water, soap or solvents, which do not enter into the product but which are required for the process. Both energy and ancillary substances must be generated, and ancillary substances must often be disposed of after use, and they therefore have their own product systems before and after the process, as shown in Figure 3.7b. Environmentally, the main and the lateral streams can be equally significant, and as the point of departure, both are included in the product system.

Comparison of solutions

Figure 3.8 illustrates a brief version of the product systems for three of the solutions for the service "maintain tidy haircut for one year" from Figure 3.2, namely the three combs in different materials. These solutions are the most uniform of those shown in Figure 3.2, but their product systems and

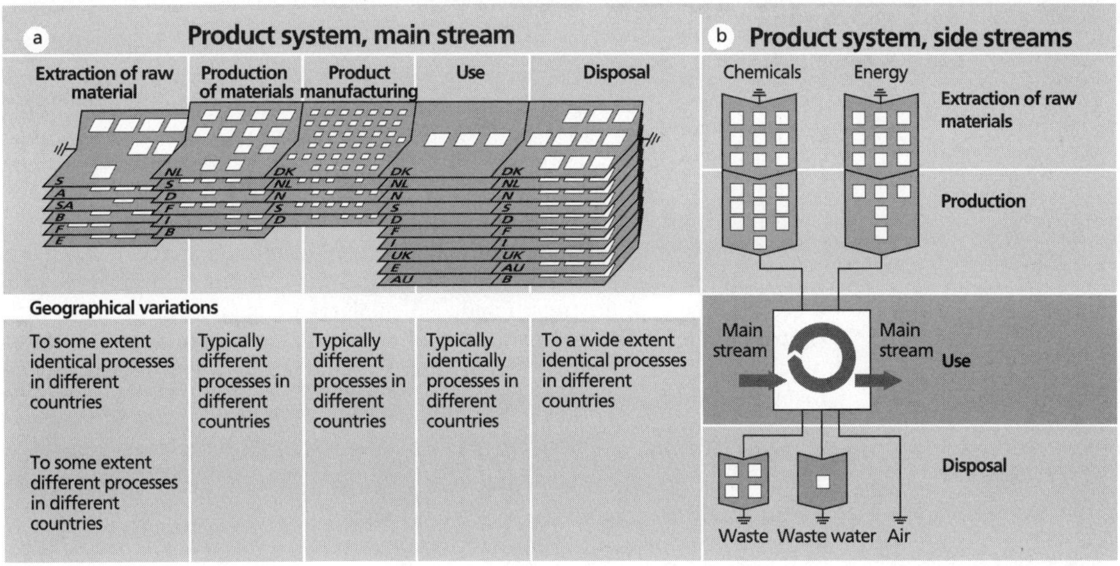

Figure 3.7. Detailed model of the product system, including main and lateral streams and geographic variation

therewith their potential environmental impacts are nevertheless very different.

Firstly, their life spans are different. The example assumes one plastic comb to provide the defined service, one wooden comb and half a steel comb. The first major differences lie here. Secondly, their product systems are very different in the first two stages, and in disposal, the materials have very different fates, and therewith different potential impacts in landfills and incineration and in nature.

Characteristics of the environmental assessment of products

Assessing the many processes in a product system can be a major task requiring large quantities of data. But it must still be possible to undertake an assessment within the time available for it. This shows that the nature of the task differs from that for the previously known assessment of an individual emission. The task is not to know all details and to assess each individual process and emission with great precision, but on the contrary to create an overview of the whole product system: an overview which has not previously existed.

Site-specific information

The large number of processes excludes the possibility of knowing all localities and site-specific conditions concerning the emissions in the

Figure 3.8. Simplified outline of the product system for three solutions for the service "maintain tidy haircut for one year"

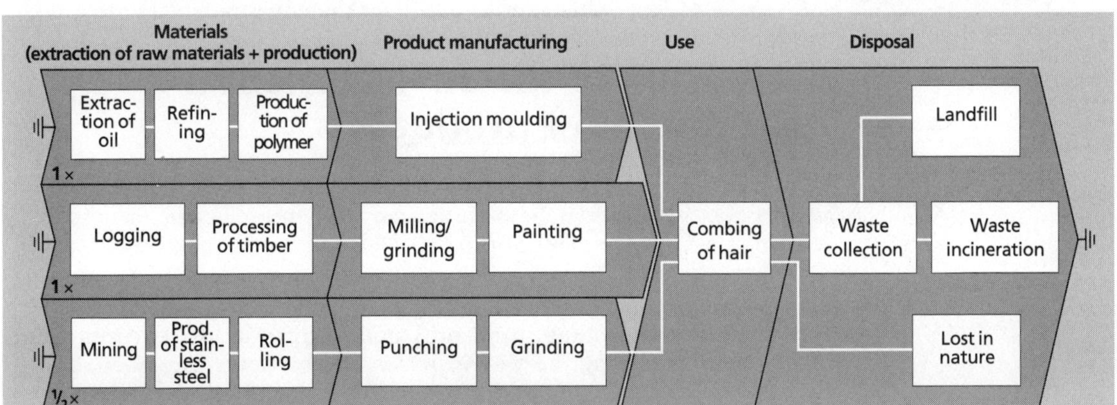

product system. In other words, variation in the emission over time and the way it is distributed in the recipient following emission are often unknown. In most cases the need for large quantities of data means that the task of collecting data for all actual localities in the product system would be too great. Instead, generally available data for corresponding processes must be used to represent the actual processes.

Consequently, detailed information on time and place cannot be found in most cases. The environmental assessment of products can therefore most often not include an analysis of exposure, as described previously for individual emissions of chemicals and mixtures. The starting point is therefore that the product system is taken as independent of local conditions, as suggested in Figure 3.9.

The principle of environmental assessment

An analysis of exposure of the environment and its biota to an emission is required in order to be able to predict the emission's actual effects. When this is not possible, the environmental assessment of products is therefore limited to consideration of potential impacts or effects. The principle of the environmental assessment of products can therefore be summarized as:

ΣPotential impact = Σquantity of substance · substance's impact potential

summing all environmental exchanges in the product system. Depending entirely on where and how an exchange actually occurs, the potential impact can trigger real effects to a greater or lesser degree. In other words, depending entirely on the how the product system in Figure 3.9 is actually "landed", the effects which are triggered can be different.

Figure 3.9. Initially, the product system is considered to be independent of specific local conditions

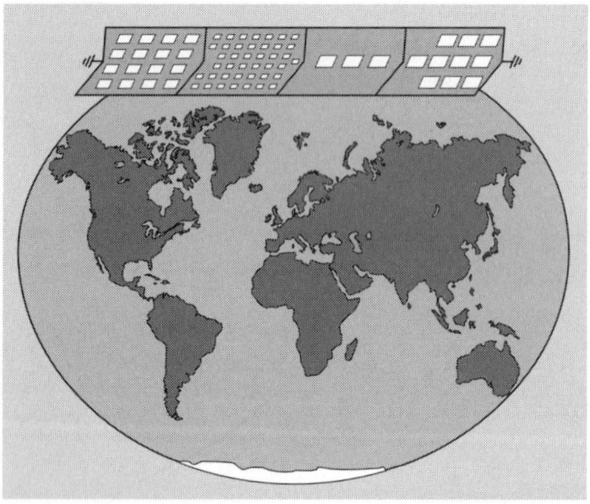

This feature of the environmental assessment of products is the most significant characteristic of the environmental assessment method. It is caused by the difficulty of generating data specific to the product in time and place, and it applies in general to all existing methods. Attention must be paid to this aspect so that results are not over-interpreted where in reality the conclusion is not unambiguous.

But this lack of accuracy in the environmental assessment is not a negative limitation. On the contrary, the accuracy in most cases is correct relative to the task of the assessment, namely to provide an overview and aid to ranking among solutions for a product which are very different in environmental terms.

Definition of the environmental assessment of products

The description of the environmental assessment and of the product and its product system can be summarized into the following definition of environmental assessment of products:

To assess a product environmentally is: to define and quantify the service provided by the product, to identify and to quantify the environmental exchanges caused by the way in which the service is provided, and to ascribe these exchanges and their potential impacts to the service.

4. Status of life cycle assessment

The generally recognized term for environmental assessment of products is *Life Cycle Assessment* or *LCA* in abbreviation. LCA is sometimes also read as life cycle *analysis,* but life cycle analysis is not a particularly good designation since an LCA always contains an element of *assessment,* namely the consideration and weighting of differing resource and environmental problems required to make a decision. The word *"analysis"* therefore signals too much objectivity. In German-speaking countries the terms *Bilanzierung, Ökobilanzierung, Umweltbilanzierung* or *Produktlinienanalyse* are often used, and in French-speaking countries the word is *ecobilan.*

This book uses only the term "life cycle assessment" or "LCA", with "environmental assessment of products" as an occasional synonym.

Historical development

The first examples of environmental assessments of products were carried out on packaging and published at the end of the 1960s and the beginning of the 1970s in the USA. They were called "Resource and Environmental Profile Analyses" (REPAs) and focused primarily on energy consumption, resource consumption and generation of waste, in accordance with the focus in the environmental debate of the time. At that time there was still too little knowledge on processes' emission of environmentally hazardous substances and their possible impacts on the environment to permit assessment of a potential environmental impact (see, for example, Hunt *et al.*, 1974).

At the beginning of the 1980s, interest in the environmental assessment of products grew in connection with discussions on environmental impacts from various forms of packaging, and LCAs were used in several European countries to compare different beverage packagings (BUS, 1984; Lundholm & Sundström, 1985; Franke, 1984). Obtaining unambiguous results and conclusions proved to be difficult in some cases, because the database and the methods varied in the different studies. This was unsatisfactory, and it was one of the reasons why a more systematic development of the methodological basis for the life cycle assessment of products began.

From the end of the 1980s to today (1997), interest in life cycle assessment has grown very strongly, and a growing number of different and increasingly complex products and systems have been assessed. Summaries are given in Pedersen & Christiansen (1992) and "Sustainability" (1993). Different interests in society have seen a major perspective in this development of the environmental focus and they have each commenced work from their own point of view. The authorities use life cycle assessments, for example, in work on environmental action plans and in environmental labelling of products. Companies use them in product development, environmental management and marketing, and consumer organizations use them in counselling consumers.

Accompanying the growing activity within the field of life cycle assessment, more attention has been paid to the development of a methodological basis. The international scientific society of environmental chemists, SETAC (Society of Environmental Toxicology and Chemistry) started work on life cycle assessments in 1990, and within a few years it became the international forum for discussion of the methodological basis of the LCA.

Table 4.1 provides an overview of some of the most important international conferences and workshops which have been held since 1990 with life cycle assessment as their theme. SETAC's dominant position is clear from the table, which also illustrates how the items have changed from the general to the more specialized aspects of the life cycle assessment method.

The object of the SETAC workshop in 1993 in Sesimbra, Portugal, was to attempt to set common guidelines for the carrying out of an LCA, and the report from this workshop, "Guidelines for Life-Cycle Assessment: A Code of Practice", has gained the status of a common reference for life cycle assessment (Consoli *et al.*, 1993).

Table 4.1. Workshops and conferences held since 1990 on subjects related to life cycle assessment (Christiansen et al., 1995, and other sources)

Venue	Time	Organizer	Topic
Washington, DC, USA	May 90	WWF	LCA, in general
Vermont, USA	August 90	SETAC	LCA, in general
Leuven, Belgium	September 90	Procter & Gamble	LCA, in general
Leiden, Netherlands	December 90	SETAC	Methodological background
Sandestin, USA	February 92	SETAC	LCA, impact assessment
Washington, DC, USA	March 92	SETAC	
Potsdam, Germany	June 92	SETAC	LCA, impact assessment
Wintergreen, USA	October 92	SETAC	Data quality
Lyngby, Denmark	January 93	TUD	Assessment of ecotoxicity
Sesimbra, Portugal	April 93	SETAC	Guidelines of LCA
Amsterdam, Netherlands	June 93	UNEP	Applications of LCA
Brussels, Belgium	December 93	SETAC	Case studies
Leiden, Netherlands	February 94	SETAC	Allocation
Brussels, Belgium	April 94	SETAC	Impact assessment
Zurich, Switzerland	June 94	SETAC	Impact assessment
Copenhagen, Denmark	October 94	Nordic Council of Ministers	Guidelines of LCA
Brussels, Belgium	December 94	SETAC	Case studies
Florida, USA	February 95	SETAC	Index and ranking systems
Hankø, Norway	March 95	WBCSD, NEP project	Improvement assessment and applications of the LCA tool
Copenhagen, Denmark	June 95	SETAC	LCA impact assessment and toxic emissions LCA in ecolabelling
Stockholm, Sweden	September 95	IVL	Waste treatment processes in LCA
Brussels, Belgium	December 95	SETAC	Case studies

WWF:	WorldWide Fund for Nature
SETAC:	Society of Environmental Toxicology and Chemistry
Procter & Gamble:	The Procter & Gamble Company
TUD:	Technical University of Denmark
UNEP:	United Nations Environment Program
WBCSD:	World Business Council for Sustainable Development
NEP:	Nordic Project for Environmentally-Oriented Product Development
IVL:	Institut för Vatten och Luftvårdsforskning

Status of the method

Table 4.2 shows the status of the methodological development in the various phases and elements in the life cycle assessment, as described in SETAC's "Code of Practice". It is clear that most phases and elements have been determined at the conceptual level, but only the introductory components have been determined operationally.

After the Sesimbra workshop in 1993, SETAC established a number of different working groups to continue work on those parts of the methodological basis which were not yet determined operationally. Most progress has been made within the scope definition phase, where agreement has been approached on a common framework for several of the elements.

The EDIP method represents an operationally determined method which covers all phases, but there have also been earlier attempts to present detailed guidelines for the carrying out of a life cycle assessment along the lines presented in this book. The most important included:

Methodological state-of-the-art according to "Code of Practice" from SETAC

LCA phases and elements	Status
Goal definition	(✓)
Scope definition	
- Functional unit	(✓)
- Choice of assessment criteria	✓
- Time scope	(✓)
- Technological scope	(✓)
- Scope definition of the product system	✓
- Allocation	(✓)
Inventory	✓
Impact assessment	
- Impact potentials (classification and characterization)	(✓)
Global environmental impacts	✓
Regional and local impacts	÷
- Normalization	(✓)
- Weighting	(✓)

Legend

✓: Operationally specified
(✓): Conceptually specified
÷: Not specified

Table 4.2. Status of the life cycle assessment method's components in SETAC's "Code of Practice" (Consoli et al., 1993)

- The Swiss *Ökopunkte* ("Ecopoints") method, which presented one of the first comprehensive assessment methods for use in the LCA (Ahbe *et al.*, 1990).
- The CML method, a Dutch method from the University of Leiden which is often quoted in the international LCA literature, especially for its detailed treatment of the impact assessment phase (Heijungs, 1992).
- Detailed recommendations on the procedure for preparation of inventories for the product system published by the US EPA (US EPA, 1992).
- Guidelines prepared by a group consisting of Scandinavian researchers and consultants within product life cycle assessment (Lindfors *et al.*, 1995).
- The EPS method for life cycle assessment, which is specially oriented towards inclusion of environmental considerations in product development (Ryding, 1995).

Standardization

Work has been continuing since 1993 within the International Organization for Standardization (ISO) on development of international standards for life cycle assessment. A general standard for the LCA area with ISO number 14040 is expected to be finally adopted during 1997. Work is also proceeding on three more detailed standards for the different phases of the LCA:

- one for goal definition, scope definition and inventory analysis,
- one for impact assessment,
- one for interpretation.

The first of these standards will be presented for voting in 1997, the remaining two presumably later.

In May 1995 the European standards organization CEN (Comité Europeén de Normalisation) adopted a so-called CEN report, which, in contrast to a CEN standard, is for guidance only, with recommendations for the elaboration of life cycle inventories for packaging systems.

Work in both the ISO and the CEN is influenced by the fact that life cycle assessment is still a very young discipline, and that in many areas a common perception of how the task is to be carried out has therefore not yet been developed. This is also apparent in Table 4.2.

Applications of LCA

Table 4.3 gives an overview of the contexts in which life cycle assessments are used today (1997).

It is important to be conscious of the goal of the LCA. Even if there are general principles for the LCA, the concrete design of the assessment depends on the application and the type of decision it is intended to support. This is illustrated in Table 4.4, which provides some examples of how the use of an LCA makes different demands on the method with respect to the time over which the result will apply.

Table 4.3. Uses of environmental assessment of products in Denmark, 1996

Active party	Application	Example
Authorities	Community action plans	■ Incineration versus recycling of paper ■ Recyclable glass bottles versus other beverage containers ■ Ranking of industrial products
	Environmentally conscious public purchase	Cars, work clothes, canteen service, office furniture
	Consumer information	Ecolabel
Company	Establish environmental focus	■ Identification of areas of improvement ■ Product-oriented environmental policy ■ Environmental management
	Design choices	■ Choice of concept ■ Choice of component ■ Choice of material ■ Choice of process
	Environmental documentation	Environmental information to costumers
Consumer organizations and other associations of interested parties	Guidance to environmentally conscious consumption	Ecolabel
	Life cycle assessment of community actions	Ecological or conventional farming, transport systems

If, for example, the life cycle assessment is used as the basis for environmental labelling of products, it must provide an actual picture of the product type's current environmental impact. For EU environmental labels the results only have to apply for three years, since EU environmental labelling criteria are to be revised at least every third year.

If, on the other hand, it is used as a part of the basis on which decisions are made in product development, the results must in principle be valid for the entire time during which the product is expected to be in production, plus the product's life span after production ceases.

Table 4.4. Periods of validity for results in various applications of environmental assessments of products

Objective	Application	Support for decision	Time horizon T
Focus	Product development	Background for environmental specifications, design strategies, principles and rules	3 < T < 3 + P + L
	Eco labelling	Identifies important environmental properties for the product category	0 < T < 3
	Community action plans	Identifies environmentally important product groups	0 < T < L
Choice	Product development	Ongoing identification of the best choices from alternative solutions	2 < T < 2 + P + L
	Cleaner technology	Identifies the best available technology by means of LCA	1 < T < 1 + L
	Community action plans	Identifies the best community strategy for a certain problem or product	0 < T < 5
	Consumer information	Documents potential environmental impacts from a certain product	0 < T < 1

Legend
T = Time horizon for which the decision must be valid (examples)
P = Period of time during which the product is produced
L = Lifetime of the product

5. Environmental assessment and product development

The development of new products is a creative process and as such difficult to characterize within a fixed framework. But the process has some basic aspects which should be understood when environmental considerations are to be integrated and the environmental assessment method adapted to product development. Some of the aspects in product development which are most important for the practitioner of the environmental assessment are therefore described here.

The task of the designer

Some of the most important tasks of the designer is to create and to choose among solutions. When a business opportunity is recognized and the goals are determined, the designer must find a solution which fulfils the goals.

The goals for the product create demands on functions in the product: firstly, demands on the general service functions, followed gradually by demands on sub-functions in the product as its concept, structure and details are determined. This is illustrated in Figure 5.1.

The figure illustrates in simplified form some of the decision making processes in the development of two very different products, namely a ship and a refrigerator. The object of the figure is to show how the product is born as a result of a number of decisions from the general concept to the individual detail, and where in this decision making process the most significant environmentally decisions are made.

The example of development of a ship to fulfil the general function "transport between Denmark and Sweden" is not a typical product development project, but in a simple manner it illustrates the significance of the various levels in the decision making process. At the top level there are several conceptually different ways of providing the service: ship, bridge, tunnel, aeroplane ..., and in principle there is an unlimited number of solutions. The choice of "ship" gives rise to a number of other functional demands: create the ability to float, maintain course, change course, create forward motion ..., and the choice of "wind" as a means of creating forward motion gives rise to demands for new functions, and so on. The choices at each level very strongly narrow the space within which decisions can be made, and the framework for the product's environmental characteristics is narrowed at each level. Consider where the most environmentally significant decisions lie. Intuitively, it appears that the choice between ship, bridge, tunnel and aeroplane is very significant, as is the choice between wind, sun, oil and uranium as propellant. The choices at the top level thus appear to mean more in environmental terms than the choices at the bottom levels, e.g. choice of material for the mast and chemicals for protecting it against the weather.

The example of development of a refrigerator is more typical for product development in a company. The figure illustrates several branches in the decision making structure, but still only a small portion of the functions determined in the refrigerator. The most significant choice at the concept level appears intuitively to be the choice of refrigeration system, i.e. the choice of electric heat pump relative to, for example,

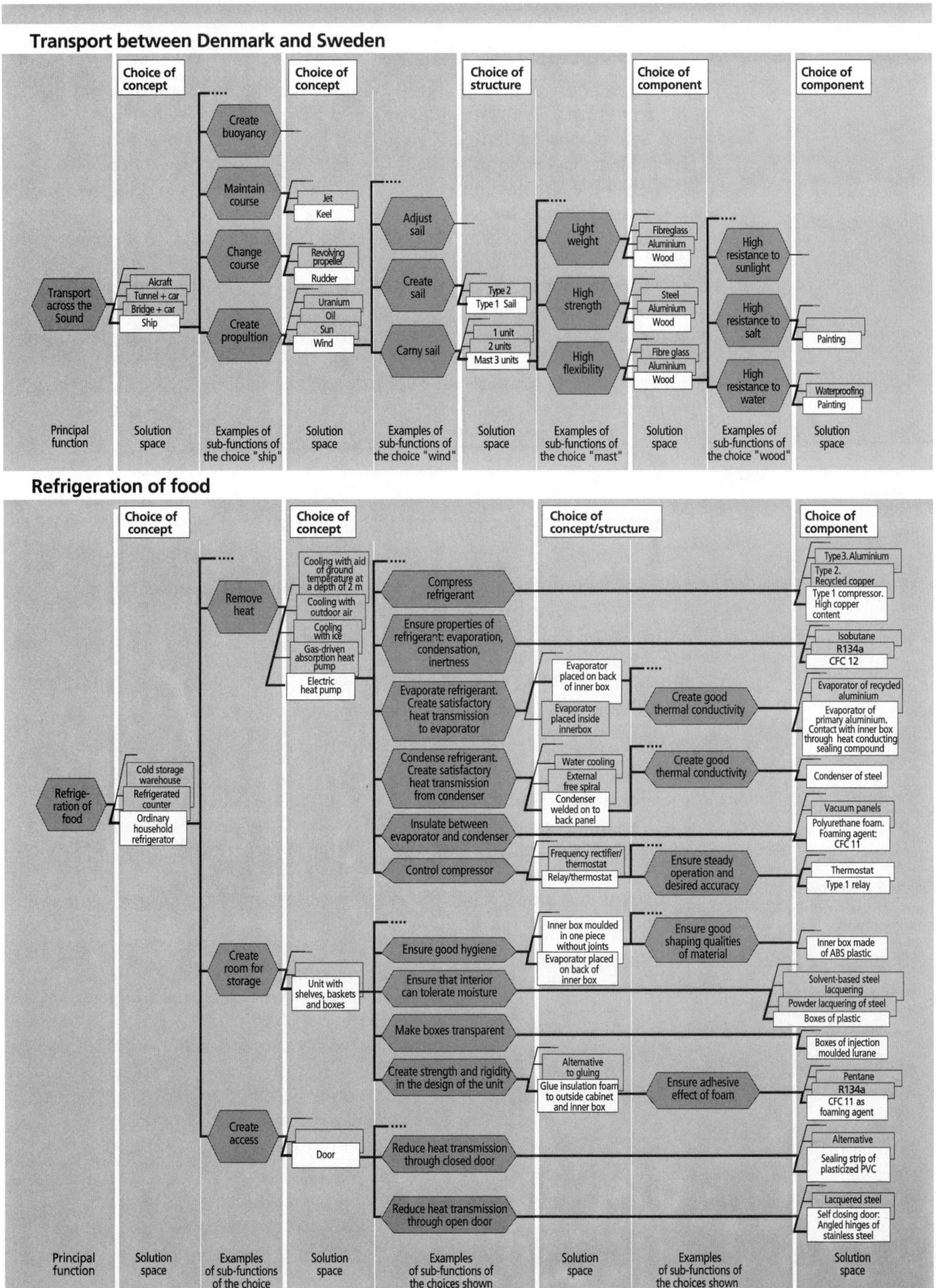

Figure 5.1. Solution space and choices in product development. Some of the choices made are shown in white. Some of the functional demands caused by the choices are shown in dark grey

cooling with outdoor air in winter or placing the unit 2 metres under ground level to make use of the permanent cold there. At the structural level, for example, choices are shown for the placing of the evaporator: considerations of good hygiene require that the evaporator is not placed inside the unit, but directly on the rear of the inner box for ease of cleaning, while considerations of good heat transfer require that the evaporator be placed inside the unit itself. Considerations of hygiene are decisive in this example. At the component level, choice of materials and processes is shown including choice of certain chemicals. In this example an environmentally decisive choice also resides at the component level, namely the choice of CFCs as refrigerant and foaming agent. Note that CFCs in the refrigerant and/or foaming can be avoided both at the concept level, for example by choosing a geothermally cooled or gas-driven unit, and at the component level by selecting alternative chemicals. The example supports the judgment that the choices at the general level are environmentally significant, but it also shows that the choices at underlying levels can be important. An environmental assessment of a refrigerator is presented in Part V, illustrating the environmental significance of many of the choices shown in Figure 5.1.

The product development procedure

The course of product development varies from company to company, but some basic aspects are common. When the first idea for a new product is conceived, the idea and the business basis are analysed, and some goals for the product are set up. The general concepts for the product are then determined, followed later by the details. In some cases, details at the component level are determined from the start, e.g. "build in aluminium" or "build without CFCs".

A picture of the development procedure is shown in Figure 5.2, by the arrow at the top of the figure. The procedure can be more or less formalized or detailed, but in the main, most companies will be able to recognize their product development in the sequence shown.

The object of the figure is to show how the decisions over the course of development narrow the space within which further decisions can be made, as described in the two examples in Figure 5.1.

At the beginning, there is in principle an infinite number of possible solutions, and at the end, when all details have been determined, one solution has been chosen which is put into production. The solutions are illustrated in the figure as product systems with grey contours under the curve. In the early phases they are not known, but in principle they exist as possibilities which can be found. The solution finally chosen is found among the many possible solutions at the beginning, as shown in black in the figure, and among these the environmentally best solution is also to be found.

Figure 5.2. Narrowing of the solution space during product development

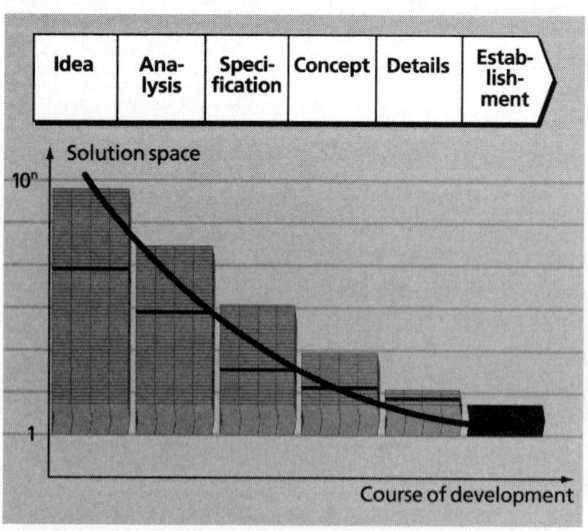

Basic requirements for the environmental assessment method

The aspects of product development shown in Figure 5.2 make some significant demands on the method of environmental assessment and on how it is used in product development. Two pivotal requirements are:

- the designer's decisions must be able to be supported early in the course of development,
- the degree of detail in the environmental assessment must be adapted to its purpose in product development.

The requirement for early support of the decisions

As the environmentally most significant decisions are most probably made in the earliest phases of product development, the environmental assessment must be able to support these decisions.

Environmentally attractive product systems and solutions in the product must be able to be identified as early in the product development as possible, so that environmental considerations can be included in the most significant decisions.

This requirement is met in the EDIP method, basically by its building on the use of reference products to represent the new product. With the aid of the reference products, the largest part of the new product system is predicted, and in this way, points of focus can be designated and goals set for the new product from the very beginning of product development. This is described in more detail in Part II.

Requirements concerning the level of detail

The actual product system for the new product is not known during product development, i.e. the variation in time and place for the product system's processes and their environmental exchanges is not known.

The closer product development comes to the establishment and operation phases, the closer the product system is to being identified. Only in the operation phase is it fully known where the processes are occurring. This is illustrated in Figure 5.3. In product development it is therefore necessary to work without full knowledge of variations in time and place, but with probabilities. This lack of knowledge about the processes' variation in time and place in product development agrees entirely with the degree of detail in the information base for environmental assessment of products. The *probabilities* in product development correspond to the *potentials* in the environmental assessment.

The environmental tasks in product development and the interplay with the designer are described more fully in Part III.

Figure 5.3. The product system's placing and variation in time and place are unknown during product development

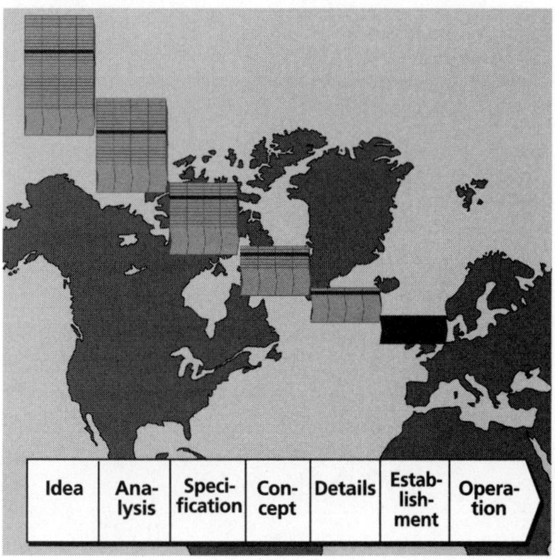

Part II. The Life Cycle Assessment Method

- *What are the method's phases?*
- *What does the method include?*
- *Which methods of assessment are used, and why?*
- *What results can the method yield, and how are they interpreted?*

6. The structure of the LCA method

This chapter provides an overview of the LCA method. It shows how the method is constructed and the types of results which it can yield.

The method's phases

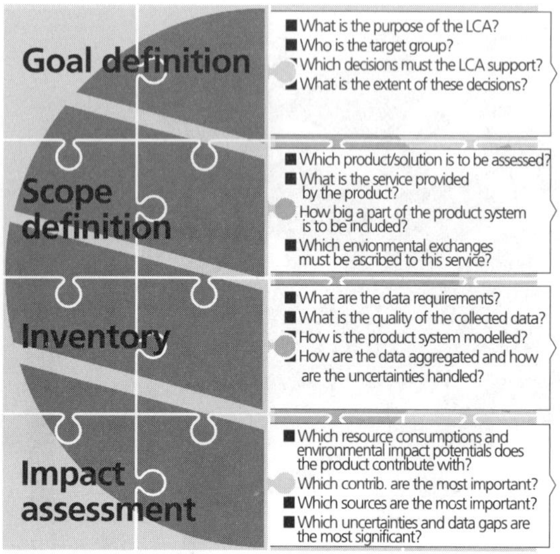

Figure 6.1. The EDIP method's phases and some of the questions answered

The EDIP method contains four main phases: goal definition, scope definition, inventory and impact assessment. The method is thus in agreement with SETAC's general guidelines for the content of a life cycle assessment (Consoli *et al.,* 1993). A procedure is assigned for each individual phase and it is supported by examples and data. The method is designed as a handbook, i.e. solutions are given for the tasks of the LCA. There can be several procedures for some of the methodological elements, and alternatives are therefore indicated after a particular recommendation. The user can thus see where the significant choices in the EDIP method are made.

The core of the EDIP method is valid for most applications, as are the basic questions which the method is intended to answer (see Figure 6.1). The type of results generated in each of the method's phases is also generally applicable (see Figure 6.2). But the concrete procedure in the method and the content of the results depend on the use to which the LCA will be put.

The EDIP method is designed for product development. The method is thus suited to the environmental assessment tasks which arise when environmental considerations are to be built into new products. These

Figure 6.2. The content of each of the method's phases

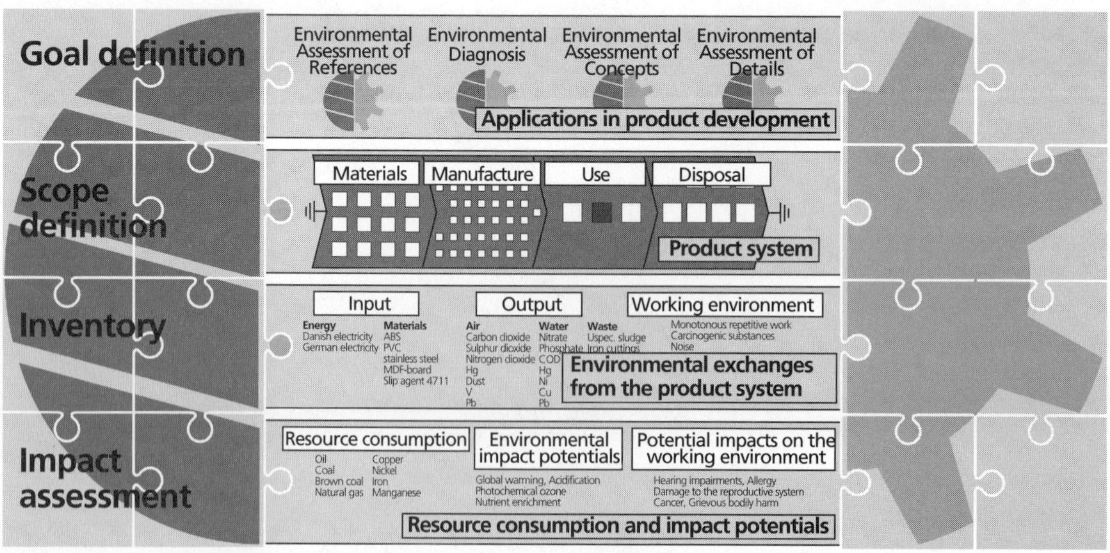

tasks can be divided into four main types called "environmental assessment of references", "environmental diagnosis", "environmental assessment of concepts" and "environmental assessment of details".

The content of each phase

The objective of the LCA is described in the *goal definition.*

How large a part of the product's life cycle is to be included in the LCA is decided in the *scope definition.* A decision is also made here on the criteria on which the product is to be assessed, and the period of time for which the results of the LCA should apply.

In the *inventory,* information is gathered on the exchanges which the product has with its surroundings via the various processes in the product system. These exchanges include both consumption and emission of substances from and to the environment and interactions between the processes and the employees who operate them.

Information from the inventory is interpreted in the *impact assessment.* Potential contributions to various environmental effects are calculated, and the potentials for impacts on the environment and the working environment and resource consumption are weighted in relation to one another in order to increase the suitability of the inventory as a basis for decisions.

Table 6.1 shows a more detailed picture of the elements of each phase.

Some typical results

Figure 6.3 shows some typical results of scope definition, inventory and impact assessment:

- Definition of the scope of the life cycle can often be presented with advantage in a part list for the product.
- The inventory provides information on the exchanges with the surroundings for the various processes, stages of the product life and the complete product system. The results are presented in tables.
- Environmental impact potentials are presented in the impact assessment in tabular form, while the comparison with the background impact and weighting is typically presented as environmental profiles, resource profiles and working environment profiles in the form of figures as shown.

The case histories (presented in the last part of this book) provide an overview of the results of the LCAs on five products.

1. Goal definition

Describes the purpose of the study, including the decisions that it is to support

2. Scope definition

2.1. The object of the study: Defines the product by its service. This includes defining a *functional unit* which describes and quantifies the service provided by the product. The results of the Life Cycle Assessment will then be expressed per functional unit. A joint reference frame is thus defined wherein various solutions can be compared. The functional unit will typically include a time factor, e.g. the lifetime of a product. The scope definition determines the *reference product(s)* which are to represent a way of providing the service. Determines the required quality for the data representing the product system of the reference product.

2.2. Assessment criteria: Defines the impact categories to be used in assessing the product, including resource consumption, environmental impact potentials and potential impacts on the working environment.

2.3. Scope definition for the product system: Determines which part of the product system is to be included in the LCA. This includes the geographical location of the product system.

2.4. Time scale: Defines the period of time within which the decisions based on the LCA are to be valid. Also defines the period of time within which the method considers the potential environmental impacts from the product-related environmental exchanges.

2.5. Technological scope: Defines the technological level that should be represented by the data included in the LCA. This will inter alia be a function of the time scale. Describes the trend analysis and the projection of data.

2.6. Allocation models: Identifies processes that contribute to more than one product. Determines the partitioning keys for the environmental exchanges from these processes. This is done in order to ascribe the environmental exchanges to each separate functional unit. Provides a guide to avoid allocation by expanding the system.

3. Inventory

3.1 Data format: Defines the data to be collected on the process and its environmental exchanges and describes how to characterize the data. Describes how the process is normalized to a unit process that can be used as a basis for modelling the product system. Defines how the relationship of the unit process to the product and the product system is defined during the modelling of products and their product systems.

3.2. Collection of data: Describes data sources and presents data networks. Provides a guide to the procedure during collection of data.

3.3. Computation model: Describes the modelling of the product system, including aggregation of the processes and handling of site-specific considerations. Reviews uncertainties and statistical calculations. Describes searching and sorting of significant parameters.

4. Impact assessment

4.1. Impact potentials: Classifies the various environmental exchanges according to the impact categories to which they contribute. Describes how environmental exchanges are converted into impact potentials. Discusses handling of site-specific aspects.

4.2. Normalization: Describes how the relative sizes of the resource consumptions and the impact potentials are assessed via comparison with a background impact. The background impact is expressed as the average impact per person, and the contribution to the various impact categories from the product is thus expressed as person equivalents.

4.3. Weighting: Describes how the magnitude of the different resource consumptions and impact potentials can be weighted: resource consumption by the scarcity of the resource, environmental impact potentials by politically targeted reduction goals, potential impacts on the working environment by the frequency of reported work injuries.

Table 6.1. The stepwise construction of the EDIP method.

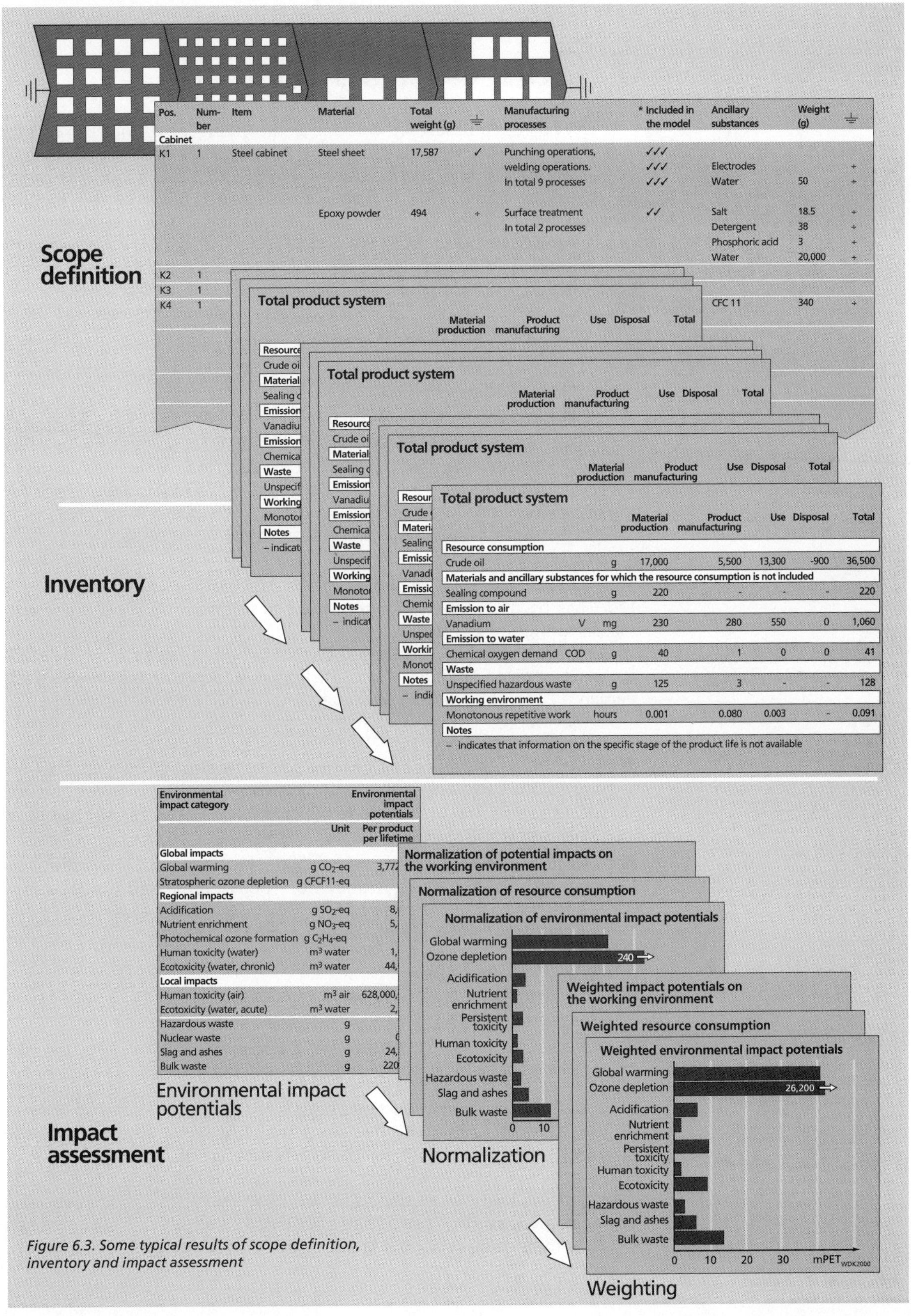

Figure 6.3. Some typical results of scope definition, inventory and impact assessment

7. Goal definition

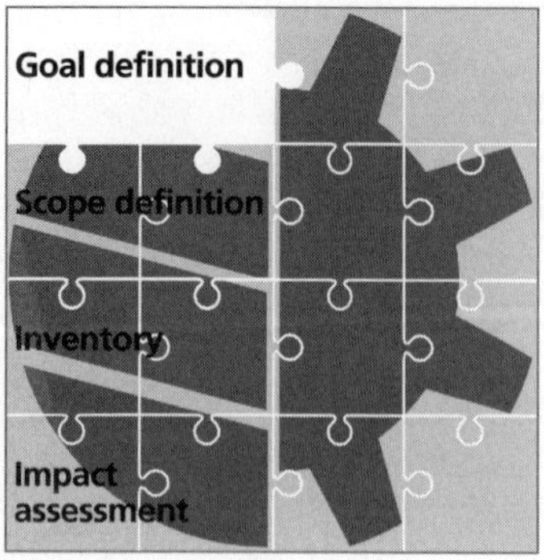

The first phase of the LCA is the fixing of its objective. This includes determining the use to which the environmental analysis will be put and at the same time assessing what it *can* and *cannot* be used for. This is important, because parts of the method depend on the objective, and because the results can be wrongly used to arrive at incorrect conclusions, if their use is not properly defined. When the LCA is presented, there must therefore also be a description of the objective, so that the user has guidelines in use of the results. The goal definition should be reviewed as an exercise which answers the questions:

- For what will the results of the LCA be used?
- What decisions can be made on the basis of the LCA?
- What is the nature and extent of these decisions? In other words, what changes will be produced in the community's activities as a result of the decisions? Which products will the new product supersede and to what extent? What changes in the surroundings will result from the new product?

The nature and extent of the decisions can only be *estimated* when the LCA has been commenced, and it may be necessary to revise these estimates during the LCA.

The EDIP method identifies four main applications of LCA in product development, and they are distributed under two main objectives, namely 1) focusing, and 2) selecting:

Focusing, using the LCA for:

- generating environmental information on the product's expected life cycle. This involves a thorough clarification of the environmental relationships for one or more selected references for the product. This task is called *environmental assessment of references.*
- identifying improvement potentials, including comparing with alternatives, designating environmental focus points and designating where they lie in the product.
  This task is called *environmental diagnosis.*

Selecting, using the LCA for:

- comparing alternative solutions at the concept level.
  This task is called *environmental assessment of concepts.*
- comparing alternative solutions at the level of details
  This task is called *environmental assessment of details.*

Goal definition consists of clarifying what the LCA can and cannot be used for, including the decisions which it must support and the environmental consequences to which these decisions can lead.

The four applications of the LCA are described with explanations and examples in Part III, "Environmental Tasks in Product Development", which is thus an in-depth treatment of this chapter on goal definition.

8. Scope definition

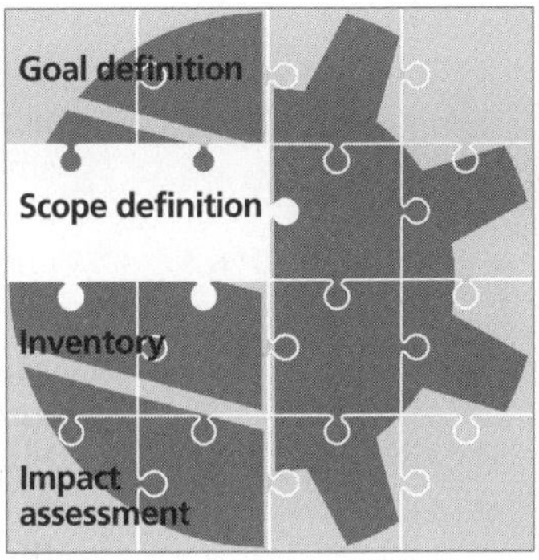

The objective of scope definition is to identify and to define the object of the assessment and to limit it to include that which is significant for the goal of the LCA. Scope definition includes:

- defining the object of the study and expressing it as the service it provides, including defining the functional unit.
- selecting one or more reference products or reference systems to represent the object of the study.
- designating the environmental assessment parameters which are important for the goal of the LCA.
- identifying the environmentally significant processes in the product system, paying respect to the goal of the LCA; setting up a model for the product system on the basis of the references selected, which will include the most significant and exclude other processes; determining the geographic framework for the product system.
- defining the time horizon for which the decisions based on the LCA are to apply, including defining the technological level which should be represented in the product system which is assessed; projecting the LCA of the reference product(s) to apply to the relevant time period; defining the time horizon under which the environmental impact should be seen.
- allocating the environmental exchanges occuring in the product system between the object studied and other user services to which the product system's processes contribute, because they also are part of other product systems.

The object of the study

If the goal of the LCA is to *establish knowledge,* the object of the study is the processes, technologies or parts of the life cycle about which information is desired. As a basis for focusing of the work, the environmentally most significant parts of the product system must be included, and the various stages of the product life must be equally represented.

In applications of the LCA which imply *comparisons* between alternative solutions, products or technologies, the object of the assessment is generally always the service provided to the user, i.e. the function which a given product has. This service must be defined and quantified. In the terminology of the LCA it is called the product's *functional unit.*

In a comparison, it is pivotal that the object of the assessment is always one and the same service for the user. Even if a given product development project is limited to looking at a sub-system in the product, e.g. development of a new door or a new compressor for a refrigerator, the LCA must look at the overall service for the user: in this example, the service provided by the refrigerator. It would be a misunderstanding only to look at the product system and environmental impacts of the door if, for example, the energy consumption in the refrigerator's use stage is altered by a new door. It would similarly be a misunderstanding only to look at the compressor's product system and energy consumption if it will alter the refrigerator's life span because of an altered control or other things.

When the LCA is used as a basis for the focusing of work, the processes in the product system must be equally represented. In comparing alternatives, their overall service for the user must be the same.

This argument also applies to development projects which focus on particular stages in a product's life, for example "design for recycling" or similar. In these cases the object of the assessment must also be the overall service for the user, and the entire product life must be included to ensure detection of whether the solutions in the new product lead to changes in the functional unit or in other stages of the life cycle.

The object of the assessment can only be narrowed down to a limited part of the product life cycle or the service from a sub-assembly of the product if, at the same time, it is ascertained that the overall service for the final user is unlikely to be changed, and that the excluded part of the life cycle is unaltered.

The definition of the scope must thus be based on a holistic view which ensures that the environmentally most significant parts of the product and the product system are equally represented. This does not mean that there may never be any short cuts in an LCA, but merely that when these short cuts are necessary, they must be based on environmentally weighted and well-considered criteria.

The functional unit

The task of the product developer is, as noted in Chapter 5, to seek solutions for given functions in the product and to compare alternatives. It is important to remember that the function of the product, or rather the service which it provides, is central, and that the reference products studied are merely selected ways of providing the service. Figure 5.1 in Chapter 5 illustrates the large number of alternatives available for some of the functions in a refrigerator.

It is therefore a good idea to define the functional unit at once. The object of the assessment in comparing alternatives will thus be clear from the start.

The functional unit must include both a *qualitative* description of the service and a *quantification.* The quantification must specify *the duration* of the service, including the product's life span.

The qualitative description must define the quality level for the service so that products can be compared at a fairly uniform quality level. The quantitative description must specify the magnitude and duration of the service.

In the comparison of alternatives, the user must experience the service as comparable for both the quantitative and the qualitative parts. It is difficult to set exact rules for what is comparable. With respect to the quantitative part, for example, it could be the view that it is not reasonable to compare a 14″ TV with a 28″ TV, while it may be permissible to compare a 20″ TV with a 22″ TV. It is a question of the buyer's selection criteria. With respect to the qualitative part, it may be considered unreasonable to compare a black and white TV with a colour TV, while it can be difficult to define precisely the fine details in the picture which the user experiences as different. Which services can replace one another for the user, and which can therefore be compared, depends on the purpose of the individual project and on the individual company.

Product	Quantity	Duration	Qualities
Egg tray	Egg packaging equivalent to the average consumption of eggs per inhabitant in Denmark	1 year	Average maximum n‰ broken eggs
Television	Reception of TV programs in colour on a 28" screen	6 hours per day for 10 years	Sharpness of image, quality of sound, number of channels, remote control
Pump	Delivering 5 m^3 water per hour with an output pressure of 1.5 bar, or a similar combination according to the pump characteristics	5000 hours of operation during 10 years	Dry running protection and self-priming
Refrigerator	200 l volume cooled to 5°C with an ambient temperature of 25°C	13 years	Accuracy in temperature control, re-evaporation of melt water from defrosting, shelves, boxes etc.
Electrohydraulic control unit for a hydraulic valve	Controlling a hydraulic proportional valve in a hydraulic plant	5 years	Error monitoring, response time, accuracy in control
Paint	Protection of 1 m^2 fir surface on an outdoor facade, facing west and exposed to sun and rain	10 years	Non-dripping, colour, durability in closed container

Table 8.1. Examples of functional units for various products.

The duration of the service is, on the other hand, often easier to define, and it will immediately be evident that it is not correct to compare a TV with an 8 year life span with a corresponding TV with a 12 year life without taking account of the difference. Similarly, the time perspective must also be included to be able to compare a paint which requires maintenance every fifth year with one requiring maintenance every tenth year. Table 8.1 illustrates examples of functional units for various product types.

Note that services with different durations may well be compared. They must merely be normalized to the same duration, for example by comparing 1.3 refrigerators with a 10 year life span with 1 refrigerator with a 13 year life span, and thus expressing all environmental impacts per 13 years.

In environmental comparisons of alternatives, the duration of the service must be the same, and the service must be experienced as comparable by the user with respect to both the quantitative and the qualitative characteristics.

Reference products

The environmental impact deriving from a new product will arise when the life cycle of the product commences. Every single encounter between the product and the community in which it exists will occasion exchanges with the surroundings and will thereby affect the environment. The object of the method used for the environmental assessment is to assist the designer to predict the environmental impacts from the encounters between the product and the surroundings.

The basis of the method is that we know these encounters, or at least most of them. The great majority of new products represent only small changes or improvements to already existing products. This means that by far the largest part of the life cycle which will belong to the new product under development is already known from existing products. It is therefore possible to predict a very large part, perhaps 80-90%, of the environmental impacts from a new product by using experience from already existing products.

One of the first tasks of the environmental assessment is therefore to select one or more existing products or technologies which can represent these anticipated encounters or processes in the product's life cycle. This is called using *references,* and it is a basic principle of the EDIP method.

Even for products which are entirely new on the market, for example the first mobile telephone or the first electric car, it is always possible to assemble a fair reference by piecing together a life cycle with the aid of various known technologies and processes.

A distinction is made between two types of reference, viz. *a service reference,* also simply called the "reference product", and *data references,* most often simply called "data".

- The service reference, or the reference product, is the general representative of the service, i.e. the complete solution selected as an example of a way of providing the functional unit.
- The data references represent selected processes or parts of the life cycle for the reference product or for alternatives to it. The task of the data references is to represent the exchanges with the surroundings from selected processes in the life cycle to be modelled.

The reference product is selected to represent provision of the service. Data references are selected as representatives of processes in the reference product's life cycle and are intended to represent the processes' environmental exchanges.

Selection of the reference product depends on the life cycle to be represented. In many cases, a single one of the company's own products can serve as reference product. In other cases it is necessary to seek several references, including products from outside the company's product range. This is the case if there are significant technological changes in the new product or if a specific framework has otherwise been fixed for parts of the life cycle which cannot be represented by the company's earlier products. The data references do not have to be from the same product category as the development product. Certain processes in manufacturing or disposal of a TV could, for example, well be represented by a PC. The task of the data references is to represent the desired product system as well as possible.

It is necessary for some of the processes in the life cycle that the data derive from a product which is very closely related to the development product. This may, for example, apply to the use process for energy-consuming products such as a refrigerator or a TV, and it can apply to certain production or disposal processes. On the other hand, production of the materials is often independent of the product into which the mate-

LER200

Data reference

Product system and process types	Data specificity: Product-specific	Site-specific	General	Data source type: 1	2	3	4	5	Comments
Production of materials									
Steel		x		x					Supplier Gram
Primary aluminium			x			x			Trade organization
Plastic			x				x		Trade organization
Product manufacturing									
Moulding of plastic	x		x	x		x			Gram: Measurements of energy consumption[a] and emissions[b] carried out by certified laboratory
Processing of steel	x			x					
Coating of steel	x			x					
Use									
Energy consumption	x			x					Continuous measurements on 120 refrigerators over a period of 1 year
Disposal									
Shredding		x	x	x					Largest scrap dealer in Denmark.
Steel		x	x	x					Danish recovery steelworks

Notes

1) Measurements 2) Computation (from mass balance considerations and input data for the process in question) 3) Extrapolation of data from similar process type or technology
4) Extrapolation of data from different process types or technologies
5) Unknown source or qualified estimate

a) Gram and a regional electricity corporation
b) Danish Environmental Center, Ltd

Product-specific data: concern prossesses specifically handling the LER200
Site-specific data: concern data from actual sites in the product system of the LER200, but the inventory of process data has not been done specifically for the LER200
General data: all others

Table 8.2. Extract from database for a reference product, refrigerator LER200 from Gram

rial will enter, and average data for material production can, and often should, be used. Table 8.2 shows a small extract from data sources for the refrigerator LER200, which was chosen as reference product for development of refrigerators at Gram. The table illustrates the data references used to represent various processes in the life cycle of the LER200. For example, the use stage and many production processes were measured directly on the LER200, while many of the other data are simply general. The picture is typical of the database for a reference product.

Selection of data references and collection of data in practice occur only after scope definition of the product system for the reference product.

Assessment criteria

A significant task for the designer is to create solutions and then to choose among alternatives. In many cases such choices imply trade-off situations, where improvements to some of the product's characteristics have to be balanced against worsening in others. In many cases this will also be the situation for the product's environmental characteristics: one alternative may have a lower energy consumption, but then lead to a greater emission of toxic substances than other alternatives.

One of the designer's tasks is to be aware of the situations implying trade-offs, and one of the environmental specialist's tasks is to be able to support the product developer with environmental assessments for choices in such situations.

The considerations to which the designer must pay attention are generally determined by the interests taken in the product and its life cycle. This also applies to environmental considerations. The circle of stakeholders is wide, covering everybody from the production employee through the user of the product and the person responsible for disposal to the authorities, lenders and interest organizations. An analysis of the attitudes and requirements of the various interested parties quickly shows that the considerations, including the environmental assessment criteria, must cover a broad field. This ensures that the relevant aspects of the trade-off situations are covered by the assessment criteria and are thus transparent in the environmental assessment.

Figure 8.1. Interrelationships between emissions, environmental impact potentials and the various impacts and their consequences

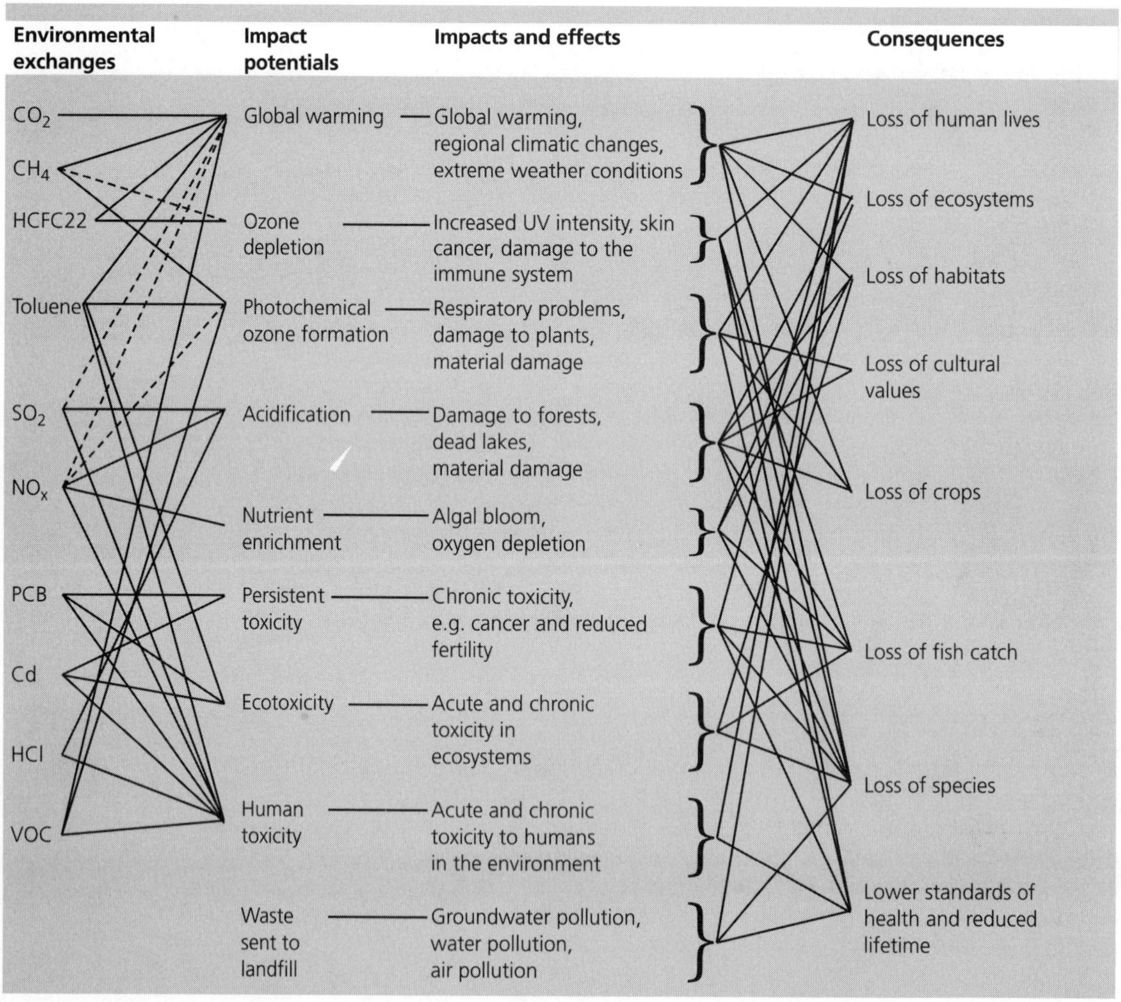

The requirement that the environmental assessment be broadly based is met by the EDIP method, which includes the following general categories of impacts:

- Environmental impacts
- Resource consumption
- Impact on the working environment

These three main categories are basically viewed as being equally important. They are handled as three different aspects, entering into each separate assessment and are not mutually weighted.

The impacts within these main categories are further divided in terms of their geographic extent into:

- Global impacts
- Regional impacts
- Local impacts

This subdivision is of significance for the final part of the assessment, where the contributions to the various impact categories are normalized and weighted, because the character and the mode of action differ for different geographical extents.

Consumption of non-renewable resources is a global impact. Environmental impacts can be global, regional and local, while consumption of renewable resources and impacts on the working environment are local, or in some cases regional.

Environmental impacts

Environmental impacts include impacts on the external environment, including human health. In the definition of what constitutes an environmental impact, it is possible to focus on impacts which lie early or late in the chain of causes and effects for various impacts on the environment. Figure 8.1 gives examples of cause-effect chains.

The impacts and consequences which arise late in a cause-effect chain are often the reason why the impact is viewed as a problem, and thus defined as being able to cause an environmental effect which has to be dealt with. But while the relevance of the impacts for the community is typically increased with progression through the chain, the cause-effect relationships also become more complex. This is evident in Figure 8.2, which illustrates a more detailed section of the cause-effect chain for acidifying substances. The more causal relationships involved between the substance's first impact on the environment and the effect under investigation, the greater is the uncertainty in the quantitative relationship between the magnitude of the emission and the extent of the effect. As noted earlier, real effects cannot be predicted in a life cycle assessment because of the lack of information on exposure of the sensitive parts of the environment.

A further consideration is the large degree of merging between the lowest parts of the cause-effect chains for different types of environmental impacts, as is evident in Figure 8.1. Ultimately, almost all impacts can cause loss of habitats, loss of species and loss of economic and cultural values.

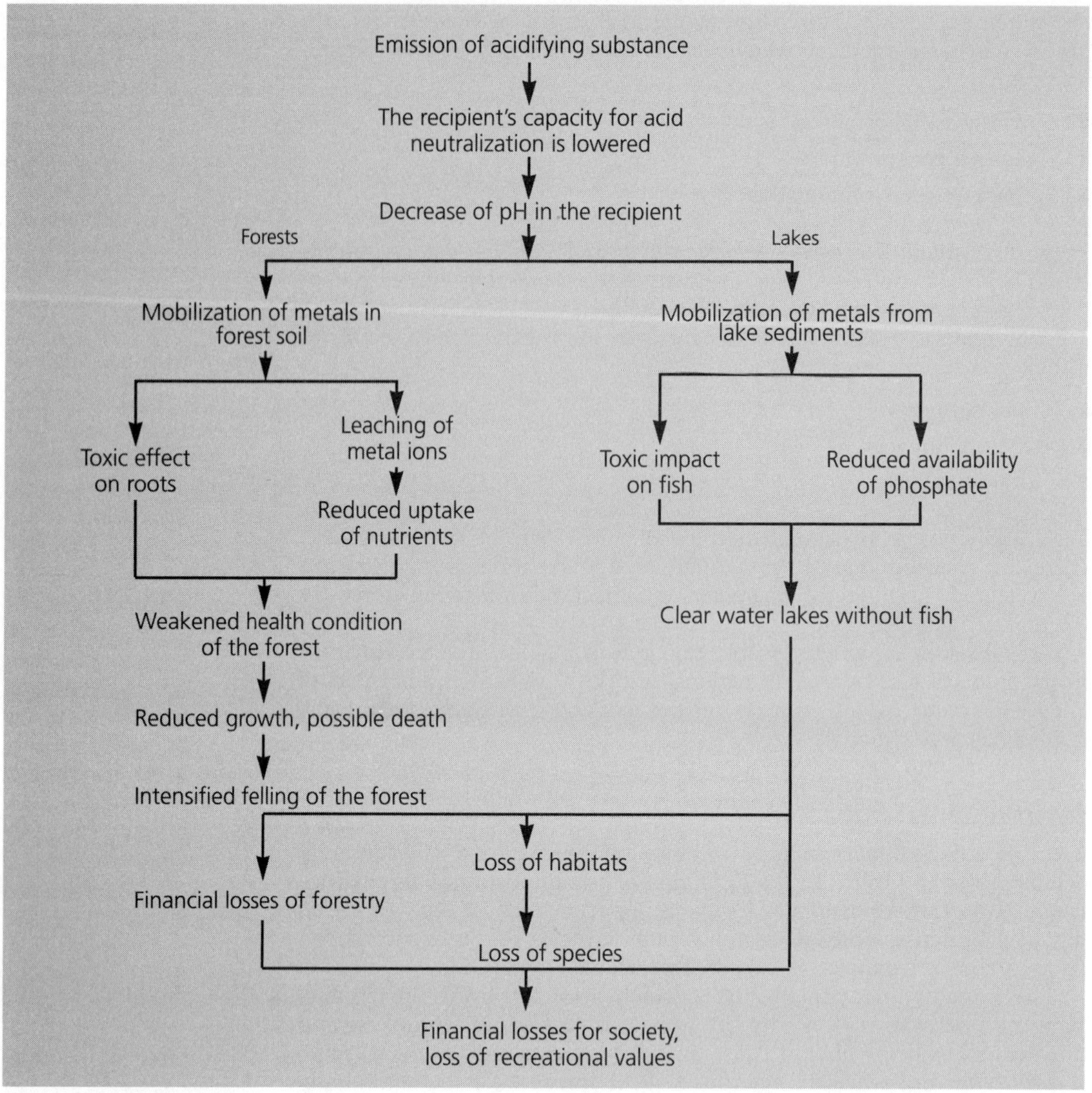

Figure 8.2. Part of the cause-effect chain for acidifying substances in the environment

Because it is not possible to predict actual effects and consequences of the environmental exchanges in the product's life cycle and in order to gain an unambiguous assessment of the impacts on the external environment, the EDIP method defines the categories of environmental impact on the basis of impacts which lie at an early stage in the cause-effect chains.

On this basis, the categories of environmental impacts shown in Table 8.3 are included as assessment criteria. When "waste for landfill" is included as a separate type of environmental impact it is because we are not yet in a position to relate the emissions from different types of landfills to the products in them, and these emissions can therefore not yet be included in an LCA. The waste is subdivided into four categories and the quantities of waste are included in the assessment to represent the possible impacts of landfilling.

	Environment	Resources	Working environment
Global	Global warming Stratospheric ozone depletion	Fossil fuel, e.g. oil, coal, brown coal and natural gas Metals, e.g. Fe, Al, Cu, Zn, Ni, Cr, Mn, Ag and Au Other minerals, e.g. lime, phosphate and salt Others	
Regional	Photochemical ozone formation Acidification Nutrient enrichment Persistent toxicity - Human toxicity from the water compartment - Human toxicity from the soil compartment - Chronic ecotoxicity in the water compartment - Chronic ecotoxicity in the soil compartment		
Local	Ecotoxicity - Acute ecotoxicity in the water compartment Human toxcity - Human toxicity from the air compartment Land filling - bulk waste (non-hazardous) - hazardous waste - slag and ashes - nuclear waste	Biomass, e.g. wood, straw and grain Water, e.g. groundwater, surface water and water for hydro electric power Others	Cancer due to chemical substances Damage to the reproductive system due to chemical substances Allergy due to chemical substances Damage to the nervous system due to chemical substances Musculoskeletal injuries due to monotonous repetitive work Hearing impairments due to noise Grievous bodily harm due to accidents

Table 8.3. The EDIP method's assessment criteria

Simultaneous contributions to several environmental impacts

All illustrated in Figure 8.1, some substances can contribute to more than one environmental impact. An example is oxides of nitrogen (NO_x), which contribute to acidification, nutrient enrichment, photochemical ozone formation and toxicity to humans via the atmosphere. Table 8.4 provides examples of other substances which can contribute to more than one of the selected environmental impacts.

As a rule, these substances should be regarded as contributing to all of the environmental impacts involved, because a contribution to one impact does not exclude a contribution to the others. This is, for example, the case for the contribution made by nitrogen oxides to acidification and nutrient enrichment. The same molecule of NO_x can contribute to acidification and at the same time to nutrient enrichment.

Table 8.4. Examples of substances which can contribute to several different environmental impacts.

Impact	HCFC 22	CH_4	Toluene	PCB	SO_2	NO_x
Global warming	x	x	x		(x)	(x)
Stratospheric ozone depletion	x	(x)				
Photochemical ozone formation		x	x			(x)
Nutrient enrichment						x
Human toxicity			x	x	x	x
Ecotoxicity			x	x		

For some environmental impacts, a contribution to one of the impacts can, however, exclude a contribution to the other. If, for example, a molecule of NO_x is inhaled by a person and thereby contributes to toxicity to humans, the molecule is removed from the atmosphere and it can therefore not simultaneously contribute to acidification or nutrient enrichment. Conversely, if the molecule of NO_x exerts an acidifying and nutrient enrichment impact in e.g. a forest, it can no longer be toxic to humans.

The contributions to toxicity and acidification, or toxicity and nutrient enrichment, are therefore mutually exclusive in principle. Simultaneous contributions from nitrogen oxides to all of the above environmental impacts are nevertheless assumed, because a negligible quantity is inhaled by humans. In practice, inhalation of nitrogen oxides therefore does not reduce the potential for acidification, and the error made is therefore entirely insignificant.

Non-chemical environmental impacts

Apart from the environmental impacts noted in Table 8.3, the product's life cycle can also cause exchanges which at a local level have a physical impact on the environment. Chapter 1 discusses various predominantly local effects of physical influences. This type of impact has not been relevant to environmental assessments under the EDIP programme and they have therefore not been implemented in the EDIP method. For certain types of products, e.g. agricultural products, it can be relevant to include impacts of physical burdens as assessment criteria in the LCA. Inclusion of the physical influences is recommended if they are judged to be relevant for the product, if not otherwise then in the form of a qualitative assessment.

Resource consumption

"Resources" are taken to mean the primary raw materials from which the materials in the product system derive: those involving energy, the construction materials and the ancillary substances. Resource consumption includes consumption of both renewable and non-renewable resources.

Renewable resources are defined as resources which can be regenerated and which will therefore not necessarily be exhausted because of human exploitation. Examples of renewable resources are plant biomass such as wood, straw and grain, or water resources such as groundwater, surface water or dammed water for use in hydroelectric power plants.

Potential impacts on the working environment	Number of trades where the problem is viewed as significant
Hearing impairments and noise causing hearing impairments	14
Noise not causing hearing impairments	4
Insufficient lighting	4
Vibrations to arm/hands resulting in "white fingers"	6
Thermal impacts	6
Musculoskeletal injuries	19
Psychological impacts	10
Chemical impacts and other disorders	18
Sick building syndrome (chemical, biological and thermal impacts)	4
Infectious diseases (chemical, biological and thermal impacts)	2
Accidents	18

Table 8.5. Extent of problems in the working environment in Denmark in a total of 20 different industrial trades (Rasmussen, 1997)

The EDIP method includes renewable resources in the inventory on an equal footing with non-renewable resources, even if their consumption does not exceed regeneration for the resource in question.

Renewable and non-renewable resources can be substituted for each other in many cases, and it can be difficult to predict such future substitutions at the time of product development, for example for fuels. For energy consumption, for example, the task of the product developer is to minimize consumption, while the task of the community in the future is to find the best possible solutions for the energy systems. When the LCA is used in connection with product development, renewable resources should therefore be included so that they can enter into the total optimization.

Non-renewable resources are resources which are not regenerated, or which are regenerated so slowly that the rate of regeneration is without practical significance for their available quantity. Fossil resources such as oil, coal and gas and all metals are examples of non-renewable resources.

Table 8.3 gives examples of resource consumptions included in the inventory.

Impact on the working environment

The working environment is affected by all the influences to which a human being is exposed in his or her work. These influences can be of both positive and negative significance for personal safety and health. Influences with negative significance for safety and health are called impacts on the working environment.

Impacts on the working environment are an entirely local phenomenon, and they are often an integral part of the processes occurring in the product's life cycle. The person in the working environment is exposed to influences of a physical, chemical, biological or psychosocial nature which can lead to nuisance or damage to health.

Table 8.5 provides an overview of the extent of a number of impact categories on the working environment in Denmark. In practice it is not possible to include all types of influences in an LCA. The spectrum is simply too large, and many of the influences cannot be related to individual processes or products, but depend on the way in which the production is organized.

Table 8.3 lists the types of impacts on the working environment selected for inclusion as assessment criteria in the EDIP method. These types of impacts were selected partly because they are the most frequently occurring in Denmark and partly because it is possible to make an inventory of them in a manner which relates them to the processes which cause them.

International consensus

It is common to include environmental impacts and resource consumption as assessment criteria in the environmental assessment of products. In life cycle assessments carried out in countries outside Scandinavia, it is, on the other hand, rare that the working environment is included as an assessment criterion.

In assessment of the potential for environmental impact from the product's life cycle, it is common practice to base the assessment on the potential contribution of the emissions to various environmental impacts. In SETAC and among researchers within life cycle analysis there is general agreement on how contributions to the global environmental impact are to be handled, but there is continuing (1997) discussion of how potential contributions to the non-global types of effects can best be presented. There is also still some divergence in the selection of types of impacts, especially for emissions which contribute to ecotoxicity and toxicity to humans and for physical environmental impact.

In assessment of resource consumption, focus is generally placed on the non-renewable resources, and if renewable resources are included at all as assessment criteria, it is only in those situations where they are used in a non-sustainable manner, i.e. where the rate of consumption exceeds regeneration of the resource.

The EDIP method is in agreement with international practice on which assessment criteria are selected, except for the inclusion of impact on the working environment and use of renewable resources. In certain areas it has been necessary to clarify the definition of assessment criteria, and above all it has been necessary to develop operational methods of assessment and a database for the assessment for a large number of them. This is discussed in further detail in Chapter 10.

Scope definition of the product system

The object of the assessment has been selected, and a reference product has been designated. The task now is therefore to set up a model for the product system for this reference product, and to assess the environmental exchanges expected to occur in this product system.

But the product system contains so many processes that it is necessary to rank them and to limit their number before detailed information is collected.

In short, defining the scope of the product system means including the most significant processes from an environmental perspective, and excluding the insignificant ones, so that the data collection can be focused on the most important. But this involves a paradox, because it means that it must be possible to know what is most significant before data are collected and assessed.

The scope definition of the product system therefore actually consists of an introductory rough environmental assessment or a *screening,* which is used to plan the data collection for the actual environmental assessment. The screening is based on the environmental specialist's experience and on the instructions in the EDIP method.

The scope definition requires assessment of the product system before data for the assessment are collected.

The first step consists of establishing an overview of the product system for both the nature and the number of processes, and as far as possible of the geographic location of the processes (see Figure 8.3). As illustrated in the figure, the nature of the product system differs in the life cycle stages to and including product manufacturing and the stages following manufacturing.

In the extraction of raw materials, production of materials and product manufacturing, the number of different types of processes is very great, and each process typically occurs in only one or a few locations. For use and disposal of the product, the number of different types of processes is, on the other hand, modest, but the geographic variation can be very great for products which are exported. This sets different conditions for how the overview of the product system is established, and the product system is therefore also divided in the middle, as shown in the figure.

For simple products which consist of few components and few materials, the overview can be arranged as a flow diagram, where the processes are indicated as boxes (see the principle in Figure 3.8 in Chapter 3). This form is seen in many of the LCAs carried out on packaging. But as soon as the product is more complex and has a large hierarchy of subassemblies, it is difficult to establish an overview in a flow diagram, especially for processes to and including product manufacturing. In such cases, the product system should instead be modelled in a structure as shown in Table 8.6.

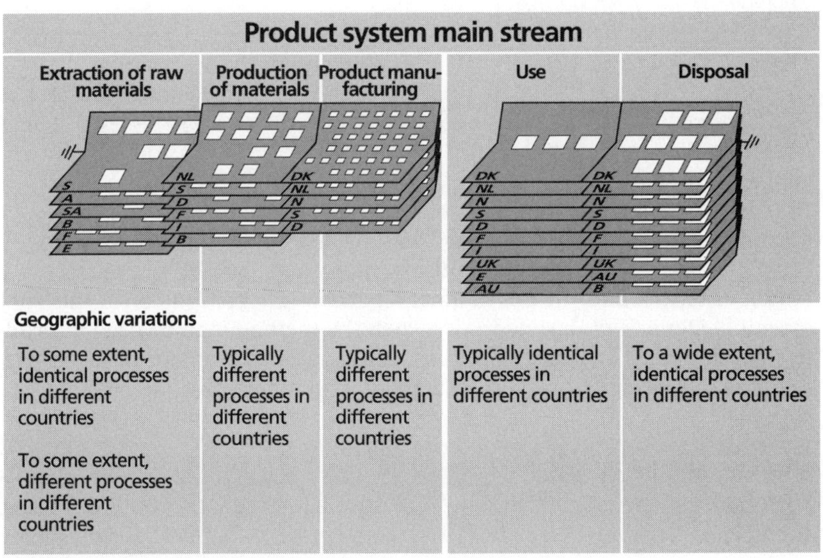

Figure 8.3. It is useful to divide the product system in two during scope definition

The parts list. The product system to and including product manufacturing

Most companies have accurately specified the material content and the manufacturing processes for each product in a so-called parts list for the product.

The parts list shows a hierarchy of the product from the top level, which is the product itself, to the bottom, which is the materials and components which the company purchases. All materials and processes are typically contained in a parts list, which also often includes the consumption of ancillary substances used in the processes. The parts list is structured in a hierarchy which corresponds to the product's structure and to the sequence of processes as illustrated by Table 8.6. The parts list is usually computerized. If not, computerization is recommended for ease of subsequent computing, searching and sorting.

For complex products, the product system is identified to and including manufacturing on the basis of the parts list, from which the scope definition is also made. The parts list is therefore central to the method. If the reference product or parts of it are not found in the company's own product range, creation of a parts list for the missing parts is recommended.

Depending on which data have already been included in the parts list, certain information will probably have to be added for use in the method, while other information can be omitted. There must therefore be a transformation from the parts list already found for the product in the company to the parts list to be used for the LCA. The information required in the parts list for use in the LCA is given in Table 8.6, which illustrates a part of the hierarchical product structure for a refrigerator, the LER200, which was used as reference product by Gram in the EDIP programme. For the sake of the overview, transport processes have been omitted from the table.

Four significant elements in the parts list structure should be emphasized: *1) hierarchical construction, 2) use of key units, 3) process sequence, and 4) grouping of small items.*

Hierarchical construction.

The parts list in Table 8.6 reflects the product's hierarchy via numbering of the levels in the product. The cabinet unit in the example (1.1) is composed of cabinet (1.1.1), plastic box with evaporator (1.1.2) and PUR foam (1.1.3), and the assembly occurs in two processes followed by a foaming process, as shown under manufacturing processes for the cabinet unit.

The processes and materials can be cited in this structure in the sub-assembly into which they enter, and by assigning the sub-assembly a unique name or ID number, this identification can be used later as a search criterion, and the environmental exchanges computed for this sub-assembly alone. This is a necessity when the method is to be used in product development because a rapid modelling and assessment of the consequences of replacing total sub-assemblies with others must be possible.

Use of key units

The quantity of all materials, processes and ancillary substances must be specified by their key unit, i.e. the unit to which the environmental exchanges can best be related. The concept of a key unit is considered more fully in Chapter 9. For materials and ancillary substances, the key unit is always the gram, but for processes it can be highly varied, and it must therefore be specified for each process.

Process sequence
The sequence of manufacturing processes in the parts list follows the sequence in the processing of the material or the component. The first process occurring for the steel plate used in door construction (1.2.1) when it enters the factory is the punching. This is followed by welding and the final surface treatment, where the epoxy powder is applied and where detergent, phosphoric acid and salt are used as ancillary substances. The sub-assembly 1.2.1 is then ready and can be installed, foamed and cleaned together with the door's other sub-assemblies.

Grouping of small items
Small items of the same material can be gathered into groups before scope definition. Table 8.6 shows a number of examples under "miscellaneous". Typical examples are O-rings, screws and electronic components such as condensers, resistors, transistors, etc.

The company's parts list for the product must be transformed into a form which can be used in the LCA, and it then forms the basis for scope definition of the product system to and including the manufacturing stage.

The product system during use and disposal

The use and disposal stages can be set up in a parts list structure in the same manner, and this should be done to retain the same form of presentation for the entire product system. Table 8.7 shows an example.

Both use and disposal of the product vary from country to country, and all countries which are significant export markets should therefore be represented separately. For products exported to a large number of countries, the table form provides the clearest overview. The columns have the same meaning as in Table 8.6 for the product system to and including manufacture. The column "fraction" here is the *percentage* of the products which pass through the process in question. In the theoretical example in Table 8.7, 20% of the products are sold and used in Denmark. Disposal of a total of 5% of the products is to landfills in Denmark, while 10% are incinerated and 5% are shredded. The disposal route for the material fractions after shredding is then seen. The columns "volume" and "unit" specify the full "volume" of process entering into the product system for one product expressed in the process's key unit, e.g. *10 years of use, 400 kgkm* of transport and *6800 g* of steel recycling. Multiplying "fraction" and "volume" thus gives the correct volume for the complete product system.

The overview of the product system in Table 8.6 to and including manufacturing and in Table 8.7 for use and disposal are now the basis for scope definition. Space can be provided for marking the scope in some columns next to "materials", "processes" (manufacturing, use and disposal, and transport) and "ancillary substances", as shown in the tables in the columns marked with * or ⏚. In a table like this, an overview of the scope definition can easily be gained in practice. The three columns cover both the main flow and the lateral flows in the product

Item			Material					
					Quantity			
No.	Number	Name	Name	Unit	μ	σ/μ	D	⏚
1		LER200						
1.1	1	Cabinet unit						
1.1.1	1	Cabinet						
			Epoxy powder	g	494	0.01	logN	✓
1.1.1.1	1	Shroud	Steel sheet	g	8,570	0.01	logN	✓
1.1.1.2	1	Back panel	Steel sheet	g	7,000	0.01	logN	✓
1.1.1.3	1	Base incl. slide bar						
1.1.1.3.1	1	Sub-base	Steel sheet	g	370	0.01	logN	✓
1.1.1.3.2	1	Base	Steel sheet	g	1,270	0.01	logN	✓
1.1.1.3.3	1	Angle slide bar	Steel sheet	g	52	0.01	logN	✓
1.1.2	1	Plastic box incl. evaporator						
1.1.2.1	1	Plastic box	ABS sheet	g	3,320	0.01	logN	✓
1.1.2.2	1	Evaporator	Aluminium	g	1,340	0.01	logN	✓
1.1.3	1	PUR foam	Polyalcohol	g	2,800	0.01	logN	✓
			Isocyanate	g	1,800	0.01	logN	✓
		etc...						
1.2	1	Door						
1.2.1	1	Door casing	Steel sheet	g	4,300	0.01	logN	✓
			Epoxy powder	g	160	0.01	logN	✓
1.2.2		Door lining	ABS sheet	g	1,800	0.01	logN	✓
1.2.3		PUR foam	Isocyanate	g	860	0.01	logN	✓
			Polyalcohol	g	614	0.01	logN	✓
1.2.4		Handle	Aluminium		260	0.01	logN	✓
		etc...						
1.3	1	Accessories						
1.3.1		Wire shelf with large back	Steel wire	g	1,646	0.01	logN	✓
			Epoxy powder	g	40	0.01	logN	✓
		etc...						
1.4		Miscellaneous						
1.4.1		Joint filler	"Thermogrip"	g	45	0.05	logN	÷
			Contact sealing compound	g	117	0.05	logN	✓
			White powder	g	100	0.05	logN	÷
1.4.2	10	Remaining steel items	Steel	g	20	0.05	logN	÷
1.4.3	5	Remaining ABS items	Regranulated ABS	g	47	0.05	logN	✓
1.4.4	6	Nylon items	Polyamide	g		0.05	logN	✓

Table 8.6. Left half. The product system to and including product manufacturing. Extract from a parts list for the refrigerator LER200, Gram. Some processes are omitted and some figures are hypothetical

Manufacturing processes						Ancillary substances					
		Volume						Quantity			
Name	Unit	μ	σ/μ	D	*Included in model	Name	Unit	μ	σ/μ	D	⩲
Assembling	s	2.300	0.05	logN	✓						
Assembling	s	2.000	0.05	logN	✓						
Foaming	kg	4.6	0.005	logN	✓	CFC 11	g	340	0.01	logN	✓
Spot-welding	m	6.0	0.01	logN	✓						
Coating	m^2	3.1	0.01	logN	✓						
						Detergent	g	38	0.01	logN	÷
						Phosph. acid	g	3	0.01	logN	÷
						Salt	g	19	0.01	logN	÷
Punching	m	3.0	0.01	logN	✓						
Punching	m	0.002	0.01	logN	÷						
Bending	m	1.2	0.01	logN	✓						
Spot-welding	m	1.2	0.01	logN	✓						
Cutting	m	0.6	0.01	logN	✓						
Plunging	m^2	0.001	0.01	logN	✓						
Punching	m	0.07	0.01	logN	÷						
Stamping	m^2	0.001	0.01	logN	✓						
Cutting	m	0.5	0.01	logN	✓						
Bending	m	0.02	0.01	logN	÷						
Assembling	s	480	0.01	logN	✓						
Cutting	m	5.0	0.01	logN	✓						
Vacuum forming	kg	4.0	0.01	logN	✓						
Subcontractor					✓						
Assembling	s	300	0.05	logN	✓	CFC 11	g	214	0.01	logN	✓
Foaming	kg	1.7	0.005	logN	✓						
Cleaning	s	60	0.05	logN	✓						
Punching	m	1.2	0.01	logN	✓						
Welding	m	0.05	0.01	logN	✓						
Coating	m^2	1.0	0.01	logN	✓	Detergent	g	2	0.01	logN	÷
						Phosph. acid	g	0.2	0.01	logN	÷
						Salt	g	6	0.01	logN	÷
Vacuum forming	kg	1.8	0.005	logN	✓						
Trimming	m	3.6	0.05	logN	✓						
Extrusion	kg	0.26	0.01	logN	✓						
Bending	m	0.6	0.01	logN	✓						
Cutting	m	0.6	0.01	logN	✓						
Spot-welding	m	0.1	0.01	logN	÷						
						Detergent	g	1.3	0.01	logN	÷
						Phosph. acid	g	0.1	0.01	logN	÷
						Salt	g	0.8	0.01	logN	÷
Mostly standard components					÷						
Injection moulding	kg	0.05	0.05	logN	✓						
Mostly standard components					÷						

Table 8.6. Right half. Legend: *μ = mean,* *σ/μ = coefficient of variation,* *D = distribution function*
**, ⩲ = columns for marking of scope*

system, and the scope definition is made by going through each single material, each process and each ancillary substance, and deciding whether or not they are to be included:

- For the material: is the material's product system back to the extraction of resources from nature to be included?
- For the process: is it to be included?
- For the ancillary substance: is its product system back to the extraction of resources from nature to be included?

Criteria for scope definition

As noted, the paradox of the scope definition is that it must designate the environmentally most significant parts of the product system before the environmental assessment has been carried out. In theory it is of course impossible. In practice, it simply means that the environmental specialist uses his or her experience for an introductory environmental assessment and then proceeds from this. The scope definition is then gradually adjusted in an iterative process as environmental information on the product system is collected, and the significance of the individual processes in the product system is clarified.

The criteria for scope definition are based on the process's resource consumption and potential impact on the environment and the working environment. However, in practice it is not feasible to analyse the contributions to all resource consumptions and categories of impacts for every single material, process and ancillary substance. It must be remembered that, at this point, it is a question of a delimitation and not of the actual environmental assessment. The latter will only be carried out after the scope definition. The criteria should therefore be simplified into a broader grouping:

- Consumption of materials, including the resource content of the materials.
- Energy consumption. Covers energy sources and energy-related environmental impacts.
- Consumption and emission of chemicals. Covers especially toxicity.
- Other, especially physical impact on the environment and impacts on the working environment not covered by the other three groups.

Materials, processes and ancillary substances are revised on the basis of these considerations, and in each of the scope definition columns in the parts list they are marked with a ✓ or a - to indicate whether or not they are to be included.

The more detailed criteria for scope definition are covered in Section 21.3.3 in Part IV of the book, the Toolbox.

Model ready for computation within the product system

The following inventory and computations in the LCA are made on the basis of the model for the product system established as the result of the scope definition. It is therefore useful to computerize the parts list and the scope definition, using the same program as for processing of data for the inventory for computations in the LCA, so that the model for the product system is ready to process when the scope definition is finalized.

					Processes						Ancillary substances					
							Volume						Quantity			
No.	Item	Loca-tion	Material/ sub-assembly	Fraction	Name	Unit	μ	σ/μ	D	*Incl. in model	Name	Unit	μ	σ/μ	D	⏚
17.1	Whole product	DK	The total product	0.20	Use	year	10	0.30	logN	✓						
				0.20	Maintenance	number	2	0.10	logN	✓						
				0.20	Transport	km	30	0.10	logN	✓	Petrol	g	2.100	0.30	logN	✓
				0.05	Transport	kgkm	800	0.10	logN	✓	Diesel	g	80	0.20	logN	✓
				0.05	Landfill	g	40,000	0.10	logN	✓						
				0.10	Transport	kgkm	1,000	0.10	logN	✓	Diesel	g	100	0.20	logN	✓
				0.10	Incineration	g	40,000	0.10	logN	✓						
				0.05	Transport	kgkm	400	0.10	logN	✓	Diesel	g	40	0.20	logN	✓
				0.05	Shredding	g	40,000	0.10	logN	✓					logN	✓
			Steel	0.05	Transport	kgkm	1,020	0.20	logN	✓	Diesel	g	102	0.20	logN	✓
				0.05	Steel recycling	g	6,800	0.10	logN	✓	Aluminium	g	3		logN	÷
			Copper	0.05	Transport	kgkm	250	0.20	logN	✓	Diesel	g	25	0.20	logN	✓
				0.05	Copper recycling	g	840	0.10	logN	✓	Oxygen	g	400		logN	✓
			Aluminium	0.05	Transport	kgkm	150	0.20	logN	✓	Diesel	g	15	0.20	logN	✓
				0.05	Aluminium recycling	g	1,250	0.10	logN	✓	Various salts	g	110		logN	÷
			Plastic, glass and others.	0.05	Transport	kgkm	1,230	0.20	logN	✓	Diesel	g	123	0.20	logN	✓
				0.05	Landfill	g	30,800	0.10	logN	✓						
17.2	Whole product	D	The total product	0.5	Use	year	10	0.30	logN	✓						
				... etc.												

Table 8.7. The product system for use and disposal. Hypothetical figures. Legend as in Table 8.6. The coefficient of variation is for "fraction x mean".

The computer program developed for the EDIP method (Danish Environmental Protection Agency, 1997) is able to import a parts list, and the transformation of the parts list, modelling and scope definition of the product system can take place in this program. The modelling and the calculations are covered in Chapter 9.

Delimitation in time

The overview established for the product system of the reference product reflects the situation as it is at the time of inventory. This means, for example, that the current distribution of sales in countries and the current disposal routes in the various countries are the basis of the established model of the product system. When an inventory of the environmental exchanges from the processes in the product system after scope definition is to be made, current data from existing processes are also used. It is often difficult to generate even current data for all processes, and much of the data will be several years old.

Time horizon for the validity of the LCA

The decisions made on the basis of an LCA during product development must, however, be valid for some time. Several years will often pass, up to 4-5 years, from the time to which the information in the LCA refers until the product is put into production and its life cycle commences. The product manufacturing period may be 5-10 years, and the product's life span 10-15 years or more. Depending on which product and which processes in the product system are involved, this means that the period of validity for the decisions can be 5-30 years into the future from the time of assessment of the reference product.

Figure 8.4 shows examples of the chronological placing of the life cycle stages of various products, namely an electro-hydraulic control unit for a hydraulic valve, a refrigerator, two TVs, a high pressure cleaner and two chairs. It is clear that there are major differences in the time horizons within which the decisions will have to apply. The LCA must be valid 5-30 years into the future for the four energy-consuming products and for still longer for the two chairs.

The information entering into the LCA should therefore be projected as far as possible to apply to the period in which the decisions have to be valid. See the section on technological scope definition.

Time horizon for the impact of the exchanges with the environment

A general reason for including environmental criteria in the development of new products is to take account of the conditions on earth which future generations will inherit. In principle there is therefore no upper time limit for which impacts are considered relevant.

There is a very different time horizon for the impacts which the exchanges from the processes in the product system can cause. Some impacts will arise immediately after occurrence of the exchange, while for others there will be delays of many years. Acidifying gases and hydrocarbons will be able to contribute to acidification and photochemical ozone formation for a few days after their emission. But the substances which are breaking down the ozone layer today were emitted 2-50 years ago, and the person who suffers cancer today may have been exposed to the

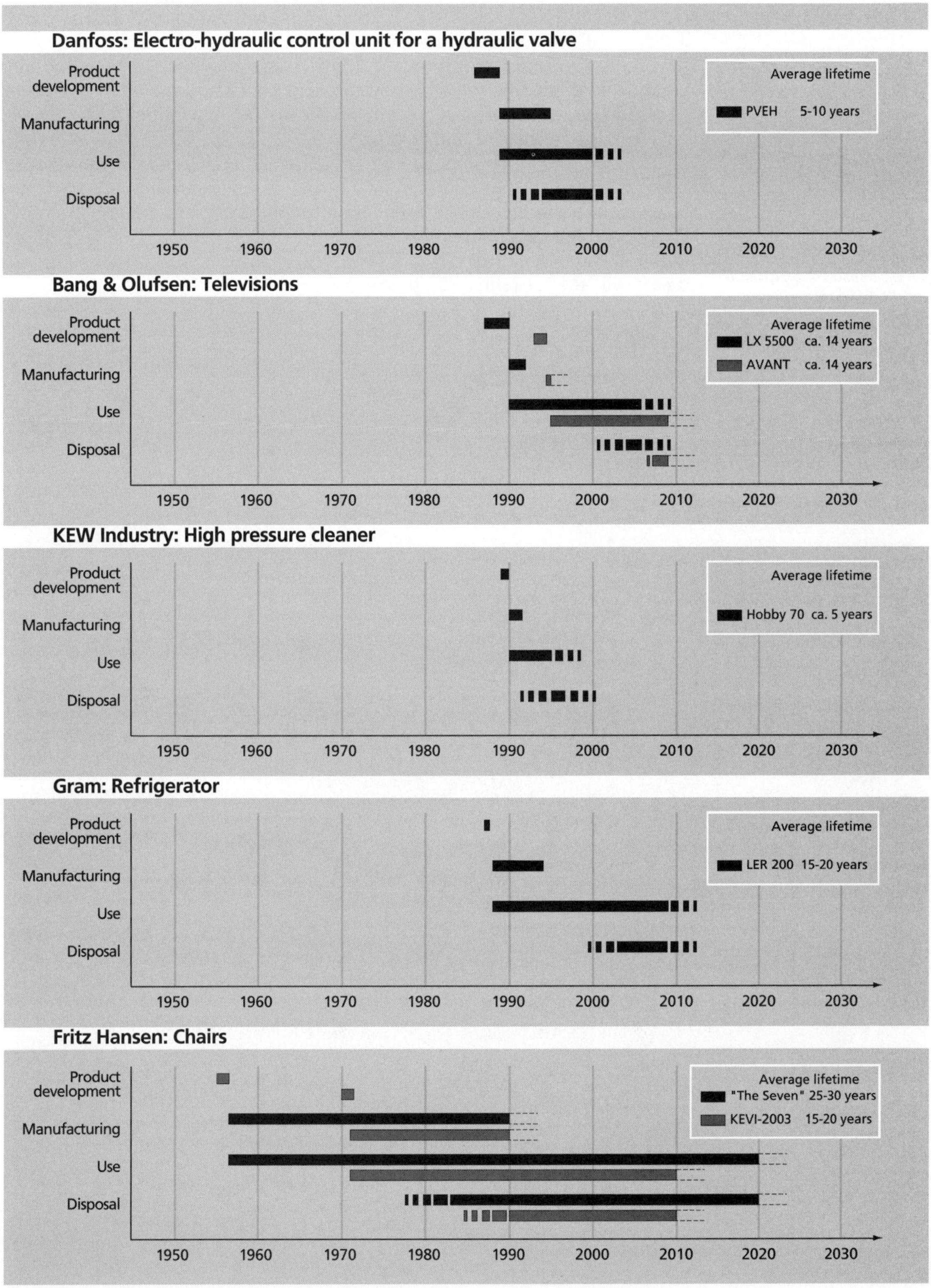

Figure 8.4. Time span for product development, product manufacturing, use and disposal for various products

causal agent 10-20 years ago. Table 8.8 shows the time horizons for a selection of impact categories.

In principle, all resource consumption and potential impacts on the environment and the working environment should be included. In practice, however, it is of course impossible to include impacts which we have no opportunity of anticipating.

An example of conditions which cannot be anticipated is the fate of many materials in landfills and in the natural environment. We have no idea of the ultimate fate and the time involved for e.g. plastic, because the speed of breakdown is too slow to have been observed. The method therefore includes "waste" as a category of impact in its own right, so that it will at least reflect the problem.

Table 8.8. Time horizons for various impacts of an exchange with the environment

Environmental impacts	**Time horizon**
Global warming	As long as the gas is present in the atmosphere, between zero and hundreds of years
Ozone depletion	As long as the gas is present in the atmosphere, between 2-5 years and several hundred years
Photochemical ozone formation	From hours to a few weeks, depending on the reactivity of the gas
Acidification	From days to years
Nutrient enrichment	From days to years
Persistent toxicity	Months for the most toxic substances and most sensitive organisms, and decades for the most persistent substances
Human- and ecotoxicity	From hours for acute toxicity of the most toxic substances from point source emissions to decades for chronic impacts from substances that are discharged continuously or at intervals.
Waste	From weeks for airborne emissions from landfills, to centuries for the degradation of the most long-lived substances and percolation to the groundwater
Resource consumption	**Time horizon**
Non-renewable resources	For the irreversible consumption of fossil resources: millions of years
Renewable resources	Depends on the regeneration rate of the resource
Impacts on the working environment	**Time horizon**
Cancer	From 1-2 decades, depending on the specific type of cancer
Damage to the reproductive system	From months to decades, depending on the type of damage
Allergy	From days to years, depending on the strength of the substance and the sensitiveness of the person
Damage to the nervous system	From months to decades, depending on the type of damage
Hearing impairments	From months to decades, depending on the type of damage
Musculoskeletal injuries	From years to decades, depending on the type of work
Accidents	From days to decades, depending on the type of damage

Definition of technological scope

To be able to project the LCA to apply to the period in which the product is on the market, the trends in technological developments must be known for the processes which comprise the product system. Important trends should therefore be investigated so that it is possible to select a technological level in the data base for the assessment which will reflect the future as accurately as possible.

Trend analysis

Figure 8.6 illustrates some developmental trends in environmentally significant key figures for a number of processes and environmental exchanges involved in the life-cycles of many products. The figure shows that many of the most important key figures change quickly with time, and that it is therefore important to take account of technological development when the LCA is used as a basis for decisions.

Projection of the most important key figures

It is neither realistic nor relevant to investigate technological trends for all processes in the product system. It is sufficient to designate the most important processes and parameters and to investigate the development in these. In practice this means that the LCA of the reference product is initially carried out on current data. The most significant processes and key parameters for the product system are then designated on this basis, and the trend analysis is carried out by projecting these. The projection consists of assessing the future magnitude of the most important key figures, changing these in the modelling of the reference product and then repeating the calculations in the LCA.

Figure 8.5 shows an example of a trend analysis for selected processes and other parameters in a product system. Note that only processes and key figures in the product system of the reference product are projected. The product technology itself must not be projected, because the purpose of the projected LCA of the reference product is to be used as the basis for assessing and ranking priorities for changes in the product.

The figure shows some of the key figures which will be environmentally significant for many products: production of the ordinary construction materials, energy systems, disposal routes, degrees of recycling of materials in the community, etc. The developments in these quantities are of considerable significance for how development of new products is focused.

The figures in the table are examples, but they reflect realistic values. The projection should be equally balanced for the entire product system for all key figures which change significantly. The LCA can otherwise be distorted.

Figure 8.5. Examples of projection of some important key figures for a product system

		Reference product system	Estimates for next generation
The material stage	**Year**	**1993**	**2010**
Steel:			
■ energy consumption	MJ/kg	25	20
■ content of recycled steel	%	20	30-40
PE:			
■ process energy consumption	MJ/kg	20	15
Aluminium:			
■ energy consumption	MJ/kg	170	120
■ content of recycled aluminium	%	0	20
The manufacturing stage	**Year**	**1993**	**2010**
Man-hours per product	hours	8	7
Overhead energy per product	MJ	250	200
The use stage	**Year**	**1993**	**2020**
Production of electricity and heat in Denmark:			
■ energy efficiency	η	54	60-65
■ coal (MJ)	%	91.6	~50
■ oil (MJ)	%	2.9	~5
■ gas (MJ)	%	3.6	~15
■ renewable energy (MJ)	%	1.8	~30
German electricity: ...etc...			
The disposal stage	**Year**	**1993**	**2030**
Disposal routes, Denmark:			
■ landfill	%	20	5
■ incineration	%	40	10
- energy efficiency	η	60	75
■ recycling of steel content	%	40	90
■ degree of steel recycling in society	%	56	80
■ recycling of copper content	%	90	95
■ degree of copper recycling in society	%	60	90

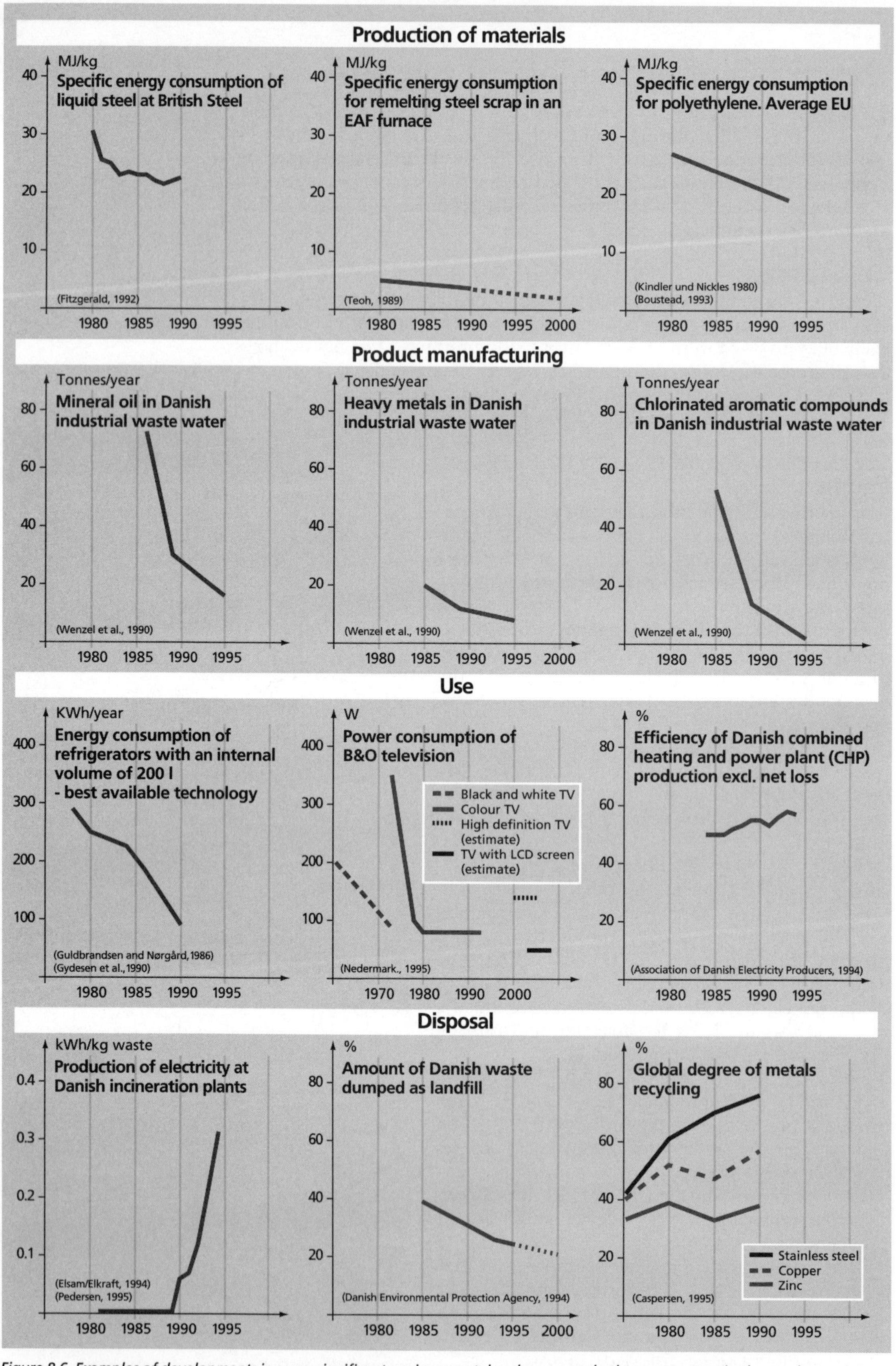

Figure 8.6. Examples of developments in some significant environmental exchanges and other parameters in the product system

Allocation

As noted previously life cycle assessment of products can be defined as:

defining and quantifying the service provided by the product, identifying and quantifying the environmental exchanges caused by the way in which the service is provided, and ascribing these exchanges and their potential impacts to the service.

It is central to the task of LCA that it concerns the service or the functional unit for the product. This means that it is also pivotal to identify the processes and environmental exchanges which are caused by this service. The "white processes" in the product system in Figure 8.7 are a consequence of the "dark grey process", i.e. the use process or the service provided to the final user.

The picture in Figure 8.7 is simple. There is only one use process, and all other processes are a consequence of this process. All exchanges from all processes in the product system can therefore be fully ascribed to this use process.

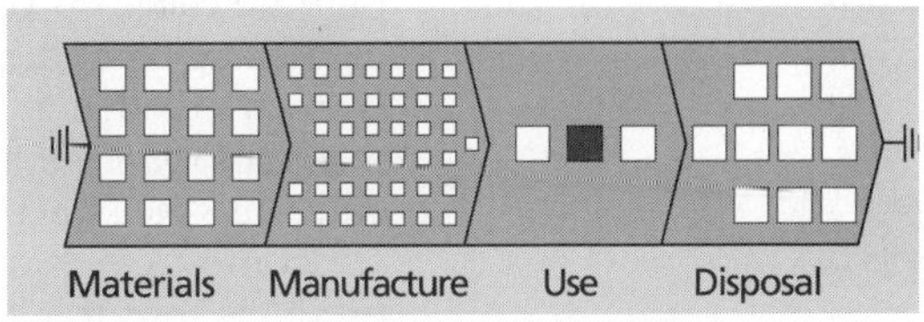

Figure 8.7. Diagrammatic outline of the product system

The picture is, however, never so simple in reality. There are always some processes in the product system which contribute to more than one use process. To which service or which use process should the exchanges from these processes be ascribed? A co-generation power plant produces both electricity and heat. How much of the coal consumption should be ascribed to the electricity?

This task is called allocation, and it is one of the most essential tasks in LCA.

Figure 8.8 shows the product system for a product in a presentation of the interrelationships with other product systems. The figure shows that many inputs into a product system derive *from* other product systems, and that many outputs from the product system continue *to* other product systems. This means that many of the processes in the product system belong equally to other product systems, and these processes' environmental exchanges must therefore be divided among

Figure 8.8. The product system's interrelations with other product systems

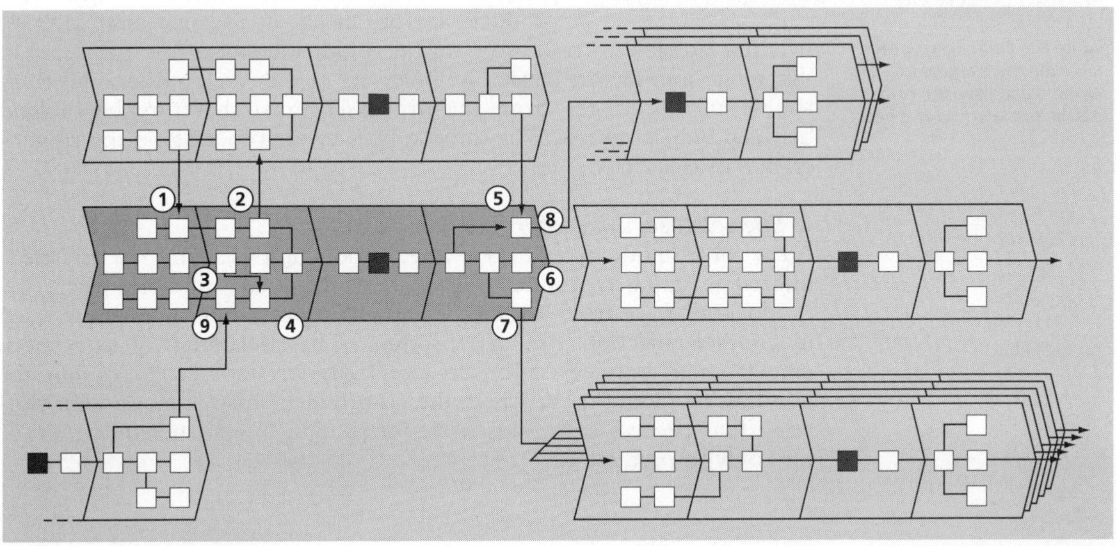

the services to which they contribute. In principle there are two different ways whereby a process can contribute to more than one service, viz.:

- When several products, called *co-products,* simultaneously arise from or enter into one process.
- In the case of *recycling* of materials or sub-assemblies from one product.

As illustrated in the figure, there are several variants under each of these two categories.

Co-products

A process, or more correctly a range of processes, can often have several outputs which subsequently result in a service for a final user. These outputs are called co-products and they must share the environmental exchanges belonging to the joint range of processes. Note that a range of processes is involved, that is, several successive processes which further process a material or a unit, and where two or more products are produced from or enter into the final process. See the examples in Figures 8.9-8.11.

In most cases, all of the preceding processes are required for both co-products, but in certain cases, some of the processes will be for the benefit of one product only. The following examples show some variants of co-product allocations in which the processes which are used by two or more products, and which must therefore be allocated, are designated.

1) Ordinary co-products

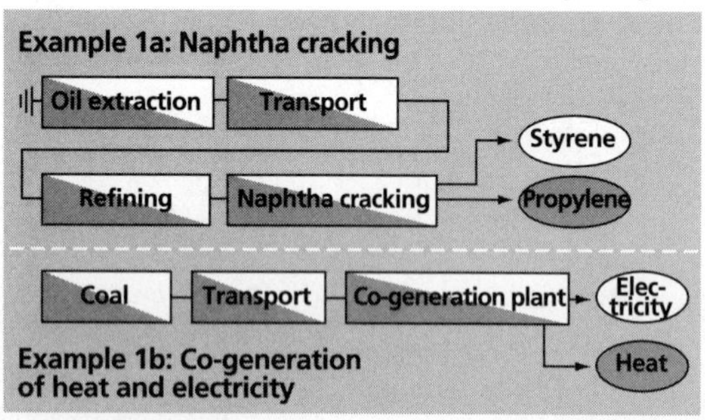

Figure 8.9. Ordinary co-products from the cracking of naphtha and from the co-generation of electricity and heat

Figure 8.9 presents two examples of ordinary co-products, namely styrene and propylene, after the cracking of naphtha, and electricity and heat from a co-generation power plant. As illustrated in the figure, all processes in the sequence in both examples contribute to both products.

For the cracking of naphtha, the process range from extraction of oil to production of naphtha is a precondition for both the styrene and the propylene, and the cracking itself, where the two products are produced, is also one joint process. It is not possible to resolve or further subdivide any of the processes in this range into factors caused by only one of the two products. All environmental exchanges from the processes must therefore be divided between both products. The same applies to the processes in the co-generation of electricity.

2) Special case: utilized production waste

A special case of co-products is when production waste from a process is utilized in another product. Figure 8.10 shows an example, where the production waste from an injection moulding of a unit in plastic is used in another injection-moulded product. The preceding oil extraction, cracking and polymerization are common for both products, but the injection moulding itself, where the co-products arise, is not a joint process. This process is not necessary for product 2 and therefore contributes only to product 1.

3) Special case:
production waste utilized internally

The case of co-products numbered 3 in Figure 8.8 corresponds to example 2 in Figure 8.10, but in Figure 8.8 it is a question of *internal* recycling of e.g. the intakes from an injection mould, i.e. recycling *within the same product system.* The processes shown thus do not contribute to several services, and it is therefore not an allocation task.

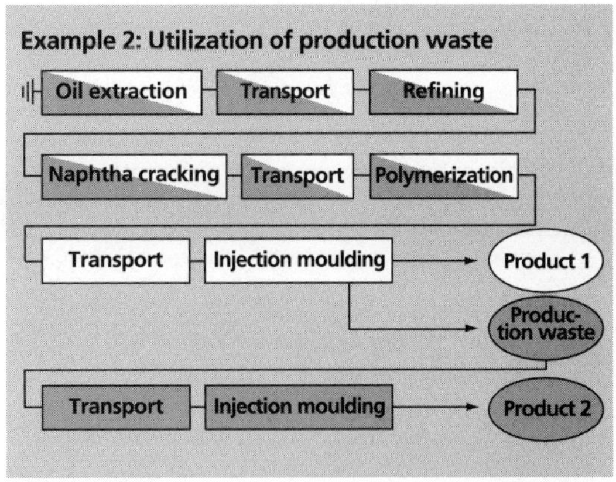

Figure 8.10. Co-products arising when production waste from injection moulding of plastic is utilized in another product system

4) + 5) Special case: overhead

For many processes in the product system there are a number of joint facilities or "fixed environmental costs", which are designated as "overhead". Two examples of overhead are shown in Figure 8.11. One illustrates overhead in the form of joint facilities in a factory such as lighting and heating; the other illustrates joint facilities and emissions in a waste incineration plant. During production, the factory's lighting and heating are joint to all products produced, as are all activities in administration. The exchanges from these facilities and their preceding product systems must therefore be allocated among the products produced in the joint locations and which draw on the joint administration.

In a waste incineration plant or in other waste treatment plants, the lighting etc. corresponds to joint facilities which must be shared. See Figure 8.11, showing that the electricity production for operation of the waste incineration plant is joint. But the environmental exchanges from the plant are not all joint. Certain emissions, such as CO_2 and SO_2 to the air or heavy metals in slag and ashes, can be traced directly back to the individual product, and must therefore not be shared according to an allocation key, but merely calculated as a function of the product's content.

Other emissions are "fixed environmental costs" ascribed to the general operation of the plant. They can, for example, be dioxins, which are not a function of the composition of the incinerated product, but which are formed as a consequence of the incineration process itself, and which the product must therefore carry a part of if waste incineration enters as a part of the product system.

Figure 8.11. Co-products which share overhead from joint facilities

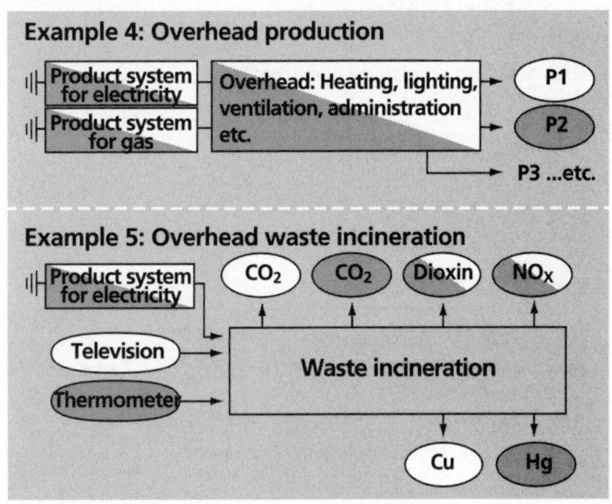

Recycling

Recycled materials can often enter into the product system, and parts of the material content of the product can similarly enter into other product systems after the product is disposed of. In the same manner, sub-assemblies in a product can be reused as spare parts in other products, as is known from e.g. the car industry.

In recycling situations such as these, many of the processes in the product system will be joint to the various user situations,

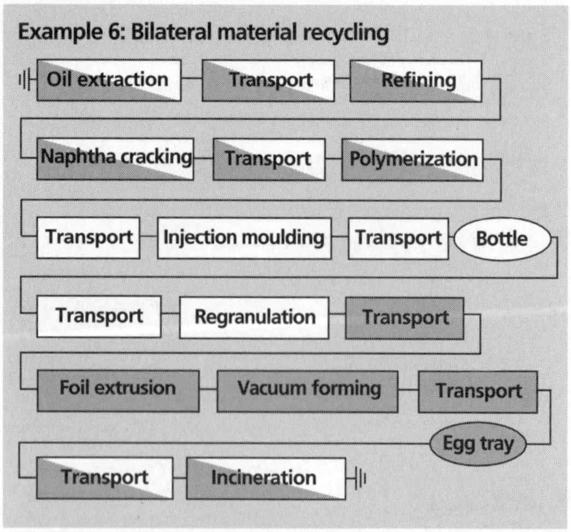

Figure 8.12. Recycling of a PET bottle into a PET egg tray

namely the processes which only occur once in a recycling system, e.g. raw material extraction and final disposal. If, for example, a motor is used in several products, the motor's copper consumption must be shared between each use process.

The following examples illustrate some different variants of recycling. Processes which are in common, and which must therefore be allocated, are designated:

6) Bilateral material recycling

Bilateral material recycling is distinguished by both the contributing and the recipient product being known. Allocation in this case is done between two known products. Figure 8.12 shows an example of the product system for packaging for beverages, namely a PET bottle, which was collected for regranulating and recycling of the PET into egg trays. The processes to and including polymerization and the final disposal are joint. The bottle is then ascribed 100% of its own manufacturing processes, including associated transport and the regranulation process which readies the raw material for the next user. The subsequent processes are then ascribed to the next user until the final incineration, which is again a joint process.

As illustrated, the recovery process itself, which is here a regranulation, is attributed to user 1, and this is done consistently in the EDIP method. The explanation for this is twofold. Firstly, the recovery process is not a once-only process. It arises in each stage of recycling in situations with several repeated recoveries, which is most often the case (see Figure 8.13 and following figures). Secondly, user 1 is responsible for delivering the material to user 2 in the same state as it is delivered to him or her by the joint processes of material production. The material in this example is delivered to user 1 as a granulate by the joint process. For user 2 to be ascribed a share of the joint process, user 2 must therefore receive the material in the same state.

Figure 8.13. The EDIP method's handling of the recovery process in allocation

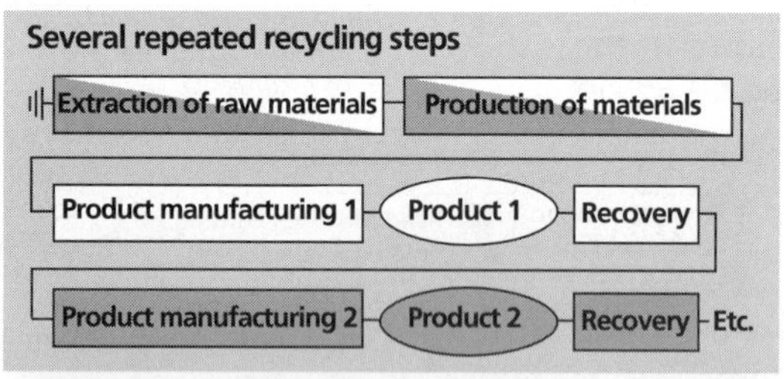

For these reasons, the recovery process and associated transport are ascribed fully to the user who delivers the material for recovery, i.e. user 1 in this example. Further explanation is given in Wenzel (1997).

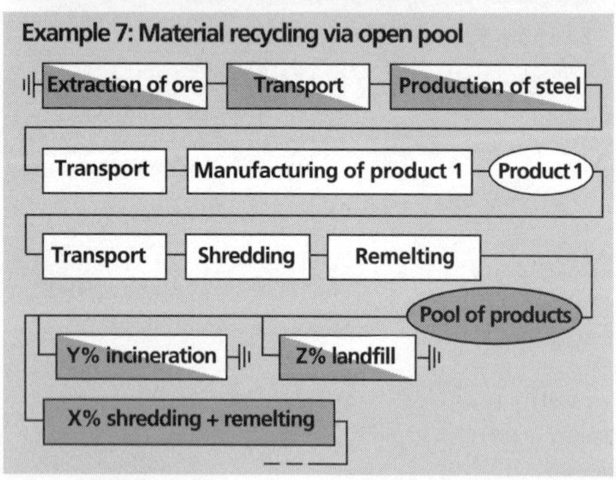

Figure 8.14. Recycling of steel via the societal pool

7) Material recycling via open pool

In most cases the recipient product is not known when the material is sent for recycling. In the typical situation, the product's material fractions are separated and sent for processing, after which they are sold on the market as recovered materials. They can therefore enter into a large number of different products, the identities of which are unknown.

This situation is typically the case for e.g. metals and paper. An example for steel is shown in Figure 8.14.

8) Utilized calorific value

In some European waste incineration plants, the heat produced is utilized by coupling the plant to a district heating network. There is, furthermore, continuing extension of power and heat co-generation at many of the plants.

Materials which have a calorific value, and which therefore release heat on incineration, thus contribute to more than one service in all the product systems in which incineration of waste forms a part. Figure 8.15 shows an example for a television set, where one of the sub-assemblies is produced in injection-moulded plastic.

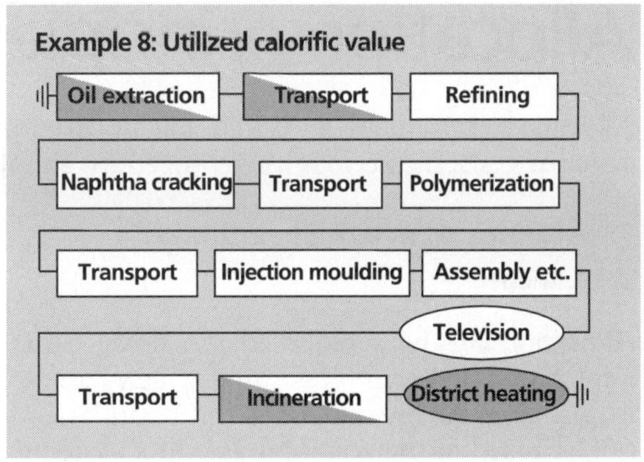

Figure 8.15. Utilization of the plastic's calorific value in incineration of waste

As the figure shows, only the initial oil extraction and transport plus the combustion contribute to the district heating. All other processes are ascribed solely to the television.

9) Reuse of components

An example of direct reuse of components or sub-assemblies in products is given in Figure 8.16, which shows reuse of a hub cap for a car. It is evident that all processes to and including manufacturing of the hub cap contribute to both use processes and are therefore joint, and the final disposal is accordingly also a joint process.

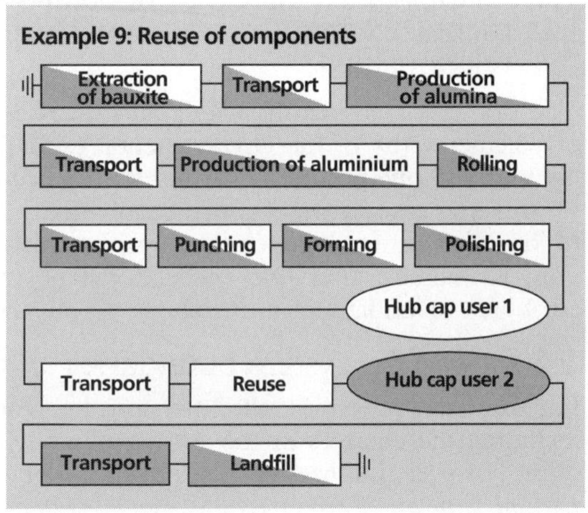

Figure 8.16. Reuse of a hub cap

The examples of allocation shown here for both co-products and recycling situations illustrate the most important variants of allocation. Their objective is to illustrate the allocation task and to show the first part of the task, the designation of joint processes. Note that there is no description here of how the environmental exchanges from the joint processes are to be allocated. This is the second part of the task.

The difference between Figure 8.15 for *utilized calorific value* and Figure 8.16 for *component reuse* provides a good impression of what the first task of designating joint processes involves, and why there are large differences in which ones are joint.

In the following section, the principal procedure of the second task, i.e. *how* to allocate and which types of allocation keys to use, is briefly explained. The detailed procedure of allocation is subsequently described and exemplified in the Toolbox in Part IV of the book. The allocation principles are further discussed and justified in Wenzel (1997), including comparisons with alternative procedures.

Allocation principles in the EDIP method

Allocation introduces an additional uncertainty, and it is an advantage to avoid allocation if possible. In general, the following procedure should therefore be followed:

1. **Attempt to avoid allocation**
2. **Allocate**

Allocation can be avoided in two ways: either by dividing the processes which are only apparently joint into smaller process units which can be attributed unambiguously to the individual products in question, or by extending the boundaries for the product system under study to include alterations caused in the adjoining product systems.

1. Attempt to avoid allocation

1.1 Clarify cause-effect relationships and ascribe processes and their environmental exchanges to each of the products in question on the basis of these relationships

The first task is to identify true joint processes and eliminate false ones. This includes identification of whether some of the environmental exchanges from the joint processes can be ascribed unambiguously to one product, or whether they are true joint exchanges.

In some cases a process can be divided into smaller process units, and it may thus be possible to identify an unambiguous relationship with one of the products in question, thus avoiding allocation. Remaining after this exercise will be true joint processes and exchanges which cannot be attributed unambiguously to a single product.

1.2 Extend the system boundaries

For true joint processes, allocation can be avoided by instead extending the system boundaries and including the changes in the surroundings which will result from the decision made. This procedure, however, implies that the object under study is a *change,* i.e. the LCA must apply to a *comparison with an alternative,* and the *magnitude* of the change must be known, e.g. "100,000 TVs incinerated instead of being dumped".

When an assessment is to be made of whether, from an environmental viewpoint, the above 100,000 TVs should be dumped or incinerated, one can choose either to allocate or to extend the system boundaries. The difference between the two procedures in this example will be as follows:

Allocation will mean calculating the environmental exchanges separately for each of the two disposal scenarios and then comparing them. For the incineration scenario, the joint exchanges from oil extraction and incineration are partitioned between the TV and the district heating as shown in the example in Figure 8.15. The resulting exchanges (and potential impacts) after allocation can then be compared with the exchanges (and potential impacts) computed for a dumped TV. Before comparison, the figures for one TV could for both product systems be multiplied by 100,000, but it would be of no significance to do so.

Extension of the system will mean that it is the *difference* between the two scenarios which is assessed, i.e. the change in going from the one scenario to the other. In other words this implies an assessment of which consequences will result in the surrounding of the product system, when going from one disposal scenario to the other. Extension of the system will include the alternative heat production based on e.g. natural gas, which is superseded when 100,000 TVs are burned in a waste incineration plant with utilization of energy instead of being dumped. In other words, extension of the system will include the environmental exchanges from the heating produced by the natural gas by subtracting them in the total calculation of the consequences of a change in disposal route for these TVs. In this case it is of significance that the scenario concerns 100,000 TVs, because this waste volume is essential for the concrete scenario to be used in relation to the capacity of incineration plants, district heating networks, etc.

System extension makes the LCA more narrowly adapted to the concrete scenarios which are set up. If the decision, with a high degree of certainty, concerns precisely these scenarios, the system extension should therefore be used, in order to provide the most accurate picture of the consequence of the decision. But if the scenario *cannot* be predicted with certainty, it can be a mistake to extend the system boundaries. If, for example, the heat of combustion of the TVs in question, when they are disposed of in 20 years, at that time does not at all supersede natural gas, but biogas or wind-generated electricity, then it is a serious mistake to assume in the LCA that natural gas will be superseded. An allocation can therefore be preferable if there is uncertainty regarding which new scenario will apply, because the allocation is less targeted and therefore more broadly applicable. Alternatively, a number of different scenarios covering the whole spectrum of possibilities can be included.

System extension requires:

- that the object of the study be a *change,* i.e. a comparison between alternative scenarios,
- that the nature and extent of the change which will actually occur as a consequence of the decision which the LCA is to support can be predicted with a fair degree of certainty, and
- that data are available for the adjoining systems in question.

It is often difficult in product development to predict the changes in adjoining product systems which will be caused by choosing one alternative over another, because the product systems in question are some distance in the future, and because it is difficult to predict the extent to which the new product will supersede the reference product or other products.

2. Allocate

The task of allocation is to partition the environmental exchanges from the joint processes in a fair manner. The allocation must make it possible to compare the individual product with alternatives and ensure that such a comparison reflects, as well as possible, the consequences to which a decision made on the basis of the comparison will actually lead.

2.1 General principle of allocation

The general principle of allocation is: *Express the joint environmental exchanges per functional unit.*

Expressing the exchanges per functional unit is the primary task of the LCA, and it therefore continues to be the primary principle in the allocation.

For the processes which contribute to more than one service, i.e. the joint processes, the task is therefore also to partition their environmental exchanges between the services to which they contribute. The objective of the partitioning is to identify the portion of the exchanges "caused" by the functional unit under study.

The services to which the joint processes contribute are, however, by nature often different, i.e. they cannot be expressed in the same type of functional unit. If this were the case the task would be simple, namely merely dividing the quantity of each joint exchange by the number of the functional units to which the joint process from which it arises contributes.

The task of allocation is therefore to find a total and uniform measure for all the services to which the joint processes contribute, such that it becomes possible to partition the joint exchanges to each of the services to which they contribute.

This total and uniform measure is defined in the EDIP method as the total *utility value* of the products, materials or sub-assemblies to which the allocation applies. The principle of allocation is thus:

When partitioning joint environmental exchanges between the products/services in question, they follow the substance, the material or the sub-assembly in proportion to its utility value.

As far as possible, the utility value is determined on the basis of a technical criterion, and if this is not possible, an economic criterion is used.

2.2 Technical criterion for utility value: grade

A technical criterion can, for example, be used in those cases where the allocation applies to co-products in the same material or products which use the same material via a material recycling. A precondition for being able to determine the utility value on the basis of a technical criterion is that it is carried by the same properties of the material for the various products. If this condition is fulfilled, the utility value is determined on the basis of a technical assessment of the essential material properties and it is called the *material grade.*

Basis for grade as partitioning key

A material can be used a number of times before its essential material properties, i.e. its material grade, are destroyed or reduced to a degree which makes further use impossible. The material grade is reduced by each use.

For each time the material grade is reduced by x%, a need for x% new joint processes and environmental exchanges is born. The joint exchanges are therefore ascribed to the user who reduces the material grade in proportion to the loss of material grade. If a material of high grade is used, the user therefore carries a correspondingly large share of the joint exchanges, but if the material is passed on to the next user without loss of material grade, the joint exchanges are also transferred without their being ascribed to the user in question. However, if the material is *not* passed on to a subsequent user, the joint exchanges are ascribed to the user who causes the material to be taken out of use, in proportion to the material grade which the material possessed before the use in question. In both cases the joint exchanges are thus ascribed to the user in proportion to the loss of utility value of the material.

Ascribing the joint exchanges in proportion to the relative loss of material grade caused is analogous to the writing off of an initial economic value of a capital goods item in proportion to the loss of value during use.

2.3 Examples of allocation on the basis of grade

Material recycling

Material recycling, either bilaterally or via an open pool, is an example of allocation situations where the partitioning key is the material grade. The allocation itself between the two products which enter into the recycling is made on the basis of the relative *loss* of material grade.

Co-products in the same material

Co-products in the same material are another example, as the joint exchanges from the preceding process range follow each co-product in proportion to its *quantity · relative material grade.*

Note that this is completely analogous to the recycling situation, where the joint exchanges also follow the material in proportion to its relative material grade until disposal or recycling occurs, and the loss of material grade can be identified. When the co-product is eventually disposed of and the material may be recycled, a loss of material grade is correspondingly identified.

The utilized production waste from the injection moulding in Figure 8.10 is an example of a co-product allocation made on the basis of this partitioning key. Note that the production waste goes through one injection moulding more than the primary product, and as there is a small loss of material grade in the injection moulding itself, the material grades of the two co-products are therefore not entirely the same. The correct partitioning key is therefore quantity multiplied by relative material grade.

Co-products in the form of different types of energy

A special example of co-products is the co-generation of electricity and heating, where both electricity and heating arise in the same process, as shown in the example in Figure 8.9. Allocation in this case is made similarly to co-products in the same material, and the partitioning key is correspondingly quantity multiplied by relative grade, where the grade of the energy is here defined as its content of exergy.

Component reuse
In the recycling of components/sub-assemblies in a product, both the material's utility value and the value which the production processes have added to the component/sub-assembly are taken into account. In this case the joint exchanges from raw material extraction to and including material production are ascribed in proportion to the material's relative loss of material grade, while the joint exchanges from the product manufacturing processes are ascribed on the basis of an economy-based partitioning key. See below.

2.4 Economic criterion for utility value: the product's relative price

Economic criteria are used when the allocation concerns co-products in different materials, or when the utility value does not reside in the same properties of the material. The product's economic value is used as an approximation for the utility value, and the relative price for the substances, sub-assemblies or materials to which the allocation applies is recommended as the partitioning key. Note that relative price and relative profit margin are identical for true co-products because all inputs and thus all variable costs of the co-products are in common.

Basis for relative price as partitioning key
The price is taken as a practical expression of the utility value, i.e., a product or a *service* which costs a user £17 is perceived by the user as having the same utility value as another service which costs £17.

The product's economic value can also be taken as an expression of the *reason* for the existence of the process, or more correctly the process range. If there is only one product as output from a process, the relative price (= relative profit margin) is thus 1.0. The economic value of this single product is the entire reason why the process range occurs. If there are several products as output from a process, the relative price (= relative profit margin) is a way of partitioning the "cause" of the existence of the process range. If one product accounts for 10% of the earnings from the process and another for 90%, the reason for the existence of the process range is partitioned in the ratio 1:9. This is considered a fair principle of allocation because economic considerations determine the existence of an industrial activity.

2.5 Examples of allocation on the basis of economic criteria

Co-products in different materials
In the case of co-products in different materials, it is difficult to define a utility value on the basis of technical material properties, because the relevant properties are different from one product to the other. If the allocation cannot be avoided, the economic partitioning key is therefore recommended. The example for co-production of styrene and propylene in the cracking of naphtha in Figure 8.9 is a case of this type.

Co-products where the utility value is not carried by the material
In the example with incineration of waste in Figure 8.11, the joint exchanges are partitioned among the products which are incinerated. This is similarly a case of co-product allocation, where the material properties in many cases are without significance and where allocation on the basis of an economic criterion is the only option. In some cases, the heat of combustion can be a technical material property used as partitioning key.

Component reuse
In the reuse of components or sub-assemblies in a product, the joint exchanges from the product manufacturing processes must be ascribed on the basis of an economy-based partitioning key. The key used is the component's relative loss in price, the material price not being included, i.e. attention is paid only to that part of the price which derives from the increase in value in the product manufacturing processes.

Material recycling where the material properties are not preserved
In the recycling of materials in situations where the material basically changes character, the material grade resides in widely different properties, and a loss of material grade therefore cannot be defined. An example is utilization of the material's calorific value in incineration as in Figure 8.15. Economic values can be used in such cases or, if possible, an extension of the system.

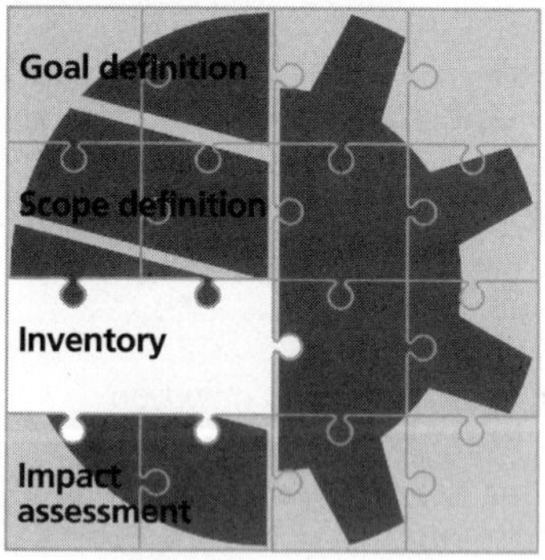

9. Inventory

The objective of the inventory is to collect environmentally relevant information from the processes identified during the scope definition and included in the model of the product system. This includes:

- generation of unit data and the setting up of unit processes
- inventory of the environmental exchanges from the complete product system or for parts thereof
- presentation of the information in a transparent form

Data collection follows the same pattern for all processes. This pattern is called the "data format", and it specifies the information which must be collected for a process. It is an advantage if all data requirements are specified from the start because time is then saved for both the person carrying out the LCA and for the provider of data.

Data format

The data format contains three categories of information, viz.:

- a description of the process,
- an inventory of the process's exchanges with the environment, and
- a characterization of the information, i.e. a delimitation of the process and a description of the origin and quality of the data.

An important goal of data collection is to establish a database which can be used on an on-going basis for LCAs at the company. This means that data for a process must be stored in a form which makes reuse possible in the subsequent modelling of other products. This is called normalizing the process or transforming it into a unit process.

The unit process

The relationship between a process and its environmental exchanges must be clarified, i.e. to which key unit are the exchanges most correctly related? For materials, the correct unit is typically the *weight* of the mate-

Table 9.1. Examples of key units for processes in a product system

	Key parameter		
Process	**Energy consumption**	**Consumption of materials and ancillary substances**	**Discharges**
Anodizing	Surface area	Surface area	Surface area
Production of recycled steel	Weight	Weight	Weight
Electroplating	Surface area	Surface area	Surface area
Milling	kg removed material	kg removed material	kg removed material
Powder coating	Residence time in cabinet	Surface area	Residence time in cabinet
Haulage by lorries > 16 tonnes	Km*kg + driving mode + speed	Km*kg + driving mode + speed	Km*kg + driving mode + speed
Production of electricity	1 kWh	1 kWh	1 kWh

rial, and the exchanges are expressed per kilogram. But for a painting process or a plating process it is often the *surface area*, and for other processes it can be *time* used or something else. When the best process unit is identified, data are expressed as a function of this unit. In this way the process can be generalized and used as a building block in the modelling of product systems, also later for products other than the current one.

Table 9.1 shows process units for various process types. An unambiguous procedure cannot be set up for identification of the key unit, but a fairly usable picture can be drawn for several categories of processes, as shown in the table.

Raw material extraction and material production: the typical key unit is the material's *weight.*

Electricity production: the typical key unit is *1 kWh* of electricity generated.

Heat production: the typical key unit is *1 MJ* of heat output.

Transport: the typical key units are *km · loaded weight* and *speed* combined with *type of transport* (e.g. urban driving vs. country driving).

Production processes: there is no typical key unit. It will vary greatly, depending on the type of process, and the company itself is best able to identify it for its own processes. The company often uses a key unit for the process for financial calculations, and in many cases this will also be the correct key unit for the environmental exchanges.

Figure 9.1 illustrates an example of a unit process. The figure shows an aggregated process for "phosphatized steel", because by such an aggregation, nearly all types of exchanges can be illustrated. The process has been created by aggregation of data for two unit processes, viz. production of steel and phosphatizing of steel, and the object of presenting it is to provide examples of the types of information reported for a unit process.

Description of the process

The description of the process must specify the operating conditions for the process which underlie the data for the exchanges. It must make it possible to understand the underlying circumstances if data for the process will be used later for other LCAs.

Figure 9.1 shows an example of the information required in the description of the process. It contains two elements, namely a description of the unit process and a description of the position of the process in the product system. The latter is not described on establishment of the unit process, but only when it is used for modelling of a particular product system.

When the results of the LCA are to be used in product development, there is often a need to seek and sort the various contributions to resource consumption and potential impacts. The various fields in the process description must therefore be seen as "active", i.e. a computer program must be able to search on them. The concepts site characterization and region characterization which appear in the figure are reviewed in Chapter 10 under impact assessment because they concern the environmental interpretation of data from the inventory. But note here that during modelling of the product system, they must be specified in the individual unit process for use in later computations in the LCA.

The process description for transport processes and use processes is somewhat different from the other processes in the product system with respect to the nature of the information to be collected.

Description of the process

The unit process

Unit process and family ID	ID-number	Process key unit	1 kg
Process name	Phosphated steel		
Data reference period	1992-1993	Updated:	26.11.95
Projected to:	xxxxx	Reference person	HW

Operational data: Example for phosphatizing

■ Process cycle:
Continuous operation of the phosphatizing cabin 16 hours a day, 5 days a week.
Spraying of units in counter-stream: 1) soaping, 2) phosphatizing, 3) rinse, 4) rinse.
Continuous overflow of rinse water directly into sewerage system and batch-wise replacement of baths 1) and 2) after 3-4 months. The batch is precipitated chemically and the sludge taken to the Danish public plant for treatment of chemical waste. Drying of the units in drying cabin after last rinse.

■ Operating data:
Processing times: phosphatizing 8 min and drying 8 min. Flow: 200-300 m^2/hour.
Temperatures: bath 1) 50-55°C, bath 2) ca. 45°C, bath 3) and bath 4) cold, pH: 4.5-6.0.
Chemical precipitation of batch: calcium hydroxide and ferrous sulphate to pH 7.

■ Working environment:
No observations of bacterial problems from aerosols from showering with water in the standing baths.

When modelling a product system

Geographic position	Germany and Denmark	Phase of the product system	Materials

Component description: Example for phosphatizing

■ Component characteristics:
Weight: 18080 g, surface area: 2.16 m^2, geometry: "box" with only three sides, 60 cm x 120 cm on all sides

■ **Process modification:**
Flow: 123 items per hour = 2224 kg/h
No modifications compared to description of unit process above

Site characterization for calculation of impact potentials

Impact	Global		Regional						Local					
Impact ID	1	2	3	4	5	6	7	8	11	12	13	14	15	16
Site factor (default)	*1*	*1*	*1*	*1*	*1*	*1*	*1*	*1*	*1*	*1*	*1*	*1*	*1*	*1*

Region characterization for normalization and valuation

Impact category	Resource comsumption			Environmental impacts			Impacts on the working environment
	Non-renewable	Renewable Water	Wood	Global	Regional	Local	Local
Reference area (default)	*The World*	*DK*	*The World*	*The World*	*DK*	*DK*	*DK*

Figure 9.1. An aggregated unit process for "phosphatized steel"

Environmental exchanges

Environmental exchanges		Unit	Allo-cation factor	Statistic			Characterization of data					
				Volume	Coeffi-cient of variation	Func-tion of distr.	1	2	3	4	5	6
ID-number	Name		(f)	(μ)	(σ/μ)	(D)						
Input												
Electricity												
ID1	Danish electricity	kWh	1	5.1	0.25	logN	1	2		1990	xx	
ID2	German electricity	kWh	1	0.04	0.25	logN	1	2		1990	xx	
Thermal energy												
ID3	Natural gas, North Sea	kg	1	0.004	0.1	log N	1	1		1990	xx	
ID4	Fuel oil	kg	1	0.07	0.1	logN	1	1		1990	xx	
Materials												
ID5	Ferromanganese	kg	1	0.007	0.1	logN	3	1		1984	xx	
ID6	Ferrosilicon	kg	1	0.0006	0.1	logN	3	1		1984	xx	
ID7	Iron ore + pellets	kg	1	1.53	0.1	logN	2	1		1975	xx	
ID8	Scrap iron	kg	1	0.11	0.1	logN	1	1		1975	xx	
Ancillary substances												
ID9	Calcium carbonate	kg	1	0.17	0.1	logN	3	1		1975	xx	
ID10	Metallurgic coke	kg	1	0.46	0.1	logN	3	1		1975	xx	
ID11	Phosphatizing agent	g	1	0.53	0.1	logN	1	1		1993	xx	C
ID12	Degreasing agent	g	1	0.25	0.1	logN	1	1		1993	xx	C
Output												
Emissions to air												
T	CO_2	g	1	1650	0.1	logN	2	2		1990	xx	
T	SO_2	g	1	1.22	0.1	logN	1	2		1990	xx	
T	CO	g	1	16.00	0.1	logN	1	2		1990	xx	
Emission to water												
ID13	SS	g	1	0.94	0.5	logN	1	3		1993	xx	
ID14	COD	g	1	5.55	0.1	logN	1	3		1993	xx	
ID15	Tot-P	g	1	3.84	0.4	logN	1	3		1993	xx	
ID16	Unspecified detergents	g	1	9.16	0.4	logN	1	3		1993	xx	C
Emissions to soil												
Waste												
T	Ferruginous furnace slags	g	1	150	0.1	logN	1	1		1990	xx	
T	Unspecified waste	g	1	140	0.1	logN	1	1		1990	xx	
ID16	Unspec. heavy metal + sludge	g	1	131	0.1	logN	1	1		1994	xx	
Potential impacts on the working environment												
T	Monotonous repetitive work	mh	1	0.04	0.5	logN	2	1		1993	xx	
T	Noise	mh	1	0.18	0.1	logN	1	3		1993	xx	
T	Damage to the nervous system	mh	1	0.12	1.0	logN	1	3		1993	xx	
T	Allergy	mh	1	0.008	1.0	logN	1	3		1993	xx	
T	Damage to the repro-ductive system	mh	1	0.06	1.0	logN	1	3		1993	xx	
T	Cancer	mh	1	0.001	1.0	logN	1	3		1993	xx	
T	Accidents	number	1	$16x10^{-6}$	0.1	logN	2	1		1993	xx	

Notes

1. Data source type: 1-5
2. Data representativeness of the process: 1-5
3. Representativeness of the process/technology for the purpose in modelling: 1-4
4. Age of data: state year
5. Reference: state a number that refers to a central reference library in the database
6. Confidentiality: indicate with a C

Characterization of data is described in the "Toolbox", Section 22.3.3

Figure 9.1 (continued)

Characterization of data

Scope definition of the process

- **Extent of the process:** For primary steel, the process includes extraction of iron ore, concentration of the ore into pellets and refining in a blast furnace. All materials, energy consumptions and ancillary sustances are "followed back to nature", i.e. all processes from raw material extraction to production of steel are included. The steel contains 11% scrap metal. The energy for shredding of the scrap metal is included. For the phosphatizing the full process is included. Energy consumptions are "followed back to nature" but for ancillary substances this was not possible.
- **Co-products:** There are no co-products.

Characterization of data

- **Data gaps:** Bacterial counts in standing baths should be made.
- **Data source:** The information is based on data from a single steelworks (Frees, 1996). For steel production there was no information on ancillary substances and alloying agents. Consumption of these is therefore taken from another source for production of steel.
- **Representativity of data:** Data are average data for a year.
- **Representativity of the process:**
- **Statistical database:** An interval was given for the aquatic emissions from the phosphating within which the data lay. The interval is taken as an expression of the coefficient of variation. The other coefficients of variation are estimated on the basis of the EDIP method's guidelines.
- **Mass balance:**

In	g	Out	g
Metallurgic coke	460	Carbon dioxide, CO_2	1700
Lime, $CaCO_3$	168	Sulphur dioxide, SO_2	1
Ferromanganese	7.3	Carbon monoxide, CO	16
Ferrosilica	0.7	Slags including	
Iron ore incl.		30 g bound oxygen	150
460 g bound oxygen	1534	Waste	140
		Steel	1000
Scrap iron	110		
Oxygen from the air	724		
Total	3004		3007

With respect to the quantity of oxygen bound in the iron ore and slag, it is assumed in the mass balance that the composition is 100% Fe_2O_3.

Technological development

- **Trends:** Within the last 10 years, energy consumption in the production of primary steel has fallen by ca. 20%. The reduction for the next 10 years is estimated to be ca. 10% only, as much has already been done to reduce energy consumption.
- **Projection of data:** not done.

Figure 9.1 (continued)

For transport processes, the process description primarily comprises type of transport, distance, load, nature of the transport (e.g. urban driving, country driving, etc.) and speed.

For use processes it is a question of pattern of use, type of maintenance, transport scenario during maintenance, and life span for the product and for components replaced during the use stage.

The procedure of the process description is presented in Section 22.3.1 in the Toolbox.

The process's environmental exchanges

The data format in Figure 9.1 illustrates most of the possible types of exchanges. These are:

- Inputs in the EDIP method are divided into six types, viz. electrical energy, thermal energy, transport, materials, ancillary substances and resources.
- Outputs are divided into aerial emissions, aquatic emissions, emissions to soil, and waste.
- Impacts on the working environment are divided into the seven types of impacts implemented in the EDIP method.

Figure 9.1 shows how the inventory of the process's exchanges itself specifies the relationship to the adjoining processes in the product system. The column "ID number" specifies with a "T" whether the exchange is *terminal,* i.e. whether it comes directly "from nature" or is conveyed directly to the final recipient in nature. For all exchanges not marked with a "T", a name and/or ID number is given to designate the preceding or next process in the product system. For output, it is typically a question of waste water treatment, waste disposal or recovery.

Note that the unit for the non-terminal exchanges must always be the key unit for the process specified by the exchange. This means that a non-terminal input to a process is a *"number of key units"* of another unit process, e.g. *0.04 kWh German electricity,* and a non-terminal output similarly draws on a "number of key units" of another unit process, e.g. *5.55 g COD to a municipal waste water treatment plant* (ID14 in Figure 9.1). It is important to maintain this discipline in filing the data in the unit process. Otherwise, the data cannot be used in a later modelling and subsequent calculation. Terminal inputs and outputs are always kept in one and the same unit, e.g. *gram* as shown in the figure, and impacts on the working environment are a multiple of hours, here *millihours,* or a *number* for the impact type "accidents".

For each individual exchange it is possible to specify an allocation factor for the case where the process in the actual product system is a joint process. As explained in Chapter 8 under allocation, each individual exchange must be able to be allocated, and the factor must therefore be able to be assigned in this way. The allocation factor is set to 1 as a default, corresponding to no allocation, and only when the process is used in a given modelling is a decision made on whether the factor must be altered.

All exchanges in the calculation tool should be given a unique identification number. This is a precondition for avoiding double entries in the database, and exchanges should always be established in and taken from a central list, when data are entered in new processes. There are various ways of checking for double entries for chemicals, for example by using the so-called CAS numbers. Searching by CAS numbers in the database should be possible. Further discussion of the data discipline required in the management of large quantities of data in a database is beyond the scope of this chapter. It has been one of the most significant tasks in the construction of the EDIP method's PC tool (Danish Environmental Protection Agency, 1997).

Statistical description of the environmental exchanges
The uncertainty in the data must be described together with the figures themselves. The data format is such that this description of uncertainty can be done in a simple way and at the same time a way which permits easy estimation of the uncertainty in the total results for the product system in a computer program. The description of the uncertainty does not require a large quantity of figures, but merely that the coefficient of variation for the exchange be estimated. The coefficient of variation is the standard deviation divided by the mean. It specifies how much the exchange can deviate to each side of the mean.

It is often found in LCAs that it is only possible to obtain one figure for a given exchange from a process. In most cases it is not possible to obtain a large quantity of data on which a standard deviation can be calculated. But it is often possible to learn, from the data provider or a specialist in the process in question, approximately how much the figure varies in the form of "±10%" or "±20%". This "±x%" can be regarded as the coefficient of variation and used in the unit process together with the figures themselves for the exchange, as given in the data format in Figure 9.1. For the subsequent calculation of the total uncertainty for the product system, the statistical distribution which the exchange follows must also be determined. Selection of a log normal distribution is recommended as default, with the possibility of altering this to a normal distribution if there is a justification for this. Section 22.3.2.2 in the Toolbox provides a guideline for estimating coefficients of variation of various types of exchanges.

The assessment of uncertainty can also be carried out manually. But then it is unrealistic to carry out an actual calculation which includes variation in all the individual processes and exchanges. Manual assessments of uncertainty are limited to qualitative assessments and estimates, and selected calculations for a few of the most significant processes and key parameters.

The statistical calculations for the complete product system are covered later in this chapter.

Data characterization

A very significant part of the inventory for a process is characterization of the process and the data which describe it. This involves:

- delimitation of what is covered,
- description of data gaps and of the quality of the data, and
- description of the technological development for the process.

The data format in Figure 9.1 shows an example of characterization of data for the aggregated process "phosphatized steel". Note that the data are firstly characterized for the individual exchange by using so-called data quality indicators. These provide an overview of how data for the individual exchange are collected, how representative they are, and how old they are. More details of the data characterization are then given in the text field in the last part of the data format for the unit process.

There must be a description of what the process covers, preferably with a flow diagram which shows which individual operations are included and which are not. If more than one product arises from the process, this must be stated, and data for the partitioning key for allocation must be collected.

Further details can be given in the data characterization of how the statistical uncertainty is determined. It is also a good idea to extend the description of the data quality with a mass balance of inputs and outputs as shown in Figure 9.1, as this will often reveal the biggest errors in the data in terms of quantity.

Finally, the technological development for the process should be described for the environmentally most significant exchanges.

The procedure used to characterize data and to assign quality indicators is described further in Section 22.3.3 in the Toolbox.

Data collection

Data collection is the most time-consuming task in the LCA of products. In general it involves a lot of work to establish qualitative and quantitative information concerning the many processes in a product system. But when this information has been collected, it will provide a base of fundamental environmental knowledge for the company, and inventory of subsequent products will be much simpler. Data collection for the first product may thus well take several months, while subsequent products can be finished in a much shorter time. Experience has shown that when the database is established and the method in general established in a good computer tool, usable assessments can be carried out in the space of a few hours to a few days, depending on the objective.

Data sources

The data collection can build on data from various types of data sources. These can be divided into four main categories, as shown in Table 9.2.

Updated measurements or calculations of the environmental exchanges from the actual processes in the product system represent the largest degree of accuracy and specificity for the product. But it is neither feasible nor necessary that data for all processes in the product system have this specificity. In practice it is necessary to base a large part of the inventory on available existing data, and for some processes in the product system it is even desirable that data represent a broad average.

1.	**Electronic databases:** Several databases are described in Hemming (1995)
2.	**Literature data:** e.g. scientific papers and public reports, LCA reports
3.	**Unreported data:** From manufacturers, laboratories, authorities and others
4.	**Measurements and/or computations:** Where data are non-existent or should be improved

Table 9.2. Various types of data sources

When using data from the literature or electronic databases it is most important to be critical of the sources and to ensure that the data used concern the relevant processes, that they are properly updated, and not least that system delimitations, allocations, etc. are in agreement with the scope definitions and the allocation rules that apply for the product system to be modelled.

Section 22.1 in the Toolbox discusses planning of the database for the product system and indicates which type of data should be collected for the various processes.

Data providers

Experience with data collection shows a particular pattern of where information on the various processes in a product system are to be found. The picture which emerges is tied to the stages of the product's life, as illustrated in Figure 9.2.

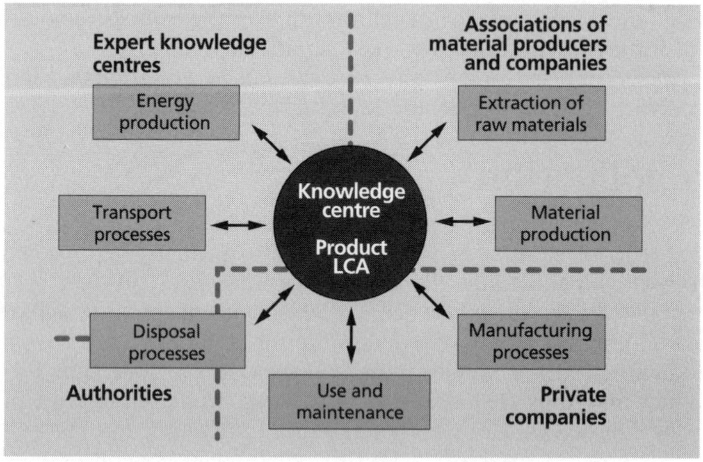

Figure 9.2. Data suppliers and networks

Raw material extraction and material production
As illustrated in the figure, knowledge of raw material extraction and material production is typically based in the larger associations of material manufacturers. Examples are the European Association of Aluminium Producers, EAA, and the Association of Plastics Manufacturers in Europe, APME. Large raw materials suppliers such as the oil companies also often have data for their raw materials.

Product manufacturing processes and use processes
Information on manufacturing and use processes is generally to be found within the companies themselves. International databases on manufacturing processes exist (see overview in Hemming, 1995). There is, however, a trend towards rapid obsolescence in such data for manufacturing processes, and in many cases the use of such databases will be avoided for fear that the data are obsolete. Furthermore, it is often very difficult to see the process conditions underlying the data.

Disposal processes
For disposal processes such as incineration and landfilling, information is typically available from the authorities and to some degree also from research institutes. For recovery processes, information is available on an equal footing with manufacturing processes with the actual recovery companies.

Transport processes and energy systems
Data on transport processes and energy systems are typically available at specialized information centres.

Data networks
In place of static databases, which often contain obsolete data, *data networks,* as outlined in Figure 9.2 and described here, will be the future

way of ensuring availability of data for use in LCA of products. Each individual expert body in a data network must handle and update its part of the database for the product system. This involves a division of responsibility as outlined in Figure 9.2. Each body in the network has an intrinsic commercial interest in participating, and this is a precondition for ensuring the viability of the network and therewith the availability of data.

Procedure

A procedure for data collection is outlined in Table 9.3, which offers some general recommendations. There are large differences in the availability of input data, output data and data on the working environment, and the procedure will vary depending on the data type required.

Input data

Input data are the most readily available. Energy consumption, material consumption and consumption of chemicals are usually registered by the company and appear in parts lists for the product and purchase lists for chemicals. Similarly, the time required for operators of the process is generally known and expressed per product unit for use in cost calculations. The latter enters into the basic data for the assessment of the working environment.

For companies with a very uniform production, energy consumption per product unit can be calculated on the basis of the company's total energy consumption. In cases with non-uniform production, energy consumption must usually be measured or estimated for each individual process.

Output data

Output data, i.e. data for emissions to air, water and soil and data for quantities of waste, are often difficult to find. The difficulty is partly due to the fact that data for emissions are not available at the companies and partly to the fact that the data which are available are difficult to allocate

1. **Literature information retrieval:** Data are extrapolated from similar processes described in available literature. Knowledge centres are contacted for relevant available information: papers, books, reports, LCA reports, electronic databases and data networks.

2. **Questionnaire:** Data are sought from contractors etc. Personal contacts are established at the relevant companies. Questionnaires are sent out and the contact is followed up. A visit will often be necessary to assist the contractor and avoid misunderstandings.

3. **Calculations:** Processes and data are examined in order to reveal to what extent the emissions can be calculated as described in the next paragraph. Calculations are carried out.

4. **Measurements:** Measurements are carried out for processes where this is judged to be desirable or necessary, and/or because sufficient data cannot be obtained by other means.

Table 9.3. General procedure for data collection

to the individual product. Which of these problems arises typically depends on the size of the company and the composition of the company's production.

Large companies will often have a complex production. There will be many measurements and data for the emissions, but the productions will often be complex and will include many different products. Measurements and data typically concern *joint facilities,* i.e. joint waste water emissions and joint emissions to air, and it can be difficult to allocate data to the individual products.

Small companies will more often have a more homogeneous production of uniform products. Data will thus more easily be allocated to the relevant product, *but data will typically be lacking* for emissions and exposure of process operators.

This presentation is of course simplified, but it describes the nature of the existing problems.

The problem with availability of output data can, however, be avoided to some degree. Given that *input data* are more readily available, the inventory should permit output data in certain cases to be *calculated* on the basis of input data via simple mass balance observations. This will be entirely adequate in many cases, and in certain cases even better than data from direct measurements of the emissions. In reality the latter will apply to all emissions where the input is very well defined, and where there can be no doubt about the route of emission, i.e. whether the emission is to water, air, soil or whether it ends as solid waste.

Section 22.4 in the Toolbox elaborates further on this and provides guidelines for the data collection.

Data on the working environment

The procedure for collection of data for impacts on the working environment is different for the various types of impact. Cancer, reproductive damage, allergy and damage to the nervous system are consequences of chemical impacts, and the inventory is therefore based on assessments of chemicals. Damage to the musculoskeletal system is a consequence of *monotonous repetitive work* and the inventory is based on analyses of the work process. Hearing impairment is a consequence of the impact of noise and the inventory includes measurements of noise. To these must be added an inventory of *work accidents,* which is based on statistics of reports.

For each of the seven types of impact, the time is computed for exposure of the employees to an effect.

Table 9.4. Impact criteria in the inventory of impacts on the working environment

Impacts categories	Impact criteria
Chemical impacts	Skin contact or respiratory exposure exceeding the threshold value by more than 10%
Impacts from noise	Exceeding 80 dB(A)
Impacts from monotonous repetitive work	Pattern of movements repeated twice or more per minute implying locked or uncomfortable working positions
Risk of accidents	Accidents resulting in more than one day of absence from work

Impact criteria

The EDIP method sets a limit to when an exposure is regarded as having an impact. This limit is called the impact criterion, and it is shown for each type of impact in Table 9.4. The exposure is assessed for each of the types of impacts, for each individual process, and calculated as an impact if it exceeds the impact criterion for the type of effect.

For the chemical impacts, an assessment is made of whether there are substances in the process, exposure to which can cause one of the effects: cancer, reproductive damage, allergy or damage to the nervous system. One should address both substances emitted and by-products formed in the process. If there is an exposure to substances with one or more of the above effects, an impact is defined as exposure to the substance by inhalation of concentrations exceeding 10% of the existing threshold value or as exposure to the substance via skin contact.

For the effects of noise and monotonous repetitive work, a similar assessment is made of whether the impact criteria in Table 9.4 are exceeded.

If an impact criterion is exceeded, the entire time of the process is calculated as a contribution to the impact time for the type of effect in question.

The risk of accident in the process is determined as the frequency of accidents leading to more than one day's absence from work.

Effects on the working environment computed as impact time or accident risk can be added for all processes in the product system within each type of impact.

Inventory of impacts on the working environment

Two different forms of data are used for assessment of the impact on the working environment for a process or a collection of processes.

- Process-specific data computed for the individual processes which are included in the product system.
- Sector data computed as the average for the relevant type of process within the sector.

Process-specific data are typically used for the processes occurring at the company commissioning the LCA and possibly for processes at first level subcontractors. Process-specific data for a process are the impact hours within each of the different types of impact on the working environment. Collection of data is done in cooperation with the company's working environment department or the associated health service.

Sector data must often be used for processes with producers of materials, producers of standard components, transport processes, electricity production and disposal processes. Sector data are computed as the number of reported work injuries within each of the types of impacts seen in relation to production in the sector, e.g. the number of injuries reported within the different types of impacts per kWh of electricity. Sector data and process-specific data cannot, therefore, be compared or added before the process-specific data are converted to potential reported work injuries, as described in Chapter 10. But aggregation can then occur in the same way as for potential environmental impacts and resource consumption.

Calculation model

When the database is established as unit processes, a model is set up for the product system on the basis of the scope definition, i.e. the processes are pieced together like building blocks in a model, and the inventory is then carried out on the basis of data in this model.

Termination of a process and aggregation of the product system

Data are aggregated in the inventory across the processes in the product system to summed values for each environmental exchange. The result is a summary of all terminal exchanges, i.e. all inputs directly from nature (resource consumption), all outputs directly to the recipient (air, water and soil) and all impacts on the working environment. The task involved in aggregation is to "follow" all non-terminal exchanges back to the terminal input or forward to the terminal output, and for the processes passed through on the way, to sum all terminal exchanges over the entire product system separately for each type of exchange. In the EDIP method this is called "terminating" the exchanges. Figure 9.3 illustrates an example of termination of a portion of the exchanges from a single process, viz. production of recycled aluminium.

Figure 9.3. Diagrammatic outline of terminalization of the environmental exchanges from the unit process for recycled aluminium

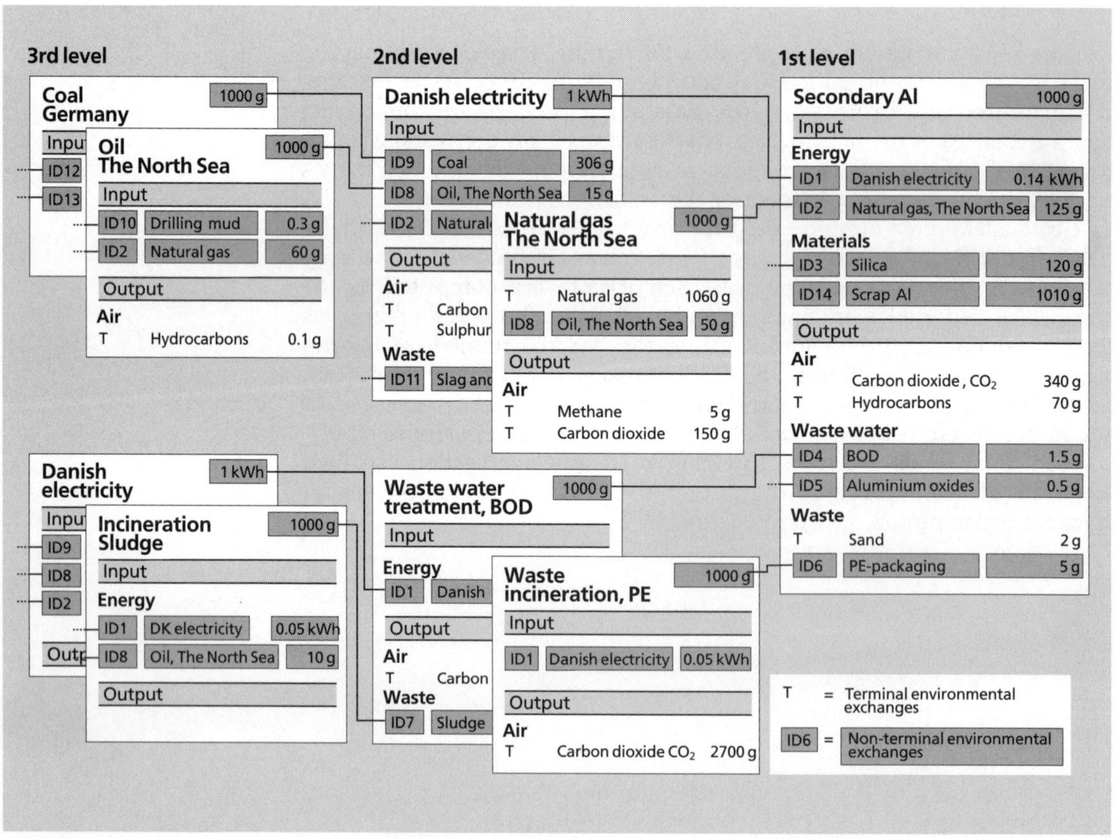

Inventory per functional unit

The object of study in the LCA is the functional unit, and after aggregation, the terminal exchanges with environment must therefore be expressed per functional unit. The inventory of each individual terminal exchange can be expressed:

$$Q_i = T \cdot \sum_{up} Q_{i,up} + \frac{T}{L} \cdot \sum_{p} Q_{i,p}$$

where

Q_i = sum of terminal exchanges (i) computed per functional unit
T = duration of the functional unit (years)
L = life span of the product (years)
$Q_{i,p}$ = the terminal exchange from the process (p) computed for the number of key units of the process entering into the product system; (p) designates all processes other than the use process
$Q_{i,up}$ = the terminal exchange per annum from the use process (up) including the processes specified by all of the use process's non-terminal exchanges

If the life of the functional unit is defined as the life span of the product, then T = L and the expression in the formula becomes simpler. But products with different life spans must often be compared, and $\sum Q_{i,p}$ must then be modified by T/L as shown in the formula so that the products can be compared on the basis of the same service.

Aggregation can be done for the whole product system or only for selected processes, a stage in the product life, a sub-assembly in the product or similar as required in the actual LCA. But irrespective of the level of aggregation, the exchanges should always be expressed per functional unit. If this is not done, what they represent must be specified very carefully in order to avoid unequal comparisons.

The product system is infinite

The example for recycled aluminium shows that the product system quickly develops branches. Each of the non-terminal inputs has in itself a number of exchanges which must be terminated, and it will soon become clear that the non-terminal exchanges will continue to increase in number as one regresses. Infinite loops will also arise during termination of the exchanges. The example for recycled aluminium contains such a loop: the recovery process for aluminium includes *Danish electricity,* which includes *North Sea oil,* which in turn includes *drilling mud* used for the extraction, and *aluminium* goes into the drilling mud. And North Sea oil is also used in the extraction of aluminium.

As the example shows, it is therefore not possible to terminate all exchanges. The more levels one goes back, the more non-terminal exchanges in practice arise. This paradox of course derives from the fact that one is also regressing in *time* when an exchange is followed back. In analysing the birth of one product in this way, industrial activity can in principle therefore be followed back to its genesis, if data were available for following the exchanges back. But this is of no interest.

The magnitude of the exchanges of course becomes smaller and smaller the more levels one goes back, and they will mean less and less. An intelligent delimitation is therefore necessary.

The example for recycled aluminium shows only the termination for a part of the exchanges, and only for one process. It is easy to imagine the complexity of the task for the product system of a complex product, e.g. a TV, and the importance of delimiting the task is clear. It is similarly clear that it is necessary to use a computer program to be able to handle the calculations within a reasonable time.

In the PC program developed for the EDIP method, the termination is automatic when the product system is modelled by the unit processes established in the database. The program also delimits the exchanges on the basis of an automatic calculation of their relative significance, after they have been followed several levels back.

Site or region-specific considerations

When the exchanges are subsequently to be interpreted in environmental terms, it can be desirable to include special site-specific considerations, e.g. if some processes occur in a more sensitive environment than others. The procedure is described in the presentation of the impact assessment in Chapter 10. It must be pointed out here that at the stage of the inventory, however, if it is desired to include site-specific aspects in the calculations of the potential environmental impact of the exchanges, this must be done *before* the exchanges are summed for the processes as shown in the above formula. The exchanges must in this case be transformed into potential environmental impacts separately for each process, taking the site-specific aspects of the process into account, and the potential environmental impacts from the various processes can subsequently be summed. This procedure is described in Chapter 10 in the section on site-specific considerations, and a formula for the aggregation and the calculations is given, which in this case is used instead of the above calculation model.

The PC program developed for the EDIP method can handle both procedures.

The inventory itself must under all circumstances be presented separately before it is transformed during the impact assessment.

Searching and sorting

The LCA results in aggregate values for the environmental exchanges and their potential impacts for the entire product system expressed per functional unit. These aggregate values provide a picture of the total contribution to various types of impacts, and they show which types of impacts are the most significant in the actual weighting. Such an overview is of interest. But in product development it is often even more interesting to see which *sources* of the various exchanges and potential impacts are significant in the total picture.

It is therefore a requirement in the modelling and the inventory of the product system that it be possible to seek and sort on the various processes and exchanges, so that the sources of the contributions can be traced and designated.

Some questions can, for example, be: From where does the contribution to the greenhouse effect derive? Or the potential ecotoxicity?

- from the materials in total?
- from the metals in total?
- from the steel?

- from the production stage?
- from the use stage?
- from the disposal stage?

- from the electronics?
- from the chassis?
- from the base plate?

- Process name and ID number
- Reference period of data
  - current or projected process
- Geographic location
- Stage of the product life
- Specially user-defined process groups, e.g. classified according to the MECO principle
- Sub-assembly in the product hierarchy
- ID number of environmental exchange
- Age of data
- Data quality indicator
- IDs of resource, environmental impact and impact on the working environment in all elements of the LCA method

Table 9.5. Significant search parameters

The environmental specialist must be able to answer these types of questions, and it is thus necessary that the results can be easily sorted and divided into the various sources. In practice this requirement also means that the modelling and the inventory must be done by a computer program.

Table 9.5 shows some significant parameters on which it is important to be able to search. As is clear, this corresponds to the parameters specified in the data format for the unit process.

Simulation of the total uncertainty

When the LCA enters into decisions of major strategic and economic significance for the company, it is important that it includes an assessment of uncertainty.

The uncertainty assessment can be carried out manually by identifying the processes and the exchanges of greatest significance for total resource consumption and potential impacts, and then estimating a coefficient of variation for these exchanges and assessing what the variation means for the total result. If necessary, individual calculations can be performed manually on the most significant key figures to gain an idea of how much the uncertainty means in the total picture.

A computer program enables automation of the uncertainty analysis. The strength of the computer program is that it can handle a large number of calculations quickly, which means that the variation in all exchanges in all processes in the product system can be included without the person carrying out the LCA having to spend further time on it. This therefore means both savings in time and a better result.

The most suitable procedure for estimating the total variation of a variable calculated as the sum of a very large number of variables, each with its own variation, is to simulate it. The so-called Monte Carlo method is used for this purpose.

	SO_2 emission (g)		
	μ	σ/μ	D
Process 1	4	0.1	logN
Process 2	2	0.4	logN
Process 3	5	0.1	logN
Process 4	17	0.3	logN
Process 5	8	0.5	logN
Σ Product system	36	?	

Table 9.6. Find the total variation in the SO_2 emission in the product system

The Monte Carlo method

In brief, the task of the uncertainty analysis is to find the variation in the sum of a number of variables, each of which has its mean, coefficient of variation and statistical distribution. Table 9.6 shows a simple example of this task, namely finding the variation in the summed SO_2 emission from five processes.

As Table 9.6 shows, the five processes have different means and coefficients of variation for the SO_2 emission. It is assumed here that all exchanges follow a log normal distribution. The mean of the sum of the emissions is not difficult to find. It is simply a matter of adding the means of the individual emissions. The variation in the sum of the emissions is, however, more difficult to calculate, and in more complex cases than that shown here, it is impractical to find the variation in the aggregate exchanges analytically, i.e. by expressing it in a formula such that it can be calculated.

But it is very simple to find the aggregate variation via a simulation, and this is the Monte Carlo method's concept and the strength of the computer program. Conceptually, the simulation is very simple, as shown here in the example with the SO_2 emission in Table 9.7.

Table 9.7. Variation in the total SO_2 emission is found by simulation in the Monte Carlo method

			Generation of random numbers within the logN distribution									
	μ	σ/μ	1	2	3	4	5	6	7	8	9	10
Process 1	4	0.1	3.8	4.5	4.2	3.9	4.4	3.6	4.2	4.0	3.7	3.8
Process 2	2	0.4	1.5	3.8	1.9	2.2	2.9	0.9	3.0	1.2	1.0	1.8
Process 3	5	0.1	5.4	5.7	3.9	4.7	5.5	4.6	4.4	5.2	5.0	5.8
Process 4	17	0.3	12.1	23.4	20.2	17.0	10.2	15.9	18.2	14.3	20.5	18.2
Process 5	8	0.5	4.3	12.7	5.7	8.4	9.8	11.7	16.5	3.8	2.8	4.2
	$\bar{x}_\Sigma$	$S/\bar{x}_\Sigma$	Σ_1	Σ_2	Σ_3	Σ_4	Σ_5	Σ_6	Σ_7	Σ_8	Σ_9	Σ_{10}
ΣProduct system	36.05	0.20	27.1	50.1	35.9	36.2	32.8	36.7	46.3	28.5	33.0	33.8

- For each of the five processes a random number for the SO_2 emission is generated within the specified distribution. One result is then calculated for the sum of SO_2 emissions for each of the five processes (result no. 1, $\sum_1 = 27.1$).
- The figure for the SO_2 emission is then altered in each of the five processes. The alteration is made by generating a new random number for the SO_2 emission in each of the five processes.
- The sum of the five SO_2 emissions is now calculated again (result no. 2, $\sum_2 = 50.1$). Generation is then done on new random numbers for all of the actual processes, and a new sum is calculated etc. until there are 'n' values for the sum (here n = 10).
- Mean and standard deviation are then calculated for the n values.

Thus, ten different random numbers for the SO_2 emission in Table 9.7 are generated for each process. In each case the number is generated within a population of numbers specified by the actual mean, the coefficient of variation and the distribution function for the emission. The ten numbers generated for the SO_2 emission from process 1 thus derive from a population of numbers with a mean of 4 and a coefficient of variation of 0.1, and which follow a log normal distribution.

The sum of the SO_2 emissions from the five processes is then calculated a total of ten times, and the mean and standard deviation of these ten values are calculated. The coefficient of variation is thus derived for the total emission of SO_2 from the five processes, which in this example represent the entire product system.

The simulation technique is simple and provides a correct estimate of the total variation in the sum of the individual exchanges. This requires calculating capacity because the inventory must be made many times to obtain a good estimate of the variation, typically 10-1000 times depending on the nature of the decision which must be supported. But the procedure is problem-free with a computer and does not require man-hours. The EDIP method's PC program is capable of such simulations.

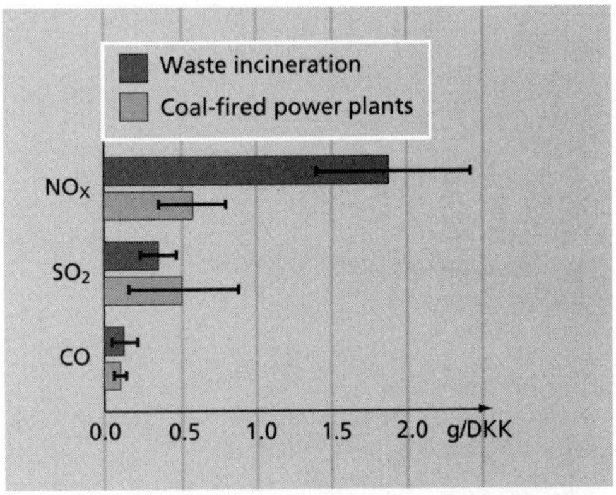

Figure 9.4. Example of presentation of confidence intervals in comparison of results (Hoffman et al., 1995)

Figure 9.4 shows an example of how the result of a simulation of uncertainty can appear. The example shows mean and confidence interval for a number of emissions from a waste incineration plant and a coal-fired power plant. It is evident that there is a significant difference in only one of the exchanges, namely NO_x, while the differences for the others lie within the uncertainty with which they were determined.

In decisions of very great significance, the presentation of alternatives should include uncertainty estimates as shown here.

Corresponding presentations can be given in the assessment step, where the uncertainties in impact equivalency factors and normalization references are included in the simulation. This option is also included in the EDIP method's PC program, but the database for the uncertainty in the various assessment factors is not yet established.

The result of the inventory

Data from the inventory comprise an independent result of the LCA and should be discussed separately, before they are translated into weighted resource consumption and potential impacts in the subsequent impact assessment phase. In the transformation of data into potential environmental impacts which occurs in the impact assessment, some of the information is lost and the impact assessment itself should therefore always be supplemented by a discussion of the exchanges at the inventory level.

In the presentation of data from the inventory, the data can be aggregated at an arbitrary level according to the use for which the inventory is required. It may be that a presentation is desired for a specific process, a component, a stage in the product life, or for the total product system.

Table 9.8 illustrates an example of an inventory of the environmental exchanges from the product system for a high pressure cleaner, divided into stages of the product life.

			Production of materials	Manufacturing of products	Use	Dis-posal	Transport	Total
Resource consumption								
Crude oil		g	2,670	90	4,920	-20	290	7,950
Natural gas		g	2,040	60	3,580	-1	20	5,700
Coal		g	1,860	1,800	17,700	20	1	21,400
Brown coal		g	290	-	13,200	-0.5	-	13,500
Uranium ore		g	0.05	-	1.80	-0.0001	-	1.9
Water for hydroelectric power		litres	340	80	7,250	-0.10	0.0005	7,670
Aluminium	Al	g	580	0.01	0.9	-160	0.02	420
Iron	Fe	g	1,220	0.004	0.3	-230	0.02	990
Copper	Cu	g	460	-	-	-50	-	410
Manganese	Mn	g	9	-	-	-1	-	8
Nickel	Ni	g	7	-	-	-	-	7
Zinc	Zn	g	9	-	-	-	-	9
Calcium carbonate	$CaCO_3$	g	220	0.008	0.6	-0.005	0.04	220
Quartz	SiO_2	g	460	-	-	-30	-	430
Sodium chloride	NaCl	g	130	0.02	2	-0.003	0.04	130
Soft wood		g	1,670	6	840	-0.05	0	2,520
Groundwater		litres	-	5	-	0.3	-	5
Surface water		litres	-	0.1	-	0.006	0	0.1
Unspecified water		litres	50	0.01	41,300	-0.04	60	41,400
Ancillary substances for which resource consumption is not included								
Fatty alcohol (C12-18) ethoxylate		g	-	-	430	-	-	430
Secondary alkane sulphonate, Na salt		g	-	-	80	-	-	80
Diethylene glycol monobutyl ether		g	-	-	140	-	-	140
Nitrilo acetic acid, 3-Na salt		g	-	-	260	-	-	260
Quarternary coconut fat amine ethoxylate		g	-	-	110	-	-	110
Difatty acid isopropyl ester dimethyl ammonium methosulphate		g	-	-	20	-	-	20
Fatty alcohol ether sulphate, Na salt		g	-	-	80	-	-	80
Emissions to air								
Carbon dioxide	CO_2	g	11,200	5,300	86,600	1,800	650	105,600
Carbon monoxide	CO	g	30	0.9	50	9	2	92
Nitrogen oxides	NO_x	g	60	20	350	0.7	8	440
Sulphur dioxide	SO_2	g	80	30	770	-0.05	3	880
Nitrous oxide	N_2O	g	0.3	0.3	7	0.003	0.01	8
Unspecified particles (dust)		g	60	3	40	-8	0.4	95
Hydrocarbons	HC	g	60	40	420	0.3	0.4	520
Volatile organic compounds	VOC	g	100	0.09	2	-0.09	1	100
Styrene		g	0.007	-	-	-	-	0.007
1-methyl-2-propyrrolidone		g	-	0.0007	-	-	-	0.0007
Unspecified methyl styrene		g	-	0.002	-	-	-	0.002
Dioxin		mg	0.0004	$2x10^{-8}$	$1x10^{-5}$	0	-	0.0004
Arsenic	As	mg	0.3	0.05	8	0	0	8
Cadmium	Cd	mg	0.07	0.005	1	$2x10^{-5}$	0	1
Copper	Cu	mg	0.4	0.4	8	0.004	0	9
Mercury	Hg	mg	0.1	0.06	2	0	2	4
Manganese	Mn	mg	0.2	-	-	-	-	0.2
Vanadium	V	mg	30	4	230	0.02	0	264
Fatty acid (C14-18)-2-ethyl ester		g	-	-	1	-	-	1
Terpine alcohol		g	-	-	0.08	-	-	0.08
Diethylene glycol monobutyl ether		g	-	-	0.01	-	-	0.01

Table 9.8. Example of presentation of the inventory for a high pressure cleaner

			Production of materials	Manufacturing of products	Use	Dis-posal	Transport	Total
Emissions to water								
Chemical oxygen demand	COD	g	7	0.0006	0.04	-0.002	0.003	7
Total nitrogen	tot-N	g	0.1	0.0007	0.2	-0.00006	0.0008	0.3
Total phosphorus	tot-P	g	0.01	-	-	-	-	0.01
Fatty alcohol (C12-18) ethoxylate		g	-	-	20	-	-	20
Secondary alkane sulphonate, Na salt		g	-	-	70	-	-	70
Diethylene glycol monobutyl ether		g	-	-	150	-	-	150
Nitrilo acetic acid, 3-Na salt		g	-	-	70	-	-	70
Quaternary coconut fat amine ethoxylate		g	-	-	6	-	-	6
Fatty alcohol ether sulphate, Na salt		g	-	-	20	-	-	20
Difatty acid isopropyl ester dimethyl ammonium methosulphate		g	-	-	20	-	-	20
Hydrocarbons		g	0.60	0.001	0.1	-0.0004	0.006	0.7
Unspecified oil		g	0.3	0.002	0.2	-0.00001	0	0.5
Phenol		g	0.0003	0.00003	0.003	0	0	0.003
Arsenic	As	mg	10	-	-	-	-	10
Cadmium	Cd	mg	5	-	-	-	-	5
Chromium	Cr	mg	20	-	-	-	-	20
Copper	Cu	mg	30	-	-	-	-	30
Mercury	Hg	mg	0.0006	-	-	-	-	0.0006
Manganese	Mn	mg	0.05	-	-	-	-	0.05
Nickel	Ni	mg	20	-	-	-	-	20
Zinc	Zn	mg	30	-	-	-	-	30
Emissions to soil								
Nitrilo acetic acid, 3-Na salt		g	-	-	30	-	-	30
Fatty alcohol ether sulphate, Na salt		g	-	-	15	-	-	15
Diethylene glycol monobutyl ether		g	-	-	20	-	-	20
Secondary alkane sulphonate, Na salt		g	-	-	15	-	-	15
Fatty alcohol (C12-18) ethoxylate		g	-	-	50	-	-	50
Quaternary coconut fat amine ethoxylate		g	-	-	70	-	-	70
Difatty acid isopropyl ester dimethyl ammonium methosulphate		g	-	-	1	-	-	1
Waste								
Unspecified hazardous waste		g	0.07	-	-	-	-	0.07
Unspecified waste containing heavy metals		g	9	-	-	-	-	9
Unspecified nuclear waste		g	0.006	-	0.3	0	-	0.3
Unspecified chemical waste		g	30	-	-	-	-	30
Unspecified slag and ashes		g	100	140	3.400	1	0.7	3,600
Unspecified bulk waste		g	710	1,000	10,300	10	0.7	12,000
Working environment								
Monotonous repetive work		hours	-	0.36	0.006	-	-	0.366
Hearing impairment inducing noise		hours	-	0.38	0.08	-	-	0.46
Substances causing damage to the nervous system		hours	-	-	0.003	-	-	0.003
Allergenic substances		hours	-	0.022	0.013	-	-	0.035
Substances causing damage to the reproductive system		hours	-	0.022	-	-	-	0.022
Carcinogenic substances		hours	-	0.022	0.009	-	-	0.031
Accidents		number	-	14×10^{-6}	3×10^{-6}	-	-	17×10^{-6}

Notes

- Indicates that information regarding the specific stage of the product life is not available.

0 Indicates that the value is very small compared to that of the other life cycle stages.

Table 9.8. (continued)

10. Impact assessment

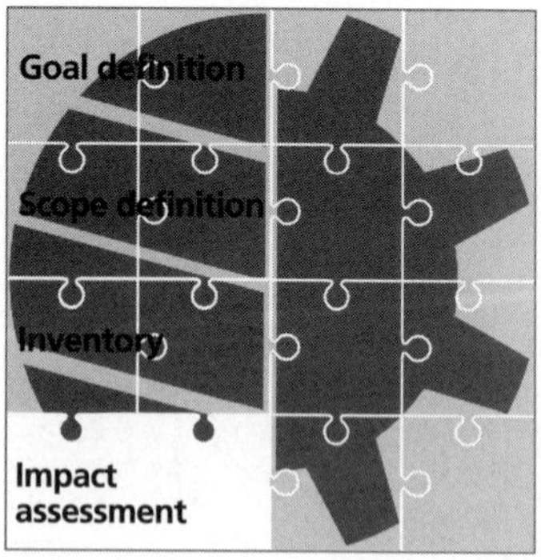

The inventory provides collected data for environmental exchanges for the entire product system. Some of the exchanges are environmentally important, others are of no significance.

For the LCA to be able to support decisions, the data in the inventory must be interpreted. The interpretation must be based on the available background knowledge of the environment, resources and working environment, and it must show which of the exchanges are significant, and how great their contributions can be.

An example based on data from an inventory is shown in Table 10.1, which illustrates a small excerpt from the inventories for two alternative solutions to a CFC-based refrigerator. The refrigerators are distinguished from each other in the use of different refrigerants and different foaming agents in their insulation. Both refrigerators represent non-ozone-depleting alternatives to CFCs. One solution uses R134a as both refrigerant and foaming gas. R134a is an HFC - a non-ozone-depleting halocarbon. The other solution uses isobutane as refrigerant and pentane as foaming agent. Isobutane and pentane are hydrocarbons. As is evident from the table, assessing which alternative is the better from an environmental viewpoint is not entirely straightforward on the basis of the available inventories, even for so simple an example as this one.

As a rule, the alternatives which are compared will involve the use of different materials and processes. Different resources are therefore consumed, and different substances emitted to the environment. The inventory can include information on emissions of many different substances, where some of the substances will be hazardous to the environment,

Substance	Unit	Pentane/isobutane	R 134a
Emissions to air			
Carbon dioxide	g	850,000	997,000
Carbon monoxide	g	1,000	1,030
Nitrous oxide	g	50	59
Sulphur dioxide	g	4,570	5,360
Nitrogen oxides	g	3,100	3,640
Lead	g	0.38	0.41
Cadmium	g	0.001	0.001
Chromium	g	0.02	0.02
Copper	g	0.06	0.08
Vanadium	g	0.95	1.05
Hydrocarbons	g	5,660	6,790
Methane	g	14	17
Isobutane	g	6	-
Pentane	g	126	-
R 134a	g	-	737
Emissions to water			
BOD	g	1.11	1.12
H^+ (hydrogen ions)	g	2.33	2.43
NH_4^+-N	g	0.22	0.22
NO_3^--N	g	0.09	0.09
Phenol	g	0.009	0.01

Table 10.1. Inventory of emissions in the product system for refrigerators with different refrigerants and foaming agents (taken from the case histories in Part V)

while others may be innocuous. In the example in Table 10.1, one alternative thus leads to emission of R134a, while the other leads to emission of isobutane and pentane. The R134a alternative also has a slightly greater energy consumption in the use stage, leading to greater emissions of a number of substances from the electricity generation.

To be in a position to decide which alternative is the more environmentally friendly, it is necessary to interpret the inventory, by weighting the various exchanges according to the magnitude of the environmental impact to which they can lead, based on existing knowledge of the substances' hazards to the environment and the scarcity of the resources.

In the EDIP method, interpretation of the inventory is done by assessing the magnitude of the contribution which each individual exchange will be able to make to known environmental impacts. The inventory is thus transformed into a list of resource consumptions and potentials for impacts on the environment and the working environment.

The EDIP method's assessment tool treats impact categories within the three main groups

- environmental impacts,
- resource consumption,
- impacts on the working environment,

and in the EDIP method's environmental interpretation tool, transformation of the inventory comprises three elements in accordance with the general framework in both SETAC and ISO recommendations:

1. *Calculation of potential environmental impacts for emissions* - how much do the emissions contribute to the various types of environmental impacts?
2. *Normalization* - how great are the resource consumptions and the potentials for impacts on the environment and the working environment relative to the impact from society's activities as a whole?
3. *Weighting* - which of the resource consumptions and the potential impacts are the most important?

After the impact assessment, uncertainty and sensitivity analyses are carried out.

Calculation of environmental impact potentials

The impact potentials for the product are the sums of the impact potentials for the emissions occurring throughout the product system:

∑Impact potentials = ∑Quantity of substance × substance's impact potential

Whether the potentials trigger actual effects depends inter alia on how the emissions occur and the concentrations of substances to which they contribute in the environment. These conditions are rarely known in product LCAs.

The EDIP method enables calculation of the potential contributions to the types of environmental impacts shown in Table 8.3. Calculation of the potential contributions from an emission to an environmental impact requires a knowledge of how strongly the substances emitted contribute to the type of environmental impact in question.

The substance's impact potential is expressed in the form of an equivalency factor. Substances which can contribute to more than one type of environmental impact have an equivalency factor for each type of impact.

In ISO and SETAC recommended terminology for LCAs, the word "classification" is used for the review of the inventory that aims at deciding the type of impact to which the emissions can contribute. The subsequent calculation of the emissions' environmental impact potentials is called "characterization". In the EDIP method, these two concepts are united under the designation "calculation of environmental impact potentials".

Method

If the emission of a substance (i) has the magnitude Q_i, and if the substance's equivalency factor for the environmental impact category (j) is called $EF(j)_i$, the emission's potential contribution $EP(j)_i$ to the environmental impact (j) is calculated as follows.

To calculate the total contribution to the impact category seen over the entire product system, the contributions of all individual emissions are summed:

$$EP(j)_i = Q_i \cdot EF(j)_i$$

Table 10.2 provides a list of the abbreviations for the various environmental impact categories.

$$EP(j) = \Sigma EP(j)_i = \Sigma(Q_i \cdot EF(j)_i)$$

Environmental Impact	Abbrevation (j)
Global warming	gw
Stratospheric ozone depletion	so
Photochemical ozone formation	po
Acidification	ac
Nutrient enrichment	ne
Ecotoxicity	et
Human toxicity	ht
Persistent toxicity	pt

Table 10.2. Abbreviations for environmental impact categories

Choice of equivalency factor

Table 10.3 shows equivalency factors for potential contributions from the most important greenhouse gases to the environmental impact category global warming, gw. The equivalency factor expresses the substance's strength measured relative to a reference substance, and this method of expressing the equivalency factor is common for most types of environmental impacts. For global warming the reference substance is carbon dioxide, CO_2, and the impact factors thus express the substances' potential impacts as grams of CO_2 equivalent per gram of substance. When methane has an impact factor of 25, it means that emission of 1 g of methane contributes as much to global warming as the emission of 25 g CO_2.

Substance	Formula	GWP g CO_2/g substance 20 years	100 years	500 years
Carbon dioxide	CO_2	1	1	1
Methane	CH_4	62	**25**	8
Nitrous oxide	N_2O	290	320	180
CFC11	$CFCl_3$	5000	4000	1400
CFC12	CF_2Cl_2	7900	8500	4200
HCFC22	CHF_2Cl	4300	1700	520
HCFC141b	$CFCl_2CH_3$	1800	630	200
HCFC142b	CF_2ClCH_3	4200	2000	630
HFC134a	CH_2FCF_3	3300	1300	420
HFC152a	CHF_2CH_3	460	140	44
Tetrachloromethane	CCl_4	2000	1400	500
1,1,1-trichloroethane	CH_3CCl_3	360	110	35
Chloroform	$CHCl_3$	15	5	1
Methylene chloride	CH_2Cl_2	28	9	3
Carbon monoxide	CO	2	2	2

Table 10.3. GWP values for time horizons of 20 years, 100 years and 500 years (Albritton et al., 1995a)

The table shows equivalency factors expressing the gases' strengths relative to that of CO_2 seen over three different time horizons, 20 years, 100 years and 500 years. It is evident that the relative strength depends on whether the gases' contributions are viewed over a short or a long time span. For some of the gases, e.g. nitrous oxide, the strength increases with a longer time span. For other gases, e.g. methane, the strength decreases with a longer time horizon. The time horizon selected for the equivalency factors will therefore affect the result of the impact assessment. In the LCA, it is common to select a time horizon of 100 years for calculation of global warming impact potentials.

For an emission of 550 g methane, 15,000 g CO_2 and 10 g CO, the global warming potential for a time horizon of 100 years is determined thus:

$$\Sigma EP(gw) = 550 \cdot 25 \text{ g } CO_2\text{-eq} + 15.000 \cdot 1 \text{ g } CO_2\text{-eq} + 10 \cdot 2 \text{ g } CO_2\text{-eq}$$
$$= 28.770 \text{ g } CO_2\text{-eq}$$

Site-specific considerations

One of the characteristics of the LCA is that there is only a very general knowledge of site-specific conditions for most of the processes in the product system, and in many cases there is no site-specific information at all. Some of the processes can occur at many different places in the world, e.g. depending on choice of supplier of materials or on how the finished product is distributed among various export markets. As described in Chapter 3, it is therefore generally not possible to include exact information on site-specific conditions, and as a consequence of this, it is also not possible to make a quantitative assessment of the exposure of those specific parts of the environment which receive the substances emitted.

Experience with environmental assessments of products has, however, revealed a substantial need to attempt to modify the LCA to include some sort of exposure assessment. For emissions from certain processes it is quite obviously unreasonable to calculate a potential contribution to an impact category, if it can be said with certainty that the emission cannot trigger an effect where the process is occurring. This can, for example, be the case for the impact category human toxicity when toxic but short-lived substances are emitted at a place where humans are not exposed to them. Or it can be the case for emission of SO_2 from ocean-going ships, where it can be said with certainty that this SO_2 will never be able to contribute to acidification because it is absorbed by the sea, which, because of its enormous volume, is not sensitive to acidification.

There has been some discussion internationally on how the LCA can include a type of *generalized assessment of exposure* which can remove such quite obviously unreasonable calculations of contributions to potential environmental impacts. The EDIP method includes a site factor which may be used to modify the calculation of the potential environmental impact.

The EDIP method offers the option of including a site factor, SF, which represents a generalized knowledge of site-specific conditions. The value of the LCA in supporting decisions is thus increased, and the method is pre-adapted to future developments within the area.

$$\mathbf{EP(j) = \sum_{p}\sum_{i} EP(j)_{p,i} = \sum_{p}\sum_{i}(Q_{i,p} \cdot EF(j)_i \cdot SF(j)_p)}$$

where 'i' is the individual substance, 'p' is the individual process and 'j' is the impact category in question.

Note that the site factor is bound to the environmental impact category and to the process, and not to the individual substance.

As is evident in the formula, calculation of the potential environmental impact must occur separately for each process, before emission from the processes can be summarized in the aggregation for the product system. For each individual process, the contributions to the individual impact categories for all substances , 'i', are therefore calculated first. The impact potentials are then subsequently summed for the processes 'p'. When site-specific aspects are included, the emissions cannot, therefore, be summarized for the processes before the calculation of impact potentials, as is otherwise done.

The magnitude of the site factor

The magnitude of the site factor can reflect both the way in which the emission in question is occurring (e.g. from a power station chimney, from the exhaust pipe of a car, or indoors) and the sensitivity of the environment in the area where the environmental impact from the emission is occurring. It therefore depends both on the process and on the type of environmental impact to which the potential contribution is calculated. Currently (1997) the site factor's primary task will be to remove quite obviously unreasonable contributions to impact potentials. If it is clearly improbable that the emission will be able to trigger real effects, SF is thus assigned the value 0.

If no information on site-specific conditions is available, SF is assigned the value 1. This corresponds to the normal practice for LCAs. It means that the conservative viewpoint is adopted, that the full potential impact should be calculated in the LCA unless different information which points to the opposite is available.

The magnitude of the site factor can be regarded as the *probability* that real environmental effects will occur as a consequence of the type of emission in question. It will still not be possible to predict real effects in specific recipients, and it is thus still a matter of *potentials* for environmental effects. But by use of a generalized exposure assessment in the form of a site factor, the LCA provides a more accurate picture of the impacts to which the product will be able to contribute.

The object of introducing the site factor is thus to increase the ability of the LCA to provide intelligent support to environmental decisions.

Examples of use of site factors

Table 10.4 gives some examples of situations where the site factor for various types of environmental impacts is assigned the value 1, and of situations where it should be assigned the value 0, or at least a value smaller than 1.

Impact category (j)	SF(j) = 1	SF(j) = 0 or SF(j) < 1
Acidification	Sensitive land areas - e.g. Sweden and Norway	Robust areas - e.g. the sea during sea transport
Toxicity via air	Exposure probable - e.g. road transport	Exposure unlikely - e.g. power plants, aircraft transport
Acute ecotoxicity	Critical exposure probable - e.g. lakes and streams	Critical exposure unlikely - e.g. the sea
Photochemical ozone formation	Areas with NO_x - typically inhabited areas	Areas "without" NO_x - typically sparsely populated areas, e.g. deserts and the sea
Nutrient enrichment	N-limited systems - typically coastal waters and the sea	P-limited systems - typically lakes
Nutrient enrichment	P-limited systems - typically lakes	N-limited systems - typically coastal waters and the sea

Table 10.4. Examples of choice of site factors for various types of environmental impacts

Use of site factors in the calculation of environmental impact potentials has consequences for the subsequent normalization and weighting. If a particular type of emission, for example, is cancelled by assigning it a site factor of 0 in the calculation of impact potentials, the same type of emission should be cancelled in the calculation of normalization references and weighting factors unless its magnitude is insignificant at the societal level and therefore plays no role in the magnitude of the normalization reference. Correction of the basis for normalization and weighting therefore requires a detailed insight into the inventories and calculations which underlie the normalization references (see Hauschild and Wenzel, 1997g).

Note that in comparing different products, normalization and weighting must be carried out with the same factors. At the same time, the use of site factors should always be reported together with the results of normalization and weighting.

A common database for the use of site factors has not yet been developed, and they should therefore be used with care, and only in the following cases:

1) As a part of the sensitivity analysis, to clarify the significance of site-specific aspects. This means that the LCA is carried out both with and without use of the site factors.
2) When unambiguous conclusions can be drawn on the basis of the environmental impact potentials before normalization and weighting, that is, in situations where all the environmental impact potentials are lower for one of the alternatives after use of site factors, thus implying that there is no need to carry out normalization and weighting.
3) When the site-specific conditions are of considerable significance for the potential environmental impacts for the product under assessment, but of only very little significance for society's background contribution to the impact.

Resource consumption

In the EDIP method, resource consumptions are computed as consumptions of pure raw materials, for example pure iron and not iron ore. This means that the so-called terminal inputs from the inventory express consumption of pure resources.

These resource consumptions appear individually as assessment parameters in the EDIP method, and no further aggregation or conversion to "potentials" is made, as is done in the conversion of emissions to potential environmental impacts.

The resource consumptions from the inventory thus enter directly into the normalization at the next step in the assessment method. The consumption RC(j) of resource 'j' is simply the summarized quantity of the resource, Q_i:

$$RC(j) = Q_i$$

For renewable resources it is relevant to take account of site- or, more correctly, region-specific aspects, and the possibility of doing this is opened in the normalization and the weighting.

Potentials for impacts on the working environment

As described in Chapter 9, the impacts on the working environment are measured for each individual process in the product system as duration of impact, Q_i, within each of the seven impact categories (j) shown in Table 8.3. The inventory includes only duration of exposure for processes where the impact is above a certain level. In this way, the inventory step already includes an assessment of whether there is an exposure, and if so, of the exposure's potential for causing impacts. The exposure in the working environment is therefore already expressed in the inventory as impact potentials, and exposure and site-specific aspects are included.

The potential for impacts on the working environment is therefore identical to the computed duration of impact within each of the impact categories.

$$WP(j) = Q_i$$

Interpretation of impact potentials and resource consumptions

In comparing potential impacts and resource consumptions from alternative products, one alternative will sometimes prove to have a lower contribution to all impact potentials and resource consumptions than the other. The calculation of impact potentials has thus fulfilled a significant objective, as it is now clear which alternative is preferable from an environmental perspective.

However, it will often be such that one alternative has the lowest impact potentials for certain impact categories, while the other has the lowest impact potentials for other impact categories. In this situation, the various impact potentials and resource consumptions must be assessed relative to each other. Are any of them greater and more serious than others?

This is the situation in Table 10.5, which shows the environmental impact potentials for the two refrigerators for which Table 10.1 presented the inventories.

Environmental impact potentials from refrigerators			
Impact categories	**Unit**	**Pentane/isobutane**	**R 134a**
Global warming	g CO_2-eq.	870,000	2,270,000
Ozone depletion	g CFC11-eq	0	0
Photochemical ozone formation	g C_2H_4-eq	101	63
Acidification	g SO_2-eq	6,820	8,000
Nutrient enrichment	g NO_3^--eq	4,380	5,150
Ecotoxicity, water, chronic	m^3 water	44,000	44,000
Human toxicity, water	m^3 water	1,610	1,610
Ecotoxicity, water, acute	m^3 water	2,160	2,160
Human toxicity, air	m^3 air	563,000,000	613,000,000
Hazardous waste	g	185	185
Nuclear waste	g	0.05	0.05
Slag and ashes	g	20,300	24,300
Bulk waste	g	165,000	194,000

Table 10.5. Environmental impact potentials for the product system for refrigerators with different refrigerants and foaming agents (from Part V of the book)

The pentane/isobutane solution clearly leads to the lowest impact potentials for global warming. This is attributable especially to the fact that R134a, a powerful greenhouse gas, is used in the other refrigerator.

The potential impacts for nutrient enrichment, acidification and human toxicity via air and the quantities of slag and bulk waste are marginally lower for the pentane/isobutane refrigerator as a consequence of a slightly lower energy consumption. On the other hand, the potential for photochemical ozone formation is clearly greater for the pentane/isobutane alternative than for the R134a refrigerator, because both pentane and isobutane can cause photochemical ozone formation when the substances escape from the insulating foam or the compressor. There is no difference between the two refrigerators for the other impact categories.

In order to decide which alternative is preferable from an environmental viewpoint, it is necessary to weigh the potentials for the various environmental impacts against one another. Is it possible that the pentane/isobutane alternative's somewhat greater contribution to photochemical ozone formation should be accorded a greater weighting than the R134a alternative's greater contribution to the other impacts? A mutual comparison of the magnitudes of the potentials is unreasonable without further consideration, because they concern different effects and are calculated in different units. The global warming potential is calculated as g CO_2-equivalents, the ozone formation potential as C_2H_4-equivalents, and these values are not comparable.

Part of the comparison therefore involves relating the potential impacts to a common scale, a common reference, which can show what is large and what is small relative to a known impact. This need is also illustrated in Table 10.6, which shows a presentation of potential environmental impacts for the product systems for seven different products or activities.

For each product or activity, the inventory has been transformed into potential contributions to ten different types of impacts and four types of

	Unit	Refrigerator LER200	Television LX 5500	Compact water supply system JetpaQ	High pressure cleaner Hobby 70	Control unit PVEH[1]	Annual consumption of milk cartons[2]	One year of motoring[3]
Global impacts								
Global warming	g CO_2-eq	3,720,000	1,340,000	1,550,000	108,200	53,100	2,460	5,860,000
Ozone depletion	g CFC11-eq	600						
Regional impacts								
Acidification	g SO_2-eq	8,000	14,000	17,600	1,200	500	26	33,600
Nutrient enrichment	g NO_3^--eq	5,200	7,900	8,500	600	660	41	61,200
Photochemical ozone formation	g C_2H_4-eq	60	190	73	4	52	3	43,900
Human toxicity (water)	m^3 water	1,600	8,300	24	19,700	6.8		
Ecotoxicity (water, chronic)	m^3 water	44,000	3,700	116	13,200	41		
Human toxicity (soil)	m^3 soil				200			
Local impacts								
Human toxicity (air)	m^3 air	628,000,000	806,000,000	936,000,000	680,000	50,000,000	2,860,000	4,850,000,000
Ecotoxicity (water, acute)	m^3 water	2,200	200	6.4	4,300	2.2		
Hazardous waste	g	750	2.8	1,300	90	9.5	0.046	
Nuclear waste	g	0.05	3.0		0.3	0.0014	0.01	
Slag and ashes	g	24,300	50,800	57,500	3,700	400	-4.2	5,000
Bulk waste	g	220,000	197,000	185,000	12,700	2,800	590	5,000

Notes

1: Is used in e.g. cranes for controlling the hydraulics

2: Consumption by an average inhabitant in Denmark

3: Exhaust fumes from driving a petrol-driven van 15,000 km in an urban area

Table 10.6. Environmental impact potentials for various products and activities (from the case histories in Part V and other sources)

waste. On the basis of the table it is possible to compare the products by each impact category. Most of the products are active, that is, they consume energy in the product's use stage. For these products there are large contributions to the types of environmental impacts attributed primarily to energy production: global warming, acidification, and to some extent photochemical ozone formation, human toxicity (air) and nutrient enrichment. Some of the products emit environmentally hazardous chemicals during their life cycles. This results in greater contributions to human toxicity and to aquatic ecotoxicity and, if the chemicals are CFCs, to ozone depletion.

It is not possible on the basis of the table to assess which of the environmental impact potentials are great and which are small. It is also not an easy matter to decide which of the products or activities cause the greatest total impact on the environment. How large is the average environmental impact from the LX 5500 television relative to the environmental impact of refrigerating one's food in an LER200 refrigerator throughout its entire life span or receiving one's milk supply over a year in milk cartons?

To be able to compare the environmental impacts from different products or activities, it is necessary to compare them against a common reference. This is the object of normalization.

Normalization

Normalization has two purposes:

- To provide an impression of the relative magnitudes of the potential impacts and resource consumptions.
- To present the results in a form suitable for the final weighting and decision-making.

In normalization, the impact potentials and resource consumptions which have been determined are compared with an impact which is common for all impact categories, and of which the consequences for the environment, resources and working environment are known. In this way an impression is gained of which potential impacts are large and which are small, seen in relation to the known reference impact.

As normalization references, the EDIP method uses the resource consumption and the potential impacts which society imposes on the environment and the working environment each year.

Even if some of the future consequences of society's current impact on the environment and the working environment are unknown, we still have an idea of how serious the situation is for each individual impact category, and this knowledge is used in the subsequent weighting.

Method

In normalization, the impact potentials or resource consumptions (P(j)) are divided by the corresponding normalization references. The normalization references are calculated on the basis of an inventory of all of society's activities over a period of time which must be the same as the duration of the product's service, as defined in the functional unit. If the functional unit defines the duration of the service as T years, the normalization reference is therefore expressed as T·R(j), where R(j) denotes the normalization reference for 1 year. The normalized impact potentials and resource consumptions, NP(j), are calculated as follows:

$$NP(j) = P(j) \cdot \frac{1}{T \cdot R(j)}$$

For the environmental exchanges which apply on a global scale, the EDIP method uses the total global impact as the normalization reference. For the other impacts of regional or local effect, the basis of the method is the total Danish contribution to the impact.

Global impacts are used for non-renewable resources and for contributions to the global impacts global warming and stratospheric ozone depletion. Danish impacts are used for the other categories. The method does, however, admit the possibility of including region-specific aspects in the normalization, and thus using other normalization references if there are special reasons for doing so. See the following two sections.

	Unit	The world/EU Total	per person	Denmark Total	per person	Reference
Environmental impact potentials						
Global warming	kg CO_2-eq/year	$4.59\cdot10^{13}$	8.700 [1]	$9.74\cdot10^{10}$	19,000 [4]	[1]
Ozone depletion	kg CFC11-eq/year	$1.05\cdot10^{9}$	0.20 [1]	$3.92\cdot10^{6}$	0.84 [4]	[2]
Acidification	kg SO_2-eq/year	$2.65\cdot10^{10}$	82 [2]	$7.12\cdot10^{8}$	140 [4]	[3]
Photochemical ozone formation	kg C_2H_4-eq/year	$5.05\cdot10^{9}$	16 [3]	$1.01\cdot10^{8}$	20 [4]	[4]
Resource consumption						
Oil	kg oil/year	$3.13\cdot10^{12}$	590	-	-	[5]
Coal	kg coal/year	$3.04\cdot10^{12}$	570	-	-	[5]
Iron	kg iron/year	$5.44\cdot10^{11}$	100	-	-	[6.7]
Copper	kg copper/year	$8.81\cdot10^{9}$	1.7	-	-	[7]
Impact potentials on the working environment						
Carcinogenic substances	Impact hours/year	-	-	$1.2\cdot10^{8}$	47	[8]
Allergenic substances	Impact hours/year	-	-	$1.6\cdot10^{8}$	62	[8]
Noise and hearing impairments	Impact hours/year	-	-	$1.3\cdot10^{9}$	514	[8]
Monotonous repetitive work	Impact hours/year	-	-	$3.1\cdot10^{8}$	118	[8]

Notes

1: Global 1990, 2: EU 1990, 3: EU 1985, 4: DK 1990

References

[1]: Hauschild and Wenzel, 1997d, [2]: Hauschild and Wenzel, 1997e, [3]: Hauschild and Wenzel, 1997b, [4]: Hauschild and Wenzel, 1997a, [5]: British Petroleum Company, 1992, [6]: British Geological Survey, 1991, [7]: World Resources Institute, 1992, [8]: Rasmussen, 1997.

Table 10.7. Examples of society's impact on environment, resources and working environment, globally and regionally

Normalization references

Table 10.7 gives values for background contributions to several impact categories calculated for a year for the whole world, the EU and Denmark.

Region-specific considerations

It is important for the subsequent weighting that the normalization references correspond to an impact for which the magnitude and contribution to effects in the environment and the working environment and to depletion of reserves are known.

It is important that the impacts are summarized for the area which actually contributes to the current condition of the environment as it is experienced.

For the global impacts it makes no difference where the environmental exchanges occur. But for the regional and local impacts, only the exchanges occurring within the regions or local areas in question contribute to the current and future condition of the environment there. Emissions of greenhouse gases thus contribute to global warming, irrespective of where in the world they occur. But the Far Eastern or North American emissions of substances which can contribute to acidification have no influence on the degree of acidification of forests and lakes observable in Europe today.

On this basis, the normalization and the subsequent weighting should ideally be carried out with normalization references and weighting factors which, for each individual emission, are representative of the region where the emission is occurring. The PC program developed for the EDIP method (Danish Environmental Protection Agency, 1997) is prepared to include such region-specific normalization and weighting, but the requisite database is not available. As default values, the EDIP method therefore uses Danish normalization references and weighting factors based on Danish environmental policy for the regional and local environmental impact categories and for the working environment.

Time dependence
The impact of society's activities changes with time, for example as a consequence of changes in consumption and living standard and as a consequence of targeted initiatives against the worst impacts. Figure 10.1 illustrates the trends in global consumption of copper, global emissions of ozone-depleting gases and Danish emissions of acidifying gases. It is evident that even for the short period shown in the figure, developments are quite different for the three types of impacts. It is therefore necessary to choose one and the same reference year for all impact categories within each of the three main groups environment, resources and working environment, to ensure that the scale is indeed common for all impact categories.

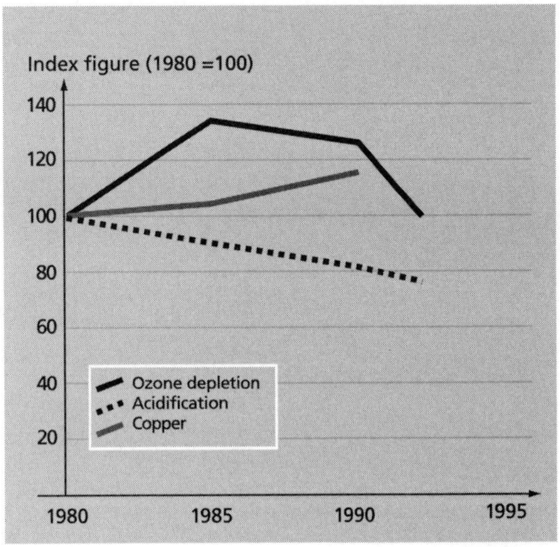

Figure 10.1. Developments over the last 15 years in global consumption of copper, in ozone-depleting potential of global emissions, and in potential for acidification for Danish emissions (World Resources Institute, 1992; Albritton et al., 1995b; UNECE, 1994)

The EDIP method currently operates with 1990 as reference year for normalization, and normalization references are therefore written as $ER(j)_{90}$ for environmental impact categories, $RR(j)_{90}$ for resource consumption, and $WR(j)_{90}$ for impacts on the working environment.

The normalization reference should be updated on an on-going basis, e.g. every fifth year, together with the weighting factors.

The person-equivalent
The global impact will always be much greater than the impact from a particular region, irrespective of the type of effect to which the impact contributes. This is illustrated for some of the impact categories in Table 10.7, and is of course attributable to the fact that the global impacts are the result of the activity of many more people than the impacts from the region in question. Use of global impacts as the normalization reference for the global impact categories and regional impacts for the regional and the local impact categories will thus give an imbalance in the normalization, and it will result in global impacts from the product system coming to appear much less than the other impacts, because they are compared with the activity of the population of the entire world, while the others are compared only with the activity of e.g. the Danish population.

To correct this bias and ensure that the set of normalizing references constitutes a common scale for all impact categories, irrespective of whether they are global or regional, the normalization references are calculated as the background impact over the course of one year per person in the area for which the impact is computed. This gives the normalization references the unit "impact potential *per person* per year" for each individual impact category.

$$\text{Normalization reference} = \frac{\text{Impact potential or resource consumption for the area in question in the reference year}}{\text{Population of the inventorized area in the reference year}}$$

Table 10.7 illustrates examples of impacts calculated per person for some impact categories and resource consumptions. It is clear that in 1990, the average Dane contributed significantly more than the average citizen of the world to the global impact categories, about twice as much to global warming and over four times as much to ozone depletion. The table also shows that there can be large differences in the nominal size of the impacts for the various impact categories and resource consumptions. The annual global average contribution to global warming was thus 8,700 kg CO_2-equivalents per person in 1990, while the contribution to ozone depletion in the same year was 0.2 kg CFC11-equivalents per person.

When the impact potentials are normalized, they are expressed in person-equivalents (PE), i.e. fractions of the contribution to the impact deriving from the average person.

On normalization, all potential impacts and resource consumptions thus assume the same unit, and it is possible to compare their magnitudes. At the same time, the normalized potential impacts of the product are expressed in a comprehensible unit as they can be viewed relative to one's own average contribution to the impact.

The person-equivalent is illustrated in Table 10.8 and Figure 10.2. The table shows some average annual resource consumptions and impact potentials for environment and working environment for the product system for a refrigerator. The refrigerator's impact potentials are compared to the impact potentials of the background load from an average person. The refrigerator's impact potentials and resource consump-

Category	Unit	1 refrigerator per year	1 person per year	Per thousand
Environmental impact potentials				
Global warming	g CO_2-eq/year	286,000	8,700,000	32
Ozone depletion	g CFC11-eq/year	48	198	240
Acidification	g SO_2-eq/year	615	138,000	4.6
Nutrient enrichment	g NO_3^--eq/year	400	264,000	1.5
Hazardous waste	g/year	750	20,700	2.8
Resource consumption				
Crude oil	g/year	2,810	591,000	4.8
Coal	g/year	24,000	573,000	42
Aluminium	g/year	36	3,370	11
Nickel	g/year	0.85	177	4.8
Impact potentials on the working environment				
Noise that induces hearing impairments	hours/year	0.0982	514	0.19
Allergenic substances	hours/year	0.0378	62	0.61
Carcinogenic substances	hours/year	0.00831	47	0.18

Table 10.8. Examples of normalized environmental impact potentials, resource consumptions and potential impacts on the working environment for a refrigerator

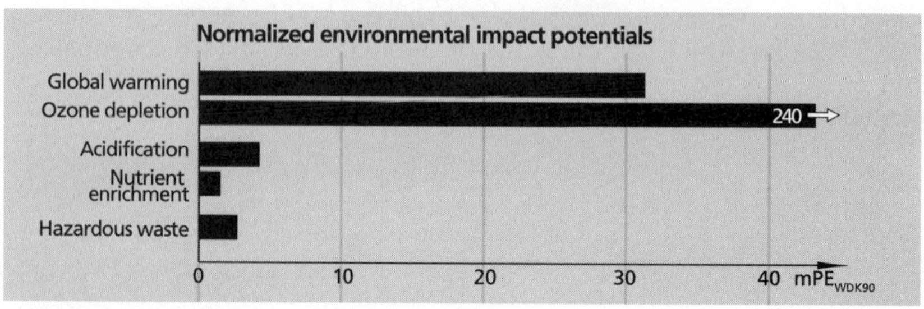

Figure 10.2. Normalized environmental impact potentials for a refrigerator. Same figures as in Table 10.8

tions can thus be expressed in parts per thousand of a person's average contribution to the various impact categories. One part per thousand of a person equivalent is a milli-person-equivalent, mPE, and the normalized impact potentials in Table 10.8 are therefore expressed in mPE.

In Figure 10.2, the environmental impact potentials are shown in a histogram as mPE. This is the way of showing the normalized environmental impact potentials in the EDIP method.

Normalization of environmental impact potentials

The impact potentials of the actual emissions per person from society's activities in the reference year 1990 are used as normalization references $ER(j)_{90}$ for the environmental impact potentials. The normalized environmental impact potentials are calculated on this basis as:

$$NEP(j) = EP(j) \cdot \frac{1}{T \cdot ER(j)_{90}}$$

On normalization, the environmental impact potentials are expressed in the unit PE_{WDK90}, i.e. person-equivalents (PE) calculated as the potential environmental impact of an average person in the world (W) or in Denmark (DK) in 1990, depending on whether the impact category is global or regional/local.

Normalization references for the environmental impact potentials are determined on the basis of an inventory of the reference year for all of society's emissions which can contribute to the type of environmental impact. When the magnitudes of the emissions have been determined, they are converted to environmental impact potentials, as described earlier in the section on calculation of environmental impact potentials. Table 10.9 gives an example of a calculation of the normalization reference for an environmental impact category on the

Table 10.9. Calculation of normalization reference for the environmental impact category photochemical ozone formation

Source	Average POCP $g\ C_2H_4/g\ VOC$	Danish VOC-emission kt VOC/year	Photochemical ozone-formation potential $kt\ C_2H_4/year$	Ref
Energy production				
Power plants	0.4	10.930	4.372	[1]
Road transport	0.4	95.032	38.013	[1]
Transportation, other	0.5	9.978	4.989	[1]
Energy production, other	0.5	3.417	1.709	[1]
Other activities				
Oil refining and distribution	0.4	8.1	3.24	[1]
Painting	0.5	25.50	12.75	[1]
Production of chemical and other uses of solvents	0.3	8.32	2.50	[1]
Cattle farming	0.007	262	1.834	[2]
Other farming	0.4	2.04	0.816	[1]
Landfills and sewage treatment	0.007	120	0.840	[2]
Total VOC			71.42	
CO	0.04	737.706	29.508	
Total			**100.928**	

References

[1]: Fenhann and Kilde, 1994
[2]: Fenger et al., 1990

basis of a total inventory for society of the emissions which contribute to this impact. The example shown is for the impact category photochemical ozone formation. The normalization reference for this impact category is the total potential for photochemical ozone formation of the Danish emissions of volatile organic compounds, VOC, and carbon monoxide, CO, in 1990. Equivalency factors can be found in the literature only for individual substances, not for groups of substances such as VOCs. It is therefore necessary to group the activities according to the composition of their VOC emissions and to calculate average equivalency factors for each of these groups. Calculation of the normalization reference for photochemical ozone formation is given in Hauschild & Wenzel (1997a).

For other impact categories, the work involved in determining normalization references is more comprehensive. Examples are aquatic and terrestrial ecotoxicity (Hauschild et al., 1997a). Separate inventories are prepared here for the potential ecotoxicity from:

- discharges from individual industries with separate emissions,
- discharge or spreading of metals and detergents with water and sludge from municipal sewage treatment plants,
- agricultural use of pesticides,
- air-borne emissions from electricity production,
- air-borne emissions of dioxins from incineration of waste, and
- all other emissions of the metals lead, mercury and cadmium.

It has been possible for all impact categories to obtain estimates of the background contribution. It has been necessary for some impact categories to use figures for emissions in the years before or after the reference year 1990.

Normalization of resource consumption

The normalization references for use of non-renewable resources are based on total global consumption of the resource, because resources are traded on one common world market. The area from which the resource originates is therefore irrelevant from the viewpoint of reserves. In the subsequent weighting, there is thus no need to take account of where non-renewable resources come from. The normalization references for consumption of a non-renewable resource are therefore found as the duration, T (years), for the service defined in the functional unit, multiplied by the annual consumption of the resource in 1990, $RR(j)_{90}$, computed per person in the world. The normalized resource consumption NR(j) is thus:

$$NR(j) = RC(j) \cdot \frac{1}{T \cdot RR(j)_{90}}$$

On normalization, the resource consumptions are expressed in the unit PE_{W90} - person-equivalents (PE) calculated as the resource consumption for an average person in the world (W) in 1990.

For renewable resources, it is decisive in the subsequent weighting whether consumption exceeds regeneration of the resource. This can vary from region to region and from local area to local area. The normalization references for renewable resources should therefore be based on the total consumption in 1990 in the relevant area, computed per person in the area. For wood, which is a renewable resource, it has not, however, been possible to find regional inventories for consumption. In the EDIP method's normalization element, wood must therefore be handled for the time being as if it were a non-renewable resource, i.e., wood consumption is normalized with the average global wood consumption per person in 1990.

Normalization of potentials for impacts on the working environment

Society's total impact is summarized for each individual type of impact on the working environment in a form which can be used directly in normalization. Impact potentials for the working environment are normalized with the background impact for Denmark. The exposure time of the impact is computed per employee per year in Denmark within each individual impact category for the years around 1990. For potential impacts on the working environment, this value WR(j) is used, multiplied by the duration T (in years) of the service, as the normalization reference. A person-equivalent is thus derived for impacts on the working environment, corresponding to the impact to which an average Danish worker is exposed during a year. On this basis the normalized potential for impacts on the working environment, NWP(j), is calculated as:

$$NWP(j) = WP(j) \cdot \frac{1}{T \cdot WR(j)_{90}}$$

On normalization, the potentials for impacts on the working environment are expressed in the unit PE_{DK90} - person-equivalents (PE) calculated as the average duration of exposure to the impact per employee in Denmark (DK) in 1990.

Interpretation of normalized potential impacts and resource consumption

On normalization, the impact potentials and resource consumptions are expressed as person-equivalents, PE_{W90} or PE_{DK90}, and it is now possible to compare:

- The impact from alternative products, from entirely different products, in fact from any activity.
- The relative magnitude of the contributions to the individual impact categories or resource consumptions within each of the main groups environment, resources and working environment.

With due regard for the difference between the types of impacts within the three main groups environment, resources and working environment, it is also possible to compare the magnitude of the contributions between the main groups in order to gain an impression of whether the product is particularly severe on the working environment or the environment, or whether it is the resource consumption which has the greatest impact relative to society's total activities. If a comparison across main groups produces results which are other than those expected, it can mean that significant environmental exchanges have not been included for one or more of the impact categories.

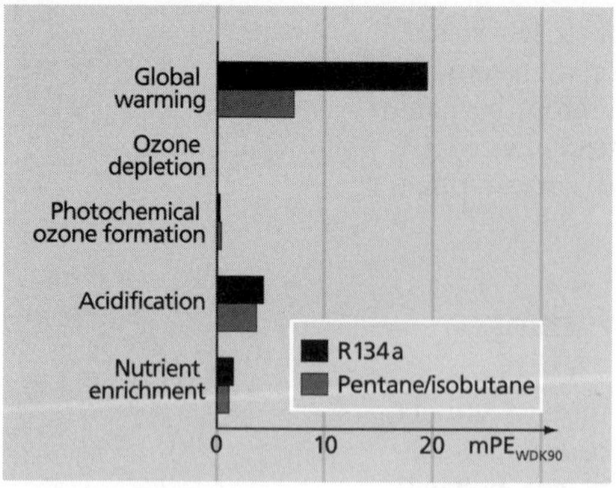

Figure 10.3. Normalized environmental impact potentials for the product system for refrigerators with different refrigerants and foaming agents (from Part V of the book)

In addition,

on normalization, the magnitudes of the potential impacts and the resource consumptions are expressed in a unit which it is simple to understand, namely fractions of the annual impact from an average person.

Figure 10.3 shows the normalized environmental impact potentials for the product system for the two refrigerators for which Table 10.1 previously gave the inventories. The impact potentials are normalized with the background loads for the reference year 1990. If the reference year had been another year, the relative magnitudes of the normalized impact potentials would have been different (cf. Figure 10.1). It is therefore important to select the same reference year for all impact categories, and the reference year used should be stated clearly.

After normalization, it is clear that for both refrigerators, the potential environmental impact is greatest for global warming. The contributions here are 20 mPE_{W90} from the R134a alternative and a bare 10 mPE_{W90} from the pentane/isobutane alternative. In other words, the refrigerators' contribution to global warming corresponds to 10-20‰ or 1-2% of an average global person's impact in 1990.

The contributions for the other impact categories are significantly less, and there are only small differences between the two refrigerators. For photochemical ozone formation there is, however, a significant difference, in that the normalized potential for the R134a alternative is 60% less, although this is difficult to distinguish in Figure 10.3. From an environmental point of view there is much to be said in favour of the pentane/isobutane alternative. The final decision depends, however, on how the potential impacts for global warming and photochemical ozone formation in particular are weighted relative to each other, i.e. which environmental impact potentials are considered the *most important.*

Figure 10.4 on the opposite page illustrates normalized environmental impact potentials, resource consumptions and potential impacts on the working environment for the seven products or activities presented with their environmental impact potentials in Table 10.6.

Among the seven examples, it is one year's running of a car (15,000 km of mainly urban driving for one person in a delivery vehicle), which makes by far the greatest contribution to almost all environmental impacts and resource consumptions. This is the case even though the inventory has only included the environmental exchanges from the vehicle's *use stage*, i.e. the exhaust gases and fuel consumption. Several of the normalized impact potentials from driving the vehicle are close to 1000 mPE or 1 person-equivalent. For photochemical ozone formation, the normalized environmental impact potential is close to 2.5 person-equivalents. A person driving this much (in a delivery vehicle in the city) contributes 2.5 times the average person's total photochemical ozone formation potential just by driving. The impact from this driving is thus significantly greater than the average impact from one person, partly because a delivery vehicle is more polluting than an average passenger car, partly because 15,000 km of urban driving per year is a high figure, and finally because it is common for more than one person to travel in a car for part of the time.

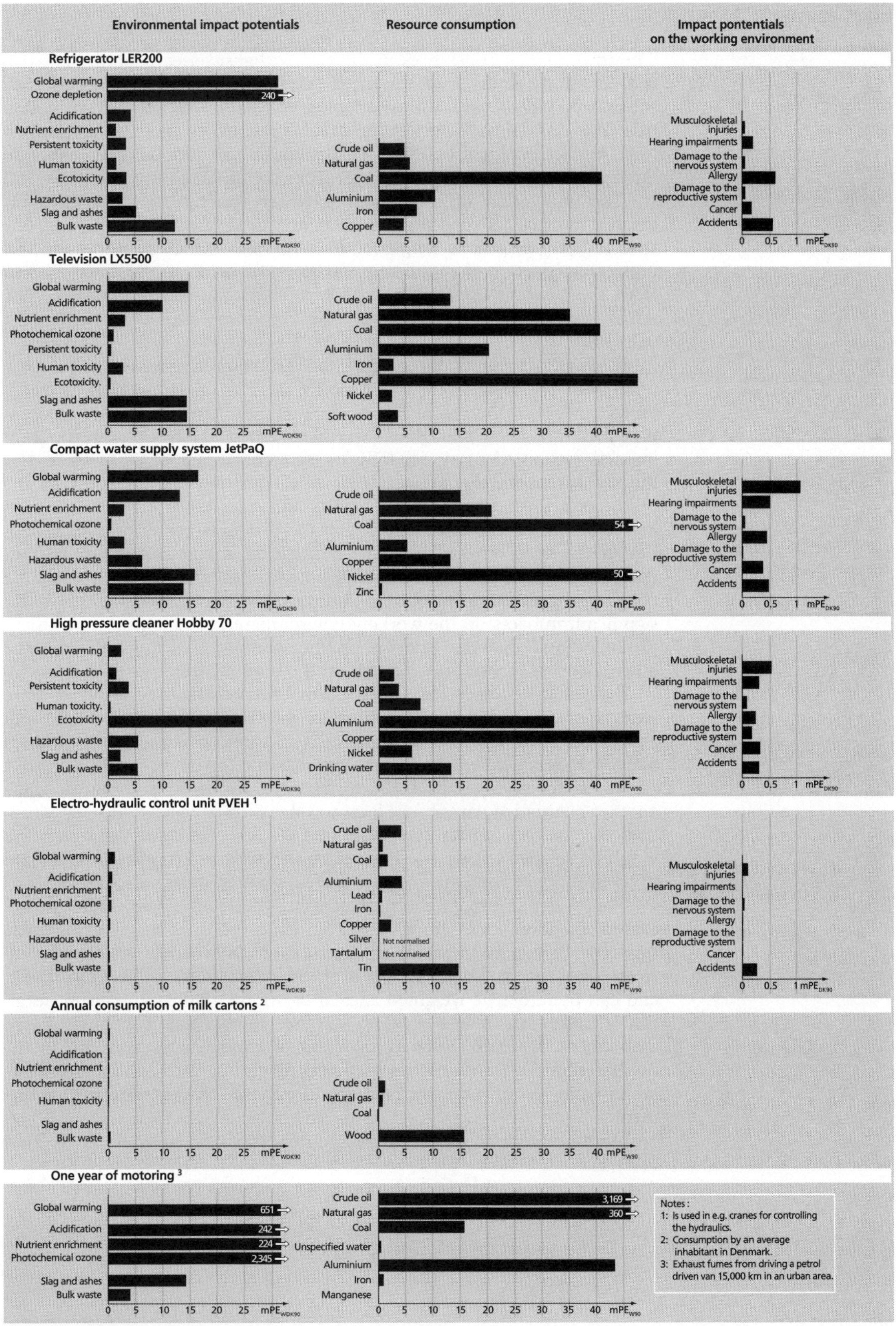

Figure 10.4. Normalized resource consumption and impact potentials for various products and activities (from the case histories in Part V and other sources)

The inventory of substances in the car's exhaust gases is based on average emission factors, and these provided no information on individual substances which would have enabled calculation of ecotoxicity potentials. For this impact category, the high pressure cleaner has a relatively large impact potential because of its spreading of various detergents and other cleaning agents with the water from the cleaner's use stage.

The only other impact category where driving does not have the greatest normalized contribution is ozone depletion. There is only one special group of gases - long-lived halocarbons - which contribute to this impact category, and significant emissions of halocarbons do not occur from the product system for products other than the refrigerator. The refrigerator contributes to the potential for ozone depletion with almost 0.25 PE.

If the car driving is disregarded, the normalized environmental impact potentials generally lie below 20 mPE or 2% of an average person's impact. Most of the products consume energy, and their greatest potentials therefore lie within global warming and acidification: categories of impacts attributed predominantly to the combustion of fossil fuels. For the same reason, the greatest resource consumption for most of the products is the consumption of lignite and coal, both of which are used in production of the electricity which the products use during operation. For the working environment, the normalized impact potentials are generally lower than the environmental impact potentials and the resource consumptions. This is attributed particularly to the fact that there are no significant impacts on the working environment in the use stage for these products, and that the impacts on the working environment from the other stages are therefore "diluted" in the total picture.

Even if it becomes clear with normalization which contributions are large and which are small relative to the background load, there will always be situations where judgment is required in comparison of alternative products, where one alternative has the lowest normalized impact potentials for some impact categories or resource consumptions, while another alternative contributes least to others. In such a "trade-off" situation, the various impact categories and resource consumptions must be weighted relative to one another before they can be compared. Are some of them more serious than others, and how much more serious?

This weighting must aim at expressing the relative seriousness within each of the three main groups, environment, resources and working environment. It can be difficult enough to decide whether acidification is worse than ecotoxicity, but both of these impacts affect the environment and can therefore be weighted against each other with some justification. But it makes much less sense to discuss whether acidification is worse than use of the earth's limited resources of copper, and the EDIP method therefore does not attempt to assign weightings between environmental impacts, resource consumptions and impacts on the working environment.

Normalization is considered in more detail in Hauschild & Wenzel (1997f).

Weighting

Normalization assists in assessing which of the potential impacts are large and which are small, by placing them in relation to the impacts from an average person in 1990. But even if the potential impacts for two different impact categories are equally large on normalization, this does not automatically mean that the two potential impacts are equally serious. To be able to compare the potentials for the various impacts, an assessment must first be made of the seriousness of the impact categories relative to one another.

The mutual seriousness of the impact categories is expressed in a set of weighting factors with one factor per impact category within each of the main groups environment, resources and working environment.

The weighting can then be made by multiplying the normalized impact potential or resource consumption, NP(j), by the weighting factor, WF(j), associated with the impact category or resource consumption in question, and the weighted impact potential or resource consumption WP(j) is therefore calculated as:

$$WP(j) = WF(j) \cdot NP(j)$$

$$\Rightarrow \quad WP(j) = WF(j) \cdot \frac{1}{T \cdot R(j)} \cdot P(j)$$

where P(j) is either EP(j), RC(j) or WP(j).

Weighting of environmental impact potentials

The weighting factor for an environmental impact must reflect the seriousness of the effect potentially being caused by the impact and the possible consequences of this effect relative to the other environmental effects. The seriousness depends on several characteristics of the environmental effect and the actual condition of the environment. Some of these characteristics are of a purely scientific nature and they should be evaluated in the same way, irrespective of the use to which the LCA will be put. Examples of scientific characteristics include:

- How far is the actual impact status from critical threshold values for the impact in the areas affected by the emission?
- What types of consequences will result if these threshold values are exceeded?
- Will the environmental damage be reversible?
- How great an area will be affected by the damage?

Other characteristics are of a more normative or political nature, and depend on the context into which the LCA will enter. If the LCA is to be used by a company to prioritize an initiative in relation to a particular market, normative questions could, for example, be:

- What is the perception of the seriousness of the environmental effect among potential buyers of the product or among others of the company's interested parties?
- What is the perception among various opinion-leaders in the community in general?

If the LCA is to be used by the authorities, for example to set criteria for assigning an environmental label for this product type, normative questions could be:

- What current political plans of action exist for reduction of the impacts which lead to the environmental effect?
- What international regulations exist in the area?

The EDIP method's weighting factors were not developed as described above, on the basis of purely scientific weightings followed by more normative weightings. The basis of knowledge of the individual impact categories is simply not good enough to enable such sharp definition of a purely scientific and a more normative weighting. When the above considerations concerning weighting are discussed here, it is to indicate that it is this type of consideration which in principle should be expressed in the weighting factors.

The EDIP method instead seeks to fix the weighting factors so that they:

- as far as possible for each individual environmental impact reflect the scientific circumstances and the relevant normative circumstances in one,
- reflect the official societal priorities as well as possible,
- give the company the best possible assurance of how the impact categories will be ranked by the authorities in the future.

This is achieved by using political environmental targets within the field of each individual impact category as a basis for the weighting.

Weighting on the basis of political environmental targets

The authorities' regulation of society's impact on the environment has naturally enough been focused on those activities which have the greatest environmental impacts. International agreements and national plans of action have set reduction targets for society's impacts on the environment. Today there are thus politically determined targets for reductions in the most significant contributions to all of the impact categories treated by the EDIP method.

When targets are set for reductions in society's environmental impact, this is based on considerations of how serious the consequences of the impact can be, and the costs which will be associated with reducing them. The considerations include such issues as:

- What damage to the environment can be observed today as a consequence of the impact?
- What damage to the environment can be expected as a consequence of the impact, and what environmental consequences can result in the short and the long term?
- What costs will this damage impose on society?
- What technological possibilities are available for preventing and repairing the damage?
- Is the public aware of the environmental effect?
- How will the planned measures against the impact affect the national and the international economies and employment?

As a rule, scientific research on an environmental effect would have been going on for a long time before plans of action are initiated or targets for reduction adopted. Plans of action and targets for reductions thus usually have a substantial scientific background.

For the individual environmental impact category, the political setting of reduction targets therefore implies a balancing of scientific, technical and political considerations. No conscious balancing of the seriousness of this environmental impact is made relative to the seriousness of the other impacts to which the environment is exposed. But the targets for reductions are set within society's total economic framework for environmental improvements, and the initiative regarding individual substances and groups of substances is therefore indirectly ranked in relation to the total environmental measures.

On this basis, the political setting of reduction targets can be considered a result of a decision-making process similar to that which should underlie the determination of weighting factors for the environmental impact categories.

The situation is such that

- the authorities have set reduction targets for the most significant of the emissions which contribute to the impact categories entering into the EDIP method,
- many authorities desire a product-oriented environmental policy as a central part of their environmental administration, and thus as a significant means of achieving the reduction targets which have been set.

This argues strongly for determining the weighting factors on the basis of political reduction targets. The authorities will not be able to defend environmental priorities for the product policy other than those which can be deduced from the reduction targets, for in such case the reduction targets from action plans, agreements, etc. and the product-oriented environmental policy will each pull in its own direction. But by using this way of determining weighting factors, the authorities have a powerful instrument to ensure that companies have the correct environmental priorities when developing new products. If the societal reduction targets are changed for certain impacts, the changes can be transferred accordingly to the corresponding weightings in the product-oriented environmental policy. The weighting factors must therefore be adjusted at intervals, for example every fifth year together with the normalization references, to keep them in accordance with the current reduction targets.

In determination of weighting factors for the individual impact categories, the EDIP method is based on the existing Danish political targets for reduction of various categories of environmental impacts.

Political targets for reductions are normally set for individual substances or groups of substances and not for total contributions to environmental impacts. For example, there is a target in many countries for reduction of society's emissions of CO_2, which is the most significant greenhouse gas, but not for a reduction in society's total contribution to the global warming impact, which is also attributable to substances other than CO_2. The reduction targets for individual substances can, however, be translat-

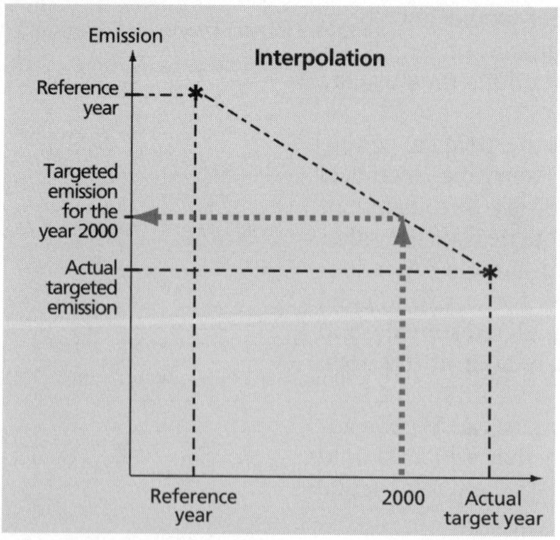

Figure 10.5. Determination of emission targets for the year 2000 by linear interpolation

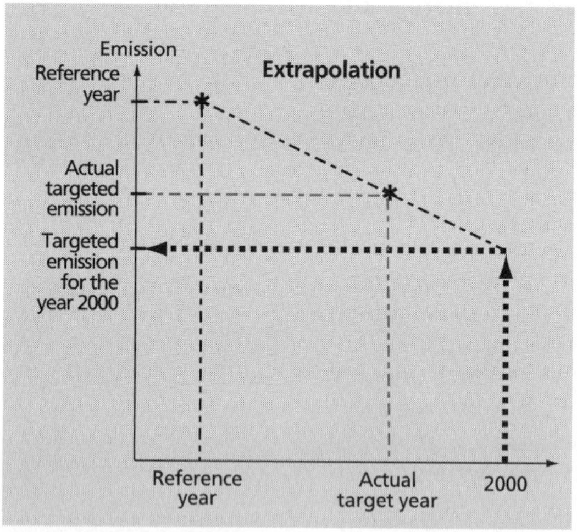

Figure 10.6. Determination of emission targets for the year 2000 by linear extrapolation

ed into reduction targets for environmental impacts with the aid of the equivalency factors in the same way that the inventory of environmental exchanges for a product system can be translated into environmental impact potentials.

As a rule, the reduction targets are formulated such that society's emissions of a substance or a group of substances in the selected target year may amount at most to a certain percentage of the emissions in a reference year. But reference year and target year vary for the various substances and groups of substances, depending on the time when the reduction targets are set, and also on the desirable and realistic time frame for achievement of the reductions.

To give a uniform treatment of all environmental impact categories, the reduction targets are harmonized in the EDIP method to apply to the same period for all environmental impact categories before they are used as a basis for calculation of the weighting factors.

Because of the requirement which has arisen for the environmental assessment of products, it can be expected that more work will be directed towards the basis of weighting in the future.

Method

The year 2000 was chosen as the common target year, while 1990, the normalization references' inventory year, was chosen as the common reference year.

An estimate is made of the magnitude of the target emission if the target year had been 2000. The estimate is made by linear interpolation if the actual target year is beyond 2000 and by linear extrapolation if the target year is before 2000, as shown in Figures 10.5 and 10.6.

The weighting factor WF(j) for the type of environmental impact category (j) is defined as the environmental impact potential of the emissions in the reference year 1990, $ER(j)_{90}$ divided by the environmental impact potential of the target emissions in the year 2000, $ER(j)_{T2000}$.

$$WF(j) = \frac{\text{Environmental impact potential of emissions in 1990}}{\text{Environmental impact potential of target emission for the year 2000}}$$

$$\Rightarrow$$

$$WF(j) = \frac{ER(j)_{90}}{ER(j)_{T2000}}$$

The weighting factor thus expresses by how much the normalization reference must be reduced by the year 2000 to be in accordance with the efforts expressed by the reduction targets for the environmental impact in question. The sharper the reduction targets, the greater the weighting factor for the environmental impact.

The choices of target year, reference year and method of interpolation are not objective. They could be selected differently, and this will be of significance for the relative magnitudes of the individual weighting factors. The target year must, however, lie a suitable number of years in the future, so that the weighting factors provide a certain assurance of how the impact categories will be ranked relative to one another when the product eventually enters the market.

Table 10.10 below shows the calculation of target emissions for the year 2000 for three environmental impact categories.

With the chosen definition of weighting factor, the weighting can be seen as a type of normalization with the target emissions in the year 2000 as normalization reference, as the weighted environmental impact potential WEP(j) is determined as:

$$WEP(j) = WF(j) \cdot NEP(j)$$

$$\Rightarrow WEP(j) = \frac{ER(j)_{90}}{ER(j)_{T2000}} \cdot \frac{EP(j)}{ER(j)_{90}} \cdot \frac{1}{T}$$

$$\Rightarrow WEP(j) = \frac{EP(j)}{ER(j)_{T2000}} \cdot \frac{1}{T}$$

In other words, the weighted environmental impact potential for the product is expressed as a percentage of the person-equivalent which can be expected in the year 2000 if society's plans for reduction are achieved.

The unit is $PET_{WDK2000}$, which stands for person-equivalent based on *target emissions* in the year 2000. WDK stands for the weighting of global impact categories on the basis of the accepted global contributions in the year 2000, while the regional and the local impact categories are

Table 10.10. Estimation of target emissions for the year 2000 for individual substances and resulting contributions to environmental impact

Substance (i)	Targeted reductions %	reference year	target year	"Target for the year 2000" kt/year	$EF(j)_i$	$EP(j)_{i,2000}$	$ER(j)_{T2000}$	Unit	Area	Ref.
Global warming									World	[1]
CO_2	30	1988	2005	22,200,000	1	22,200,000		kt CO_2-eq/year		
CFCs	100	1986	1995	0	various	0		kt CO_2-eq/year		
Methane	-	-	-	351,000	25	8,780,000		kt CO_2-eq/year		
Nitrous oxide	-	-	-	7,230	320	2,310,000		kt CO_2-eq/year		
CO	-	-	-	996,000	2	1,990,000		kt CO_2-eq/year		
- total							**3,420,000**	kt CO_2-eq/year		
Acidification									Denmark	[2]
SO_2	80	1980	2000	89	1	89		kt SO_2-eq/year		
NO_x	30	1987	1998	186	0.7	130		kt SO_2-eq/year		
NH_3	-	-	-	141	1.88	265		kt SO_2-eq/year		
- total							**484**	kt SO_2-eq/year		
Photochemical ozone formation									Denmark	[3]
VOC (all)	30	1985	2000	127	0.42	53.4		kt C_2H_4-eq/year		
CO	-	-	-	707	0.04	28.3		kt C_2H_4-eq/year		
- total							**81.7**	kt C_2H_4-eq/year		

References:
[1]: Hauschild and Wenzel, 1997d
[2]: Hauschild and Wenzel, 1997b
[3]: Hauschild and Wenzel, 1997a

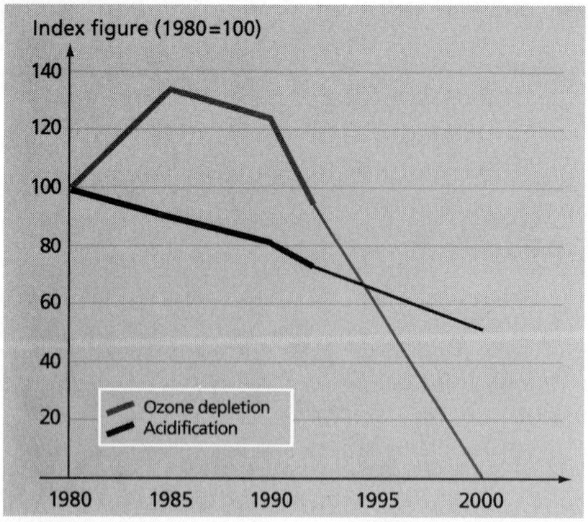

Figure 10.7. Reduction targets for the environmental impacts leading to ozone depletion and acidification (Hauschild & Wenzel, 1997e; Hauschild & Wenzel, 1997b)

weighted on the basis of the accepted contributions in Denmark. The word "accepted" should not be taken too literally. It is not supported by statutory requirements, but by national and international conventions and plans of action for the extent of reductions by the year 2000. Figure 10.7 gives examples of the development which can be expected in society's contribution to ozone depletion and acidification if the reduction targets are achieved.

Region-specific considerations

The EDIP method admits the option of including region-specific factors in the weighting of regional and local impact categories. Normalization references and weighting factors are available today only for Denmark and for certain impact categories for the EU, but the method is open to future developments within the area.

From an environmental viewpoint, the weighting should take account of region-specific factors so that the non-global impact potentials are weighted with the acceptable person-equivalent for the area which they affect. The environmentally acceptable person-equivalent is equivalent to the environmental latitude, i.e. the average impact which is acceptable for a person in a sustainable society. When the potential environmental impacts are weighted in this way, they are expressed as fractions of the environmental latitude available in the places where the environmental impacts occur.

The product system can have ramifications in many places on earth, and a region-specific weighting carried out on the potential environmental impacts will therefore require a comprehensive file of environmentally acceptable person-equivalents. A more operational solution could be that for each impact category, all countries on earth are divided up into a limited number of sensitivity classes based on the countries' population densities, technological level and the sensitivity of their natural environments. The task is thus reduced to determining environmentally acceptable person-equivalents for each of these sensitivity classes.

In the absence of regionally determined values for the environmentally acceptable person-equivalent, the second best solution could be politically set reduction targets for environmental impacts in the various regions. The level of environmental ambition reflected by environmental policy is, however, highly diverse in different parts of the world, and this will introduce some uncontrollable distortions in the weighting. For this reason, weighting factors based on Danish environmental policy as described above are being used for the time being.

Weighting of resource consumption

The weighting factor for a resource must reflect how scarce the resource is relative to consumption of it, i.e. the size of known reserves relative to annual consumption of the resource. This is called the "supply horizon" and it provides a measure of the resource's scarcity.

The supply horizon for a non-renewable resource is the number of years for which current consumption of the resource can continue before known reserves are exhausted. The scarcer the resource relative to con-

Non-renewable resources:

$$\text{Supply horizon} = \frac{\text{Known reserves}}{\text{Annual global consumption}}$$

sumption, the shorter is the supply horizon. For non-renewable resources the supply horizon can be long or short, but it is always finite. If there is consumption of the resource, reserves will not last forever.

For renewable resources which are not consumed faster than their rate of regeneration, it is meaningless to speak of a supply horizon. If the resource is not plundered, it can in theory be used forever, i.e. the "supply horizon" is infinite. But there can be regions where current use exceeds regeneration. This is the case for forests in large parts of the earth's tropical regions, where the original forest is being cleared and the forested area reduced year by year as illustrated in Table 1.2. A more local example is consumption of water in Denmark. In Denmark's capital city area, the groundwater resource is being overused because of extraction of drinking water, and use of groundwater should therefore be weighted more heavily than if it occurred in Ringkøbing County in western Jutland, where extraction of groundwater is much less than regeneration of groundwater.

If a renewable resource is consumed more quickly than it is regenerated, the supply horizon can be defined as the number of years which will elapse before the resource is exhausted at the current ratio between present reserves and the difference between rate of consumption and rate of regeneration. The supply horizon for renewable resources is thus defined on the basis of actual consumption in the local areas where consumption is occurring:

Renewable resources where consumption exceeds regeneration:

$$\text{Supply horizon} = \frac{\text{Known reserves}}{\text{Annual consumption - annual regeneration}}$$

Renewable resources where consumption does not exceed regeneration:

Supply horizon = Infinite

Method

The EDIP method defines the weighting factor for resource consumption as the reciprocal of the supply horizon for the resource, calculated on the basis of consumption, computed reserves and any rate of regeneration in the reference year 1990.

$$WF(j) = \frac{1}{\text{Supply horizon for resource (j)}}$$

For renewable resources, the weighting factor is defined on the basis of the total consumption where the process is occurring. If the resource is not used faster than it is regenerated, the supply horizon is infinite and the weighting factor is therefore zero.

For non-renewable resources, this definition of the weighting factor means that consumption, R(j), of resource (j) in the product system is compared against reserves of the resource in question at the weighting. The weighted resource consumption, WR(j), is found as:

$$WR(j) = WF(j) \cdot NR(j)$$

$$\Rightarrow \quad WR(j) = \frac{RR(j)_{90}}{\text{Known reserves of (j)}} \cdot \frac{R(j)}{RR(j)_{90}}$$

$$\Rightarrow \quad WR(j) = \frac{R(j)}{\text{Known reserves of (j)}}$$

The weighted resource consumption is given the unit "person-reserve", PR_{W90}, i.e. the proportion of known reserves per person in the world in 1990, where 'W' stands for the world. The weighted resource consumption is typically expressed in mPR_{W90} and the product's consumption is thus expressed in parts per thousand of known person-reserves in 1990.

Weighting of potentials for impacts on the working environment

The weighting factors for the impact categories for the working environment should reflect the seriousness of the individual types of impacts, as do the weighting factors for the environmental impact categories. The seriousness of an impact category for the working environment is determined by

- the probability that an impact actually results in a work injury, and
- the seriousness of the consequences of such a work injury for the employee.

Method

In the EDIP method, the weighting factor is calculated for each individual type of working environment impact as the relationship between the number of work-related injuries reported for persons exposed to the type of impact in question in Denmark and the total time of exposure in Denmark to the relevant type of impact in the years around 1990.

$$WF(j) = \frac{\text{Reported injuries for impact category (j) per annum in Denmark}}{\text{Total annual time of exposure to the impact category (j) in Denmark}}$$

The weighting factor expresses the annual average number of work injuries per employee in Denmark per annum for the impact category in question.

The weighted potential for impacts on the working environment, WWP(j), is determined as:

$$WWP(j) = WF(j) \cdot NWP(j)$$

$$\Rightarrow WWP(j) = [\text{Reported injuries per annum per employee for impact category (j)}] \cdot \frac{\text{Exposure time for impact category (j)}}{\text{Exposure time per employee per annum for impact category (j)}}$$

With this definition of the weighting factor, the weighted potential for impacts on the working environment can be interpreted as a measure of the number of work injuries within the relevant category of impact to be expected as a result of the product system. In the weighting, the potentials for impacts on the working environment are expressed in the unit RWI_{DK90}, reported work injuries based on inventories for Denmark in the years around 1990. Note that the weighting factor with this definition includes only the first of the two aspects noted at the beginning of the section. The seriousness of the consequences of various types of work injuries is difficult to quantify, and it is not attempted in the EDIP method. This is discussed in the section on interpretation of potentials for impacts on the working environment.

Section 23.10 of the Toolbox shows the EDIP method's default values for normalization references and weighting factors. It is important that accurate records are kept for the points in the LCA where there are deviations from the default values.

Interpretation of weighted impact potentials and resource consumptions

Environmental impact potentials and the environmental space determined by political targets

In the weighting, the environmental impact potentials are expressed as proportions of the person-equivalent corresponding to the policy target for environmental impact in the year 2000. The weighted environmental impact potentials are measured in $PET_{WDK2000}$, which stands for the person-equivalent based on target emissions for the world (W) or Denmark (DK) in the year 2000.

Chapter 1 introduced the "environmental space" as the environmental impact available on average to each person in a sustainable society. In the same way, the "environmental space determined by policy targets" for the year 2000 can be defined as the impact which we on average may cause if the targets for reductions are to be fulfilled. The size of the latitude determined by environmental policy targets will gradually approximate the size of the environmental space as the environmental policy targets approximate the requirements for sustainability.

A contribution to acidification of 1 PET_{DK2000} is equal to the average annual impact of acidifying substances which can be expected per Dane in the year 2000 if present reduction targets for emissions of acidifying substances are fulfilled. Thus, 1 PET_{DK2000} can be regarded as the politically-determined environmental space for each individual impact.

This figure is immediately comprehensible. If one buys a car with a global warming potential of 500 $mPET_{W2000}$, half of one's "ration" for

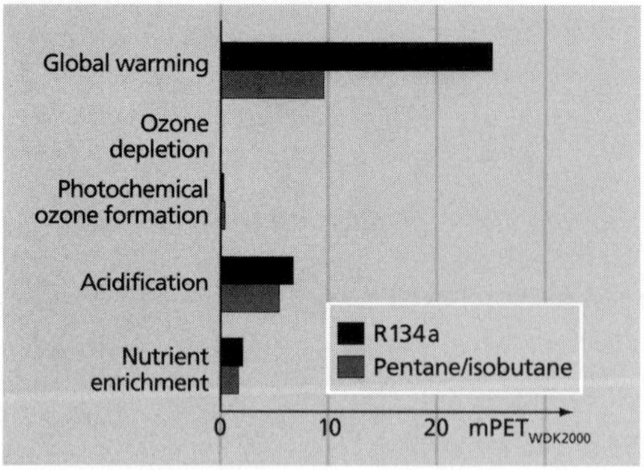

Figure 10.8. Weighted environmental impact potentials for the product system for refrigerators with different refrigerants and foaming agents (from Part V of the book)

the contribution to global warming has been used for all of the time during which the car is in use.

As discussed above, the weighted environmental impact potential is suitable for supporting decisions because it represents the product's potential impact relative to the best prediction of the community's future priorities within the environmental field.

Figure 10.8 shows the weighted environmental profiles for the two refrigerators with various alternatives to use of CFCs as refrigerant and foaming agent. It is evident in the figure that the weighting with factors based on the political reduction targets does not change decisively in the picture from the normalized environmental profiles in Figure 10.3, but the weighting of the seriousness of photochemical ozone formation against global warming and the other environmental impacts is now included in the figures. From an environmental point of view, the pentane/isobutane alternative is preferable to the R134a alternative.

It is also evident that the weighted global warming potential is the largest impact potential for both refrigerators, and that for the R134a alternative it amounts to ca. 25 $mPET_{W2000}$, corresponding to 2.5% of the user's environmental policy latitude in the year 2000 if the refrigerator serves only one person.

Resource consumption and known reserves

In the weighting, resource consumptions are expressed relative to the average global reserves per person in 1990, and they are expressed in the unit PR_{W90}, PR for person-reserve, W for world because the reserves are computed for the entire world, and 90 because the reserves are computed as known quantities in 1990.

These figures are also immediately comprehensible. For example, with a figure of 10 mPR_{W90}, it is possible to buy 100 products of the relevant type, and the "ration" of known reserves has thus been used for the entire future for all generations, i.e. also that portion of the known reserves which were otherwise available for one's children, grandchildren and subsequent generations.

In the weighting of resource consumption, a number of factors apart from the scarcity of the resource can be considered as supplementary weighting criteria:

- Irreversible versus reversible consumption: fossil fuels are burned and disappear entirely as resources, while metals, for example, are only made inaccessible in landfills with some possibility of future recovery. If the consumption is reversible, the weighting can take account of the magnitude of the energy consumption required for regeneration of the resource, for example on a later recovery of the resource from a landfill.
- Usability: oil has more possible uses than other fossil resources such as coal and gas.
- Substitutability: certain metals can be more difficult than others to replace and to do without for the industrialized world, because they are essential to certain applications.

The significance of the above three circumstances is difficult to quantify, and it is not attempted in the EDIP method. The weighting of resource consumption only includes the scarcity on the basis of consumption relative to the size of reserves, and the various resources are accorded an equally large value. In presentation of the weighted resource consumption, care must be taken to ensure that the decision maker knows the weighting principle, and has been presented with the above "missing" weightings.

It should also be noted that the size of the reserves is not absolute, but depends to a high degree on the market price of the resource and on its strategic significance. The more expensive or important the resource, the more intensively it is sought, the more inaccessible the places on earth where it is sought, and the greater is the environmental impact per unit of resource extracted. The known reserves of crude oil have thus been more or less constant over the last few decades despite a continuing large consumption of the resource.

In spite of this situation, the supply horizon defined on the basis of the size of known reserves is the best measure we have for the scarcity of the resource, and thus the best option at present for a weighting factor for resource consumption.

Potential impacts on the working environment and reported work injuries

In the weighting, a potential for impacts on the working environment is expressed as the number of reported work injuries to be expected as a result of the potential. The expectation is based on the relationship which can actually be confirmed between impacts on the working environment and the frequency of reports of work injuries in Denmark. The weighted potentials for impacts on the working environment for the product can thus be interpreted as a measure of the number of work injuries to be expected under Danish conditions as a consequence of the product system. The weighted potentials for impacts on the working environment are expressed in the unit RWI_{DK90}, reported work injuries in Denmark.

In the weighting of potentials for impacts on the working environment, there are also several other considerations which should be accorded weight, possibly as supplementary weighting criteria:

- As the weighting factor is defined above, it does not take account of the fact that the various types of injuries are not equally serious.
- The frequency of reporting can vary for the various categories of impacts on the working environment. Allergy is often recorded immediately after the impact which triggered it, and it is therefore easier to identify as an impact on the working environment than cancer, which is only observed years after the impact which triggered it. This means that the impact categories with the lowest relative frequency of reporting are weighted too lightly with the weighting factor defined above. For these impact categories, a greater number of

work injuries must be expected than predicted by the weighted working environment profile.

- For the types of effects recognized only long after the working environment impact which triggered them, the weighting factor does not reflect the impact in 1990, but for e.g. carcinogenic effects the impact 10-20 years earlier. It can be expected that the total impact on the working environment has changed since then. If the exposure to these impacts on the working environment has decreased in Denmark over the last few decades, the corresponding impact categories are weighted too heavily. If the exposure has increased, they are weighted too lightly.
- When using sector data, the impacts on the working environment are computed directly as reported work injuries. These can include region-dependent distortions in the weighting of various impacts, as the frequency of reporting will be lower for occupational diseases in many developing countries than in industrialized countries.

The significance of these considerations is difficult to quantify, and it has not been attempted in the EDIP method. In the presentation of the working environment profiles, care must be taken to ensure that the decision maker knows the weighting principle and is informed of the existence of these other weighting criteria.

Figure 10.9 illustrates the weighted environmental impact potentials, resource consumptions and potential impacts on the working environment for the seven products or activities presented in their normalized profiles in Figure 10.4.

If ozone depletion, which has a very high political priority, is disregarded, the weighting factors for the environmental impact categories lie between 1 and 3. This is not enough to introduce significant changes into the picture already drawn by the normalized environmental impact potentials in Figure 10.4.

There are, however, very large differences in the weighting factors for resource consumption and potentials for impacts on the working environment. These differences reflect the fact that some resources are much more scarce than others, and that some types of impacts on the working environment result in reported work injuries much more frequently than others.

The weighting of resource consumption in Figure 10.9 means that attention is directed to a higher degree to consumption of metals such as nickel, copper and tin, and it reduces the significance of aluminium consumption. For the TV, consumption of copper corresponds to 1.2‰ of the person-reserve, unless the TV's materials are recovered on disposal.

Consumption of copper is thus more serious than consumption of coal for the production of the electricity used by the TV. A large normalized consumption of water for the high pressure cleaner disappears in the weighting, because a renewable resource is involved which for the areas where the cleaner is used may on average be assumed to be consumed more slowly than it is regenerated.

Hearing impairment, injury to the musculo skeletal apparatus and accidents are all working environment effects which are relatively simple to relate to an impact on the working environment, and they gain significance relative to the other impact categories of the working environment in the weighting.

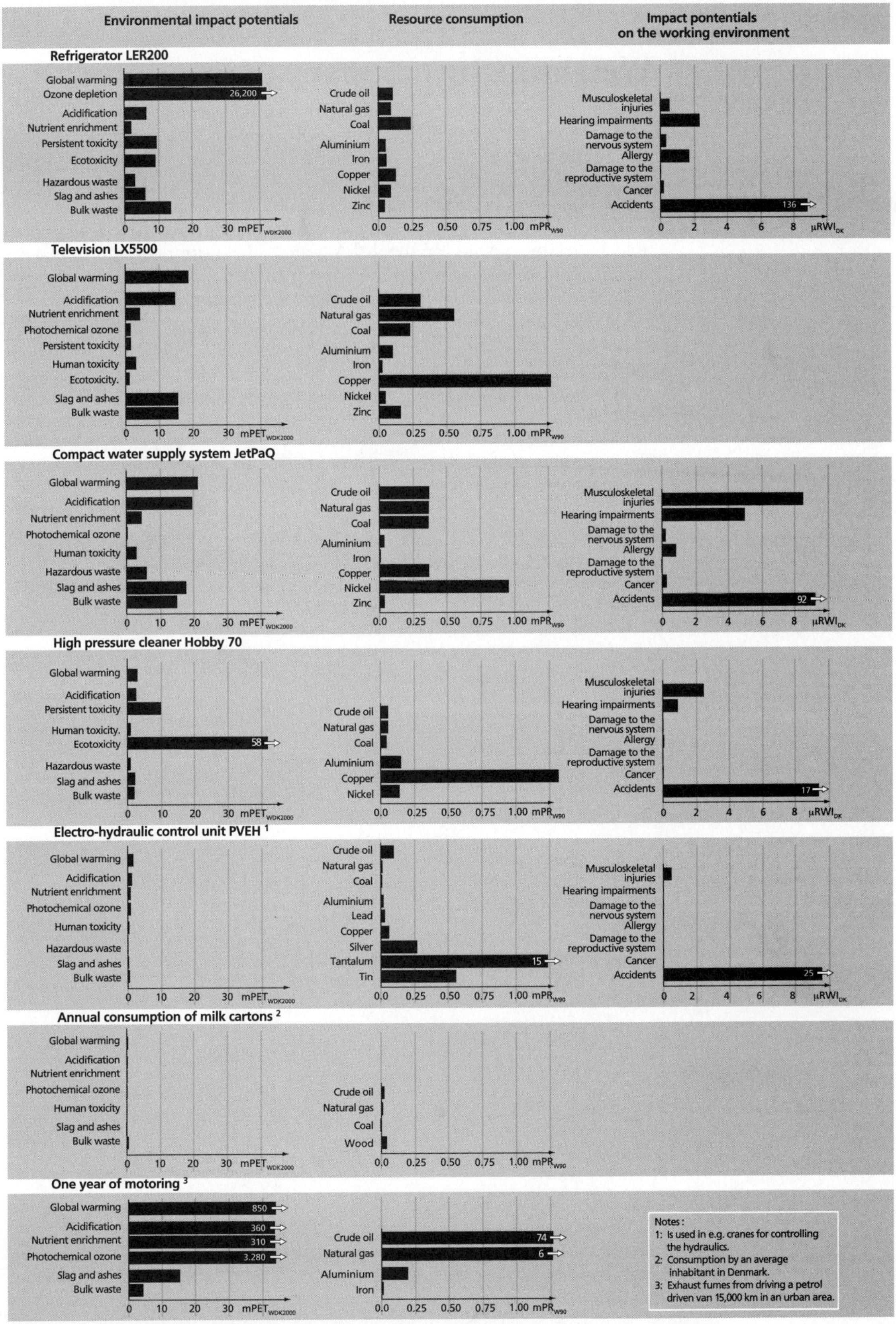

Figure 10.9. Weighted resource consumptions and impact potentials for various products and activities (from the case histories in Part V and other sources)

Weighting by company policy

As a final step in weighting, the EDIP method permits a further weighting based on the company's environmental policy. This may be particularly useful when the method is applied in product development or in the company's environmental management. The company's environmental policy must then be expressed in a set of weighting factors.

When a weighting expressing company policy is included, it is possible to let the results of the LCA reflect the priorities determined by management in the company's environmental policy. The LCA thereby becomes a powerful tool in the company's environmental management.

At the same time, the LCA ceases to be in agreement with the authorities' priorities. This will affect public acceptance of the use of company-specific weightings.

If a final company policy weighting is included, it is therefore important:

- That the profiles for environment, resources and working environment are always shown both with and without the company policy weighting.
- That account is taken of where the company policy weighting enters, and how it affects the result.

11. Uncertainty and sensitivity analysis

An analysis of the *uncertainty* in the LCA consists of estimating how much the results can vary. The uncertainty analysis covers three types of elements, viz.:

- The model of the product system, i.e. the assumptions concerning *which* processes enter into the product system and to how great an *extent* they enter, including assumptions on distribution of sales, patterns of use, disposal routes and life spans.
- The processes' environmental exchanges, i.e. the actual data for inputs, outputs and impacts on the working environment.
- The assessment factors, i.e. the equivalency factors, normalization references and weighting factors.

The uncertainty analysis thus consists of estimating the uncertainties in the scope definition, the inventory and the impact assessment, and the resulting uncertainty in the total result of the LCA.

A *sensitivity analysis* consists of assessing the sensitivity of the total results and conclusions to the uncertainties in the data used. The sensitivity analysis identifies the model assumptions, processes and environmental exchanges of greatest significance for the total result, and these data are designated the *key figures* of the LCA. The significance of the uncertainty in the key figures is examined, for example by assuming alternative values for them and investigating how these will affect the total result of the LCA and the conclusions which can be drawn. Furthermore, an investigation of the significance of the uncertainty in the most important assessment factors for the total result can be made.

The uncertainty and sensitivity analyses are carried out after the assessment phase of the LCA, thus making use of the knowledge of the various environmental exchanges' potential contributions to impacts included in the assessment phase, and it is thus possible to decide which exchanges are the most significant and which are without significance in the total picture.

The EDIP method includes a number of choices and decisions, presented in full in Chapter 13. Some of the choices are of a conceptual nature while others concern details within the selected concept.

An example of a conceptual choice is the decision on basing the assessment phase on an impact category method where the emissions are converted to potential contributions to various categories of environmental impacts, as defined on the basis of basic types of environmental effects. A detailed choice within the framework of this conceptual choice is the recommendation to use equivalency factors for global warming calculated for the substances' contributions over a time horizon of 100 years.

Changing the conceptual choices really means choosing a different method of LCA. The general conceptual choices in the EDIP method are therefore taken as nominal, while the EDIP method's detailed choices for each of the phases of the LCA are included in the uncertainty and sensitivity analyses.

Uncertainty analysis

The uncertainty analysis is carried out for each of the phases, scope definition, inventory and impact assessment.

Scope definition
The following choices from the scope definition phase enter into the uncertainty analysis:

- scope definition of the product system, including assumptions concerning the model of the product system,
- choice of allocation model,
- definition of technological scope, including projection of key figures.

In scope definition, a model is set up for the product's total life cycle, and an assessment is made of how much the product system draws on the processes which enter into it. The assumptions made in the construction of the model are encumbered with uncertainties. These uncertainties are assessed and noted together with the processes' scope in a parts list structure for the total scope definition as proposed in Chapter 8, and enter into the total calculation of uncertainty in the inventory as described in Chapter 9.

There are also elements of the scope definition which are not apparent in the parts list structure, but which can be key figures for the product in environmental terms and which must therefore be subject to an uncertainty analysis. This applies, for example, to patterns of use for the product and the life span of the product.

Inventory
The uncertainty in the individual environmental exchange is estimated in connection with the inventory analysis for all processes in the product system, and a statistical distribution is assumed for the magnitude of the exchanges. On this basis, the uncertainty in the total inventory for the entire product system can be estimated for each type of exchange by a Monte Carlo simulation, as described in Chapter 9.

Impact assessment
The subsequent assessment phase also introduces uncertainties in the result of the LCA via the use of

- equivalency factors,
- normalization references, and
- weighting factors.

The uncertainties in the *equivalency factors* can be attributed to both doubts concerning how the equivalency factors are to be defined and uncertainty in the models and assumptions used in their calculation.

For some of the impact categories there is some international agreement on the definition of the equivalency factors. This is the case for consumption of non-renewable resources and the global impacts global warming and stratospheric ozone depletion.

For other impact categories there is continued discussion on how the equivalency factors are to be defined or on which modelling assumptions are to underlie their calculation. On this basis, some equivalency factors can vary several hundred percent from one procedure to another.

There is no systematic estimate of the uncertainties in the EDIP method's equivalency factors, normalization references and weighting factors, and the uncertainty analysis of the assessment phase is therefore not implemented. It can, however, be asserted that among the equivalency factors, factors calculated especially for toxicity-related impact categories will be encumbered with large uncertainties.

The normalization references are based on available statistics for Danish or global environmental exchanges and are relatively certain with the exception of the normalization references for the impact categories related to ecotoxicity and human toxicity.

The uncertainties in the normalization references are transferred to the weighting factors, but if the definition of the weighting factors is taken as nominal, including choice of method of extrapolation, calculation of the weighting factors does not lead to further uncertainty.

Sensitivity analysis

The significance of missing data for processes in the product system and the significance of the uncertainties found in the uncertainty analysis for the key figures in the LCA are assessed in the sensitivity analysis.

The sensitivity of the LCA to changes in a key figure are investigated by assigning different values to the key figure in question and calculating the total result of the LCA while all other assumptions are unchanged. The variation in the key figures is chosen on the basis of the uncertainty with which they are encumbered.

It can be a time-consuming task to perform the sensitivity analysis manually if there are many key figures to be varied or if some of the key figures enter into many of the processes in the product system. The latter will, for example, be the case for products which use electricity in the use stage, because the chosen scenario for electricity production will contain key figures for such products.

If the LCA is made with the aid of a computer, it is simple and not particularly time-consuming to carry out a sensitivity analysis, as the product system is merely simulated for the new value of a key figure and the calculation is then repeated automatically and without spending time on it. Example 11.1 provides an example of a sensitivity analysis for modelling assumptions on life span and pattern of use carried out with the aid of a computer program.

Modelling assumptions

In reviewing the individual choices made during the scope definition, an assessment is made of whether they will be significant for the impact of the total product system. This applies to both the decision on which processes are to be included in the product system and assumptions concerning distribution on, for example, distribution of sales in the product's export markets, patterns of use, distribution among various routes of disposal and assumptions on the product's life span. If a choice or an assumption proves to be of major significance for the total result, the choice is considered carefully, and the variation intervals in the result due to other choices is simulated by a computer program. The arguments for the choice made are given in the LCA report.

Missing data
If there are processes in the product system where it has not been possible to generate data for the environmental exchanges, a realistic order of magnitude for them is estimated and an assessment made of whether the total result of the LCA and its conclusions could be significantly altered by these data if they were available.

Data for individual processes
In a review of the entire inventory, focus is on the processes which, in light of the results of the impact assessment, can be seen to contribute significantly to the total results for the product system. The sensitivity analysis is concentrated in the first instance on these processes, and an assessment is made of whether the uncertainty in their environmental exchanges is of significance for the total result of the LCA and for the conclusions drawn on the basis of it.

The sensitivity analysis thus implies a new review of the scope definition and the inventory in the light of the results of the impact assessment. As described, the purpose of the sensitivity analysis is to clarify the sources of variation and to assess their relative significance. The most significant sources of variation are also the key figures for which it is most important to make any projection.

Assessment factors
After the sensitivity analysis of the scope definition and the inventory, a sensitivity analysis of the equivalency factors and any normalization and weighting factors in the assessment phase is carried out. In the same way as for the key figures, the assessment factors with the greatest influence on the total result are identified. The uncertainty in the most important assessment factors is estimated as accurately as possible, and an estimate of its significance for the uncertainty in the total result of the LCA is made.

The results of the sensitivity analysis are included in the discussion of the results of the LCA and of the certainty of the decisions made on the basis of the LCA.

The case histories in Part V provide examples of uncertainty and sensitivity analyses for scope definition and inventory in the LCA of five different products.

Uncertainty and sensitivity analysis stage by stage

The following section provides several examples of the factors in the various stages of the product system which can be of significance in the sensitivity analysis.

The raw material stage
The raw material stage often lacks an inventory of environmental exchanges from the life cycle of chemicals from extraction of raw materials to manufacture of the product. If the chemicals enter into the product in large quantities or if substances which are particularly hazardous to the environment are emitted during production of the chemicals, this lack can be of considerable significance. But it will often be the case that the emissions in the early life- cycle stages of the chemicals are insignificant relative to the emissions in their use stage.

As a rule, it is difficult to obtain data for impacts on the working environment from the raw material stage, and this lack can be of significance if some of the raw materials are produced under stressful conditions or with a large consumption of manpower.

The product manufacturing stage
The significance of choice of allocation principle is investigated for:

- significant joint environmental exchanges in the product manufacturing stage, and
- overheads to be allocated to various products.

The use stage

There are a number of assumptions for the use stage which, depending on the nature of the product, can be of major significance for the result of the LCA and which should therefore be investigated:

- distribution of sales,
- patterns of use,
- the lifetime of the product,
- efficiency of the community's electricity production,
- types of fuel for the community's electricity production,
- degrees of treatment in the municipal sewage treatment plant,
- ....

For products which use energy or chemicals in the use stage, these exchanges will often be among the largest and most significant for the product system. It is therefore important to carry out a sensitivity analysis for the scenarios which underlie the inventory of energy and chemical consumption in the use stage. Example 11.1 shows such a sensitivity analysis.

Not only the magnitude of the product's electricity consumption, but also the choice of production scenario for electricity can play a large role in the total environmental impact, and the inventory should therefore be calculated with various alternative scenarios.

The disposal stage
Among the assumptions for the disposal stage which can be of significance for the result, and which should therefore be investigated in a sensitivity analysis, are:

- the assumed distribution of the product among different disposal routes,
- the degree of efficiency of the utilization of energy released on incineration,
- the percentage of the various materials in the product for which recycling is expected,
- the average degree of recovery for materials from the society pool.

1. The dimensioning criteria for the operating time of the television, stated by Bang & Olufsen, is 6 hours a day for 10 years corresponding to 21,900 hours. This choice is highly significant, and the consequences of two alternative use scenarios are therefore investigated:
2. Danish average: Investigations carried out by B&O based on installation of timers on 800 televisions in Danish homes show that they are used 4 hours a day on average. Based on enquiries at Danish B&O retailers, the LX 5500 television is assumed to have an average life span of 12.8 years.
3. English average: Based on a similar investigation on the English market it is assessed that the television on average is used 5.3 hours a day and has a life span of 8.5 years.

Use scenario	Daily operating time, hours/day	Life span years	Total operating time hours	Total stand-by time hours
B&O dimensioning criteria	6	10	21,900	65,700
Danish average	4	12.8	18,688	93,440
English average	5.3	8.5	16,443	58,017

Assuming that the television uses electricity based on an average European production, the use scenarios will result in the following resource and environmental profiles.

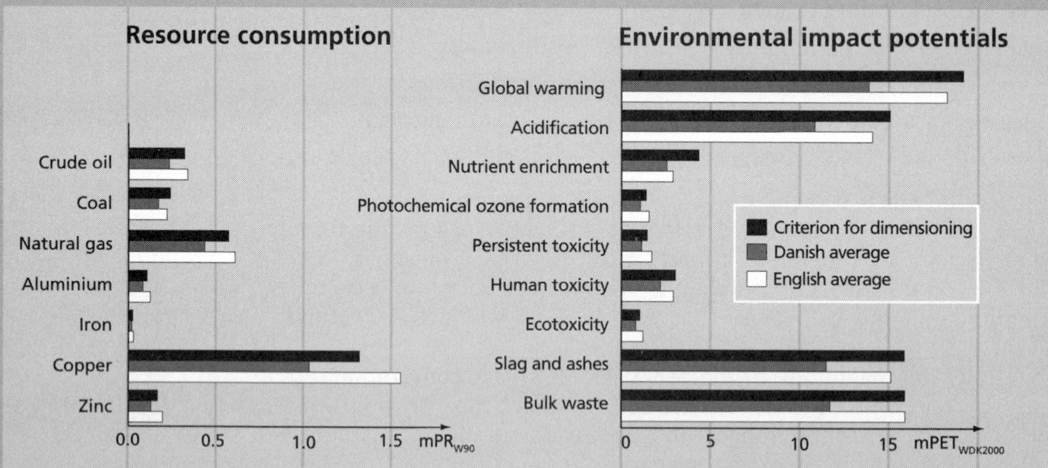

Example 11.1. Sensitivity analysis for assumptions on patterns of use for television (see also the case histories in Part V)

The product system in general

Apart from the assumptions described within each individual stage of the product system, there are some more general choices and assumptions which can be of significance for the result of the LCA:

- choice of allocation models,
- assumptions on extension of the system.

The points where the product system is delimited by allocation or a system extension has been made are assessed with respect to whether other choices or assumptions could be reasonable, and what these would mean for the outcome of the LCA.

12. The MECO principle

The MECO principle is used to structure the work of the LCA and to systematize and simplify its results.

The principle consists of dividing the assessment into four areas in accordance with the underlying causes of the product's environmental impacts, viz.

- Materials
- Energy
- Chemicals
- Others

from which the name "MECO" is derived. The principle enters into most of the method's phases and applications and is a type of "hands on" principle in the EDIP method.

The first three areas are self-explanatory, while the fourth, "Others", covers the "odds and ends" which should enter into every good systematic study. It contains various elements, depending on the part of the LCA in which the MECO principle is used.

The basis for dividing into these four areas is given in Table 12.1.

Table 12.1. Basis for systematizing the LCA after the MECO principle

1. Together M, E, C and O represent all the terminal environmental exchanges	Materials, energy and chemicals represent all terminal inputs to the product system. All terminal outputs are caused directly or indirectly by elements from one of these three groups. The group "Others" cover impacts on the working environment not related to inputs or outputs, e.g. monotonous repetitive work, noise and work accidents.
2. M, E, C and O each represent their own type of resource consumption and impact potentials	Material consumption represents one type of resource consumption, typically reversible consumption. Energy consumption represents another kind of resource consumption, typically irreversible. Together, M and E represent all resource consumptions in the product system. Furthermore, the energy consumption represents a group of impact categories due primarily to emissions from combustion processes. Consumption of chemicals represents another group of impact categories. Together, E and C represent the major part of the environmental impacts, arising in the product system. "Others" represents the impacts on working environment that are not covered by M, E and C.
3. M, E, C and O represent typical areas of improvement in the product development	The designer is accustomed to working with energy, materials and chemicals savings and substitutions. These are represented by the groups M, E and C in the MECO principle. Properties of the product, such as lifetime and ease of disassembly, which the designer is also accustomed to working with, belong to the group "Others". An environmental initiative in the design phase is often targeted at one of the areas represented by M, E, C or O.

Representation of resource consumption and impact categories

A significant reason for classification after the MECO principle is, as indicated in Table 12.1, that the four areas M, E, C and O each represent their categories of resource consumption and potential impacts, which to some degree are complementary. This can be used for a systematic arranging and simplification of the results of the LCA. Table 12.2 shows which of the four areas especially covers the various assessment parameters.

Focus areas in product development

The principal object of introducing the MECO principle is to make the LCA easier to understand and to relate to the product.

This is of particular relevance when the results of the LCA are to be used as inspiration or as the basis for making a decision in product development.

The LCA results, for example, in the conclusion that photochemical ozone formation is the most significant environmental impact potential for the product system. This information can, however, be difficult for the designer to use if it is not translated into characteristics of the product itself. What the designer can affect with his or her decisions is the use of materials and chemicals and the use of energy in various places in the product system, and such factors as life span, the pattern of use, and the disposal of the product.

Table 12.2. Assessment parameters in the LCA method covered by M, E, C and O

	Environmental impacts	Resource consumption	Impacts on the working environment
Materials	Bulk waste Slag and ashes	Resources used in materials Mainly reversible consumption	
Energy	Global warming Photochemical ozone formation Acidification Nutrient enrichment Bulk waste Slag and ashes	Energy carriers, especially fossil resources and wood Mainly irreversible consumption	
Chemicals	Ozone depletion Photochemical ozone formation Persistent toxicity Ecotoxicity Human toxicity	Resources used in the production of chemicals	Impacts related to chemical exposure: cancer, damage to the reproductive system, allergy and damage to the nervous system
Others			Monotonous repetitive work, noise, work accidents

For the designer to utilize the knowledge generated by the LCA, the product characteristics which the designer can control are therefore identified, and for this purpose a classification after the MECO principle provides a good overview.

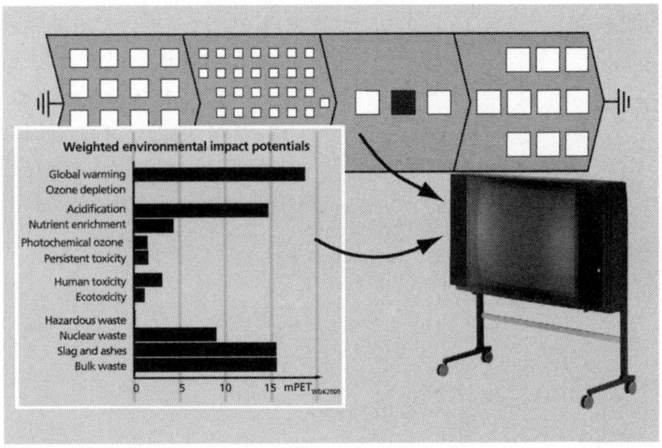

Figure 12.1. The LCA is translated into characteristics of the product

The materials, the energy and the chemicals which enter into the product system are to some extent present in the product. They are either directly present, for example as construction materials or paint, or they are indirectly present because the chosen solutions in the product specify their involvement in the product system.

Translation of the LCA into characteristics in the product is illustrated in Figure 12.1.

The MECO principle is used *inter alia* in the LCA of the reference product and in the environmental diagnosis, where there is clarification of how much various changes in material consumption, energy consumption, consumption of chemicals and "others" (e.g. life span) can mean in environmental terms for the product type. When this overview is established via the systematization of the results of the LCA after the MECO principle, both the environmental specialist and the product developer have a good feeling of proportions and priorities for an environmental initiative for the product. This facilitates communication of the product's environmental qualities and is thus of major significance for environmental work during product development.

The MECO principle is not only directed towards identifying areas of initiative in product development. It can be used with advantage in most of the EDIP method's components to simplify the overview and

Table 12.3. Use of the MECO principle in various applications of the EDIP method

	Application of the MECO principle	Described in
In product development		
Environmental assessment of references	During searching and sorting on the causes for resource consumption and impact potentials	Chapter 15
Environmental diagnosis	During simulation of environmental consequenses of changes in a product or its product system	Chapter 16
Environmental specification	M, E, C, and O are used as headlines in the setting of targets	Chapter 17
Environmental assessment of concepts	As a rough assessment of a concept	Chapter 18
In all applications of the EDIP method		
Scope definition	As a rough preliminary assessment, used when defining scope	Chapter 8
Sensitivity analysis	When examinating the significance of data gaps and uncertainties	Chapter 11

structure the work in the LCA. Table 12.3 provides an overview of the places where the MECO principle is used in the method.

The MECO principle can only be used together with the LCA method and cannot replace an LCA. The LCA is required to gain an adequately detailed insight and to rank different alternatives in the environmental initiative.

13. Choices in the EDIP method

During an LCA of a product it is necessary on several occasions to make choices, for example in the scope definition of the product system, in choice of allocation principle and in choice of normalization references and weighting principle. There is often no unambiguous "correct" choice in these situations. An argument can be made for several of the alternatives, but one of them must be chosen so that the LCA can be carried out.

Some choices are very general and concern the method's concept itself. Other choices concern alternatives within the selected concepts.

The choices can have decisive influence on the result of the LCA, and it is therefore important that the environmental specialist consider and investigate their significance. Chapter 11 includes a description of how the uncertainty and sensitivity analyses are carried out within the conceptual framework of the EDIP method.

The EDIP method's choices and recommendations are described in various places in the descriptions of methods in Part II and in the Toolbox in Part IV. The choices are summarized in tabular form in Table 13.1 to give the environmental specialist a complete overview of the method's special distinguishing features and an opportunity to change some of the choices if there are special reasons for doing so.

The choices made in the LCA should always be described in the report on the LCA.

Phase of the method	Choices within the EDIP method	Own choices
Goal definition	■ The method including the four main applications is dedicated to product development. ■ The core of the method is also suitable for other applications.	
Scope definition Object of the assessment	■ The method is based on the use of reference products.	
Assessment parameters	The method operates with the following assessment parameters: ■ Environmental impacts: global warming, ozone depletion, photo-chemical ozone formation, acidification, nutrient enrichment, ecotoxicity in water, soil and sewage treatment plants, toxicity to humans via air, water and soil, and finally waste for landfills divided into: bulk waste, hazardous waste, slag and ashes and nuclear waste. ■ Resource consumption: non-renewable resources and renewable resources. ■ Impacts on the working environment: chemical exposure leading to: cancer, damage to the reproductive system, damage to the nervous system and allergy. Hearing impairments from exposure to noise. Musculoskeletal injuries from monotonous repetitive work. Grievous bodily harm from accidents.	
Scope definition of the product system	■ The method recommends the hierarchical structure of the parts list as a starting point for the scope definition for complex products.	
Time scope	■ Projection of data to the period for which the decision must be valid is recommended. ■ No upper time limit for the impact of the environmental exchanges.	
Technological scope	■ Performing a trend analysis on the most important key figures and projection of these in the LCA is recommended.	
Allocation models	The method recommends the following procedure: ■ Try to avoid allocation, - Clarify the cause-effect relationschips and ascribe the environmental exchanges on the basis on these. - Extend the system boundaries. Recommended if the consequences of the decision to be supported by the LCA are predictable. ■ Allocate, - Identify processes that contribute to more than one service and allocate their environmental exchanges to each of the services in question. - If it is possible, divide the joint environmental exchanges, using a technical assessment of the utility value of the co-products, materials or sub-assemblies that are part of the allocation. - If a technical assessment is not possible: divide the joint environmental exchanges using an economic assessment of the utility value of the co-products, materials or sub-assemblies that are part of the allocation.	
Inventory Data format	■ Resource consumption is expressed as 100% pure resource, not as e.g. ore. ■ Impacts on the working environment are expressed as time of exposure for the process on the basis of a defined impact criterion or based on sector statistics for reported work injuries. ■ The data are filed in unit processes, based on an environmental key unit.	

Table 13.1. The choices in the EDIP method

Phase of the method	Choices within the EDIP method	Own choices
Inventory		
Data format	■ The variation and the uncertainty in the collected data are described by an estimate of their coefficient of variation and their statistical distribution and by applying data quality indicators. ■ The data format contains the possibility of including site and regional dependent characterization, to be used in a subsequent calculation of the environmental impact potentials and region-dependent normalization and weighting.	
Collection of data	No special method-related choices.	
Computation model	■ Aggregation of the environmental exchanges can be done before the calculations of impact potentials, provided that site- or region-dependent considerations are not included. ■ Otherwise, aggregation must be done after the calculations of impact potentials. ■ Aggregation of data variability for the exchanges into a total variation is recommended, using a simulation according to the Monte Carlo method.	
Assessment		
Environmental impact potentials	■ The method operationalizes use of the impact categories noted in the scope definition of the assessment parameters by giving methods for determining impact equivalency factors. At the same time, impact equivalency factors are given for the most common substances. ■ The following choices have been made for calculation of impact equivalency factors: - time horizon: 100 years for global warming, - time horizon: infinite for stratospheric ozone depletion, - low background level of NO_x and a time horizon of four days for photochemical ozone formation, - own method for ecotoxicity and human toxicity. ■ Use of site factors is defined by the method, but has not yet been operationalized with the necessary substantiating data.	
Normalization	■ Actual yearly impacts are used as normalization reference. ■ 1990 has been chosen as reference year. ■ Selection of the world as reference area for global impacts and non-renewable resources is recommended. ■ Denmark or the EU are recommended as reference areas for regional and local impact categories and consumption of renewable resources. ■ Use of region-specific normalization is defined by the method, but has not yet been operationalized with the necessary underlying data. ■ The normalization references are expressed per person in the reference area, and the result of the normalization is expressed in person-equivalents.	
Weighting	■ Use of region-dependent weighting is defined by the method, but is yet not operationalized with the substantiating data.	
■ Environment	■ As a starting point, weighting factors based on Danish political reduction targets are recommended. ■ The targets for reduction are harmonized to the year 2000 by linear interpolation or extrapolation. ■ The weighting factors are calculated as actual impacts divided by target impacts.	
■ Resources	■ Weighting factors are based on the supply horizon.	
■ Working environment	■ Weighting factors are based on the frequency of reported work-related injuries in Denmark.	

Table 13.1 (continued)

Part III. Environmental Tasks in Product Development

- *What are the environmental tasks in product development?*
- *What are the questions for the environmental specialist?*
- *How do the environmental specialist and the designer work together?*
- *How can the questions be answered?*

14. Cooperation with the designer

The environmental considerations involved in product development should be handled in the same way as all other considerations such as costs, quality, feasibility of manufacture and style. Environmental considerations should be an integral part of the designer's normal routines, assigned a status corresponding to the competitive potential which it can contribute to the product. Environmental considerations should thus not enjoy any special status. In attempts to develop environmentally friendly products, environmental considerations must not be allowed to overshadow and displace other considerations: for in this event the product will not sell, and its potential environmental gains will thus never be realized.

The significance of environmental considerations for the product's competitiveness must, however, be conceived broadly, and not seen simply in terms of the customer's current priorities. The customer's understanding of the environment and ranking of priorities must be projected into the period in which the product is expected to be marketed, and the influence of all other interested parties on the product's and the company's competitiveness must be considered.

The tasks of the designer

The designer is used to dealing with a range of considerations in the development of a new product, and environmental considerations therefore do not introduce any fundamentally new types of tasks. As illustrated in Figure 14.1, the tasks of the designer may be divided into four general categories, namely focusing, specification, synthesis and verification. It is helpful to look at this classification because it provides a summary of the designer's environmental tasks and of where environmental assessments are needed. The four categories of tasks include:

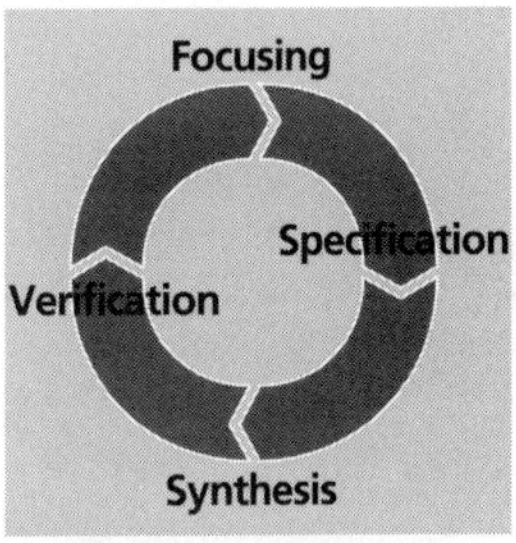

Figure 14.1. The tasks of the designer fall within four main categories

- **Focusing:** indicating what is most significant. In environmental terms, this means pointing out the most important categories of impacts for the product type and indicating where in the product system they can arise, and where in the product they can be expected to reside.
- **Specification:** determining targets for the new product. In environmental terms, this means assessing the designated focal points in commercial terms, ranking them and setting environmental objectives.
- **Synthesis:** creating the product, including its product system. In environmental terms, this means creating an attractive product life cycle, adapting the product to the systems with which the product will interact and, if possible and feasible, adapting the systems to the product.
- **Verification:** ensuring compliance with the objectives. In environmental terms, this means ensuring that the environmental objectives are fulfilled together with any other product requirements which were not formulated in the objectives.

As noted, the designer's tasks include routines for parameters other than those concerned with the environment, and the designer is responsible for including environmental considerations at the right time and place. The task of the environmental specialist is to be able to contribute the right knowledge in the right way. The object of this chapter is to provide the environmental specialist with an idea of the tasks to be expected, and how responsibility for them should be shared with the designer.

Table 14.1. Allocation of environmental tasks between environmental specialist and designer

Task	Allocation of environmental tasks	
	Tasks of the environmental specialist	**Tasks of the designer**
Focusing	**Environmental assessment of a reference product**	
analysis	■ Identify the essential impact potentials and the main causes of these	■ Identify an existing or a theoretical product which can serve as a reference for the new product
	Environmental diagnosis of a reference product	
	■ Simulate theoretical changes in the reference product or its product system and perform an LCA on these	■ Identify existing alternatives for the chosen design solutions in the reference product(s), including solutions found in competing products
	■ Perform LCAs on specific alternatives, including competing products	
diagnosis	■ Identify environmental focus points in the reference product	
Specification		**Specify environmental goals**
		■ Analyse the consumers' environmental perceptions and priorities and make a projection of this analysis
		■ Define the environmental specification for the product
Synthesis	**LCA of new products**	**Design for environment**
	■ Environmental assessment of concepts	■ Create environmentally attractive product systems
	■ Environmental assessment of details, typically in the form of changes in relation to a reference product	■ Adapt the product environmentally to existing systems in the product system
		■ If feasible adapt the system environmentally to the product
Verification	**Verify design solutions**	**Verify the product's environmental properties**
	■ If necessary perform an LCA	■ Verifiy that the environmental specification and any other requirements are met

Allocation of environmental tasks

The environmental tasks involved in each of the four categories are shown in Table 14.1. Some of the tasks are pure environmental assessments and are clearly the responsibility of the environmental specialist, while others are the responsibility of the designer.

The tasks at the left of the table are typical environmental specialist tasks, while those at the right are traditional designer tasks now simply assigned an environmental label. The task of "creating a product system" is, however, a new concept for many designers. The task of "designating environmental focal points" is a joint exercise requiring the knowledge of both the environmental specialist and the designer.

Environmental assessments under "verification" will, if they are required, be identical with environmental assessments made during the analysis phase. They will therefore not be described separately.

In many cases the environmental tasks can be placed before the development projects, e.g. as a part of the company's product planning or formulation of long-term objectives. This applies to tasks under both focusing and specification. In other cases the tasks must be undertaken in connection with a development project, and for this the environmental specialist must be available. The next section describes where the tasks occur in the development procedure.

The place of environmental tasks during product development

The development procedure is typically not formalized in smaller companies. But smaller companies will also recognize the four main categories of tasks noted here, and will benefit from support for the types of environmental tasks shown in Table 14.1. For smaller companies, the environmental tasks may have the character of once-only tasks for a family of products, after which the company can continue work using simpler guidelines which can be derived from the environmental assessments.

In larger companies, where there is a formal product development procedure, the four main groups of tasks are covered more systematically, often many times during development of the product. The initial idea can contain them in one: the analysis, the objectives, the creation of the product and the verification. Subsequently, they are elaborated in more and more detail until the product goes into production. The need to focus is most prominent in the analysis phase of the development procedure, but there must also be analysis and diagnosis later in the procedure to maintain the correct focus. The setting of targets is most prominent in the specification phase, when the product's basic specifications are determined, but the objectives are then gradually specified in more detail and adjusted as necessary as product development proceeds. Synthesis is concentrated in the development phase, where the concepts are fixed, and in the detailed design phase, where the details are selected. But potential solutions are also visualized in the earlier phases.

The tasks noted in Table 14.1 may therefore require repeating several times during the development procedure. Figure 14.2 illustrates how they typically occur.

In the figure, the tasks of the environmental specialist are shown in dark grey; those of the product developer are shown in light grey. Some companies will elect to employ and train their own environmental spe-

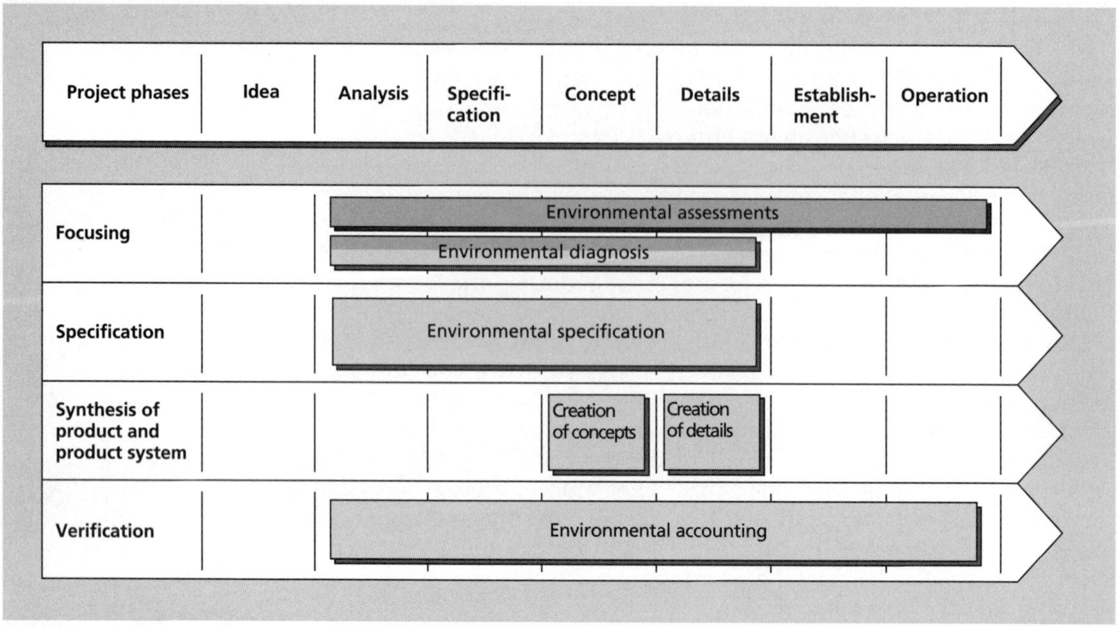

Figure 14.2. Environmental tasks in the development procedure. Modified after Olesen et al. 1996

cialist, who will then be responsible for all environmental assessments. Others will use a consultant. Others again will prefer the designer to make the environmental assessment during development, with support from an environmental specialist.

If the company elects to let the designer make the environmental assessment, a routine must be established to allow an environmental specialist on-going access to ensure that data and methods are used correctly, and that the conclusions reached are tenable. Use of the environmental assessment method should be based on the life cycle concept and on the method described in this book.

15. Environmental assessment of references

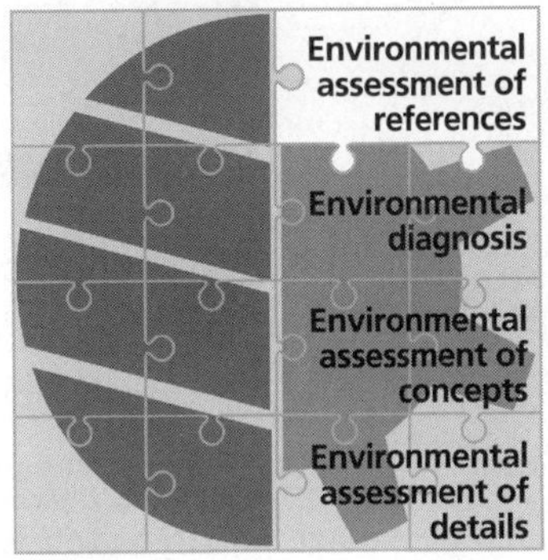

The object of assessing one or more reference products is to establish a basic environmental knowledge of the product system which can be expected for the new product. One or more existing products in the same family as the new one are therefore selected, and this/these product(s) assessed.

Environmental questions

The most important questions to be answered in the environmental assessment of reference products are given in Table 15.1. As is evident, most of the questions are answered by the basic steps in the method described in Part II of the book.

The designer is most interested in finding answers to questions 5-8, and especially in defining the most significant potential impacts and the most important sources of these, as the environmental assessment of the references in this way provides an environmental map for development of the new product. The detailing of the assessment to be carried out under question 8 is therefore the most characteristic feature in application of the method during assessment of reference products. This is therefore described in more detail and illustrated by examples. For unity of presentation, question 7 is addressed first. For the other questions, reference is made to the procedure described in the Toolbox and in Part II.

Examples of environmental assessment of reference products

What are the most significant impact potentials?

The weighted profiles must first be presented without company-specific weighting, or in other words, weighted only after the EDIP method's weighting principles as described in previous chapters. Any company-specific weightings can then be added, and new comparative figures presented. The reason for presenting the EDIP weightings alone is that the unit arrived at in the EDIP weighting is significant, i.e. the figures themselves can be interpreted as explained in Chapter 10:

- Resource consumption expressed in PR_{W90}.
- Potential environmental impacts expressed in $PET_{WDK2000}$.
- Potential impacts on the working environment expressed in RWI_{DK}.

Figure 15 illustrates an example of weighted profiles for a high pressure cleaner.

The weightings show that copper is the most significant resource consumption, ecotoxicity is

Tabel 15.1. Questions to be answered by the environmental assessment of references

	LCA related questions
1	What is the service provided by the product, i.e. what is the functional unit?
2	Which processes are caused by this service, i.e. what is included in the product system?
3	How is the product system delimited, i.e. which processes are essential?
4	Which environmental exchanges must be ascribed to the functional unit, i.e. how are allocations made?
5	What are the impact potentials of the environmental exchanges?
6	How large are the contributions?
7	Which impact potentials are the most significant?
8	Which sources of impact potentials are the most significant? - materials, energy, others? - which stages of the product life? - which sub-assemblies of the product?
9	What are the gaps in environmental knowledge of the product, and what are the most significant uncertainties?

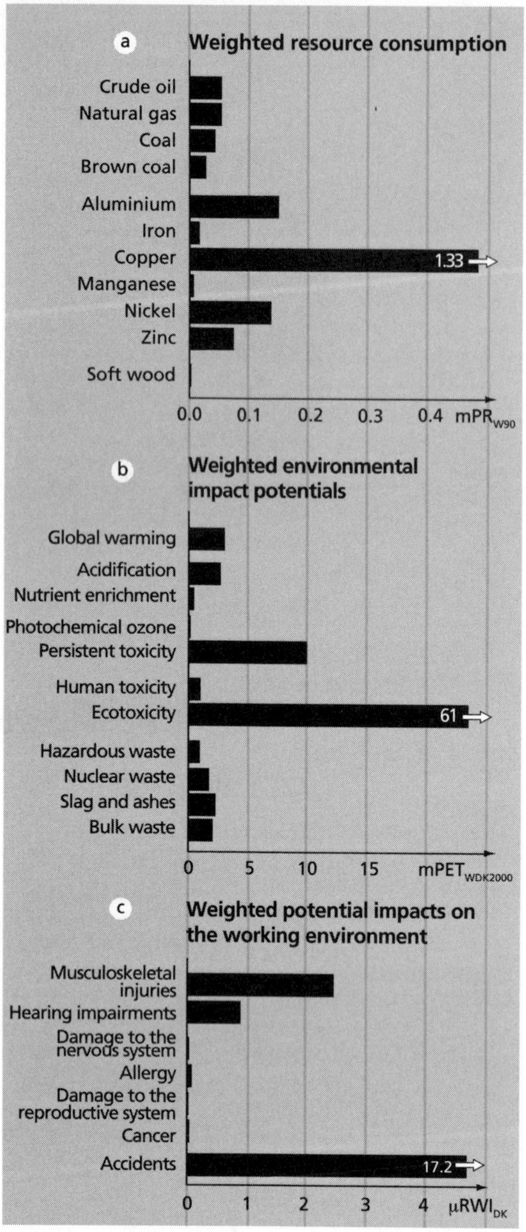

Figure 15.1. Weighted resource consumption and potential impacts for high pressure cleaner Hobby 70

the most significant environmental impact potential, and accidents are the most highly weighted potential impact on the working environment.

The three main problem areas, resource consumption, environmental impact potentials and potential impacts on the working environment, cannot be compared directly as they can in principle at the normalization step, but within each of the three areas they are mutually weighted.

For all three main problem areas, some further weightings could be included, both technical and value-oriented, and the absence of these must be brought to the decision maker's attention. Some of the weightings which can be added to the EDIP method's weightings are described in Chapter 10.

Care must be taken when presenting the weighted profiles to ensure that the decision maker is familiar with the principle of weighting and has been provided with the weighting considerations believed to be lacking.

Note that the most significant potential impacts here are designated purely on the basis of resource, environmental and working environment considerations. No account is taken of the nature of the product. But it is clear that simply by definition, certain products will be weighted most heavily on certain impact categories. In resource terms a plastic bucket is weighted most heavily on oil because it is made from oil. In environmental terms a kilogram of nitrate is weighted most heavily on nutrient enrichment because it is a nutrient; and a kilogram of pesticide is weighted most heavily on toxicity because it is a poison.

It is true that in terms of the environment, the most highly weighted potential impacts are the biggest problem within the given weightings, and the weighting must therefore also be accepted at full value. But this does not necessarily mean that the weightings represent the problems on which the product developer must focus. The nitrate content in a kilogram of nitrate cannot be reduced, but it could be that the energy consumed in making it can be halved. The point is that to some extent the potential impacts *which can be reduced* are the problem, i.e. where the product causes more pollution than necessary. See the next chapter on the environmental diagnosis.

What are the most important sources of potential impacts?

Subdivision of the reference product's potential impacts on sources should be carried out as required, i.e. according to the cause-effect relationships which the designer or the environmental specialist want to investigate. As the LCA starts with a computation of the individual environmental exchange from the individual process, it is possible in principle to investigate the source of all contributions. In practice a computer program is required to process this degree of detail and to carry out fast searches. The PC program under development for the EDIP method makes this possible.

It may be necessary to answer questions at many different levels, e.g.:

What are the most important environmental exchanges? Classification under the MECO principle

The contribution to each impact category must be able to be broken down to the contribution from the individual exchange, e.g.:

- How much of the acidification derives from:

  1) SO_2? 2) NO_x? 3) HCl? 4) NH_3?

  in other words, a breakdown of the contribution in parallel on a series of exchanges.

- How much of the potential for human toxicity derives from:

  1) all chemicals in the product system? 2) solvents?
  3) butyl acetate?

  in other words, a breakdown of the contribution into one or more impact categories at various levels in a hierarchical family of substances.

- How much of the contribution to toxic waste derives from:

  1) surface treatment? 2) electroplating? 3) chrome plating?
  4) chrome plating at a particular company?

  in other words, a breakdown of the contribution into one or more types of impact at various levels in a hierarchical family of processes.

Table 15.2. Contributions to resource consumption and potential impacts classified after the MECO principle

Materials	
Material	- consumption - disposal
Processes	
Energy	- thermal - electric
Chemicals	- ancillary substances
Others	- overhead

The results can be grouped and presented in different categories of exchanges as required. As a standard, all types of impacts are presented in a classification after the MECO principle (see Chapter 12), i.e. classified by materials and processes in the complete product system as shown in Table 15.2. This classification was made for the high pressure cleaner

Figure 15.2. Weighted resource consumption and impact potentials classified by sources after the MECO principle

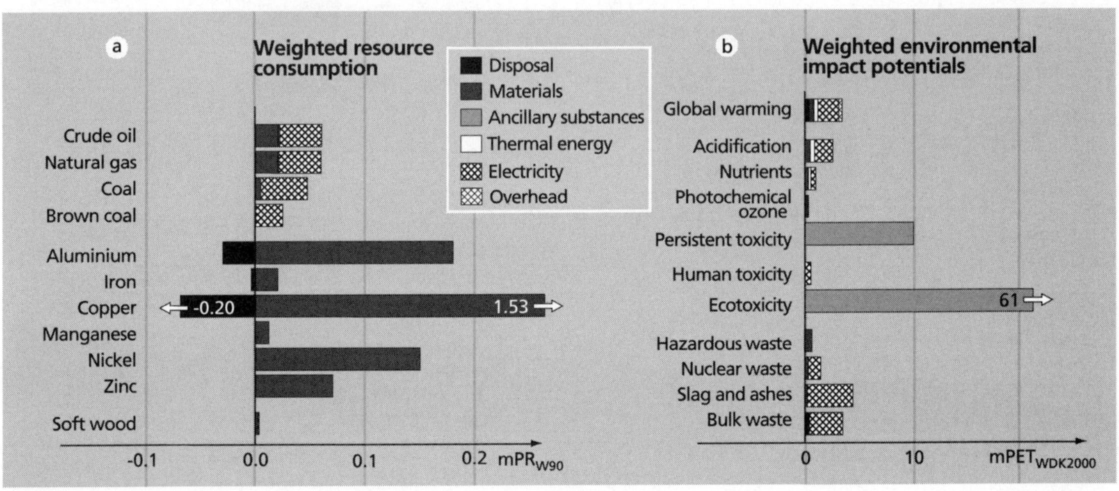

in Figure 15.2 for resource consumption and potential environmental impacts.

A breakdown into materials, energy, chemicals and "other" will provide the designer with a clear picture of the consequences of the solutions chosen in the reference product. The classification shows that M, E, C and O each have a characteristic contribution, and a pattern for the product type will be drawn which the designer gradually comes to recognize. The designer, in other words, will know the resource consumption, the environmental impact potentials and the potential impacts on the working environment typically involved when materials are selected, disposal or energy consumption planned, or processes involving consumption of chemicals are chosen.

Which stages in the product life are the most important?

An overview of potential impacts within the product system can be presented in a life cycle stage model, as done in Figure 15.3 for resources and environmental impacts. A presentation which includes the working environment can be seen in the example for Gram in Part V of the book. The transport processes are omitted because they are of little significance and can therefore be disregarded in such an overview, in any case for electromechanical products of the type studied in the EDIP programme.

The purpose of the figure is to show the significance of the stages, e.g.:

- How significant is the manufacturing stage?
- How significant is the use stage?
- How significant is disposal?

Figure 15.3. Weighted resource consumption and potential impacts classified by stages of the product life

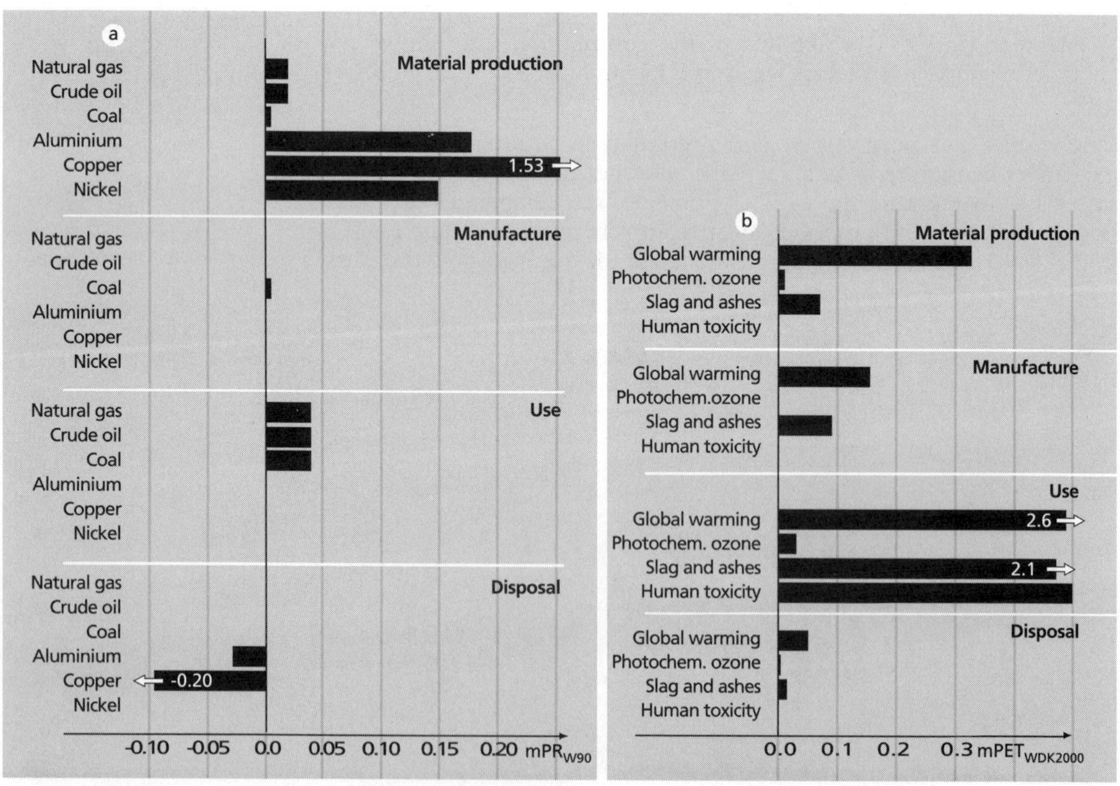

For a few selected impact potentials, the figure shows the clear dominance of the use stage for the energy-related impact categories and ecotoxicity. It also shows how small an amount of resources is recovered on disposal.

Which sub-assemblies of the product are the most important?
Renewal in a product will often consist of finding other solutions for particular components or parts of the product, e.g. for a refrigerator: new cabinet, new type of door, new compressor, new top panel, new evaporator. It is rare for the whole product to be changed completely in relation to existing products. It is therefore useful to know the potential impacts of each of the individual components and parts, so that the effects of changes can be estimated quickly. Figure 15.4 illustrates such a presentation for a refrigerator.

The presentation in Figure 15.4 contains the entire product system for each component. Only the use stage's energy consumption is omitted, because it cannot be attributed to a single component, but to a combination of many components including evaporator, compressor, condenser, insulation and control system. All other environmental exchanges in the product system can be attributed to one component.

It is important for the presentation of the contribution by component to the various impact categories that the entire product system be included as far as possible. When a component is replaced, all stages of the product life are affected, and as far as possible the total impact should therefore be presented. To be able to make this presentation of component contributions across the entire product system, certain conditions regarding modelling must be fulfilled, especially the requirement that the disposal stage must also be able to be broken down into components. As is evident from the case histories in Part V, the full component product systems are included only for the refrigerator. For the other products, the presentation includes only the product system to and including the manufacturing stage.

The numbers under the columns in the figure refer to the parts list for the product shown in the example for Gram in the case histories in Part V.

Method

The requirements in the method for environmental assessment of references correspond substantially to the description of the core of the method in Part II of this volume. Scope definition includes the following special aspects.

Reference product

The reference product must represent the new product as well as possible, and it must be selected on the basis of criteria as described in Chapter 8. A product is often selected for which the product system broadly represents a product family at the company. As an example, Table 15.3 shows Gram's range of refrigerators, all of which are represented by the LER200, the reference product selected by Gram. There are only minor differ-

Table 15.3. Examples from Gram's product range, all of which can be represented to some extent by the reference product LER200

Model	Type	Volume
K130	Refrigerator excl. freezer	116 l
K155TU	Refrigerator excl. freezer	155 l
KS200-01GT	Refrigerator excl. freezer	195 l
KS240-01GT	Refrigerator excl. freezer	240 l
KS180-01E	Refrigerator excl. freezer	168 l
KS300-01E	Refrigerator excl. freezer	271 l
KS350-01E	Refrigerator excl. freezer	323 l
KS410-01E	Refrigerator/chiller	Cool: 183 l Chill: 194 l
KF120	Refrigerator incl. freezer	Cool: 94 l Freeze: 14 l
KF135TU	Refrigerator incl. freezer	Cool: 109 l Freeze: 27 l

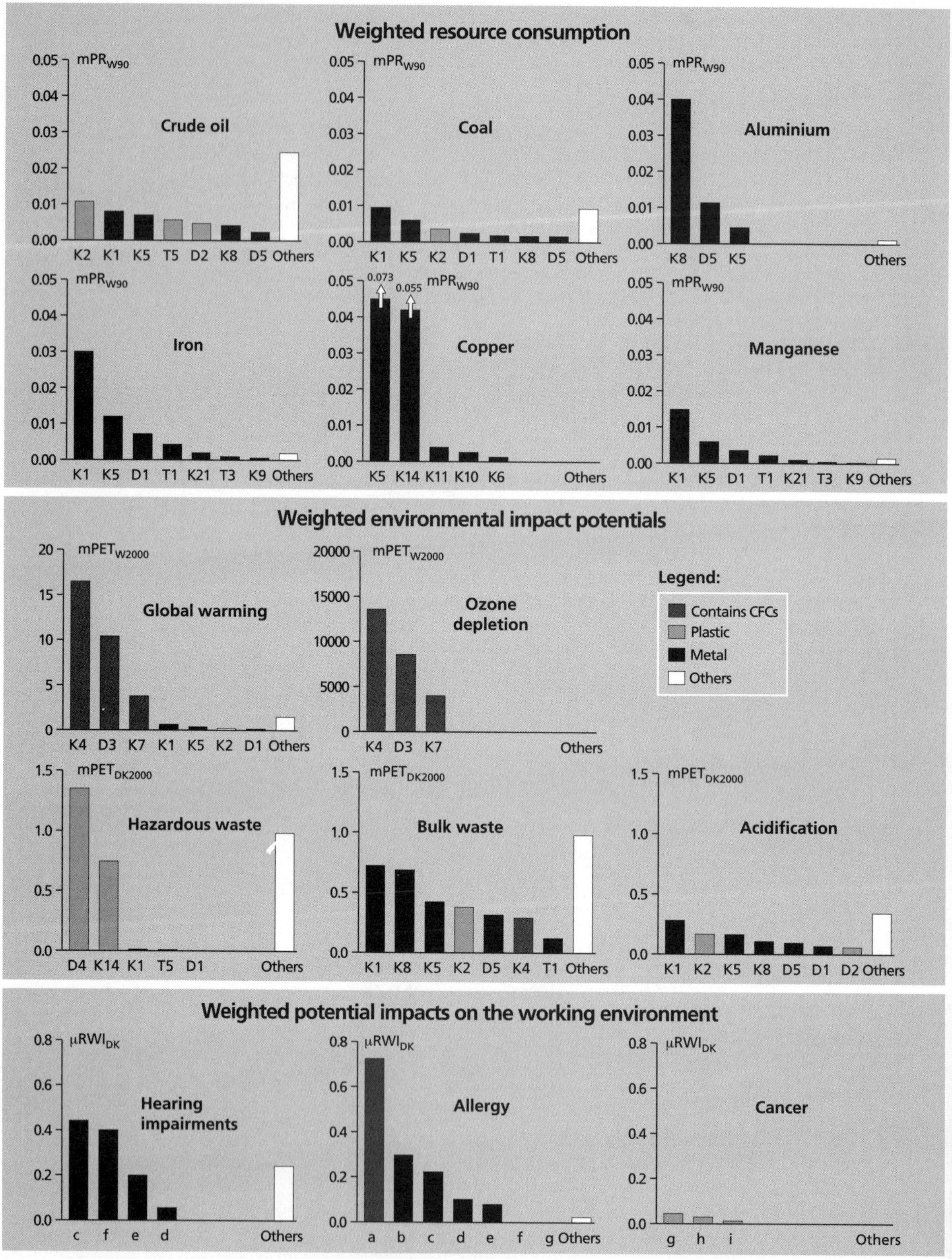

Figure 15.4. Weighted resource consumption and potential impacts classified by components within the product

ences in the product systems of the various refrigerators and the LER200 data can be used with relatively simple modifications for the others.

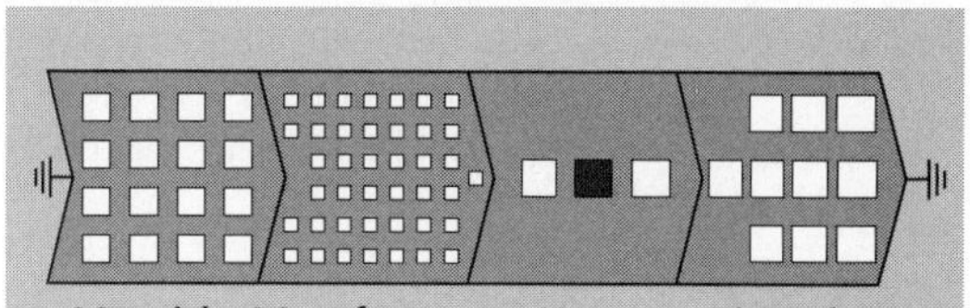

Figure 15.5. The scope definition comprises the entire product system

Scope definition of the product system
Scope definition follows the general principles. It is typical for this application of the LCA that the product system must be included in as broad and comprehensive a manner as possible, because the object is to establish a broad base of information on the product. This is illustrated in Figure 15.5.

Scope definition with respect to time and technology
The trend analysis and the projection are carried out in accordance with the principles described in Chapter 8 and in the procedure in the book's Toolbox. In this application of the LCA, the projection should be based on broad average observations. Table 15.4 illustrates some examples.

Allocation
It is not possible to extend the system boundaries during environmental assessment of reference products. The object is to acquire a basic knowledge, and the assessment is not a comparison among alternatives. It is therefore not possible to anticipate changes in adjacent systems caused

Examples for materials production:

- Expected development in technologies for the essential material production processes, e.g. specific energy consumption, consumption of additives and content of recycled materials.

Examples for manufacture:

- Expected development in average efficiency for production of electricity in the countries expected to produce the new product and expected development of air purification in the power plants.
- Expected development in average degree of purification of municipal waste water treatment plants in the countries expected to produce the new product (relevant for processes which discharge waste water, e.g. phosphatizing).
- Expected technological development for the most important manufacturing processes.

Examples for use:

- Expected development in average efficiency for production of electricity in the expected markets and expected development in air purification at the power plants (relevant for products consuming electricity).
- Expected development in the average performance of municipal waste water treatment plants in the expected markets (relevant for products which discharge waste water, e.g. high pressure cleaners).

Examples for disposal:

- Expected developments in the average distribution between incineration, land filling and recycling in the expected markets
- Expected development in the average reuse of materials for the various material fractions in the expected markets.
- Expected development in the average production of heat and combined heat and power in incineration plants in the expected markets, and expected average development in degree of air purification in the plants.

Table 15.4. Guidelines for projection of the LCA

by any later new product, and it is in any case not the task of this application of the LCA. Allocation is made in accordance with the partitioning keys in the EDIP method.

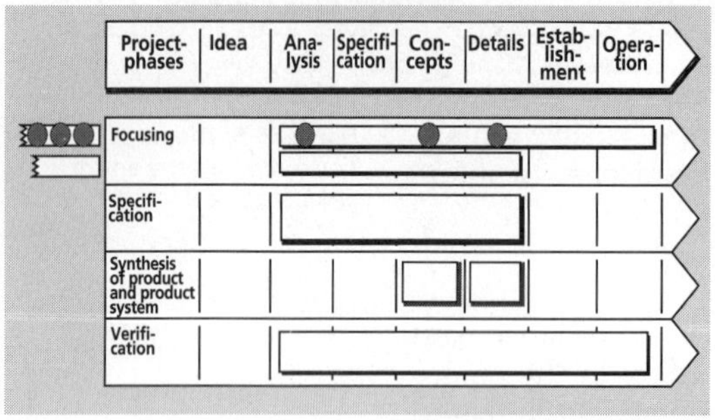

Procedure

The major work of data collection and calculations can and should be done before product development commences to prevent any bottleneck from occurring.

The LCA is reported in a brief report to the decision makers involved in product development. The first section of each example in the case histories in Part V shows such a report.

It can be useful to present the results of the LCA at a meeting or a workshop for the decision makers involved in product development. The LCA is of interest to both management and designers:

- Management can use the results as input to a product-oriented environmental policy and for long-term strategies for technology and product development.
- The designers can use the results as a basis for focusing and specification in development projects within the same product family.

The report therefore serves as background material, and the results will subsequently enter into development projects. Modelling of the reference product(s) should be done on computer so that it can accommodate any simulations of changes made later under the environmental diagnosis.

16. Environmental diagnosis

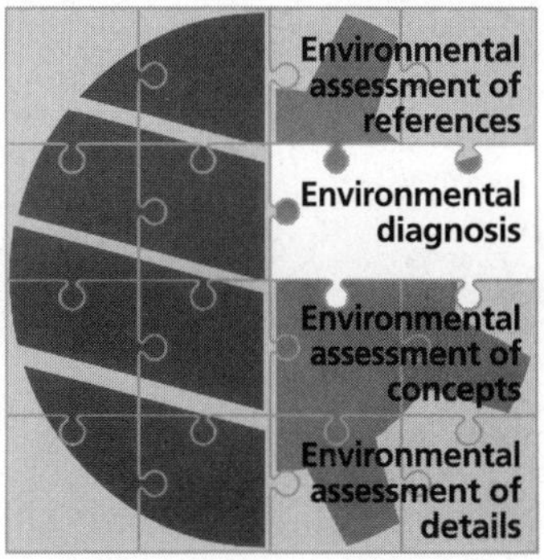

The goal of the environmental diagnosis is to designate environmental points of focus in the product: to designate which of the potential impacts demonstrated by the environmental assessment of the references should be considered problematic, and to localize where in the product the problems reside.

As noted previously, the most highly weighted potential impacts are not necessarily the most problematic for the product; to a large extent it is those which are greater than necessary. The goal of the environmental diagnosis is to identify potential improvements in the product so that, together with the environmental assessment, they can form the basis for designating the focal points.

The task of the environmental specialist is to simulate various changes in the product or its product system and then to assess these, both theoretical changes and known alternatives, e.g. competing products. The task of the designer is to supply information on known alternatives and/or new technologies, for improvement potentials can only be designated on the basis of a knowledge of alternatives.

Note that the basis for the diagnosis must be projected so that the assessments are valid for the period during which the new product is expected to be on the market.

Environmental questions

Table 16.1 lists the questions to be answered by the environmental diagnosis on the reference product.

Examples of environmental diagnosis on reference products

The environmental tasks arising from the questions in Table 16.1 are reviewed here, with illustrations of how they can be solved and how the results can be presented.

Which changes in the reference product or its product system will have significant environmental impacts?

To simulate the environmental impact of a change in the reference product or its product system will mean changing the model of the product system on which the LCA is based, and then repeating the calculations. When many simulations are to be done, it is necessary that the modelling be computerized, so that changes can be made quickly, and calculations then proceed automatically. The time required will otherwise be too long.

Many simulations have been made in the EDIP project in a prototype for the EDIP method's PC program. The example of the refrigerator LER200 in Part V illustrates 20 different simulations. The time required for the simulations is very limited when a PC program is used and when the basic data are available.

Table 16.1. Questions to be answered by the environmental diagnosis

	Questions for the environmental diagnosis
1	Which changes to the reference product or its product system will have significant environmental consequences? ■ Simulate theoretical changes and assess these. ■ Simulate known alternatives and assess these.
2	Where in the reference product do the environmental improvement potentials reside? ■ Which alternative design solutions could be found to those chosen in the reference product? Ideas, new technologies, existing alternatives. ■ Where in the product are the environmental improvement potentials located? ■ Where in the surroundings are the environmental improvement potentials located?

M		E		C	O
Material consumption (substitutions)	Disposal	Consumption of thermal energy	Consumption of electricity	Comsumption of chemicals	Others
■ Use of wood instead of plastics/steel/Al (one simulation at a time) ■ Use of recycled material instead of primary steel/Al/plastics (one simulation at a time) ■ Other relevant material substitutions	■ Landfill 100% instead of the present scenario ■ Incinerate 100% instead of the present scenario ■ Recycle 100% instead of the present scenario	■ Reduce consumption of thermal energy in selected processes by a realistic percentage	■ Reduce consumption of electricity in manufacturing processes by e.g. 25% ■ Increase energy efficiency during operation of the product to 100% ■ Reduce consumption of electricity during operation by a realistic percentage, e.g. 10% ■ Reduce consumption of electricity during standby by 100% ■ Reduce energy consumption during standby by a realistic percentage, e.g. 50%	■ Omit use of selected chemicals: paint, CFCs, lubricants etc. (one simulation at a time)	■ Reduce lifetime by 50% ■ Increase lifetime by 100% ■ Simulate different use patterns (one simulation at a time) ■ Simulate use of the product in different regions or locations (one simulation at a time)

Table 16.2. Classification of simulations after the MECO principle

It can be useful to organize the simulations after the MECO principle, as shown in Table 16.2. This implies that the simulations must illustrate the consequences of changes in choice of materials and disposal of materials and changes in the processes' consumption of energy and chemicals (ancillary substances) as well as changes in "other items", e.g. changes in life span and patterns of use.

The simulations are carried out by changing the modelling of the reference product by one item at a time, and making an LCA each time, e.g. in the form of a calculation of weighted resource, environmental and working environment profiles. The calculated profiles are then compared with the profiles for the reference.

Figure 16.1 gives examples of results of such simulations for resource consumption and environmental impact potentials. Some of the most interesting simulations from the environmental diagnosis for Gram's reference product, the refrigerator LER200, have been selected, and the simulations shown have been discussed and interpreted to show the results which they yield.

To be able to understand the simulated changes, it is necessary to know the most important assumptions in the modelling of the reference product, the refrigerator LER200:

- Use and disposal are in Denmark.
- The life span is 13 years.
- On disposal, the refrigerant is collected from 30% of the units, while the CFCs are collected from the foam for only 7.5%.
- The metals are recycled, while plastics and foam are dumped.

Only the most pertinent resource consumptions and potential impacts are presented for each individual simulation.

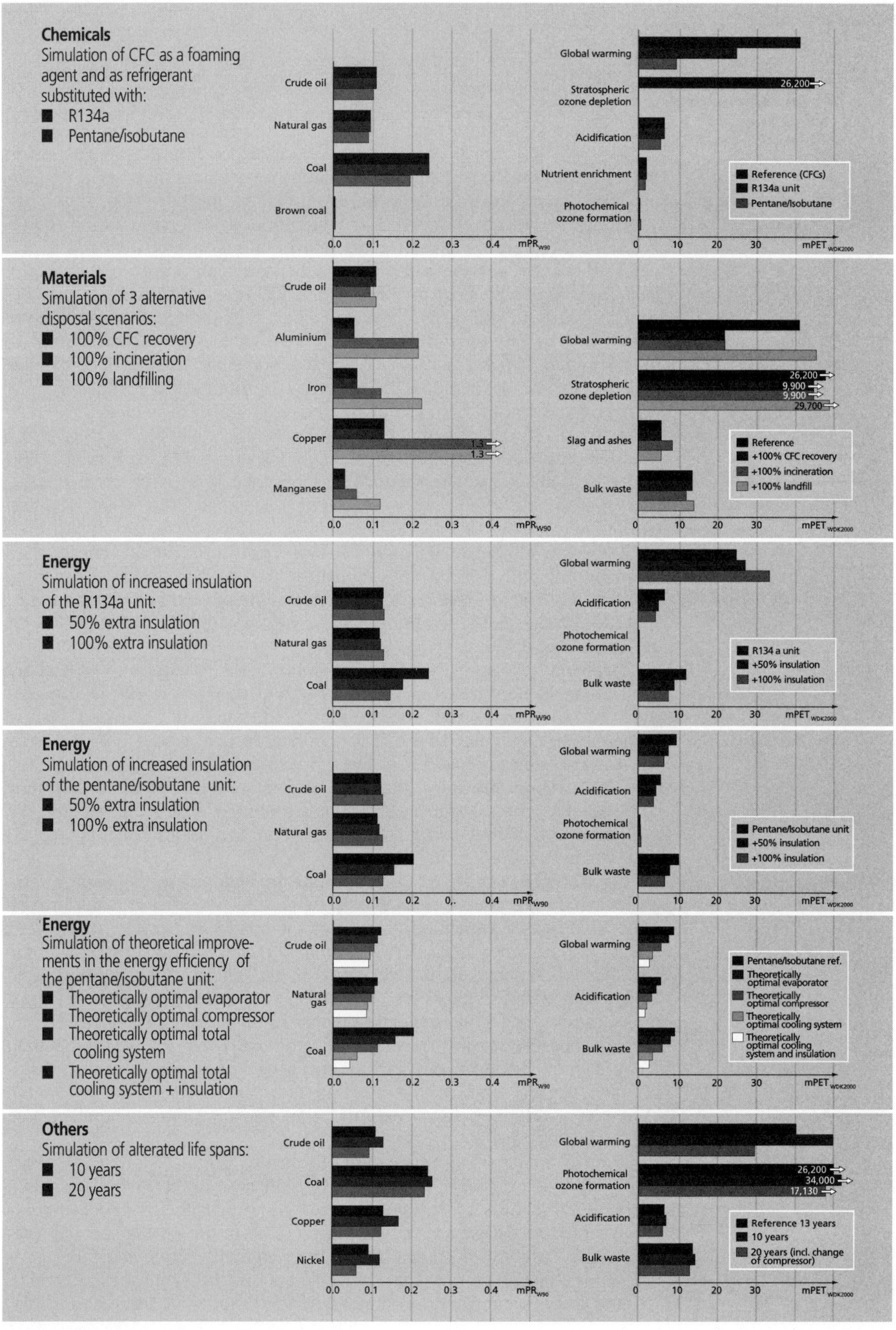

Figure 16.1. Some simulations carried out for the refrigerator LER200

The simulations of the three alternative *disposal scenarios* in Figure 16.1 show that the significance of the disposal route is very great.

One hundred percent CFC recovery on disposal yields a 50% reduction in the contribution to global warming, namely the portion accounted for by CFC emissions on disposal. A corresponding 60% reduction in the contribution to depletion of the ozone layer is attained. The remaining 40% derives from CFCs in the manufacture and use stages.

CFCs are destroyed by 100% incineration, and corresponding reductions are therefore obtained for the contributions to global warming and ozone depletion. At the same time, other fuels are substituted because heat from the incineration of waste is used. It is assumed here that the substituted fuel is oil, and the effect is visible in the form of a reduction of ca. 10% in oil consumption. Incineration reduces the contribution to bulk waste but makes a relatively greater contribution to slag and ashes because of the refrigerator's high metal content. Incineration of the refrigerator without first removing the metallic parts will have a very unfortunate effect on resource consumption, especially for copper, where it will increase consumption by a factor greater than ten. For iron and manganese the consequences are less because it is assumed that after incineration, the steel is recovered magnetically in 50% of incineration plants.

Dumping of the refrigerator has even more unfortunate effects than incineration, both for resource consumption and in terms of environmental impacts: for the metals, because iron and manganese are not reused, and for global warming and ozone depletion, because the CFCs are emitted after dumping.

The simulations of *chemical substitutions*, i.e. alternatives to CFCs, are used as an in-depth example in the method section on "Impact assessment" in Chapter 10, where the figure is interpreted. The interpretation is therefore not repeated here, but it is included because it is a precondition for the three simulations of energy savings.

Increased insulation for both the R134a and the pentane/isobutane solutions also shows large improvement potential. It is assumed that the insulation is increased externally to maintain the volume of 200 l specified by the functional unit.

For the R134a unit, the contributions are reduced to the impact categories which are energy-related only, while the potential global warming effect is increased with the thickness of the insulation because of R134a's effect as a greenhouse gas. As is evident, the increase is greatest for the last part of the insulation layer because the energy saving per unit of R134a falls with the thickness of the layer. Oil and natural gas consumption increase with increasing thickness of insulation in spite of the energy savings, because these raw materials enter into the insulating material. They do not, however, enter into Danish electricity production to a significant degree, and this is what is saved in the use stage. Danish electricity production uses coal, and coal consumption is therefore also reduced significantly.

For the pentane/isobutane unit, the same trade-off situation as for R134a is not seen with respect to the contribution to global warming. Increased thickness of insulation gives an improvement in all energy-related impacts. However, there is an increase in the potential for formation of photochemical smog, but in the weighted picture the increase is again not so significant as the improvements for other impact categories. Oil and gas consumption are increased as a function of increased insulation for the same reason as before, but to a greater extent, because energy consumption for this unit is less in the use stage than for the R134a unit. Total resource consumption is reduced with increased insulation because of the improvement in coal consumption.

The consumption of extra materials, both insulating foam and steel used to increase insulation, is of course included in the calculations illustrated in the figures.

The *energy efficiency* is 3-4% for the LER200, which is one of the world's most energy-efficient refrigerators. The energy efficiency is defined as the energy required to cool down the goods and to compensate for door openings divided by the refrigerator's energy consumption. In practice the energy required to cool down goods and to compensate for door openings is 25% of the total energy consumption, i.e. 1/4. The rest is lost through heat penetration via the insulation. At the same time, the efficiency of the refrigeration system, i.e. compressor, evaporator and condenser, is about 14%, i.e. ca. 1/7. The total energy efficiency for the unit is thus $1/4 \times 1/7 = 1/28$ or ca. 3.5% of the theoretical maximum for a fully optimized thermodynamic system (Jakobsen, 1995). Figure 16.1 shows the theoretically achievable improvements in a pentane/isobutane unit.

A theoretically optimal evaporator can increase efficiency of the refrigeration system by about 50%, and an optimal compressor can correspondingly increase efficiency by a further ca. 100%. If the entire refrigeration system is optimized, energy consumption in the use stage is lowered by the factor of seven already discussed, and if the insulation is optimized at the same time, there is a further lowering by a factor of 4. Figure 16.1 shows the consequences for resource consumption and potential environmental impacts in the refrigerator's total product system. Energy consumption for the insulating materials is *not* taken into account here, but a theoretically optimal vacuum insulation of one kind or another is assumed.

The *life span* is of considerable significance for the relative weighting of all life cycle stages other than the use stage, i.e. the material, manufacturing, transport and disposal stages. The LCAs expressed per unit of time, and the contributions from these stages are thus reversed in proportion to life span. Figure 16.1 shows the significance of two theoretical changes in the life span.

It is clear that the significance of the life span is greatest for the contribution to global warming and ozone depletion because of the very significant contribution to these impact categories during disposal. A compressor replacement is included in the 20-year life span in order not to overestimate the improvement, meaning that copper consumption is not reduced significantly in the 20-year life span compared with the 13-year life span.

The figure indicates that increased life span yields an environmental improvement. However, it is not certain that this will be the case, because refrigeration technology is developing so quickly. For example, it is expected that the refrigerator in the year 2000 will have a much smaller environmental impact during operation than in 1995. As the use stage is also expected to be decisive in future, the shorter life spans of units with today's technology may be environmentally preferable.

The simulations must fix proportions for the environmental consequences of various possible changes in the reference product or its product system, enabling identification of the directions in which the greatest potential for environmental improvement will be found within the MECO classification. This provides a basis of knowledge for the discussion on alternatives.

The selected examples from the refrigerator LER200 in Figure 16.1 give an impression of the overview which should be established.

Where do the potentials for environmental improvements reside in the reference product?

The environmental assessment of references and the environmental diagnosis have given the environmental specialist and the designer a good knowledge of the reference product and of the environmental contributions from the various parts of the product and from the various sources within the MECO classification. This overview must now be confronted by ideas and practical possibilities for alternative solutions.

In the discussion of alternatives, the MECO principle can therefore be applied now, i.e. discussions can focus on materials, scarce resources, energy efficiency, chemicals, life spans and patterns of use. These factors are readily understood in practical terms and are of immediate relevance to the product. In addition, in many cases changes will only affect one factor. The consequences of changes are clearest when they are viewed on the basis of the MECO classification, while the LCA on the reference product and the simulations remain as background knowledge which allows an assessment and a ranking of the alternatives while they are under discussion.

The discussion of alternatives must help to clarify which of the impact potentials from the assessment and which of the improvement potentials from the simulations should be brought into focus in the development project. Several aspects must be resolved before the final focusing and specification, namely:

- the environmental impact potentials and improvement potentials,
- the customers' perceptions and priorities,
- all other considerations which will be included in the product and which will give it its commercial potential: economy, quality, ease of manufacture, etc.

The environmental diagnosis describes the focal points, firstly on the basis of purely environmental considerations. The result of the environmental diagnosis then forms the basis for a commercial discussion, where environmental considerations are weighed against other considerations and the customers' perceptions and viewpoints are considered. These considerations are presented in the next chapter, "Environmental specification".

Environmental improvements of the reference product can be created in two ways which are different in principle, namely by:

I Creating other product systems
- i.e. affecting the *existence* of processes in the product system.

II Adapting the product to existing systems and processes in the product system
- i.e. affecting the product's *performance* in the processes which it encounters.

Table 16.3. Classification of means for creating more environmentally friendly products.

Method	Level in the decision hierarchy		
	Conceptual level	Structural level	Component level
I Existence of processes - create product systems	MECO	MECO	MECO
II Performance in processes - adapt product to system	MECO	MECO	MECO
- adapt system to product	MECO	MECO	MECO

Environmentally attractive product systems (I) can be created both by establishing desirable product systems and by avoiding undesirable ones. Desirable product systems can, for example, be use of renewable resources, establishment of recycling pathways or utilization of wastes. Undesirable product systems can, for example, include irreversible consumption of scarce resources, processes with heavy use of chemicals, or energy systems with low efficiency.

Better performance (II) can be created both by adapting the product to the processes and systems encountered, but also by adapting the system to the product, or mutually adapting them to each other as illustrated in Figure 16.2. An adaptation of the product to a system can, for example, mean avoiding having the product contain chemicals which are not desired in the process, e.g. plasticizers in a landfill or chlorine-containing substances in an incinerator.

To some degree there is overlap between the two principles. For example, it is possible to increase the probability of another product system (I) by adapting the product to it (II). But it will often be useful to keep the two principles separate, as this can provide inspiration for the various possibilities available for improvement of the product's environmental characteristics.

Table 16.3 illustrates the various routes to environmental improvements in a product. The table also shows that, as noted above, changes are realized at various levels in product development. Some ideas can be proposals for changes to the reference product at the detailed level, e.g. selection of another material. Others can be ideas for entirely new conceptual solutions, e.g. an entirely new control principle. The purpose of the table is to show that environmental improvements in a product are achieved in various ways.

The degree of thoroughness of the search for alternatives differs from project to project. In most cases the task is implied to some degree. For the environmental diagnosis, however, it is interesting in the first instance to keep it completely open, and to consider all conceivable alternatives to the product's basic service.

Figure 16.2 indicates such a procedure for alternative solutions for the refrigerator LER200. The figure represents an example of how alternatives are distributed in a matrix as shown in Table 16.3, and it shows some of the choices made in the LER200, and some alternatives.

Figure 16.2 illustrates selection and alternatives at the concept, structure and component levels. At the same time the figure illustrates alternative considerations concerning 1) creating other product systems, 2) adapting the product to processes in the existing reference product system, or 3) better adaptation of the existing systems to the product.

Refrigeration of food

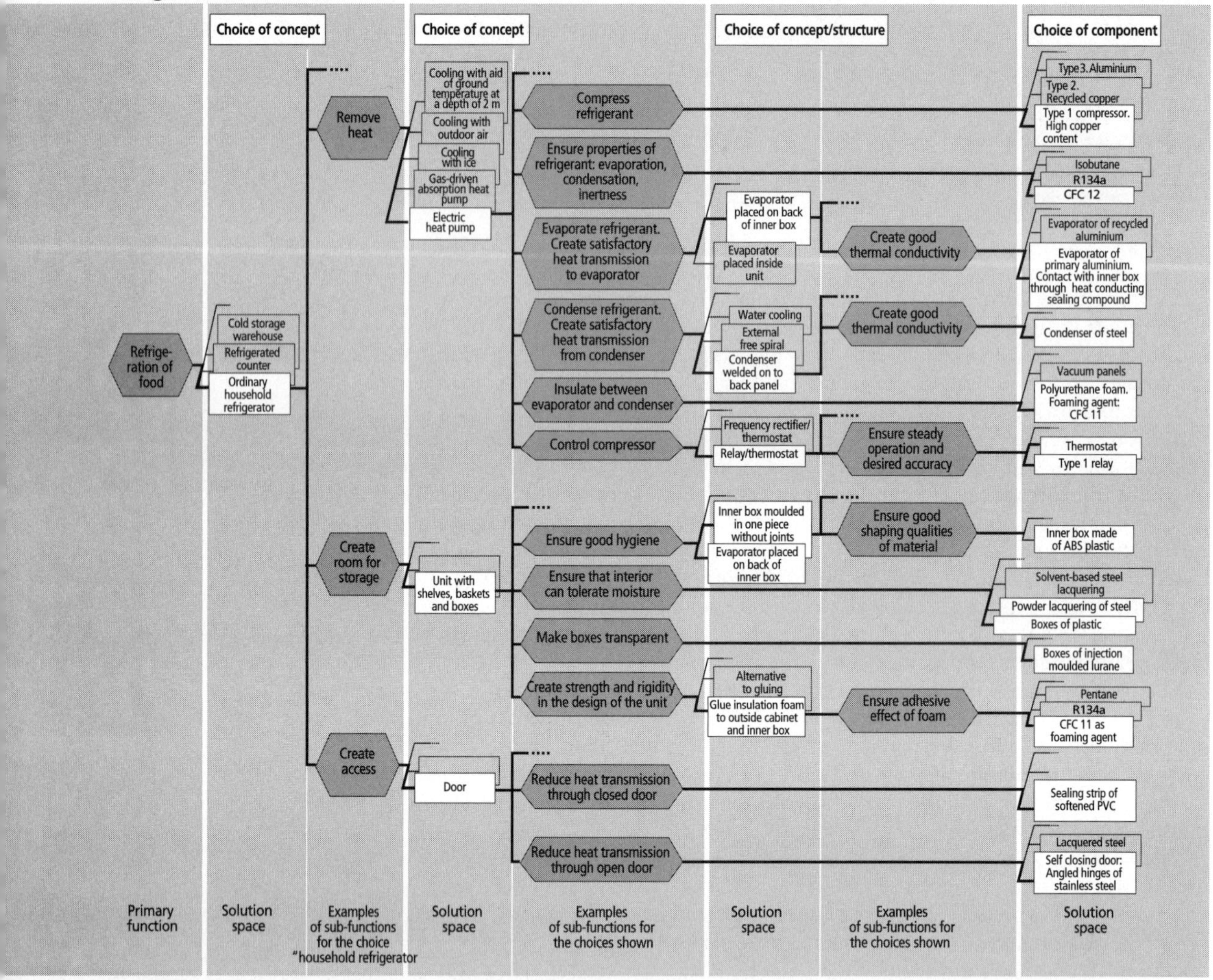

Level in the decision hierarchy

Instrument	Conceptual choice and alternatives	Conceptual or structural choice and alternatives	Component choice and alternatives
Create product systems	**Choice:** Electric heat pump **Alternative 1:** Cooling with atmospheric air **Alternative 2:** Cooling with soil temperature	**Choice:** On-off relay-controlled compressor **Alternative 3:** Continous operation of compressor by means of frequency rectifier	**Choice:** Primary copper **Alternative 7:** Steel, aluminium, recycled copper **Choice:** Primary aluminium **Alternative 8:** Recycled aluminium
Adjust product to system		**Choice** : Air-cooled condenser **Alternative 4:** Water cooling **Choice:** Sandwich construction **Alternative 5:** Avoid sandwich construction **Choice:** Evaporator placed behind inner box **Alternative 6:** Evaporator placed inside unit	**Choice:** Stainless steel **Alternative 9:** Lacquered steel **Choice:** Softened PVC **Alternative 10:** Other materials **Choice:** CFC as cooling and foaming agent **Alternative 11:** R134a or pentane/isobutane
Adjust system to product			**Choice:** Insulation, standard kitchen module **Alternative 12:** Increased amount of insulation, special kitchen module **Choice:** Shredding **Alternative 13:** Specially designed shredder system with CFC recovery

Figure 16.2. Examples of choices and alternatives classified in a matrix according to different pathways to improvement

Cooling: Electric heat pump.
A very basic conceptual choice in the LER200 is that the cooling is based on the heat pump principle, i.e. evaporation/condensation of a refrigerant and use of a compressor as known in almost all refrigerators. The energy efficiency of the best units using this principle in 1995 was only about 3-4%.

Alternative: Using outdoor air.
When the outdoor temperature drops below a certain temperature, it is possible to use outdoor air as a supplement to the existing system, *creating* a more attractive product system. In Denmark the savings potential is about 50% of the *energy consumed* in the use stage.

Alternative: Geothermal cooling.
At a depth of 2 m, the soil temperature in Denmark will seldom be higher than 5-8 °C. It is possible, by applying a shaft/lift principle, to utilize this temperature, thus *creating* a more attractive product system in the entire product use stage.

In Denmark the savings potential is about 100% of the energy consumed in the use stage, possibly at the expense of an increased *consumption of materials.*

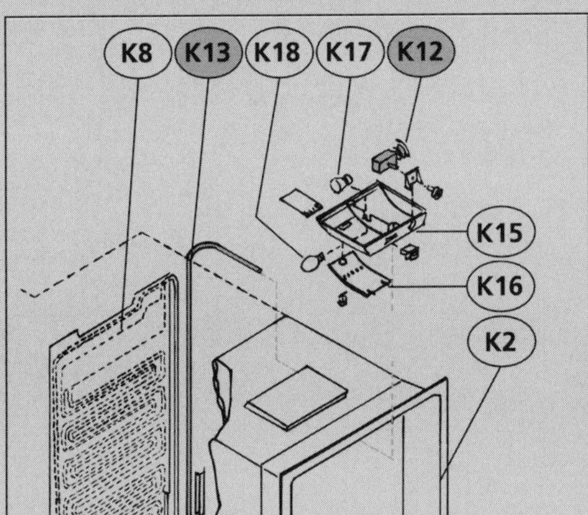

Control: On-off relay control.
In the LER200, the temperature is controlled by on-off relay control of the compressor. This results in a large consumption of energy each time the compressor is activated.

Alternative: Continuous operation.
The temperature can be regulated by controlling the compressor continuously via a frequency rectifier, creating a more attractive product system in the use stage.
If the compressor is run continuously, the savings potential will be about 30% of the *energy consumed* in the use stage.

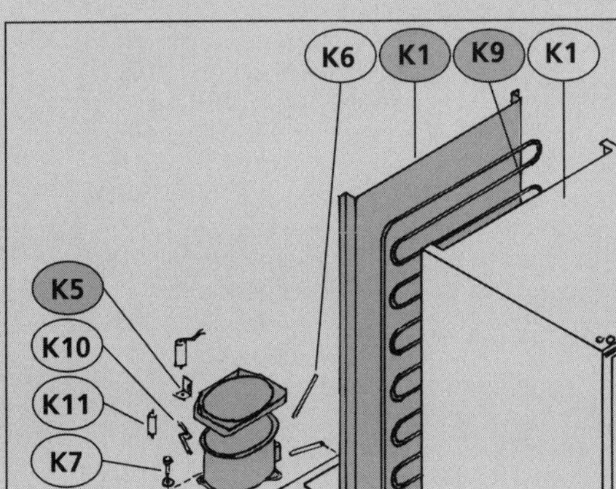

Air cooling of the back panel.
The heat produced by the compressor and the condensor is removed by heat exchange with the air. The heat exchange usually has a low efficiency due to the limited renewal of air behind the unit.

Alternative: Water cooling.
Condensor and compressor can be cooled by water. The water can then be used to preheat household water. This will increase the efficiency of the cooling system and the waste heat will be utilized, also during summer. This exemplifies the concept of adjusting the product to the surroundings in the use stage. The heat utilized will correspond to *energy savings* in the use stage of 20-25%.

Table 16.4. Examples of conceptual choices and alternatives

Table 16.5. Examples of structural choices and alternatives

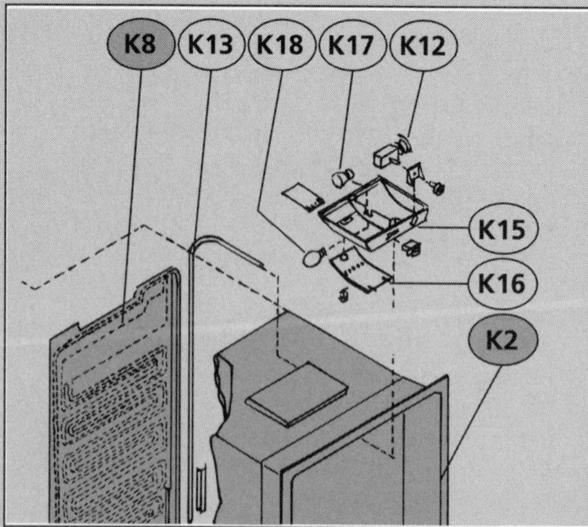

Evaporator: mounted behind inner cabinet.
In the LER200, the evaporator is placed behind the inner cabinet to simplify cleaning.

Alternative: placing inside cabinet.
The evaporator can be placed inside the cabinet, as in many other models. This will help *adapt* the product better to the use stage and increase efficiency by at least 10%. Increasing the area of evaporation will also increase efficiency by 10%. Mounting the evaporator in the top of the cabinet will allow a more uniform temperature distribution: the cold air will sink to the bottom of the cabinet and maintain some air circulation. Efficiency can also be increased by adding a small fan (1 W) to circulate air past the evaporator. Total energy consumption during the use stage could be reduced by at least 20%.

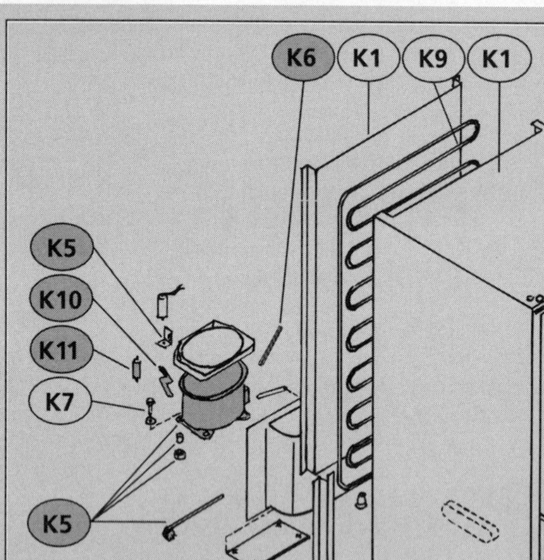

Copper.
Copper is used in the windings in the compressor, the wires, various pipes and the drying filter. As it is impossible to recover all the copper for recycling, copper consumption will be a resource problem.

Alternative: Steel, aluminium, recycled copper.
Soft iron may be substituted for the copper pipe in the compressor. The newer compressors contain less copper. Recycled copper can be used. The wiring can be made of aluminium which is not a scarce resource. From a resource point of view this will create a better product system and reduce the use of copper, a scarce resource.

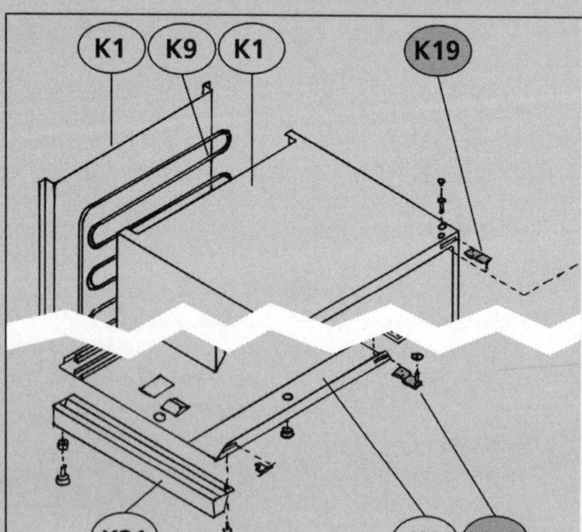

Stainless steel.
The two hinges holding the door are made of stainless steel. Using current shredding technology this alloy will probably be sent to a recycling plant for normal steel, the nickel content thus being lost in the process. Nickel then becomes one of the unit's most highly weighted resource consumptions.

Alternative: Lacquered steel.
It is possible to avoid the use of stainless steel by using lacquered steel hinges instead. This is an example of adjusting the product to the shredder technology. The result will be a *reduced consumption* of nickel but a small increased *consumption of chemicals.* This trade-off is judged to be in favour of avoiding the use of nickel.

Table 16.6. Examples of component choices and alternatives

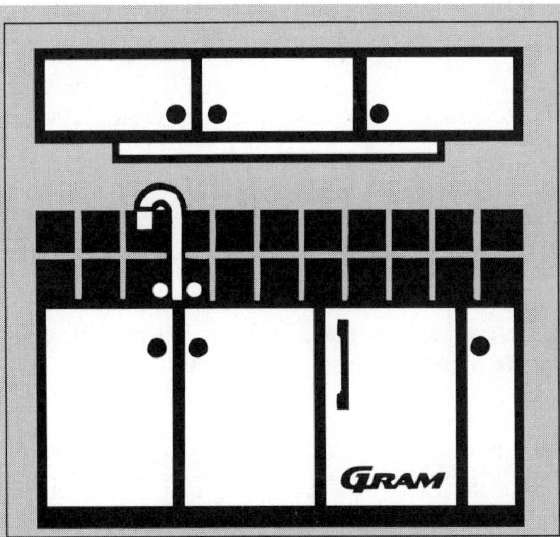

Insulation: standard module.
The insulation of the LER200 is significantly better than for other refrigerators. Increasing the thickness of the insulation while maintaining an internal volume of 200 l results in the external dimensions of the refrigerator being too large for present standard kitchen units. Increasing the insulation thickness is environmentally preferable, but will continue to cause problems as long as kitchen units demand the present external dimensions of the refrigerator.

Alternative: Special modules.
It must be possible to produce kitchen modules which allow for increased refrigerator dimensions. This will *adjust the system to the product* and is an example of the desinger's task reaching beyond his own product, i.e. by collaborating with a supplier of standard kitchen modules.

Energy consumption could be reduced by more than 50%. Another alternative could be to use vacuum panels. This would adjust the product to the system and could potentially increase insulation by a factor of two.

Table 16.7. Examples of choices and alternatives in the surroundings

For each of the categories filled out in the matrix in Figure 16.2, an example is selected to illustrate how environmental improvements can be achieved, partly at all levels in the decision-making hierarchy in product development and partly via each of the three ways. The examples are shown in Tables 16.4-16.7, where potential improvements are briefly outlined with an indication of where they reside in the product. There is no assessment of whether the alternatives are realistic commercial options or of how realistic they are in technical terms. The examples are shown partly to provide ideas and partly to show the principle involved in searching for alternatives, and they will serve as a basis for the subsequent commercial judgment and technical assessment.

For a more thorough review, refer to the case histories in Part V, where the refrigerator LER200 is diagnosed in detail and a number of points of focus are designated, and where points of focus are also given for the other four case products.

Method

Use of the LCA during the environmental diagnosis is distinguished by:

Reference product

For each simulation, the original modelling of the reference product is changed with respect to one or more processes in the product system, and each new modelling in principle amounts to a new reference. Some of the new references are theoretical and perhaps unachievable in practice, but they can serve as a basis for comparison or as positive and negative models, and as such as references for use in product development.

Figure 16.3. The scope definition comprises only the change relative to the reference product

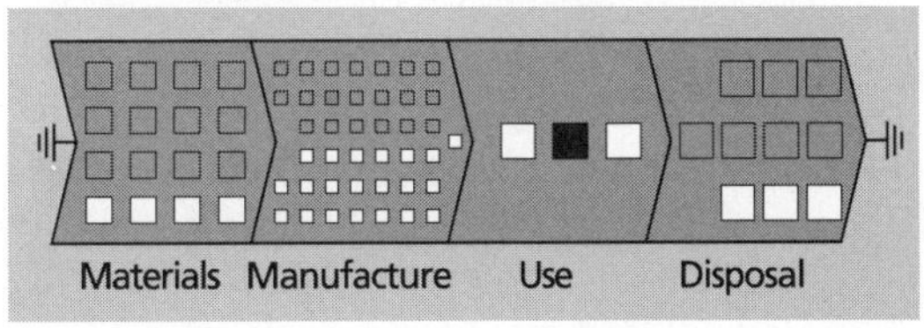

Scope definition of the product system

It is typical of the simulations that the product system is often changed only slightly compared to the model for the reference product. The task is thus merely to model a difference in relation to the reference product as shown in Figure 16.3. This procedure is also termed Δ-modelling (delta modelling). If the Δ-modelling is done by changing the modelling of the reference product, the difference can be illustrated as the reference product before and after the change, as done in the simulations in Figure 16.1.

Scope definition with respect to time and technology
The environmental diagnosis should be carried out on the reference product after a trend analysis and projection have been done. If the changes simulated are specific for particular materials or technologies, a trend analysis must then be carried out specifically for these rather than using average observations as for the reference product.

Allocation
Some of the simulations will represent specific changes which permit extension of the system boundaries. If, for example, disposal is changed from landfilling to incineration in Denmark, the substitutions of other fuels in the Danish energy system to which this will lead can be predicted with some confidence. In this case the system boundaries can be extended. Extension of the system boundaries is recommended if the displacement to which the change will lead can be predicted with a fair level of confidence; simple allocations should otherwise be used in accordance with the EDIP method's partitioning keys (see Chapter 8).

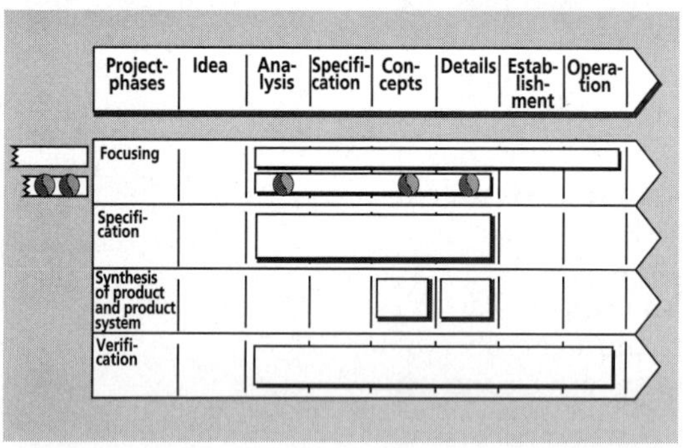

Procedure

Environmental diagnosis of the reference product forms a part of the basis for environmental specification and must therefore be carried out in good time, before the product specification is determined. Be careful to obtain data in time if some of the changes lead to processes for which the company does not possess data.

Environmental diagnosis can also be carried out on solutions for the new product during the course of development, in exactly the same way as described here. The new solution is simply modelled beforehand by changing the modelling of the reference product.

The environmental diagnosis is reported to the decision makers in product development. Examples of such reporting are given in the case histories in Part V, Section 2 in each example, in which the elements the report may contain are shown.

The results of the environmental diagnosis may also be presented at a meeting or a workshop for the decision makers in product development. Like the environmental assessment of references, the diagnosis is of interest to both management and designers.

17. Environmental specification

The objective of environmental specification is to retain the weighting of environmental considerations in the product's total commercial basis, and to determine the specific requirements for the product's properties imposed by environmental considerations.

To be able to accomplish this, the designer must know the customer's opinion and ranking of the product's environmental characteristics in relation to other qualities. The environmental considerations must be held against other considerations, and there must be an assessment of how the business as a whole is to be optimized.

Environmental considerations are often found to be pulling in the same direction as the other considerations, e.g. quality and costs, and the task of the environmental specification is therefore not necessarily one of weighing up conflicting interests.

Environmental questions

The questions to be answered in connection with environmental specification are shown in Table 17.1.

Examples of environmental specification

The tasks indicated by Table 17.1 are the responsibility of the designer. The environmental points of focus designated by the environmental assessment of references and the diagnosis are part of the basis which the environmental specialist must help to prepare, but the tasks of weighing up and of rendering the product commercially viable are the responsibility of the designer.

The designer will probably involve the environmental specialist in parts of the process. The process will therefore be reviewed below, with examples of how the tasks were solved in several companies.

Table 17.1. Questions to be answered in environmental specification

	Questions to the environmental specification
1	What is the environmental competitive framework? ■ What is the customer's perception of the product's environmental characteristics, and how does the customer rank them in relation to the product's other qualities? ■ What is the environmental focus of the competitors? ■ How does the company ensure that the answers to the above questions are valid during the period in which the product is marketed?
2	Which parameters have to be specified and what is the product's environmental specification? ■ How are the company's environmental policy and any long-term environmental goals for the product category considered in the specification? ■ How are environmental considerations weighted against other considerations in the total commercial business optimization? ■ How is the specification divided between optimizing parameters and requirements?

What are the environmental parameters of competition?

The customer's expectations and priorities should be assessed by a market analysis, and this must give a realistic picture of the place of environmental considerations in the customer's perception of the product. If environmental considerations are ranked at a very low level relative to e.g. price, quality, design, etc., this must be described in realistic terms.

Table 17.2 shows how Bang & Olufsen assess the customers' ranking of environmental characteristics, and how Bang & Olufsen's TV is placed relative to those of competitors with respect to several environmentally important product properties.

Environmental competitive framework for televisions. Bang & Olufsen

Bang & Olufsen are used to assessing the market and the customers' perceptions of the product's properties. On the basis of Bang & Olufsen's experience, product properties can be grouped into three categories relative to how customers perceive them, viz. positioning, expectations and mandatory properties. The company has studied how the environmentally important properties are distributed across these groups:

- The positioning properties form the commercial basis and the basis for the competitiveness of the product and the company. The qualities involved are those which particularly characterize and differentiate the product relative to the market and to the customer. Environmental qualities are not included here.
- Expectation properties involve those qualities which the target group and the market perceive as natural in a Bang & Olufsen product and which are not part of positioning. These include some of the environment-related qualities, e.g. low energy consumption, absence of toxic substances and long life span.
- The mandatory properties must ensure the customer a safe and trouble-free use of the product throughout its life, and ensure market access for Bang & Olufsen's products. The mandatory properties are those defined in international standards and in the company's own standards. These include safety, possibility of recycling, and stand-by power consumption.

Market analyses have shown that Bang & Olufsen's customers rank the environment fairly high in their choice of product, but the product's environmental qualities are not among the most significant. The environmental qualities are ranked with expectations and mandatory properties as shown.

Competitors' comparison for some important product properties

For some of the most environmentally significant product properties noted above, it is possible to assess where the Beovision LX 5500, Bang & Olufsen's reference product in the EDIP programme, stands in relation to competitors. The aspects involved are:

- Energy consumption in operation and stand-by
- The possibility of recycling
- Life span

Energy consumption: The Danish journal *Advice and Results* (National Consumer Agency of Denmark) made a comparison of common 28" TVs, including energy consumption. Power consumed ranged from 2 to 18 W in stand-by and from 75 to 107 W in operation. However, neither the worst nor the best combination of stand-by and operation was found in a single unit. One model with a stand-by consumption of 0.1 W was later found. The Beovision LX 5500 consumes 81 W in operation and 3 W in stand-by, and the new Beovision Avant consumes 75 W in operation and 3 W in stand-by. Bang & Olufsen are therefore at the respectable end of the scale. The company is continuing to work towards minimizing power consumption during both operation and stand-by.

Recycling: For the most critical resource, copper, consumption is assessed to be the same for all sets on the market. The differences among various televisions therefore lie in the possibility of dismantling, i.e. in how difficult it is to remove the rear cover. There are no notable differences among the various products in this field. Bang & Olufsen are working towards making dismantling of their products easier.

Life span: When the customer decides to dispose of the television, it is often because it is too expensive to repair it relative to buying a new set. Repairs cost about the same, whether the set is cheap or expensive, large or small. Cheap TVs will therefore often be thrown out before expensive sets, and will thus have shorter lives. Bang & Olufsen offer a range of spare parts and replacement modules, enabling several repairs/replacements and increasing the set's life.

Regulatory requirements

Voluntary environmental regulations are currently in force covering take-back arrangements for electronics and maximum limits for stand-by power consumption. Germany, Sweden and the Netherlands, for example, are relatively advanced in their take-back arrangements. Switzerland has already introduced voluntary maximum limits for stand-by power consumption for a large number of electronic products, and the EU is also working on their introduction.

Table 17.2. Example of an assessment of an environmental competitive framework for televisions, Bang & Olufsen.

Assessments from the market analyses do not reflect a projection of the customers' attitudes and the competitors' intentions, but these are seen indirectly in Bang and Olufsen's own rankings, which are briefly summarized in the table, and which are further described in Bang & Olufsen's long-term environmental targets for televisions (cf. Part V). Market evaluations for the other four case products are also given in Part V.

What is the environmental specification for the product?

The environmental specification for the development product is built into the product specification, with the requirements and optimization parameters which will apply for the actual product under development. But the environmental product specification for the new product is of course part of a more comprehensive environmental management, being subject both to the company's overall environmental policy and the long-term objectives which may exist for the family of products to which the actual product belongs.

Table 17.3 shows two examples of environmental product specifications, one for a mini-waterworks called JetpaQ from Grundfos, i.e. a pump with associated pressure tank, the other for the refrigerator LER200 from Gram.

The two examples show different methods of formulating environmental specifications. Grundfos has a product-oriented environmental policy which provides a framework for the basic specifications for all of the company's products. The product specification includes environmental figures as *optimization parameters* falling within the framework of the environmental policy. An extract of the environmental policy and a set of environmental product specifications are shown in Table 17.3a. Gram has formulated long-term environmental objectives for the company's refrigerators, and this is built into the product specifications of new products in step with the commercial significance of environmental considerations. As is evident from Table 17.3b, Gram formulates its environmental objectives partly as requirements (see environmental targets for energy and chemicals) and partly as optimization parameters (see environmental targets for materials). The long-term environmental objectives for refrigerators are structured in accordance with the MECO principle in the form of declarations of intent for all areas: materials, energy, chemicals and others (including the working environment).

The extent to which the environmental objectives are designed as requirements and the extent to which they are designed as optimization parameters will vary from company to company, depending on how many of the environmental decisions the company wishes to make centrally and how many are delegated to the individual product development projects. In environmental terms the two alternatives can be equally good.

Method

The LCA method is not used during environmental specification, which relies on the results of the environmental assessment of references and the environmental diagnosis.

The environmental policy of Grundfos

Grundfos was one of the first companies in Denmark to establish a product-oriented environmental policy which defines key parameters for environmental work and product development. The following is an extract of Grundfos's environmental policy:

- "Within an economically responsible framework, environmental impact and consumption of resources are to be minimized in the development of new products and processes.
- The environmental impact throughout the product's life must be assessed and described. In practice, all forthcoming products will thus be subject to a life cycle assessment, and the environmental impact shall be assessed relative to corresponding Grundfos products. New technologies shall be assessed relative to existing technologies.
- Grundfos's service department shall be able to accept and repair defective products or advise on ways of disposal of worn-out products. Grundfos shall attempt to recycle parts from worn-out or defective pumps. Attention shall thus also be paid in the design phase to designing parts for reuse or recycling."

Grundfos has recently obtained an environmental certification under the EU system ECO-Management and Audit Scheme (EMAS). The certification requires motivation of employees and subcontractors to demonstrate increased environmental awareness. Subcontractors will also be assessed on their environmental performance and if the latter conflicts with Grundfos's environmental policy, their contracts may be terminated.

Basic specifications for new pumps at Grundfos

Grundfos have performed life cycle assessments on a number of products and thus designated the most important points of focus for pumps. The environmental policy and the LCA carried out have led to focus on the following product specifications for new pumps:

- The number of composites should be limited as far as possible
- Embedding of metal in plastic should be avoided
- Optimization of design with respect to disassembly
- Optimization of the motor's efficiency
- Reduction of idling power consumption
- Reduction of energy loss from electronics
- Reduction in consumption of scarce resources, especially copper and nickel

Table 17.3a. Example of product-oriented environmental policy and product specifications in the company Grundfos

Gram: Long-term objectives for refrigerators

In general, Gram will work towards increasing implementation of life cycle assessments in product development. As a consequence of the focal points defined by previous environmental assessments, work will be directed particularly towards the following aspects:

Materials

- Choice of materials and disposal: recognition of the significance of choice and disposal of materials, both in terms of resources and the environment relative to the impacts of the manufacturing process as such, will be incorporated by the designers, i.e. choice and fate of materials in the product system will be central.
- Scarce resources: attempts will be made to reduce copper consumption in the longer term and, if possible, to replace it entirely with other materials. Alternatively, copper will be used in construction in a manner such that it becomes very easy to separate it from the other materials in the product. Attempts will also be made to avoid the use of stainless steel, or to place and label it such that its recycling in an appropriate facility is ensured.
- Plastics: attempts will be made to replace softened PVC with other materials. The production of ABS will be investigated further to identify potential problems.

Energy

Efficiency: energy consumption is also expected to be an important future competitive parameter. Attention will be paid to improving both the cooling system's performance and the insulation.

Chemicals

Foaming agents and refrigerants: attention will be paid to substitution of the greenhouse gas R134a, and alternatives will be introduced when deemed responsible relative to any inherent problems associated with their use.

Other

Working environment: attention will continue to be paid to the impact of manufacturing processes on the working environment.

Product specifications for refrigerators

Materials

Copper: select compressors with the lowest possible copper content. As far as possible, avoid copper in other components.

Stainless steel: should only be used where strictly necessary. The product's components will in any case be labelled and constructed so that stainless steel can be sent for recycling.

Primary aluminium: should be replaced by recycled aluminium or, where feasible, possibly steel.

Softened PVC: avoid unless strictly necessary.

Energy

The product must comply with the requirements of category A under EU energy labelling.

Chemicals

No CFCs may be used.

Table 17.3b. Examples of long-term environmental objectives and product specifications for refrigerators at Gram

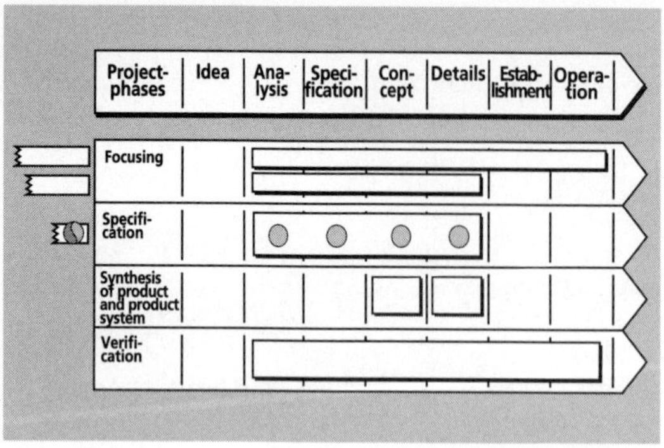

Procedure

The environmental specialist should be available when the environmental objectives are determined, and the specialist should help to ensure that the conclusions of the environmental assessment of references and the environmental diagnosis are interpreted and used correctly within the commercial framework.

When the objectives are determined as optimizing parameters, they will often be broken down and detailed throughout the course of development. At these times the environmental specialist should also be available to assist with the correct interpretation.

If the LCA is to enter into work with a product-oriented environmental policy and long-term goal definitions, it should be carried out independently of product development.

18. Environmental assessment of concepts

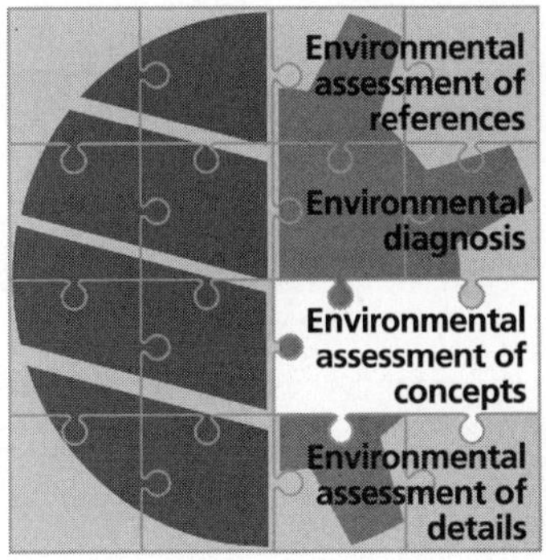

The purpose of being able to assess various product concepts is to be able to include environmental considerations early in development. The environmental assessment cannot wait until after the product's concept and structure have been determined, because then many of the most significant decisions have been made.

The task is thus to be able to support the choice of conceptually different solutions before details of how the solutions will be realized in materials and processes are known.

A major part of this task is solved with the aid of reference products for focusing and environmental specification as described in the previous sections. To a large extent, environmental specification concerns those product properties which are expected to be the most important from an environmental point of view for the new product, and a framework for the product's environmental characteristics is thus established by the specification.

The purpose of assessing solutions at the conceptual level is therefore to navigate and to optimize within the framework of the specifications.

Environmental questions

The most important questions to be answered by the environmental assessment in choice of product concept are set out in Table 18.1. As is evident, all questions here have a high element of probability: they always concern those characteristics which the solutions are *expected* to be able to represent within the latitude of the concept.

The latitude of the concept is often very large, and it would be wrong at this stage in product development to choose a given solution and model and assess all details in its product system. In all probability, this will not be the solution which is finally selected. A high degree of detail is, therefore, not required in the environmental assessment of concepts: what is required is a broad coverage of the solution space.

Examples of environmental assessment of product concepts

In many cases, a rough assessment within the MECO principle will be adequate to support the choice of concept. If necessary, the assessment can then be followed up gradually by a detailed environmental assessment as decisions on details are made. In other cases a solution can be simulated within the concept by a very rough modelling of the product system, and the environmental assessment can then be made. During the EDIP programme both options have been used, and in most cases a MECO assessment has proved sufficient.

Table 18.1. Questions to be answered in the environmental assessment of concepts

Questions to be answered in the LCA of concepts	
1	What differences from the reference product can be expected in the solutions for the chosen product concept, how great are they, and how broad is the solution space expected to be?
2	Within which of two alternative concepts can the most environmentally attractive solutions be expected to be found?
3	Is it possible within the concept to find solutions which can be expected to comply with the environmental specifications?

MECO assessments

Some examples of concept choices assessed in connection with product development in the

KEW Industri. High pressure cleaners

Concept choice: Automatic start/stop. The high pressure cleaner has a handle, where the operator can open and close the water flow to the nozzle. When the water flow is stopped the water is pumped at low pressure in an internal loop, a type of "idling".

To avoid the "idling", KEW have been working on a system that starts and stops the pump automatically, a type of hydraulic remote control. Energy consumption for idling is thus avoided, and so is mechanical wear on motor and pump during idling.

M	Practically unchanged
E	Approximately 10% reduction in the energy consumption in the use stage
C	Unchanged
O	Life increased approximately 5%. Reduced noise during use

Concept choice: Nozzle. KEW Industri have been working on a new nozzle design which in contrast to previous designs, where the jet is shaped by the mechanical design of the nozzle, utilizes a hydrodynamic principle to shape the jet. A MECO assessment of this concept shows:

M	Practically unchanged
E	Energy consumption reduced approximately 20-30% in the use stage
C	The increased cleaning efficiency of the jet is expected to replace chemical consumption to some extend
O	Lifetime of the nozzle is expected to be increased. Can not be quantified

Concept choice: Foaming system for chemicals. KEW Industri has developed a new dosage concept which produces a mixture of soap and water that foams considerably more than in previous models. The mixing is performed at the tip of the nozzle while air is added, resulting in a large amount foam. Due to the visibly increased production of foam, the consumer will reduce the amount of chemicals used for cleaning. At the same time, because the foam adheres better to the surface of the object being cleaned, the efficiency of the soap is increased. A MECO assessment showed the following changes compared to the reference product:

M	Practically unchanged
E	Practically unchanged
C	Consumption of chemicals reduced by approximately 50%
O	Unchanged

Grundfos. Pumps

Concept choice: Electronic controls. Grundfos are working with a system for electronic measuring and controlling of pumps that can result in large savings in energy consumed during operation. The electronic controls result in an increased consumption of materials and scarce resources for the electronics. Grundfos uses a so-called thick film technology. The MECO assessment is as follows:

M	Increased consumption of scarce resources
E	Up to 75% of the energy consumption can be saved in the use stage
K	Increased consumption during production of materials and manufacturing of electronics
A	Approximately unchanged

Table 18.2. Examples of a rough MECO assessment to support the choice of concepts

EDIP programme are shown in Table 18.2. The table illustrates the concept and describes the change to which the concept is assessed to be able to lead compared to the reference product. The purpose of the table is to show that the MECO assessments are often enough to decide whether the concept selected is environmentally attractive relative to the reference. Within the concept in question, solutions can then be sought which optimize the product environmentally, and environmental assessments can then, if feasible, enter into the choices of the product's structure, materials and processes.

Table 18.2 shows three conceptual choices in the development of high pressure cleaners at KEW Industri, where a MECO assessment shows unambiguous potential for improvement, i.e. there are no trade-off situations. The consequences of the changes of the product system are easily overviewed, i.e. it is clear what is changed, and that all else remains the same. For the pumps in the example from the development project at Grundfos there is a trade-off situation to be considered: the large potential for reduction in energy consumption in the use stage must be weighted against the extra resources consumed in the electronics and the possible impact of the consumption of chemicals in the manufacturing of the electronics.

The environmental assessment of Grundfos's reference product can assist in the considerations of these trade-offs. The assessment shows that energy consumption in the pump's use stage strongly dominates the weighted profiles, both for resource consumption and potential environmental impacts. This gives a feeling that the concept is environmentally attractive, but to be certain, the MECO assessment should be supplemented by an assessment of consumption of scarce resources in the electronics and the chemical consumption in the manufacturing processes.

Rough modelling and environmental assessment

A rough modelling of the product system followed by an environmental assessment is made for a concept which Danfoss has introduced in the electronics of an electro-hydraulic control unit for hydraulic valves. Danfoss used ordinary printed circuit technology, but wanted to convert to thick film technology. The modelling was carried out on the basis of a rough scope definition of the product system:

- Model resource consumption as the resource content of the materials, the printed circuit itself and the thick film.
- Disregard the manufacturing processes for electronic components.
- Model the application of conductor channels for both printed circuit and thick film.
- Take account of life span and energy consumption in the use stage.
- Model disposal.

On the basis of the environmental assessment of the reference product, which was based on a printed circuit, it was judged that this scope definition covered the most significant resource consumptions and impact potentials from the product.

The environmental assessment is shown in Figure 18.1. It is clear that there are unambiguous improvements for the thick film. At the same time, the number of electronic components is significantly smaller for the thick film, so the picture would not be altered if the manufacturing pro-

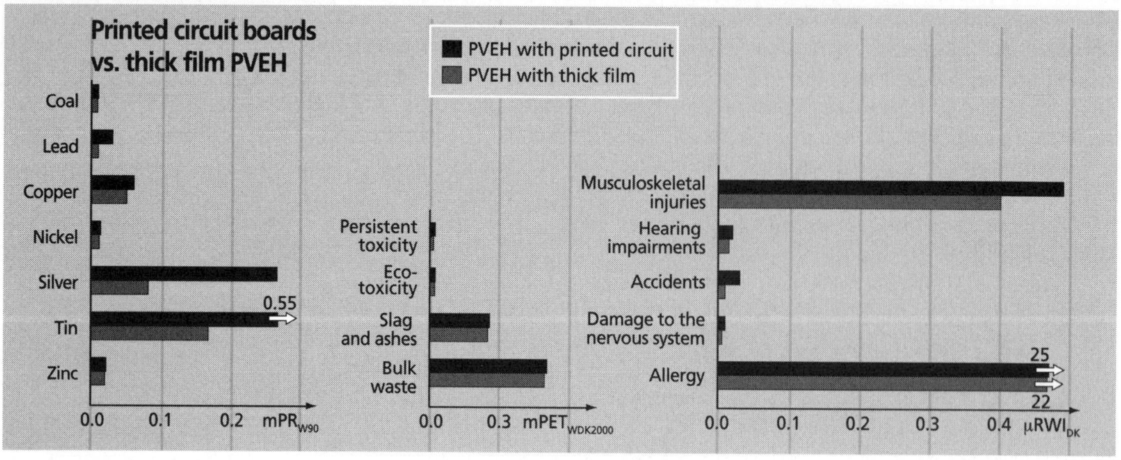

Figure 18.1. Example of an LCA of a concept on the basis of a rough modelling of the product system

cesses for these were included. Part of the improvement is attributable to the fact that the thick film is judged to alter the product's life from five to eight years.

Method

Environmental assessment in support of concept selection can be undertaken as MECO assessments. If the MECO assessment reveals a trade-off situation where gains and losses have to be weighed and the environmental preference is not immediately obvious, it is necessary to assess the concept on the basis of a rough modelling of the product system for a representative solution within the concept, followed by an environmental assessment of this. The requirements for scope definition of such modelling are:

Reference product

The original modelling of the reference product system is changed with respect to the part with which the concept is concerned. A rough modelling of the altered part of the product system is done for a representative solution within the concept in question.

Figure 18.2. The scope definition comprises a rough modelling of the difference between the reference product system and the system to be simulated

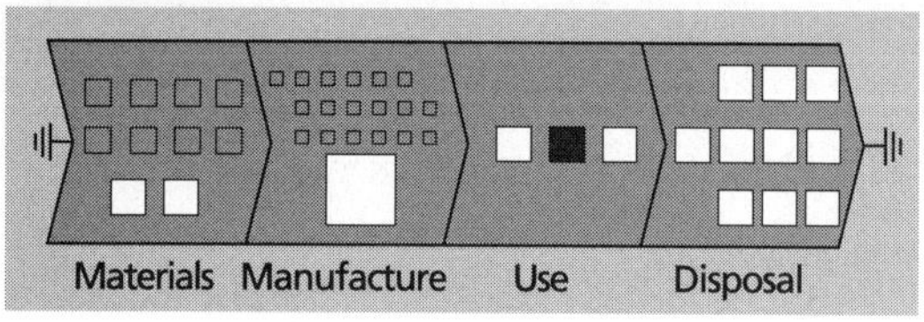

Scope definition of the product system

The modelling must be approximate and include only the most significant part of the product system (see Figure 18.2). As an example of guidelines for scope definition, some points applicable to electromechanical products are given here:

- Model material production as the product's contents, i.e. do not take account of wastage, but simply model the quantities existing in the product.

- Do not take account of energy consumption in the production processes, but only of their emissions of chemicals and also essential non-chemical-related working environment conditions. Model total chemical emissions and the impact on the working environment as one process.

- Do not include overheads.

- Model the use stage as an average for the markets (countries) which together comprise 80% of sales. Take account of the pattern of use and the product's life span.

- Model disposal on the basis of an average disposal scenario for the above countries.

- Disregard all transport.

Scope definition with respect to time and technology
The trend analysis and projection from the environmental assessment of the reference product are retained. If the concepts simulated are specific for particular materials or technologies, a trend analysis must be carried out specifically for these, and average observations as for the reference product not used.

Allocation
The system boundaries can be extended to the degree to which the changes in the surroundings caused by the new concept can be predicted. If, for example, it can be predicted with confidence that introduction of products with the new concept on to the market will lead to less incineration of waste, the system boundaries can be extended to include the energy production which will replace the energy utilized from incineration of waste. In most cases it will be difficult to predict the changes which are caused by products with the new concept coming on to the market, and allocation is used in such cases.

Procedure

The environmental assessment of concepts is carried out during product development in the concept design phase. Note that new concepts can involve processes which the reference product does not represent. There is no simple solution to generating information on new processes in the product system, but it is important to be aware of any lack of data as early as possible so that the data can be obtained.

Present the conclusions of the assessments in a brief report to the designers.

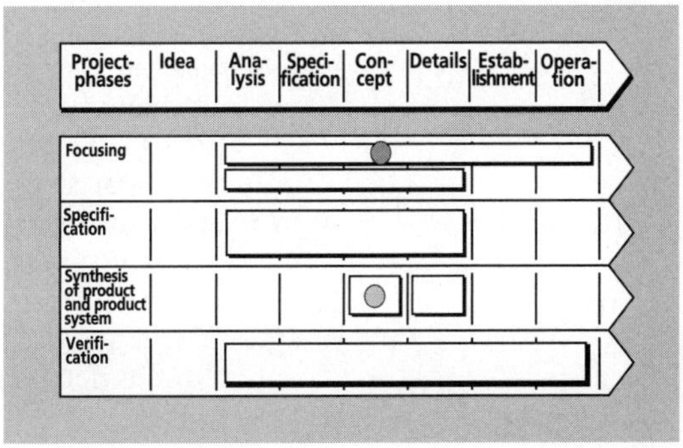

19. Environmental assessment of details

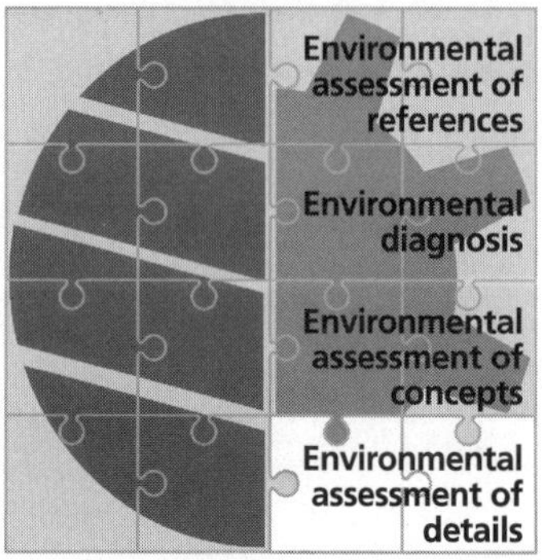

The objective of being able to assess detailed solutions during product development is to be able to navigate and optimize environmental considerations within the framework of the environmental specification and the selected concepts.

Environmental questions

The most important questions to be answered by the environmental assessment in the choice of detailed solutions are given in Table 19.1. Questions 1 and 2 in the table result in the same type of task, namely a comparative environmental assessment. This is the task of the environmental specialist. Question 3 concerns complying with the specifications. This is typically a direct comparison on product properties and does not result in an LCA. This is the task of the designer.

Questions in LCA of details	
1	What environmental differences does the solution represent compared to the reference product, and how great are they?
2	Which of two alternative design solutions is environmentally more attractive?
3	Does the solution comply with the environmental specification for the product?

Table 19.1. Questions to be answered during LCA of details

Examples of environmental assessment of detailed solutions

The task of the environmental assessment is to compare concrete detailed solutions:

- two or more different components,
- two or more different materials for the same component,
- two or more different processes or ancillary substances.

The total consequence of the change for the product system must be assessed, including the significance for efficiency in the use stage, life span and disposal. As the environmental assessment is a comparison of details, a significant degree of detail is required in the assessment.

In simple cases without trade-off situations, a MECO assessment will be sufficient, e.g. when replacing a compressor in a refrigerator with another with greater efficiency, but which otherwise has the same resource consumption and life span and the same characteristics with respect to disposal. In other cases an LCA is required.

During their development of the new television Beovision Avant, Bang & Olufsen carried out several LCAs of detailed solutions. Two of them are shown in Figure 19.1, namely a comparison of various types of materials for the TV's casing and a comparison of rear covers with and without painting.

The LCAs shown in Figures 19.1 and 19.2 are both Δ-LCAs, i.e. only those parts of the product system were evaluated which are *changed* in relation to the reference product system. As decisions on details often concern only one component or two alternative materials or processes as shown here for the television, a relatively small part of the product system will typically have to be assessed. This makes the task feasible and allows a detailed LCA to be carried out, as shown in Figure 19.1.

Δ-LCA in product development. The television Beovision Avant, Bang & Olufsen.

Choice of materials for the cabinet

In the development of the cabinet for this new television, Bang & Olufsen worked with several solutions, all expected to fulfil such demands as low cost, maintaining original style, small number of items, fast and easy assembling - and environmentally correct design.

LCAs were performed on cabinets made from the following materials:

- Cast polystyrene. Material consumption: 3.30 kg PS, of this 0.05 kg wasted.
- Wood (MDF sheet). MDF (Medium Density Fibreboard) is produced from finely ground wood chips mixed with glue and other additives. The consumption of glue was simulated in the LCA as plastic. Material consumption: 7.2 kg, of which 3.7 kg is wasted.
- Aluminium frame covered with wood. Consumption of MDF: 3.1 kg of which 1.0 kg is wasted. Aluminium consumption: 2.9 kg, of this 0.3 kg is wasted. (Assumption: 80% secondary aluminium and 20% primary aluminium).

Resource consumption and environmental impact potentials are illustrated in the figure below. Note that the profiles only show the product system for the front frame - the rest of the television is not included as it is unaffected by choice of cabinet. Paint is not included in the LCA as the same type and volume are used for the three solutions. It is assumed that the PS frame and the wood frame are incinerated after use and that the aluminium is recycled.

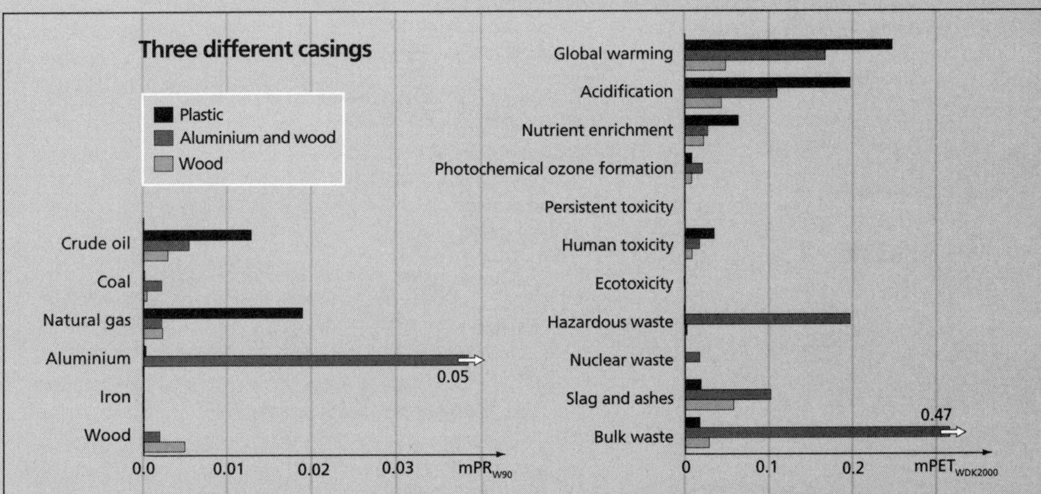

The figure illustrates resource consumption and the potential environmental impacts of the three solutions. It is evident that:

- Not unexpectedly, the aluminium cabinet consumes the most aluminium while the plastic cabinet consumes the most oil and natural gas. These consumptions are considerable relative to those for other resources.
- The wood consumption for the wooden cabinet is insignificant compared to that for the other cabinets, and in a total valuation, the wooden cabinet consumes fewest resources.
- Due to aluminium production, the aluminium cabinet contributes more to nuclear waste, slag and ashes and bulk waste than any of the other cabinets.
- The plastic cabinet contributes more to global warming, acidification and human toxicity than any of the other cabinets.
- The wooden cabinet contributes least to all impact categories except production of slag and ashes.

On this basis, Bang & Olufsen decided that a wooden cabinet is the most environmental friendly solution. The production of MDF wood results in emissions of formaldehyde and can result in problems in the working environment. This has been discussed with the manufacturer of the MDF in connection with selecting the MDF wood. The assessment is that this will not change the picture. The wooden cabinet solution was investigated further for other requirements, e.g. low cost and durability in different climatic conditions. As the wooden cabinet met these requirements, Bang & Olufsen chose this solution.

Figure 19.1. Example of Δ-LCA at Bang & Olufsen: choice of materials for the cabinet

Δ-LCA in product development. The Beovision Avant television from Bang & Olufsen

Painting of rear cover

Based on expected legislation regarding recycling of electronic products, Bang & Olufsen have been working on a concept to increase the possibilities for reusing materials in their products. Bang & Olufsen have therefore performed several experiments with reuse of plastics. These have shown that painted plastics result in recycled plastics with a lower quality as the paint lies as flakes in the plastic, causing problems with screwing into the plastic. One of the strategies in construction of the rear cover has been to avoid the use of paint.

The rear cover had to fulfil a number of demands regarding price, quality and assembly, and environmental considerations should also be included.

Attaining an acceptable surface proved to involve some problems. Several experiments were undertaken in co-operation with the Danish Institute of Technology before production methods and tools that could provide the required quality where found. The successful experiments led to the Beovision Avant being produced with an unpainted rear cover. The figure illustrates the resource and environmental profiles of a painted and an unpainted rear cover for a Beovision Avant. Note that the profiles show the total product system, but only for the rear cover - the rest of the television is not influenced by

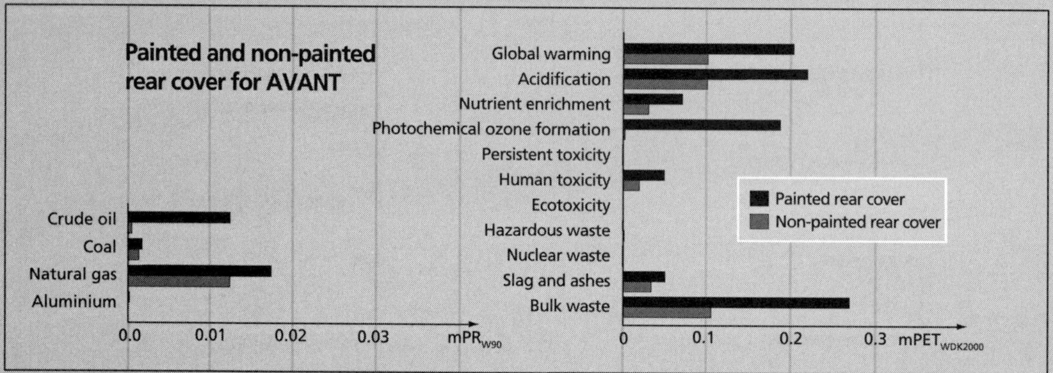

the choice. Without paint the rear cover can be recycled, and the recycling rate is estimated at 80% (the plastic manufacturer has stated that the plastic can be reused at least five times).

It is evident that consumption of oil and natural gas are the resource consumptions with the greatest reduction, as the reuse of plastics substitutes for these raw materials. In addition, energy consumption for production of new plastics is reduced. This can be seen in lower values for global warming, acidification, nutrient enrichment and the waste categories. The reduction in human toxicity and photochemical ozone formation is mainly due to the solvents in the paint not being present.

For Bang & Olufsen the changes led to savings in consumption of paint, but the tools became more expensive and the running-in more difficult, resulting in a longer period of running-in.

Figure 19.2. Examples of Δ-LCA at Bang & Olufsen: painting of the rear cover

The environmental assessment can be carried out as Δ-LCAs because the goal is to sub-optimize and support the specific decision and not to optimize the whole product. The general optimization lies partly in focusing and specification and partly in selection of concepts, and these phases of the development have been passed at the present stage.

Method

Environmental assessments in support of selection of details can be made as MECO assessments if there are no trade-offs. In case of trade-off situations where the preference cannot immediately be determined, the solution can be assessed on the basis of a Δ-LCA. A trade-off situation can occur if the solution in question comprises both deteriorations and improvements compared to the reference product or where a choice has to be made between two alternatives which both offer improvements, i.e. which improvement is the greater has to be decided.

The requirements for scope definition in a Δ-LCA are the following.

Object of the study
The object of study is the functional unit which describes the service to the user. If the change concerns only part of the product, it can be useful to define a functional unit for the sub-assembly in question as the object of study, e.g. the cabinet or the rear cover in the examples in Figure 19.1. But for all solutions, care must be taken to ensure that the general functional unit is unchanged, or the comparison with the reference will be based on false premises.

Reference product
The original modelling of the reference product system is changed for the part with which the detailed solution is concerned. The changed part of the product system is modelled in detail.

Scope definition of the product system
The modelling is delimited as follows:

- Model that part of the product system which is changed relative to the reference product (cf. Figure 19.3). The modelling should use the delimitation principles described for the method in Chapter 8 and in the book's Toolbox.

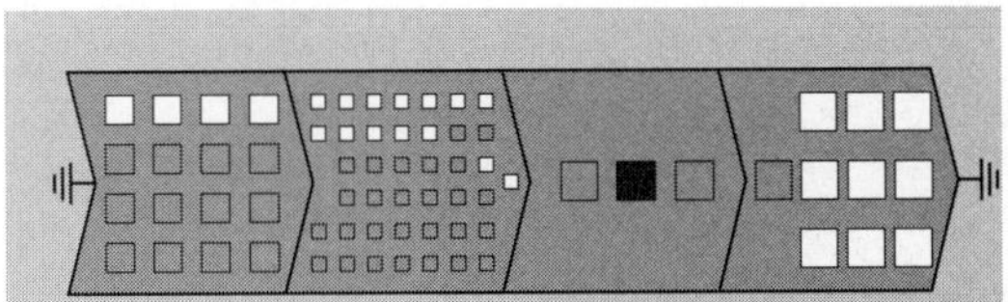

Figure 19.3. The scope definition comprises only the difference from the reference product system

Scope definition with respect to time and technology
Retain the trend analysis and the projection of the LCA of the reference product. If the details modelled are specific for particular suppliers, materials or technologies, a trend analysis must be carried out specifically for these, and average observations as for the reference product are not used.

Allocation
The system boundaries can be extended to the degree that changes in the surroundings can be predicted. In most cases it will be difficult to predict which changes are caused by products with the new solution, and in such cases a simple truncation of the product system in accordance with the EDIP allocation method is used.

Procedure

The environmental assessment of detailed solutions is carried out during product development, in the detail design phase. Note that new details can result in parts of the product system where there are no data from the reference product, so that data must be collected as early as possible.

Present the conclusions of the assessments in a brief report to the designer.

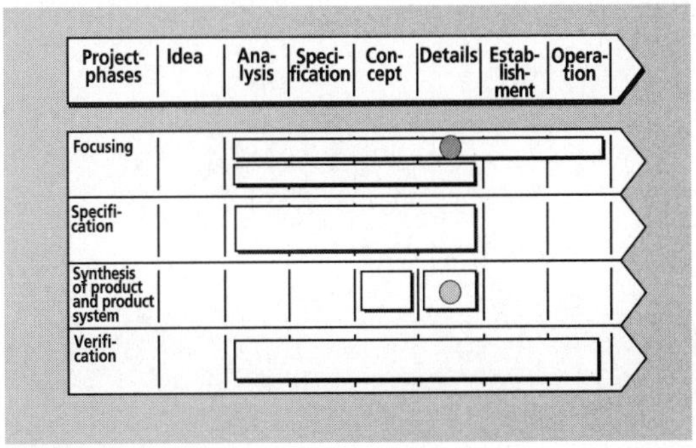

Part IV. The Toolbox

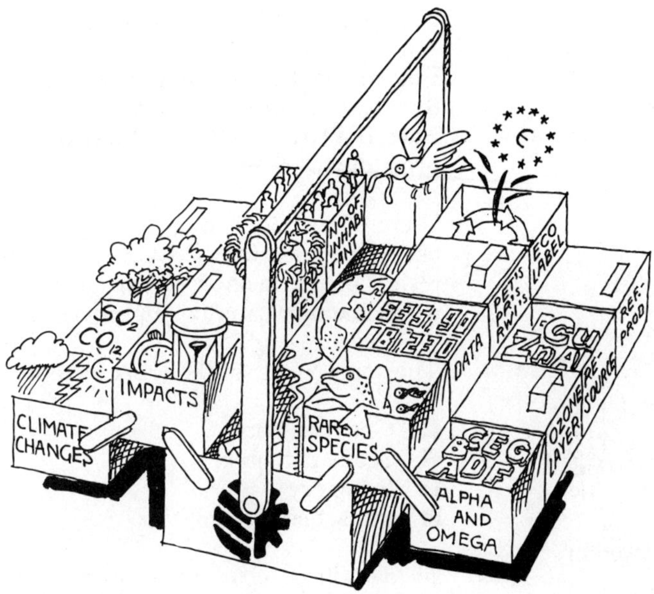

- *What is the procedure, step by step?*
- *How is the task delimited?*
- *What data must be used and how are they obtained?*
- *Which assessment factors and references must be used?*

The Toolbox is constructed as a handbook which provides detailed guidelines in carrying out an environmental assessment of a product. It contains a stepwise review of the EDIP method's elements and provides concrete instructions and tools each time a task is to be carried out or a selection made. The Toolbox is a "recipe book" in the environmental assessment of products.

The instructions are explained and elaborated in the text, but the review in the Toolbox has been purposely kept free of discussions of the selections made and of documentation for why such selections should be made. The Toolbox user will find these aspects covered in the review of the method in Part II of the book or in the more extensive notes in "Scientific Background for Environmental Assessment of Industrial Products" (Hauschild and Wenzel, 1997g).

Before the Toolbox is opened and the tools used for the first time, the user should therefore read and understand the description in Part II of the EDIP method and of the choices made during the procedure.

Users of the EDIP method should draw on the Toolbox in subsequent use of the book for the LCA tasks. Together with Part I and Part III, the review of the method in Part II will function as background knowledge which the user can draw upon via the book's index.

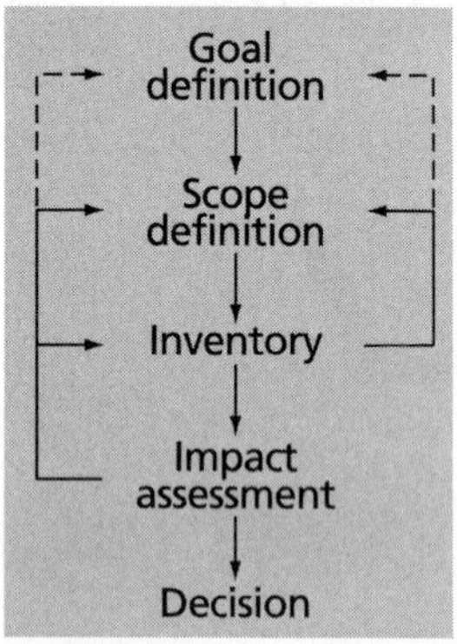

The components of the LCA method

- Goal definition
- Scope definition
- Inventory
- Impact assessment

are described separately to provide a systematic review of the method. This should not be taken as indicating that the four components are carried out in a linear sequence. Environmental assessment of products is an iterative process where some of the steps are carried out more than once as a consequence of findings made later in the sequence.

After defining the goal of the LCA a scope definition of the product system is carried out, and subsequently an inventory of the environmental exchanges from the processes involved is carried out. Work on the inventory can provide new insights into the actual appearance of the product system, and thus provide input to the scope definition which in turn may have to be modified in some points in the light of these findings.

After the inventory, an assessment of the significance of the environmental exchanges for the selected assessment criteria is done, followed by a sensitivity analysis. This makes it possible to identify the processes and assumptions of greatest significance for the results of the LCA, and therefore provide a better basis for making the final scope definition of the product system and for focusing data collection under the inventory phase on the most significant environmental exchanges.

The Toolbox compartments

20. Procedure for goal definition

- Determine what the LCA can and must be used for.
- Consider what the LCA cannot and must not be used for.
- Consider who will use the LCA.
- Consider the decisions which the LCA can and must support, and those which it cannot and must not support.
- Consider the scope of the changes to which the decisions can lead:
  - What sales figures can be expected for any new products in which the decisions are incorporated?
  - Is the new product expected to supersede other products? Which products, and to what extent?
  - What changes in the community's activities are expected from the new product, i.e. which processes will be superseded, which newly introduced, and what is the scope of these changes?

The last point is especially relevant to the allocation. It can only be estimated roughly at the time of goal definition and must be returned to later in the LCA, when adequate information has been established for the product system of the reference product.

The four different applications of the LCA in product development are:

Environmental assessment of references	Environmental diagnosis	Environmental assessment of concepts	Environmental assessment of details

Table 20.1. The four applications of the LCA in product development

The LCA depends mainly on application with respect to the scope definition, especially scope definition of the product system, technological scope definition and allocation.

20.1 Presentation of goal definition

- Present the goal definition together with the results of the LCA, with emphasis on a description of what the LCA can and cannot be used for.

Note that the EDIP method does not describe the requirements for external use of the LCA, i.e. use as environmental documentation for the product outside the company for marketing etc. The EDIP method is desirable for use in product development. All of the EDIP method's elements can be used in an LCA for external use, but a higher degree of data documentation in the presentation of results than described in the EDIP method is required, possibly followed by an external review process.

21. Procedure for scope definition

21.1 The object of the LCA

21.1.1 The functional unit

- Define the product's primary service for the user.
  If the assessment covers a sub-assembly within the product, define the service of this too.
- Define the functional unit for the product and, if relevant, for the sub-assembly.

A form as illustrated below may be helpful:

Table 21.1. Form for description of the product's functional unit

Service
Description:
Functional unit
Quantity:
Duration:
Significant qualities:

In principle, the object of the study is always the product's primary service for the final user. If only a sub-assembly of the product is assessed, e.g. the picture tube in a television, a comparison of alternatives must make it probable that the primary service is unchanged.

Duration of the service:

- Set the duration of the service to a time span defined as calendar time. As a starting point, it is recommended that duration be set to the product's life span.
- Define the product's operating time if an active product is involved, i.e. a product requiring input or creating output during operation.
- If durations other than the product's total life span are used, all environmental exchanges from the product system, apart from those attributable to the use process, must be multiplied by the ratio between the selected duration and the life span when the exchanges are calculated per functional unit. This procedure is described in Section 22.5.3. in the Toolbox.

Comparisons must be made on the basis of the same functional unit, i.e.:

- The duration of the service must be the same for all alternatives compared.
- The quantitative and the qualitative characteristics of the alternative ways of providing the service must be seen by the user of the product as being comparable.

21.1.2 Reference products

- Select one or more reference products to represent a way of providing the service and carry out the LCA on this/these reference product(s).

The reference product can be one of the company's own products, e.g. the existing generation of the product which is due for product development. It can also be a competing product or a contemplated product composed of existing parts and known technologies.

The essential point is that the reference product must form the basis on which the company wishes to build the LCA for use in product development, or the basis on which the company wishes to determine environmental policies and long-term objectives if the LCA is to be so used.

The reference product specifies the product system for which data will subsequently be collected. Planning of the data, i.e. selection of data references, is carried out when the product system is delimited. This is described under "Procedure for the inventory", Section 22.1 in the Toolbox.

21.2 Assessment criteria

Part of the basic concept for LCA of products is that it must be a holistic assessment. The method's point of departure is therefore that all significant resource consumption, environmental impacts and impacts on the working environment must be covered.

- Include all resource types, environmental impact categories and categories of impacts on the working environment implemented in the EDIP method.
- Consider whether some of the resource consumption and impact categories which the EDIP method has not implemented are significant for the product.
- If so, include such resource consumption and impact categories. Describe and assess them qualitatively, and include them in the data on which the decision will be based, together with the quantitative assessments of the impact categories implemented in the EDIP method.

Use Table 21.2 as the basis for consideration. The table shows the categories of resource consumption and impacts which SETAC's European working group on impact assessment in LCA notes at the end of 1996 as a guiding list for impact assessment, together with the impact categories which are implemented in the EDIP method.

The impact categories included as assessment criteria specify the information to be collected in the inventory. The choice of assessment criteria should therefore be made before the inventory begins.

21.2.1 Environment

The environmental impact categories implemented in the EDIP method are all results from emissions of chemicals. The method does not include a basis of assessment for smell and physical environmental effects such as noise, alteration of habitats and destruction of landscapes (see Table 21.2). But like SETAC, the EDIP method recommends that these be included if they are significant for the product. They can be included in parallel with the other criteria within the pattern which in general characterizes the method.

21.2.2 Resources

As noted previously, the EDIP method includes renewable resources in the inventory on an equal footing with non-renewable resources, even if consumption does not exceed regeneration for the resource in question. Renewable resources are included because they must enter into the total optimization of the product.

21.2.3 Working environment

Effects on the working environment are of a highly diverse nature, and the inventory requires collection of highly varied types of data. For the working environment in particular, assessment criteria must therefore be known before the inventory. Table 21.2 shows the impact categories for which the EDIP method has developed a basis of assessment. These impacts were selected because they represent the most frequently occurring working environment problems in Denmark and because electromechanical industrial products were the case products within the EDIP programme. For other types of products, e.g. foodstuffs, it can be relevant to include other types of impact categories.

Impact category	Guiding list from SETAC *	Operationalized in EDIP	Include	Resource consumption	Guiding list from SETAC *	Operationalized in EDIP	Include
Environmental impacts				**Resource consumption**			
Global							
Global warming	✓	✓		*Renewable*		✓	
Ozone depletion	✓	✓		Wood	✓	✓	
				Plants	✓		
Regional				Animals	✓		
Photochemical ozone formation	✓	✓		Groundwater	✓	✓	
Acidification	✓	✓		Water (hydro power)	✓		
Nutrient enrichment	✓	✓		Sun	✓		
Human toxicity (persistent)	✓	✓		Wind	✓		
- through water		✓		Rivers	✓		
- through soil		✓		Tidal water	✓		
- through groundwater		✓					
Ecotoxicity (persistent)	✓	✓		Add more:			
- water		✓					
- soil		✓					
				Non-renewable			
Local				Energy	✓	✓	
Human toxicity	(✓)	✓		Oil	✓	✓	
- through air		✓		Coal	✓	✓	
Ecotoxicity	(✓)	✓		Brown coal	✓	✓	
- water		✓		Natural gas	✓	✓	
Odour	✓			Peat	✓		
Physical impacts				Minerals	✓	✓	
- noise	✓			Clay	✓		
- destruction of landscapes	✓			Sand	✓		
- division of habitats	✓			Aluminium	✓	✓	
- radiation	✓			Antimony	✓	✓	
- accidents	✓			Beryllium	✓	✓	
				Lead	✓	✓	
				Cadmium	✓	✓	
Impacts on the working environment				Cerium	✓	✓	
Physical impacts				Gold	✓	✓	
Noise		✓		Iron	✓	✓	
Radiation				Copper	✓	✓	
Vibrations				Cobalt	✓	✓	
Thermal impacts				Mercury	✓	✓	
				Lanthanum	✓	✓	
Musculoskeletal impacts				Manganese	✓	✓	
Lifting				Molybdenum	✓	✓	
Monotonous repetitive work		✓		Nickel	✓	✓	
Poor working positions				Palladium	✓	✓	
				Platinum	✓	✓	
				Silver	✓	✓	
Chemical impacts				Tantalum	✓	✓	
Inhalation of hazardous substances		✓		Tin	✓	✓	
Skin contact with harzardous substances		✓		Uranium	✓	✓	
				Zinc	✓	✓	
Biological impacts							
Inhalation of biological material				Land	✓		
Skin contact with biol. material							
				Add more:			
Psycho-social factors							
Working accidents		✓					

Table 21.2. List of assessment parameters which it can be relevant to include in LCA of products, and list of parameters implemented in the EDIP method.
** From SETAC's working group on the impact assessment phase of the LCA, September 1996 (Udo de Haes, 1996).*

21.3 Scope definition of the product system

An overview of the product system is first established. This is divided into two parts, namely the product system to and including production, and the product system for use and disposal.

21.3.1 Model for the product system to and including product manufacturing

- Establish a model for the product system to and including product manufacturing.
  - Use a simple flow diagram of the product system's processes if the product is very simple.
  - Use a hierarchical process tree in table form if the product is complex with several sub-assemblies and materials.
  - If the reference product is one of the company's own products, use an existing parts list as the starting point if one exists.
  - Transform the parts list to include the information required for scope definition.
- Aggregate small items into joint groups of items of the same material before defining the scope.
- The above points should be carried out for the part of the product system and in the degree of detail required by the application (cf. description in Part III of the book).
- Set up the product system in the model used for subsequent calculations from the start. If the parts list is entered in a computer program, use the structure used in the tool in question.

The model should include the information shown in Table 21.3.

Table 21.3. Structure and information in modelling of the product system to and including product manufacturing

Item			Material						Manufacturing processes						Ancillary substances					
ID No.	Number	Name	Name	Quantity Unit	μ	σ/μ	D	⏚	Name	Volume Unit	μ	σ/μ	D	Incl. in model	Name	Quantity Unit	μ	σ/μ	D	⏚

Legend:
μ = Mean, σ/μ = Coefficient of variation, D = Distribution function
⏚ = Column for marking the scope, cf. Section 21.3.3
"Incl. in model" = Column for marking the scope, cf. Section 21.3.3

Procedure for setting up the model on the parts list form:

- The information entering into the modelling of the product to and including product manufacturing is as follows:
- Item no.: specify an unambiguous number which describes the item's place in the product hierarchy.
- Number of items: specify the number of units of the item in question. Identical or practically identical components are described once only. Enter the number in the column as a multiplication factor.
- Name of item: specify a name, which should be unambiguous even if the number is also unambiguous.
- Name of material: specify the material(s) comprising the item by an unambiguous name and/or ID number which can be used later as a search and sorting criterion.
- Material unit: specify the key unit for the material. Use grams as the standard unit.
- Material consumption: specify a mean value for consumption of material in the product and an estimate for the coefficient of variation. Assume as default that the consumption follows a log normal distribution.
- Name of manufacturing process: specify the process(es) involved in the item's production by an unambiguous name and/or ID number which can be used later as a search and sorting criterion. In principle, all processes occurring from the raw material received at the company or the subcontractor to final assembly and packaging must be specified, i.e. the existing parts list must be extended by the processes occurring at the subcontractor's. If it is impossible to obtain information on the processes, specify the best estimate or simply specify "subcontractor" in the column for processes.
- Manufacturing process unit: specify the key unit for the process.
- Manufacturing process volume: specify as an average value the volume of the process involved as a number of key units for the process. Specify an estimate of the coefficient of variation. Assume as default that the process volume follows a log normal distribution.
- Name of ancillary substance: specify the ancillary substance(s) entering into the process by an unambiguous name and/or ID number which can be used later as a search and sorting criterion. The column should include all ancillary substances. Specify the substances next to the process in which they are used.
- Unit of ancillary substance: specify the key unit for the ancillary substance. Use grams as the standard unit.
- Consumption of ancillary substances: specify as average value the consumption of ancillary substances entering into the product and an estimate of the coefficient of variation. The default assumption is that the consumption follows a log normal distribution.

21.3.2 Model for the product's use and disposal

■ Set up a model for the product's use and disposal on the same form as used for the preceding part of the product system.

Table 21.4. Structure and information in modelling of the product's use and disposal

The model should include all information in Table 21.4.

					Process						Ancillary substance					
							Volume						Quantity			
No.	Item	Locality	Material/ sub-assembly	Frac- tion	Name	Unit	μ	σ/μ	D	Incl. in model	Name	Unit	μ	σ/μ	D	⏚

Legend:
μ = Mean, σ/μ = Coefficient of variation, D = Distribution function
⏚ = Column for marking the scope, cf. Section 21.3.3
"Incl. in model" = Column for marking the scope, cf. Section 21.3.3

Procedure for setting up the model on the parts list form:

- The information entering into the modelling of the product's use and disposal is as follows:
- Location: specify an unambiguous number and name for the location in question, e.g. classify by country or use average scenarios, e.g. "EU".
- Item fraction: specify as a fraction the percentage of the total number of products sold which run through the process in question. If, for example, 20% of the product is sold in Denmark, the fraction of the use process in Denmark is 0.2.
- Item name: specify the name of the item. For products in the use stage it is typically the whole product, but for the disposal stage it can be a sub-assembly, e.g. "picture tube", if disposal is separate, or a material fraction, e.g. "steel", if materials follow their own disposal route, as is often the case.
- Process name: specify the processes which the item enters into by an unambiguous name and/or ID number which can be used later as a search and sorting criterion. In principle, all processes occurring during use and disposal must be specified.
- Process unit: specify the key unit for the process.
- Process volume: specify as an average value the volume of the process in question as a number of key units for the process. Specify the full volume for the item running through the process as the percentage occurrence of the process in the product system is already specified under fraction. Multiplying the fraction and volume will thus provide the correct description for the complete product system. Specify an estimate of the coefficient of variation and begin by assuming that the process volume involved follows a log normal distribution.
- Name of ancillary substance: specify the ancillary substance(s) entering into the process by an unambiguous name and/or ID number which can be used later as a search and sorting criterion. The column should include all ancillary substances. Specify the substances next to the process in which they are used.
- Unit of ancillary substance: specify the key unit for the substance. Use grams as the standard unit.
- Consumption of ancillary substances: specify as an average value the consumption of ancillary substances entering into the product and an estimate of the coefficient of variation. The default assumption is that the consumption follows a log normal distribution.

21.3.3 Criteria for scope definition

- Go through all material consumption, processes and ancillary substances item by item, and decide on the following:
  - For the material: must the product systyem of the material be included back to the point of raw material extraction?
  - For the process: is it to be included?
- For ancillary substances: must the product system of the substance be included back to the point of raw material extraction?
- If the product system is set up in the parts list form as shown in Tables 21.3 and 21.4, use the tables' columns designed for noting ✓ or - to indicate the scope definition.
- Use the following four considerations as a basis for scope definition:
  - Material consumption. Includes materials' contents of energy, chemicals and resources.
  - The processes' energy consumption. Covers energy resources and energy-related environmental impacts.
  - The processes' consumption and emission of chemicals. Covers especially toxicity and photochemical smog, and in special cases stratospheric ozone depletion.
  - Others, especially impacts on the working environment not covered by the other three groups and physical environmental impacts.
- Include the material, the process or the ancillary substances in the scope of the LCA if one or more of the four considerations requires it.

Criteria for scope definition for materials:

- Include the materials' product system from extraction of raw materials to and including production of materials for the heaviest items. Start by including the heaviest, then the second heaviest, and continue until x% of the material consumption is included on a weight basis. If x is 95, the product system to and including production of materials thus excludes 5% of the product on a weight basis.
- Check whether the omitted materials contains scarce resources which are not proportionally represented in the x%, and if so include the materials in question anyway.
- Check whether particularly energy-intensive materials are included in the (100-x)% omitted, and include them in any case so that approximately x% of the materials' energy consumption is also included.
- Check whether the (100-x)% of the materials hereafter omitted involve special chemicals as ingredients or during production which can contribute significantly to problems in the working environment, ecotoxicity or human toxicity, or other environmental problems specific to the chemical in question. In such cases include these materials if they are not already proportionally represented.

- Check whether the materials hereafter omitted contribute significantly to problems in the working environment such as damage to the musculoskeletal system, noise or work accidents or to other impact categories which are not covered to a greater extent by the items already included, e.g. physical damage to the landscape or effects of radiation. If so, include them.

Criteria for scope definition for processes:

- Include the most energy-intensive production processes so that an estimated y% of the total energy consumption in the product manufacturing stage is included.
- Include the processes in the use stage if the product consumes energy, and always include the processes in the disposal stage. Include the use and disposal processes for the countries which together comprise z% of the market for the product, and then extrapolate these countries to 100% by increasing their relative proportions. If desired, model use and disposal stages for the individual country or for an average scenario, e.g. the EU average.
- Include overhead consumption of energy, i.e. the consumption for heating, lighting, shared ventilation, etc.
- Include all manufacturing processes, use processes and disposal processes which involve environmentally harmful chemicals and/or emissions. The hazard assessment of chemicals is described in Section 23.3 of the Toolbox. Include the process in cases of doubt. In many cases it is feasible to include all processes involving chemicals, so that the assessment of hazardous chemicals can be postponed to the later impact assessment phase.
- Check whether the processes of the product system hereby omitted contribute significantly to problems in the working environment such as musculo-skeletal damage, noise or work accidents or other impact categories not covered by the processes already included in the product system, e.g. physical damage to the landscape or danger of radiation.

Criteria for scope definition for ancillary substances:

- Include the substance's product system from extraction of raw materials to and including production of the substance if it is significant in terms of quantity relative to the product's weight or if they contain scarce resources not already represented to a significantly greater extent.
- Include the substance's product system from extraction of raw materials to and including production of the substance if there are special emissions of environmentally hazardous substances not represented in much larger quantities in the product manufacturing process in which the substance is used. By far the greatest emissions will, however, often occur in the process in which the substance is used.
- Check whether the parts of the ancillary substance's product system hereby omitted contribute significantly to problems in the working environment, such as damage to the musculoskeletal system, noise or work accidents, or to other impact categories which are not covered by the model of the product system thus established, e.g. physical damage to the landscape or effects of radiation. If so, include these parts anyway.

The company can elect to fix x, y and z as required by the goal of the actual LCA. The guidelines in the box below are recommended as a starting point.

■ $x > 95, y > 80, z > 80$

As is evident, the scope definition requires knowledge of the energy consumption for production of materials, the materials' calorific value, the processes' energy consumption, the scarceness of the resources and the chemicals' level of environmental hazard. This knowledge is gradually established as fingertip knowledge as the environmental specialist gains experience. Until then there is no alternative to estimating and seeking specialist advice. In cases of doubt it is preferable to include too many materials, processes and ancillary substances than too few, thus establishing the requisite information during the LCA for the benefit of subsequent scope definitions.

If it becomes clear during subsequent gathering of information that some of the parts of the product system previously excluded should be included, the scope definition is of course adjusted accordingly.

21.3.4 Scope definition in the computer tool

In practice the calculations will be done on a computer as noted previously, and the setting up of the product's parts list in this tool comprises the basis for the modelling of the product system. The task consists of importing the parts list as it is, and then deleting superfluous items of information and adding others as required. In practice the scope definition is then done in the tool.

In a situation where the basic data have been established, the materials, processes and ancillary substances modelled in the parts list specify unit processes in the tool's underlying database. As described in Sections 21.3.1 and 21.3.2 this specification is always expressed as a number of key units of the unit process. When the parts list is specified and delimited according to the scope definition, the resulting model of the product system is ready for computation. The computation itself is automated in the program for both aggregations in the inventory and the calculations in the impact assessment phase.

In a situation where the data or parts thereof have not yet been established, the unit processes are simply established as "empty" processes without data, which are subsequently gradually filled in with data as they are obtained.

The EDIP method's PC program (Danish Environmental Protection Agency, 1997) is based on the product's parts list structure and builds on an underlying database. Scope definition can be done in this tool as described in brief here.

21.4 Definition of time scope

- Determine the period of time within which the decisions which the LCA has to support must apply.
- Differentiate the period of time into the various stages of the product life which the LCA concerns or for which it acts as a reference.

A table as below can be used:

Table 21.5. Form for defining time scope

	Start	Finish
	year	year
Product development		
Extraction of raw materials + production of materials		
Manufacture		
Use		
Disposal		

The time scale over which the various types of environmental impacts must be seen is determined in Chapter 23 of the Toolbox and it is therefore not described here.

21.5 Definition of technological scope

- Designate the most important environmental exchanges, processes and distribution assumptions for the reference product system. See Chapter 24 on sensitivity analysis.

The most important exchanges, processes and distribution assumptions are:

- the most important processes from an environmental point of view in the product system of the reference product and the most important exchanges from these processes,
- the most environmentally significant distribution assumptions for the product system, e.g. patterns of use, distribution of sales and distribution of disposal for the product.

Data for the most significant exchanges and distribution assumptions are called the LCA's *key figures.*

- Carry out a trend analysis on the environmental key figures in the product system of the reference product.
- Project the LCA by changing the key figures in the modelling of the reference product and repeat the calculations in the LCA.

In principle, this point in the scope definition can only be done when the LCA has been carried out on current data, as the most significant aspects can only be designated then. After the LCA on current data, a sensitivity analysis is carried out as a described in Chapter 24 of the Toolbox, and the key figures are designated.

To save time for both the provider of data and the person carrying out the LCA, it is, however, a major advantage to collect data on trends together with the other data. Inclusion of data on trends is therefore recommended for the process's most important parameters from the start, so that contact with data providers, e.g. the operator of the process and experts in the process, as far as possible need only occur once. After the sensitivity analysis, check whether it has been possible to obtain information on trends for the most important key figures, and if necessary return to the trend analysis to adjust it.

Examples of processes and distribution assumptions which are often significant in a product system are given in Figure 8.5 in Chapter 8. In most cases an available knowledge of the development over the last 10-15 years exists, and it is often possible to project the trend which will apply during the next few years.

It is recommended that advice be obtained from experts within the actual industrial processes and/or experts among the authorities who can assess the societal developments in the energy sector, the transport sector and the disposal sector.

21.6 Allocation

The question of allocation arises in two different situations, namely:

- In the case of co-products, i.e. when two or more products arise as output from a process or enter into a process, and the process thus contributes to more than one service for a user.
- In case of recycling or reuse of a material or a sub-assembly in a product, i.e. when the material or the sub-assembly enters into a new product and thus contributes to one or more other services for a user.

All cases where processes in a product system contribute to more than one service fall into one of these two categories. Whether it is necessary to allocate depends on the individual case. In both situations the following general ranking applies:

I. Attempt to avoid allocation
II. Allocate

21.6.1 Attempt to avoid allocation

There are two ways to avoid allocation, both of which should be tried before a decision on allocation is made, namely:

- identify true joint processes and environmental exchanges and eliminate false ones,
- extend the system boundaries.

21.6.1.1 Identify true joint processes and exchanges

The first step in avoiding allocation is to identify processes and exchanges which are in common only at first glance, but which on closer analysis can be assigned unambiguously to one product.

In general:

- Clarify cause-effect relationships and partition the exchanges on the basis of them, including:
- Subdivide the process into smaller units if it is possible to obtain data for a subdivision, and if doing so will make it possible to identify an unambiguous relationship to one product only and thereby avoid allocation.
- Examine each individual exchange in the process and identify whether it is a joint exchange or an exchange which can be unambiguously ascribed to an individual product.

True joint processes and exchanges which cannot be ascribed to a single product will remain. This exercise is undertaken for each of the two categories: co-products and recycling.

Identification of joint processes for co-products:

- Identify true joint processes and exchanges. A true joint process is distinguished by the fact that the process volume included in the model of the product system is for the benefit of two or more products.

For the example on utilization of production waste in Figure 8.10, it is evident that the processes from oil extraction to and including polymerization are for the benefit of both product 1 and product 2. On the other hand, the injection moulding itself and the preceding transport for product 1 are for product 1 only. Work through corresponding considerations for all of the examples in the section on co-product allocation in Chapter 8.

In most cases, all exchanges are in common if the process is in common. However, the example in Figure 8.11 on incineration of waste illustrates a process where this is not the case. The example shows how an assessment is made for each individual exchange on whether it must be partitioned or ascribed to an individual product.

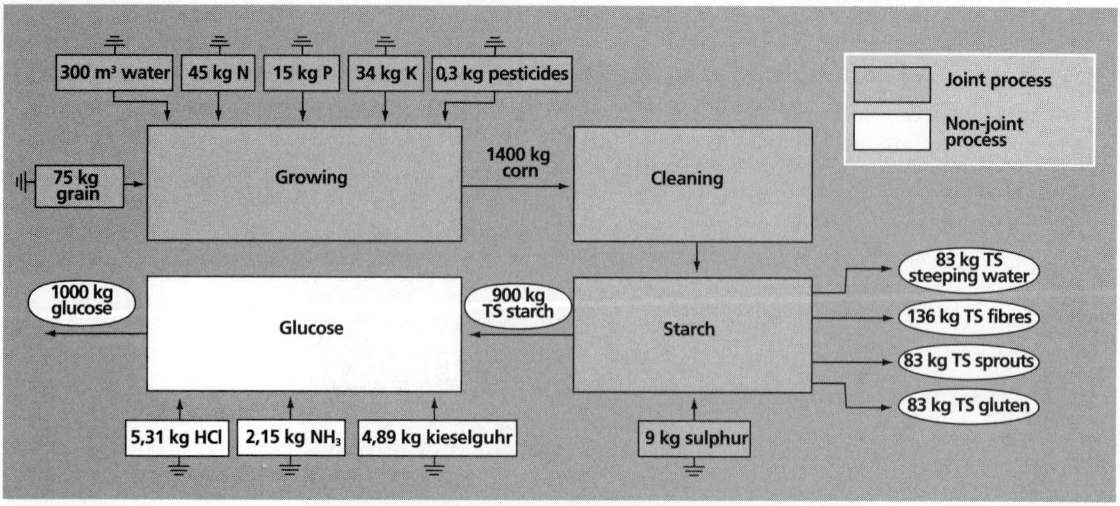

Example 21.1. Identification of joint processes for co-products in the production of glucose from maize (Frandsen & Hansen, 1995)

Identification of joint processes in recycling:

- Identify true joint processes and exchanges. A true joint process is distinguished by the fact that the volume of the process included in the model of the product system is for the benefit of two or more products.

As example 7 in Figure 8.14 shows, the processes from ore extraction to and including production of steel are for the benefit of both the product which uses new materials and the product which uses recycled materials. On the other hand, the product manufacturing processes themselves and the preceding transport are for one product only. Work through corresponding considerations for all of the examples on recycling allocation in Chapter 8.

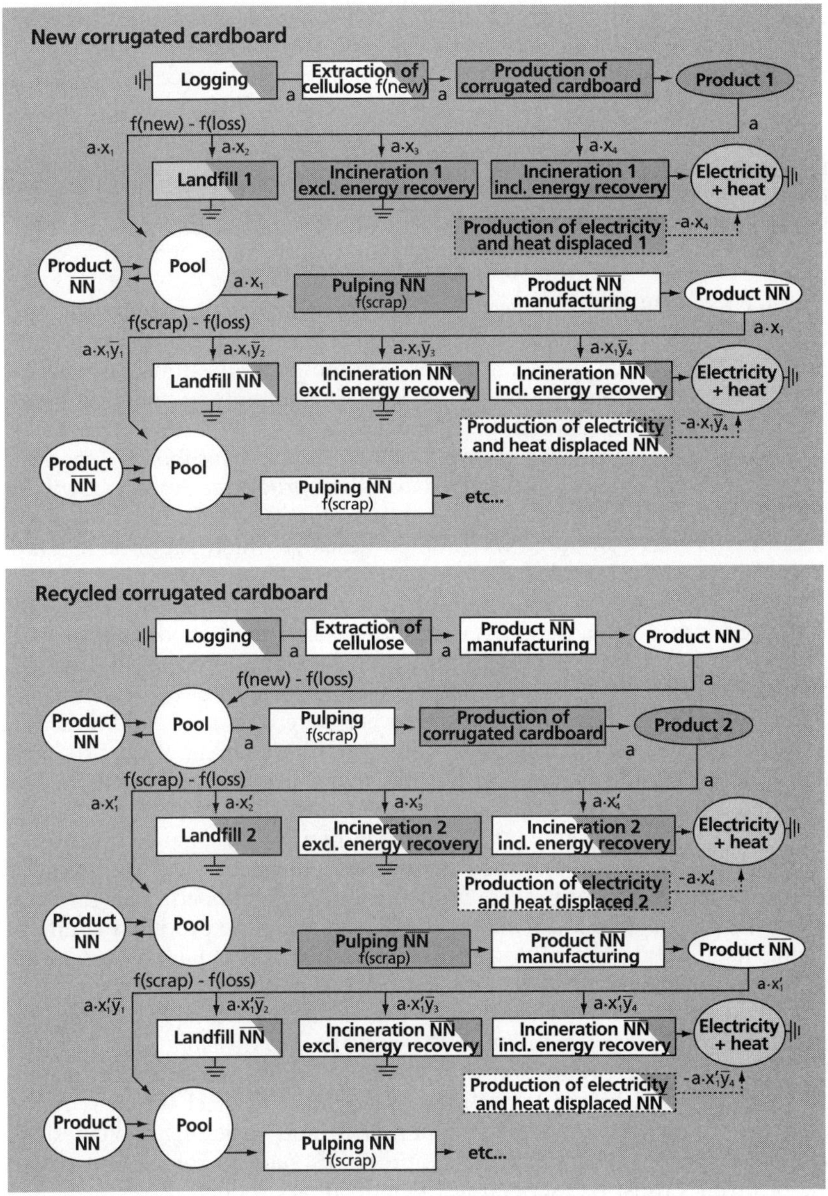

Example 21.2. Identification of joint processes in the recycling of corrugated cardboard.

In the example 21.1 for corrugated cardboard, an expansion of the system is selected for the heat utilized in incineration of waste, so that the heat does not enter as a product in the allocation.

In the figure, the joint processes are designated as those occurring only once in the recycling cascade, namely those being responsible for the terminal input and those which are responsible for the terminal disposal of the material. This implies that the user who sends the material for recycling must also be ascribed a part of the terminal disposals which are unavoidable when the material can no longer be used, namely the part corresponding to the loss of utilization value via use and recovery. The magnitude of this part is examined in Section 21.6.2.1.

Note that the recovery process itself is not designated as a joint process. The EDIP method considers it as belonging to product 1 because user 1 is responsible for the grade of the material until it is handed on to user 2 with the material grade with which it is passed on. The material is ready to be passed on to the next user only at the end of the recovery process, partly because it must be passed on in the same state in which user 1 received it from the joint processes of primary material production, and partly because a major portion of the grade loss lies in the recovery process itself. This allocation method is examined in more detail in Wenzel (1997).

The figure illustrates an example of recycling via an open pool, the most common form of material recycling. There is a very large number of different products supplying material to the pool and drawing material from it, so the terminal processes must be divided among many different products. The corrugated cardboard example illustrates the parts ascribed to new cardboard (product 1) or recycled cardboard (product 2), together with the parts ascribed to other products. As is evident in the figure, the allocation is always between one product and the average pool of products (product NN (nomen nescio)) which supplies or takes from the pool of material.

In the example, the production process for new corrugated cardboard is divided into a cellulose extraction and a subsequent production of the cardboard. Extraction of the cellulose is a once-only process which is in common, and which must therefore be allocated. But the subsequent production of the product occurs each time the material is recovered. It is not a once-only process and it is not in common. The shaping of the product, drying and pressing depend on the individual product and occur in for each recycling step. This division of production of corrugated cardboard into *cellulose extraction* and *cardboard production* for the new cardboard or *pulping* and *cardboard production* for recycled cardboard are examples of the division into smaller process units previously mentioned. This division avoids an erroneous allocation, namely allocation of the product manufacturing itself. Data are often grouped for the individual process units because they are within the same company, e.g. cellulose extraction and production of corrugated cardboard. But in most cases it is possible for the person providing the data to make an estimated division.

Division of the production of material as shown here for corrugated cardboard should be made for most materials. For steel which is recycled, ore extraction and steel production to and including molten steel are, for example, in common, while processes such as rolling and shaping of profiles must be ascribed to the individual product.

21.6.1.2 Extend the system boundaries

After designation of the true joint processes, one should further seek to avoid allocation by extending the system boundaries. The following conditions must be fulfilled before the system boundaries can be extended:

- The LCA must apply to a change, i.e. the difference between two alternative scenarios.
- The nature and scope of the change which will actually occur as a consequence of the decision, which the LCA must support, must be predictable with a fair degree of certainty.
- Data must be available for the adjoining systems in question.

If the conditions for extension of the system are fulfilled:

- Expand the model of the product system to include processes in the adjacent product systems to the extent to which they *change* in relation to the alternative, i.e. model the total difference between the two scenarios being compared and carry out an LCA on the difference.
- Create more scenarios as necessary if there is a possibility for several different changes.
- If the changes result in the volume of processes being reduced in the adjacent product systems, then summarize the environmental exchanges from these processes and deduct them in the total modelling. Similarly, add exchanges from processes of increased volume to the total modelling.

Extension of the system boundaries makes the LCA specific to a particular scenario. When the system boundaries are extended, it is therefore not the "product" which is assessed, but a concrete scenario into which it is entering. If the changes cannot be predicted with a fair degree of certainty, extension of the system boundaries should therefore be avoided, or several possible scenarios should be set up to cover the full spectrum of possibilities.

The allocation will demarcate the product system from adjacent systems. After the allocation, it is therefore the "product" which is analysed, i.e. the environmental impact which is calculated then derives from the product alone or, more correctly, from the manner in which it provides its functional unit. This demarcation makes the LCA less correct relative to a specific scenario, but on the other hand, more widely applicable as support for decisions which result in a number of different, and perhaps uncertain, scenarios.

21.6.2 Allocate

When the options for avoiding allocation have been investigated and exhausted, the exchanges from the remaining true joint processes are divided among the product under evaluation and the other products or materials or sub-assemblies in question. This takes place by an allocation in accordance with some simple principles of partitioning.

General principle:

- Express the joint environmental exchanges per functional unit.

To express the exchanges per functional unit is the general task of the LCA, and this continues as the general principle of allocation.

For the exchanges deriving from processes which contribute to more than one service, i.e. the joint exchanges, the task is therefore to partition them between the services to which they contribute. How large a portion of the exchanges to be ascribed to the functional unit under study is thus identified.

The services to which the joint processes contribute are, however, often different, i.e. they cannot be expressed in the same type of functional unit. If they could, the task would simply be to divide by the number of functional units to which they contribute.

The task of allocation is therefore to find a general measure for the service to which the joint processes contribute, so that it is possible to partition the joint exchanges between each of the products to which they contribute.

General measure for the service to which the joint processes contribute:

- Determine the total measure for the service to which the joint processes contribute as the total utility value established via the joint processes.
- If possible, determine the utility value on the basis of a technical criterion.
- Determine the utility value on the basis of an economic criterion if it is not possible to determine it on the basis of a technical criterion.

21.6.2.1 Allocation on the basis of a technical criterion for utility value: grade

On partitioning of the joint exchanges in, for example, recycling of materials, allocation is made according to a partitioning key based on a technical criterion for the material's utility value, if the utility value can be ascribed to the same properties of the material in the products into which it enters. The utility value of the material is called the material grade.

Principle of allocation:

- Determine the material's utility value, understood as its material grade.
- Ascribe the joint exchanges to the material in proportion to its relative utility value.
- Determine the loss of material grade to which the use of the material leads. Ascribe the joint exchanges to the product in proportion to the utility value lost on use, i.e. in proportion to the relative loss of material grade.
- Use the material grade immediately before manufacture of the product in question as an expression of the utility value.

Note that the joint exchanges are ascribed to the material as a function of its relative utility value. In other words, the material always carries with it joint exchanges in proportion to the relative utility value residing in the material. For co-products in the same material, the joint exchanges are partitioned at the point where the co-products arise. On recycling this means that the joint exchanges accompany the material further to the next product in proportion to the residual utility value. The product is thus ascribed joint exchanges in proportion to the relative loss of grade caused in the material, including the loss during the recovery process necessary for recycling.

Ascribing joint exchanges proportional to the loss of utility value is the correct method of allocating, because the task of the LCA is to ascribe the exchanges to the service. To allocate after loss of utility value corresponds exactly to ascribing the joint exchanges to the total volume of services to which they can contribute.

Determination of the material grade is described in more detail in the note on allocation in Wenzel (1997).

21.6.2.1.1 Allocation in case of material recycling

Allocation follows the same procedure, irrespective of whether there is a bilateral material recycling or recycling via the material pool. In almost all cases the recycling proceeds via a material pool in the community, and this example will therefore be given here.

A guideline for determination of the material grade is that at a minimum it must correspond to the utility value realized in society in practice. Example 23.1 shows what this means:

Example 21.3. Utility value realized in practice

- A product contains 100% primary steel, and after use, 100% is sent for recycling via the community pool.
- The recycling degree of steel in the pool is 50%. This corresponds to all steel in the pool being reused an average of two more times before it is finally lost. A recycling degree of 100% would mean that the material would be recycled an infinite number of times.
- The actual steel is thus used a total of three times, i.e. the steel supplied to the pool is in practice used another two times out of a total of three.
- In the actual case, the residual utility value is in practice thus proven to be a minimum of 2/3 or ca. 67% of the utility value of the primary material. The loss of material grade in the actual case should therefore not be set higher than 33%, unless specific reasons exist for this.
- The loss of material grade can be set lower if it is possible in purely technical terms to argue for a higher utility value of the recycled steel.

	f (new)	f (scrap)	f (loss)
Newspaper (DK)	1	0.50	0.17
Unspecified paper (DK)	1	0.75	0.20
Corrugated cardboard (DK)	1	0.65	0.17
Glass (DK)	1	0.75-0.80	0.10
Steel (Western Europe)	1	0.85	0.10
Steel (USA)	1	0.75	0.10
Stainless steel (Western World)	1	1.00	0.00
Copper (Western World)	1	1.00	0.00
Aluminium (Western World)	1	1.00	0.00
ABS (theoretical estimate for pure ABS regranulation, post consumer)	1	-	0.20

Table 21.6. Material grades and their losses for various types of materials on recycling

Table 21.6 gives some material grades for the types of material typically entering into recycling pools, and the loss of material grade which can be ascribed to one cycle of the material in an average product. Use of these values for material grade is recommended as a starting point. Their determination is described more fully in the note on allocation in Wenzel (1997). This will inevitably be altered in future as knowledge in the field increases.

For that part of the product subject to final disposal, the loss of material grade corresponds to the material grade before manufacture of the product, i.e. f(new) or f(scrap), according to whether a new or a recycled material is concerned. The total utility value of the material is lost when the material is removed from the cycle, and the responsible product is therefore ascribed joint processes corresponding to the material grade before manufacture of the product.

For that part of the product sent for recycling, the loss of material grade corresponds to f(loss), which reflects the loss caused by product manufacture, use, and material recovery.

Example 21.2 for new corrugated cardboard and recycled corrugated cardboard shows where material grades are apportioned in the material's product system. Example 21.4 shows how the partitioning keys based on the material grade are deducted.

The procedure for allocation with material grade as partitioning key is:

- That part of the product subject to final disposal is ascribed a part of the joint environmental exchanges, corresponding to the fraction finally disposed of multiplied by f(new) if the material is new, or f(scrap) if a recycled material is involved.
- That part of the product sent for recycling is ascribed a part of the joint exchanges corresponding to the recycled fraction multiplied by f(loss).

	Product 1 is ascribed	Product 2 is ascribed
Logging	$a \cdot (x_2+x_3+x_4) \cdot f(new) + a \cdot x_1 \cdot f(loss)$	$a \cdot (x'_2+x'_3+x'_4) \cdot f(scrap) + a \cdot x'_1 \cdot f(loss)$
Recovery of cellulose	$a \cdot (x_2+x_3+x_4) \cdot f(new) + a \cdot x_1 \cdot f(loss)$	$a \cdot (x'_2+x'_3+x'_4) \cdot f(scrap) + a \cdot x'_1 \cdot f(loss)$
Landfill 1 and 2	$a \cdot x_2 \cdot f(new)$	$a \cdot x'_2 \cdot f(scrap)$
Incineration 1 and 2 - energy recovery	$a \cdot x_3 \cdot f(new)$	$a \cdot x'_3 \cdot f(scrap)$
Incineration 1 and 2 + energy recovery	$a \cdot x_4 \cdot f(new)$	$a \cdot x'_4 \cdot f(scrap)$
Displaced production of electric and thermal energy 1 and 2	$-a \cdot x_4 \cdot f(new)$	$-a \cdot x'_4 \cdot f(scrap)$
Landfill $\overline{NN}$	$\frac{1}{1-\bar{y}_1} \cdot ax_1\bar{y}_2 \cdot f(loss)$	$\frac{1}{1-\bar{y}_1} \cdot ax'_1\bar{y}_2 \cdot f(loss)$
Incineration $\overline{NN}$ - energy recovery	$\frac{1}{1-\bar{y}_1} \cdot ax_1\bar{y}_3 \cdot f(loss)$	$\frac{1}{1-\bar{y}_1} \cdot ax'_1\bar{y}_3 \cdot f(loss)$
Incineration $\overline{NN}$ - energy recovery	$\frac{1}{1-\bar{y}_1} \cdot ax_1\bar{y}_4 \cdot f(loss)$	$\frac{1}{1-\bar{y}_1} \cdot ax'_1\bar{y}_4 \cdot f(loss)$
Displayed production of electric and thermal energy $\overline{NN}$	$-\frac{1}{1-\bar{y}_1} \cdot ax_1\bar{y}_4 \cdot f(loss)$	$-\frac{1}{1-\bar{y}_1} \cdot ax'_1\bar{y}_4 \cdot f(loss)$

Example 21.4. Example of material grade as partitioning key in allocation for recycling of corrugated cardboard

Example 21.5 illustrates a calculation of the total energy consumption in the product system processes for the recycled corrugated cardboard. The example shows how the allocation is calculated for the joint processes.

Process	Primary energy MJ/kg	Product 1 New corrugated cardboard	Product $\overline{NN}$	Product 2 Recycled corrugated cardboard	Product $\overline{NN}$
Logging	18.7	10.9	7.8	7.7	11.0
Extraction of cellulose	12.0	7.0	5.0	4.9	7.1
Corrugated cardboard, new	10.0	10.0	0	-	-
Corrugated cardboad, recycled	10.0	-	-	10.0	0
Landfill 1 and 2	0	0	0	0	0
Incineration 1 and 2 - energy recovery	0.6	0.06	0	0.04	0.02
Incineration 1 and 2 + energy recovery Displaced production og electric and thermal energy	-16.4	-4.9	0	-3.2	-1.7
Pulping $\overline{NN}$	2.0	2.0	0	2.0	0
Landfill $\overline{NN}$	0	0	0	0	0
Incineration $\overline{NN}$ - energy recovery	0.6	0.01	0.06	0.01	0.06
Incineration $\overline{NN}$ + energy recovery Displaced production of electric and thermal energy $\overline{NN}$	-16.4	-0.70	-0.34	-0.70	-3.4
Product system		24.4	9.46	20.75	13.08

Assumptions

$x_1 = x_1' = 0.5$	$x_3 = x_3' = 0.1$	a = 1 kg	f(scrap) = 0.65	$\bar{y}_1 = 0.6$	$\bar{y}_3 = 0.1$
$x_2 = x_2' = 0.1$	$x_4 = x_4' = 0.3$	f(new) = 1	f(loss) = 0.17	$\bar{y}_2 = 0.1$	$\bar{y}_4 = 0.2$

Example 21.5. Example calculations of total energy consumption in the product system processes of new and recycled corrugated cardboard

21.6.2.1.2 Allocation between co-products in the same material

The partitioning key for allocation between co-products in the same material is determined as *relative quantity multiplied by the relative material grade.*

Co-products in the same material arise, for example in the case of utilized production waste. In this case the material grades of the co-products are almost the same, but not entirely the same because there can be a loss of material grade in the manufacturing process in which the production waste arises.

21.6.2.1.3 Allocation in case of co-production of different forms of energy

Co-production of electricity and heat in co-generation power plants is probably the most frequently occurring form of co-production in product systems. The allocation is made in the same manner as for co-products in the same material, and the partitioning key is determined as *relative quantity multiplied by the relative grade* of the energy. The quantity of energy is measured as the total output of electricity and utilized heat used from the co-generation plant, measured in e.g. MJ, and the grade of the energy is measured as its content of exergy.

21.6.2.1.4 Allocation in case of reuse of a sub-assembly in the product

On reuse of a sub-assembly in a product, allocation is made for the processes which contribute to the material production on the basis of the material grade, in the same way as described for material recycling. Example: used spare parts for a car.

21.6.2.2 Allocation on the basis of an economic criterion

When partitioning joint environmental exchanges, allocation is made in accordance with a partitioning key based on economic value if it is not feasible to define the utility value on the basis of technical criteria.

Principle of allocation:

- Determine the utility value of the products among which the allocation occurs on the basis of their economic value.
- The relative price of the products in the process in which they arise is used as the partitioning key.

The price of the products in the process in which they arise is often a secret. Note, however, that a relative price for two or more products is concerned, i.e. their mutual *price relationship,* and providing this information is often unproblematic to a company.

21.6.2.2.1 Allocation between co-products in different materials

For co-products in different materials, including substances of different chemical nature, the utility value does not reside in the same properties of the product. It is therefore not possible to find a technical criterion for determination of a utility value, and it must therefore be determined on the basis of the co-products' economic value as specified.

Example 21.6 illustrates partitioning keys for the co-products in the production of glucose from maize.

	Steeping water	Fibres	Sprouts	Gluten	Starch	Glucose
Product system grain	0	0.01	0	0.08	(0.91)	0.91
Product system recovery of water	0	0.01	0	0.08	(0.91)	0.91
Product system production of N	0	0.01	0	0.08	(0.91)	0.91
Product system production of P	0	0.01	0	0.08	(0.91)	0.91
Product system production of K	0	0.01	0	0.08	(0.91)	0.91
Product system production of pesticides	0	0.01	0	0.08	(0.91)	0.91
Growing	0	0.01	0	0.08	(0.91)	0.91
Cleaning	0	0.01	0	0.08	(0.91)	0.91
Product system sulphur	0	0.01	0	0.08	(0.91)	0.91
Production starch	0	0.01	0	0.08	(0.91)	0.91
Product system HCl	0	0	0	0	0	1
Product system NH_3	0	0	0	0	0	1
Product system kieselguhr	0	0	0	0	0	1
Production of glucose	0	0	0	0	0	1

Example 21.6. Example of partitioning keys. Co-products in production of glucose from maize (Frandsen & Hansen, 1995)

The example uses the relative price of the products after the production process for starch as the partitioning key, and the co-products' relative quantity and price are multiplied so that the partitioning key can be applied directly to the exchanges of the joint processes. To express the exchanges from the joint processes per tonne of glucose, multiply the actual quantities of the exchanges by 0.91 or 1.00 as shown in the table.

21.6.2.2.2 Allocation in reuse of a sub-assembly in the product

In reuse of a sub-assembly in a product, allocation is made on the basis of economic value for the manufacturing processes which process the material and contribute to manufacture of the sub-assembly. The joint exchanges from the manufacturing processes are ascribed to the product in proportion to the *relative loss* of utility value in the same manner as for material recycling. The loss of utility value is expressed here as loss of the additional value which the manufacturing processes have created in the material as a result of processing the material into the finished sub-assembly, i.e. the loss of price is determined on the basis of the sub-assembly's price minus the material price before and after use.

21.6.2.2.3 Allocation in case of utilized calorific value

In cases where the material is destroyed, e.g. in incineration, extension of the system boundaries rather than allocation is recommended. Alternatively, allocate using a partitioning key based on economic value. As economic value of the material at the point before incineration, the price of *alternative fuel* expressed per MJ of energy output can be used. This allows for a calculation of the relative prices of the material at the point immediately before product manufacturing versus the point before incineration.

21.7 Presentation of the scope definition

- Present the scope definition together with the LCA, including:
- Describe the functional unit and explain that data for inventory and impact assessment are expressed per functional unit.
- Describe the reference product and explain the choice. Describe the database and for this purpose use a table such as Table 22.3 in Section 22.1 if required. Examples are given in five product case histories, in Part V of the book.
- Present the scope definition of the product system, but use only the scope definition actually used after finalizing the LCA, i.e. not the one planned before the inventory, but the one actually carried out. If necessary, present the scope definition as a simplified form of the parts list which need not include the hierarchical level. The case histories in Part V present scope definition of the product system for five products in simplified parts lists.
- Describe the scope definition in terms of time and technology and any projection of data.
- Explain the most significant allocations.

22. Procedure for inventory

Reference product(s) have been selected at this point in the LCA. Their product systems have been established and the scopes of these are defined, and a decision has been made regarding periods of time which the data are to represent. Joint processes and environmental exchanges have been designated to the extent to which they are clear before collection of data, and partitioning keys for the allocation have been selected. Data collection can now be planned and commenced.

22.1 Planning the data

- Consider which types of data will underlie the processes in the product system of the reference product.
- Consider whether data must derive from processes which are specific to the selected reference product or whether data can derive from processes which are not specific to the reference product.

The considerations can be noted on a form as shown in Table 22.3. Note the most significant process types and requirements for data specificity and desired source type under each stage. Table 22.2 explains the required considerations in more detail, and the case histories in Part V of the book show how the data are planned and collected for the five reference products in the EDIP programme.

Table 22.1. Typical picture of requirements for specificity of data

A typical picture of the requirements for product specificity in data collection is shown in Table 22.1.

Required data specifity	**Extraction of raw materials**	**Materials production**	**Product manufacture**	**Use**	**Disposal**	**Trans-port**	**Energy systems**
Product-specific			x	x	x	x	
Site-specific	x	x	x		x	x	x
General	x	x	x		x	x	x

The degree of specificity is in the range product-specific, site-specific, general. The concepts are defined thus:

- Product-specific data: derive from processes which directly involve the reference product, e.g. data from a manufacturing process collected while processing components specifically for the reference product, or a use process measured specifically on the reference product.

- Site-specific data: derive from the actual locality but are not product-specific, e.g. data from a steelworks, a painting box, a shredder or an incineration plant which actually enters into the specific product system, but where data are average data and do not derive from processing of the specific reference product.

- General data: are neither site-specific nor product-specific, but general data for a process type occurring in the product system, e.g. average data for production of polyethylene in Europe, average data for oil extraction, for lorry transport on highways or for electricity production in Denmark. The existence of the process in the product system and the process volumes are defined by the actual product.

The requirements for data specificity and source type are a function of how specifically the product system for the product/object studied is known and/or how specifically it is appropriate to define it:

Product system	Required specifity
Extraction of raw materials	Often, the raw material originates from large parts of the world and the precise source is not known, as in the case of metal ore. In other cases, e.g. natural gas from the North Sea, the source is well defined. For practical purposes representative averages are often used, i.e. mainly general data and only in certain cases site- or region-specific data.
Materials production	Same point as above. In many cases, however, the companies have regular suppliers , e.g. "steel from a specific steelworks" or"Norwegian aluminium". This allows the use of site-specific data.
Manufacturing	Certain processes are specific to a product, e.g. a particular surface treatment or assembly process. Others are more common, e.g. an injection moulding of plastic. For some processes, data should therefore be specific to processing of the actual product. For others, extrapolation from general data for the process is permissible.
Use	This stage is unique for the product and should, as far as possible, be product-specific.
Disposal	Same points as for the manufacturing processes described above, but with a predominance of site-specific data, e.g. incineration, shredding, and recycling of metals or representative general data. Certain disassembly processes should be product-specific.
Transport	Means of transportation and distances are specifically defined by the actual product, as is the degree of driving in urban areas, on highways etc. Data for the environmental exchanges from the transport processes must be general.
Energy systems	The product concerned specifies the site or region where the energy system is situated, while the data are general for the particular site/region.

Table 22.2. Description of requirements for specificity of data

Planning the reference data									
	Data specifity			Data source type					Comments
Product life stage and process type	Product-specific	Site-specific	General	1	2	3	4	5	
Extraction of raw materials									
Materials production									
Product manufacturing									
Use									
Disposal									
Transport									
Energy systems									

Notes

1) Measurements for the specific product or a specific process in the product system
2) Calculations for the specific product or a specific process in the product system (based on a mass balances and input data)
3) Extrapolation from data for the same process type or technology
4) Extrapolation from data for other process types or technologies
5) Estimate

Table 22.3. Form for planning of data

In planning of the data, scope definition with respect to time and technology must be remembered so that account can be taken of the fact that subsequent data projection must be possible.

- As far as possible, plan the projection of data at the same time as the planning of the data.

LCAs in product development must, as noted previously, apply several years into the future. This means that specific suppliers and localities are often unknown, and for many processes it is therefore most appropriate that the data be general. Each individual process should, however, be considered in itself as shown in Table 22.3.

22.2 Data sources and data networks

Working with LCA in product development requires easy access to data. The company should therefore join a data network, as illustrated in Figure 9.2 in Chapter 9, from the beginning.

- Establish a network of data providers. Make contact with LCA centres and gain an overview of the bodies which can provide data for the various processes in the product system for the company's products.
- Scan the data sources available in the form of literature and electronic databases; cf. e.g. Hemming (1995). Use the data in the EDIP unit process database.

22.2.1 The EDIP unit process database

The EDIP database of processes (Frees, 1996) contains data for a large number of processes occurring in the product systems of industrial products. Table 22.4 provides a summary of the contents.

	Category		
Number of processes	**1st Level**	**2nd Level**	**3rd Level**
30	Raw materials	Recovery	Ore, sand, lime, etc.
60	Materials (incl. material recovery processes)	Metals	Steel, stainless steel, aluminium, copper and zinc
		Plastics	LDPE, LLDPE, HDPE, PP, PS, HIPS, PVC, PS, ABS, PU
		Glass	
		Wood and paper	Wood, pulp, corrugated cardboard, cardboard, newsprint, etc.
		Others	Flour, beef, milk, cotton, wool, greens
100	Ancillary substances		Detergents, lubricants, acids, organic and inorganic compounds, etc.
50	Manufacture		Metal cutting, forging, moulding, degreasing, anodizing, galvanizing, plating, painting, etc.
30	Transport	Transport by road	Vans, small and large lorries, 3 scenarios for each category
		Transport by rail	electric and diesel trains
		Transport by ship	Coasters, small cargo ships, container ships, etc.
		Transport by aircraft	Medium size and large aircraft. Transportation by air is divided into "take off/landing" and "cruising"
40	Disposal	Waste incineration	Oil, paper, wood, glass and several types of plastics and metals
		Landfill	
35	Energy	Electricity	Electricity produced from natural gas, oil, coal, nuclear power, hydro power or biofuels. There are scenarios from 16 countries, geographic areas or areas of production
		Thermal energy	15 scenarios for the combustion of fuels

Table 22.4. Summary of EDIP unit processes

22.3 Data format: specification of data types

Data collection covers more than simply the environmental data for the processes' exchanges. The data to be collected fall into three categories, namely:

- Description of the process
- The process's environmental exchanges
- Characterization of the data

A procedure is described here for data collection, and a form is illustrated for the information which should be included (see Table 22.5).

22.3.1 Description of the process

- Define one and only one key unit for the process. If there are different key units for different types of exchanges, the most representative should be selected, and other existing relationships should be described under characterization of the data. In cases of doubt the key unit should be discussed with the data provider.
- Define the process's name, identity and family relationships.
- Define the reference period for data, the date of updating, the period over which the data are projected, and the person responsible for data.
- Describe operating data for the process.
- The information in the data format in Table 22.5 concerning the placing of the process in the product system:
  - description of geographic location and stage in the product life,
  - description of special conditions for the actual component/process,
  - definition of any site- and region-specific circumstances,

should only be added when the process is used during modelling of the product system.

22.3.2 The process's environmental exchanges

22.3.2.1 Normalization of the exchanges to a unit process

- Express all exchanges as a function of the process's key unit.
- Identify any co-products.
- In case of co-products: express data in the unit process for the entire process including the co-product(s), i.e. without allocation of the exchanges. Express data in the key unit for the product which is actually being assessed (the "main product"). The unit process will be used in this form in a subsequent LCA if allocation is avoided via an expansion of the system. Take particular note of the occurrence of the co-products in the scope definition of the process. Co-product allocation is carried out only when the product system is modelled.
- Identify the key unit for the unit processes which the non-terminal exchanges represent and express all non-terminal exchanges in the number of their respective key units.
- Express terminal exchanges in grams for inputs and outputs and hours or millihours for all impacts on the working environment except "accidents", which are expressed in numbers.
- As far as possible avoid double entries in the process database, i.e. avoid the possibility that one and the same exchange may be called by different names and/or ID numbers. Always establish new types of exchanges on central lists in the database, and select from these lists when exchanges are to be included in a process. Establish a check function via the substance's CAS number and an internal check of the CAS number's validity. A guideline is given in the manual for the EDIP method's PC tool (Danish Environmental Protection Agency, 1997).

The data discipline, described in brief in Table 22.5, is implemented in the EDIP process database on which the EDIP method's PC program builds.

22.3.2.2 Statistical description of data

- Enter data as mean value and coefficient of variation: use the obtained figure itself as the mean value, and estimate a coefficient of variation.
- If no other data for the coefficient of variation are available, the following values can be used initially as a rule of thumb:
  - Input: - 0.1 for consumption of resources, materials and ancillary substances, including consumption of fuels, and
    - 0.25 for electricity consumption.
  - Output: - 0.1 for CO_2 emissions calculated on the basis of input which is known with certainty to undergo complete combustion,
    - 0.1 for emissions calculated on the basis of inputs, and which are known with certainty to be emitted to the recipient in question,
    - 0.1 for quantities of waste determined by weighing or with a similarly high degree of certainty,
    - 0.5-1.0 for other emissions.
- Impacts on the working environment
  - 0.5-1.0 for chemical impacts,
  - 0.1-0.5 for noise,
  - 0.1-0.5 for monotonous repetitive work,
  - 0.1 for work accidents.
- Attempt to qualify the above estimates of the coefficients of variation by questioning the data provider or experts on the process in question. Check which processes and exchanges are of greatest significance for the result of the LCA, and attempt to qualify the estimates of the coefficients of variation for these first. Cf. the procedure for the sensitivity analysis in Chapter 24 in the Toolbox.
- Under all circumstances the rules of thumb should be assessed in each individual case and adjusted by common sense.

22.3.3 Characterization of data

- Describe the scope definition of the process as in Table 22.5.
- Describe deficiencies in data, characterize existing data, briefly describe how the coefficients of variation are estimated/calculated, and if desirable, establish a volume/mass balance for input and output. The criteria for characterization of data are given in Table 22.6.
- Describe technological trends for the process for the most important exchanges: energy consumption, significant emissions, etc.
- If the process is projected, describe how.

The unit process

Unit process and family ID		Process key unit	
Process name			
Data reference period		Updated on	
Projected to		Reference person	

Operating data:

- Process cycle:
If relevant, describe the hourly, daily, weekly and annual cycles. Describe peak loads, cleaning processes, etc.
- Operating data:
Describe operating data as e.g. flow (number of key units per unit of time), retention time, temperature, pressure and pH. Assess which are relevant to characterize the process.
- Working environment conditions:
Describe possible arrangements concerning the working environment as e.g. encapsulation of the process, ventilation, use of respiratory protection devices, ear protectors and/or protective goggles, work rotation arrangements. State number of employees working at the process and briefly describe the working environment conditions, including any symptoms stated by the employees.

When modelling a product system

Geographic location		Stage of the product life	

Component description

- Component characteristics: Describe briefly, if relevant, physical and geometric data for the item treated in the process for the product modelled: weight, surface area, volume, geometry, etc.
- Process modifications: If relevant, describe process characteristics or operational conditions unique for the specific component: flow, retention time, temperature, pressure, pH, waste percentage, etc.

Site characterization for calculation of environmental impact potentials

Environmental impact	Global		Regional						Local					
Impact ID	1	2	3	4	5	6	7	8	11	12	13	14	15	16
Site factor (default)	*1*	*1*	*1*	*1*	*1*	*1*	*1*	*1*	*1*	*1*	*1*	*1*	*1*	*1*

Region characterization for normalization and weighting

Impact category	Resource consumption			Environmental impacts			Impacts on the working environment
	Non-renewable	Renewable Water	Wood	Global	Regional	Local	Local
Ref. area (default)	*World*	*DK*	*World*	*World*	*DK*	*DK*	*DK*

Table 22.5. Data format for description of the unit process

Environmental exchanges from the process

Environmental exchange: ID number	Environmental exchange: Name	Unit	Allocation factor (f)	Statistic: Quantity (μ)	Statistic: Coefficient of var. (σ/μ)	Statistic: Distr. function (D)	Characterization of data: 1	2	3	4	5	6
Input												
Electricity												
Thermal energy												
Materials												
Ancillary substances												
Transport												
Resources												
Output												
Emissions to air												
Emissions to water												
Emissions to soil												
Waste												
Impacts on the working environment												

Notes

1. Data source: 1-5
2. Data's representativeness for the process: 1-5
3. The representativeness of the process/technology for the purpose in the modelling: 1-4
4. Age of data: state year
5. Reference: specify a number that refers to a central reference library in the database
6. Confidentiality: indicate by a C

Table 22.5. (continued)

Data characterization

Definition of the scope of the process
■ **Definition of the scope of the process:** Describe which operations are included and which are not included. Describe also which environmental exchanges are not terminated, i.e. input and output that have not been followed "back to raw material extraction" or "forward to final recipient". ■ **Co-products:** Specify possible co-products from the process. If co-products are found in the process: ■ Specify if the environmental exchanges from the process have been allocated to the product concerned. If yes, specify the allocation method. ■ Specify partitioning keys for co-product allocations, i.e. relative price or grade for each product arising from the process.
Data characterization
■ **Deficiencies in the data:** If gaps in the data are known to occur for certain environmental exchanges, briefly describe them. ■ **Data source:** If necessary, describe the data source, i.e. how have the data been generated? ■ **Representativeness of data:** If necessary, elaborate on the representativeness of the collected data, i.e. how well do the data describe the process and how representative are they of an average? ■ **Representativeness of the unit process for the purpose:** Using the process in the modelling of a specific product system, elaborate as necessary on the process's representativeness for the objective in the model. ■ **Statistical data:** Briefly, describe how the coefficients of variation for the environmental exchanges were assessed/calculated. ■ **Mass balance:** Briefly verify data by use of a mass balance describing inputs and outputs. Remember to include oxygen and nitrogen as inputs in combustion processes where e.g. CO_2, NO_x and H_2O are formed.
Technological development
■ **Trends:** If the process is essential for the total LCA, describe the technological development and trends in the most important exchanges. ■ **Projection of data:** If the process has been projected, describe how.

Table 22.5. Data format for description of the unit process

22.4 Collection of data

- Scan electronic databases and the literature for information on the processes included in the model of the product system.
- Contact LCA centres and specialist centres and ask how data are best collected.

When electronic databases and data from the literature have been scanned and the centres of knowledge asked about data, it is probable that data will still be lacking. For some processes the data will be entirely lacking and for others the data will be of poor quality.

Data grade	Data source	Representativeness of data for the unit process	Representativeness of the unit process for its purpose in modelling
1	Measurements involving the actual product or an actual process in the product system	Data represent the process cycle including long-term variations as an average based on continuous measurements or frequent random samples	Same process/material and same technology
2	Calculations involving the actual product or process in the product system	Data represent the total process cycle in including long-term fluctuations as an average of a series of random samples	Same process/material but different technology
3	Extrapolation from corresponding process or technology	Data represent the process cycle as a short-term average based on a series of random samples or one representative sample during the cycle	Different but comparable process/material and same technology
4	Extrapolation from other process or technology	Data represent the essential parts of the process cycle as a short-term average based on a series of random samples or a representative sample during the period	Different but comparable process/material and different technology
5	Estimate	Data represent an essential part of the process cycle based on one or very few random samples, or reprensentativeness unknown	

Table 22.6. Guideline for characterization of data

- Contact companies responsible for the actual processes in the product system. If feasible, use the data format in Table 22.5 and the guideline in Table 22.6 as the basis for a questionnaire.
- If necessary, hold a meeting with the data provider to check the quality of the information. Note particularly how data are ascribed to the individual product. Be careful to include overheads from the company in question.
- If necessary plan a programme of measurement to collect the data still lacking.

As noted, there is a difference in the accessibility of data for the three types of environmental exchanges: inputs, outputs and impacts on the working environment. Some guidelines are therefore given here for the procedure for procuring data within each of the three categories:

22.4.1 Input data

- Summarize material and chemical consumption per product on the basis of data in parts lists for the product and purchasing lists for chemicals. Most companies already have this information.
- Calculate energy consumption per product unit on the basis of the company's total consumption if the company's production is highly uniform. If production is non-uniform, energy consumption must often be measured for the individual process.
- For processes assessed to have an impact on the working environment, summarize time used for the process operators expressed per product unit. This is usually known because it is computed for use in cost calculations. The criteria for summarizing impacts on the working environment are covered in Section 22.4.3.

22.4.2 Output data

Identify the outputs from the process.

- Calculate outputs on the basis of input data if this will provide a more correct result than measurements.
- Obtain already existing measurements of outputs at the company.
- Calculate outputs on the basis of input data if it is not considered necessary to take measurements.
- Measure outputs if significant for the LCA and if measurements of adequate quality are not already available and outputs cannot be calculated with a fair degree of certainty on the basis of input data.
- If necessary seek expert advice within the actual process.

Example 22.1. Examples of outputs best calculated from input data

Examples of emissions best calculated from input data

Emissions to water

- Detergents: after washing, there is no doubt that the detergents are present in the waste water, but the detergent content of the water can be difficult to measure, partly because it is camouflaged by micelle formation with the substance actually washed out and partly because the measurement itself involves a high degree of uncertainty and is difficult to reproduce. A calculation on the basis of inputs and simple mass balance considerations is therefore better.

- Oil: for oil films etc. on items degreased in e.g. an aqueous alkaline process, there is no doubt that the oil is present in the water. It will often be significantly easier to obtain data on the quantity of oil film on the item before degreasing than to measure the emission with corresponding accuracy.

Emissions to air

- Volatile solvents: after painting, gluing, degreasing or similar processes involving volatile solvents, there is no doubt that the solvent is emitted to the air. It will never be possible to measure an emission of volatile solvents with total accuracy, and measurements will often significantly underestimate the emission because part of the emission will be diffuse and will escape the sampling. A calculation on the basis of inputs and a simple mass balance consideration is therefore better.

Solid waste

- As a rule, solid waste is relatively easy to summarize because it typically leaves the process in well-defined containers and the quantity per product is usually known. For many types of waste consisting of well-defined materials, a simple input–output balancing can determine the quantity of waste. For example, there could be:

- Waste from processes which remove material, e.g. chipping, sanding, etc. Weighing the item before and after the process can determine the quantity of waste.

In reality, a calculation on the basis of input data is the best procedure for all water-soluble non-volatile substances for which consumption is known, and for which it is known that the substances are added to the aquatic phase and do not alter in chemical composition during the process. This will be the case for most substances in aqueous emissions. The argument correspondingly applies to all volatile substances emitted to air and which do not alter chemically in the process.

In certain cases of calculations of emissions on the basis of input data, there will be a quantity of chemicals/materials carried further with the product. Such quantities of materials can often be determined, e.g. in determination of the thickness of layers of paint or heavy metal in galvanizing or simply by weighing the product before and after surface treatment.

For many substances, a calculation of emission data on the basis of input data can be very simple. For substances where there can be doubt concerning the emission route, the calculation can be made on the basis of a qualified assessment of the substance's physico-chemical characteristics.

As shown in Example 22.2, however, many types of emissions cannot be determined via calculations.

Example 22.2. Examples of output best determined on the basis of measurements

Examples of emissions best calculated from input data

- Reaction products: substances generated in the process can typically not be determined via calculations. Data for such emissions must be obtained from measurements.
- Input unknown: substances from undefined input can naturally not be determined on the basis of input, e.g. fat, wax, oil or similar from natural products (cotton, wool), and impurities in various raw materials, e.g. oil, coal, ore or similar which are undetermined.
- Emission route undefined: in certain cases, the emission route can be difficult to determine for well-defined substances for which input is known, e.g. volatile substances emitted to water but which evaporate from the aquatic phase to a certain extent.
- Emissions from waste water treatment plants: substances which are emitted to waste water treatment plants are usually only partly removed, and there is a partitioning of the substance between water, air and solid waste, which in many cases is best determined by measurements. In certain cases a thorough knowledge of the substance's physico-chemical characteristics and the operation of the treatment plant can, however, permit model calculations of the distribution as described in Hauschild *et al.* (1997a).

If it is judged is that measurements are required, they should be made under well-defined and representative operating conditions for the process and by a recognized sampling and analytical laboratory.

In most cases, knowledge of the process and of the actual chemicals involved will help to determine whether data should derive from measurements or whether calculations on the basis of input are preferable or acceptable.

22.4.3 Data on the working environment

The impact of a process on the working environment is calculated only if it is assessed in advance that it can contribute to some of the observed impacts (see Section 21.2.3): in other words, there is an element of assessment in the inventory for the working environment, separate from the inventory of the process's inputs and outputs.

For each individual impact on the working environment, the assessment criteria are therefore included here under the guideline for the inventory. The section on collection of data on the working environment is therefore also significantly more comprehensive than the sections on data collection of inputs and outputs.

Two different forms of data for assessment of the impact on the working environment through a product system are used.

- Process-specific data: to be used by the company which produces the product under investigation if this company produces different types of products or has many different types of processes. In this case it is difficult to use average figures from the company to describe the product's impacts on the working environment, and it is necessary to collect data for each individual process. This will typically apply to the company's own manufacturing and to some manufacturing processes which are subcontracted out.

Collection of process-specific data means that for each individual process which the product passes through, it is necessary to survey the impacts to which the employees are exposed and assess whether the exposure is significant enough to be classified as causing an impact. If this is the case, the employee's exposure time per unit produced is registered. This exposure time can later be converted to the number of work injuries which will potentially be reported because of the impact. The conversion to reported work injuries is described in Chapters 10 and 23.

- Sector-specific or company-specific data: for large companies or sectors which produce uniform types of products and have done so for a number of years, the company's reports of work-related incidents and accidents can be related to total production. The potential number of work injuries per product can thus be determined, e.g. per kWh of electricity or per ton of steel plate produced. This method was tested in the EDIP programme and used in the inventory of the impact on the working environment in Danish co-generation power plants, for oil extraction and for transport of goods.

  Sector- or company-specific data can be used for producers of materials, producers of standard components, transport, production of electricity and disposal.

22.4.3.1 Collection of process-specific data

- Collect data on the impacts on the working environment which are included in the EDIP method.

The EDIP method includes impacts on the working environment from:

- chemicals which can cause cancer, reproductive damage, allergy or damage to the nervous system,
- noise which can cause hearing damage,
- monotonous repetitive work which can cause injury to the musculo-skeletal system,
- work accidents which cause grievous bodily harm.

Collection of data should proceed in cooperation with the company's working environment department or the associated employees' health service. If data on the working environment are collected by other people, the data should be checked by an expert from the company or an employees' health service.

- Measure or calculate the working time per product or component.

For each process, register working time per product or component. This is often registered in industrial companies as it is used in cost calculations, possibly in the form of man-hours or "process hours".

22.4.3.1.1 Chemical impacts

■ For each process, register which substances are used and formed, and whether the process employee is exposed to the substances via skin contact or by inhalation of concentrations higher than 10% of the relevant threshold value.

Data on ancillary substances in the process can be collected from the parts list for the product and with the aid of user guides on working hygiene. The user guide should include the name and any CAS no. for all dangerous substances entering into the ancillary substance in quantities over a certain magnitude.

Chemicals can also be formed in or released from the process, e.g. on heating. Information on which substances might occur in the individual processes is available in the literature, and in many cases the company expert on the working environment will know which substances will occur.

In case of doubt, measurements can be made in the air by either the employees' health service or another authorized body.

Exposure to chemicals is regarded as an "impact" when the concentration in air exceeds 10% of the threshold value or if there is skin contact with the substance. If the employee reports problems which can be attributed to actual chemical impacts, the impact is similarly included. If one of these criteria for impact is fulfilled, the full working time per product for the process is recorded as "impact time".

Carcinogenic substances

■ Check in the EEC list of dangerous substances whether the substance is classified as CARC 1, CARC 2 or CARC 3. (EEC 1994).

According to the Danish Working Environment Service, there are ca. 1,250 new cases of cancer a year in Denmark attributable to conditions in the working environment, and 800 people a year die in Denmark of cancer contracted at work. Apart from chemical impacts, radiation can be a direct cause of cancer, while other effects, e.g. psycho-social, can contribute indirectly to cancer. In the EDIP method, cancer is only considered when it is a consequence of the effect of chemicals in the working environment.

A list of carcinogenic substances based on the EEC list (EEC, 1994) is given in Rasmussen (1997).

Substances causing reproductive damage

■ Check whether the substance is included in the EEC list of dangerous substances as REP 1, REP 2 or REP 3 (EEC, 1994).

Reproductive damage covers damage to ova, sperm and embryo, and effects on the hormonal system (e.g. menstrual disturbances). The damage can take the form of miscarriages, congenital deformities, premature births, stillbirths, cancer in children or mutations in the germ cells.

Very few cases of reproductive damage are reported to the Danish Working Environment Service, among other reasons because it can be difficult to say whether a given instance of reproductive damage is attributable to the working environment or whether it should be blamed on other factors.

There are different forms of impacts on the working environment known to be able to cause reproductive damage. These include the chemical impacts of certain pesticides, organic solvents, lead, certain plastic monomers, and substances deriving from the welding of metals. Physically taxing work and biological impacts can also lead to reproductive damage. The EDIP method includes reproductive damage as a consequence of the impacts of chemicals.

In their report "Chemicals in the working environment causing reproductive damage" (Hass *et al.*, 1991), the Danish institute for the working environment has surveyed substances which can be regarded as causing reproductive damage. Rasmussen (1997) presents a list of substances causing reproductive damage, based on the EEC list (EEC, 1994).

Substances causing allergy and hypersensitivity

■ Check whether the substance is classified R42 or R43 in the list of dangerous substances (EEC, 1994).

Allergy is a result of reactions to the substance by the immune system, part of the body's defence mechanism. On repeated exposure to certain substances or materials, the immune defence overreacts. The symptoms are visible in eyes, respiratory passages or the skin.

About 10-15% of the Danish population has some form of allergy, but only a limited proportion can be regarded as occupational.

Chemicals can be tested for their allergenic properties on skin exposure by a patch test, but there are no standard methods for assessing a substance's ability to cause allergy in the respiratory tract (asthma).

Rasmussen (1997) presents a list of the substances classified as allergenic substances in the EEC list (EEC, 1994).

Substances causing neurological damage

■ Check whether the substance is evaluated under the report from the Nordic Council of Ministers on substances causing neurological damage as being neurotoxic (group 1) or being possibly neurotoxic (group 2A or 2 B) (Simonsen *et al.*, 1995).

The brain consists of interconnected and highly sensitive nerve cells protected against exposure to undesired substances by a barrier. A number of substances including solvents and certain heavy metals can, however, penetrate this barrier and thereby damage the brain. The deleterious effect of solvents on the nervous system came into focus in Denmark in the late 1970s, when various painters suffered recognized occupational brain damage. Since then, a reduction in the use of organic solvents has been sought, and the annual number of cases of reported brain damage in Denmark has fallen greatly.

Rasmussen (1997) presents a list of the substances which have been assessed as being neurotoxic by a Nordic group of experts (Nordic Council of Ministers, 1995).

22.4.3.1.2 Impact of noise

- Register the noise level for all noise-generating processes. Measurements of noise are easy to make and recommended in the first instance, but the noise level can be estimated on the basis of a few rules of thumb, as given in Table 22.7.
- Only levels above 80 dB(A) are regarded as being able to cause hearing impairment.

22.4.3.1.3 Impact of monotonous repetitive work

Monotonous repetitive work is defined by:

Table 22.7. Guidelines for assessment of noise level in the workplace

Description	Experiences	Estimated noise level
Extremely loud noise	At a distance of 25 cm: shouting not understandable	> 95 dB(A)
Very loud noise	At a distance of 50 cm: shouting not understandable	90-95 dB(A)
Loud noise	At a distance of 1 m: loud speech not understandable	85-90 dB(A)
Disturbing noise	At a distance of 1 m: normal speech not understandable	80-85 dB(A)

- the same work movements are carried out continually,
- the work movements are repeated very frequently during a significant part of the working day,
- a work cycle is less than 30 seconds, or the same movements are repeated more than 50% of the time during the cycle (Danish Working Environment Service, 1994).

Consider, for example, the soldering of components on to a printed circuit, where the work with each printed circuit is a work cycle. Measure the time from the start of one circuit to the start of the next. If, for example, there are ten solderings on each circuit carried out with the same movement, the work will be considered monotonous and repetitive, even if the work cycle itself is over a minute.

The monotonous repetitive work must occur over at least half an hour to be classified as having an impact on the musculoskeletal apparatus.

- Check the length of the work cycle and note whether the same work operation is carried out more than 60 times within 30 minutes.

An assessment of whether the individual process involves monotonous repetitive work is made by an expert in working environmental issues.

22.4.3.1.4 Work accidents

In Denmark, work accidents should be reported to the National Labour Inspection if they cause injury to one or more employees leading to more than one day's absence from work. The frequency of accidents, measured as the number of accidents per million work-hours, reflects the department's or the company's general attention to the working environment, and is therefore a good assessment criterion.

- Register the number of work accidents causing absence from work for more than one day.

22.4.3.2 Collection of sector data

For assessment of the impact on the working environment in life-cycle stages outside one's own production, it is possible to use the number of reported work accidents per unit produced, transported or disposed of. Such data are not immediately available. It is usually necessary to calculate the number on the basis of the annually produced quantity and number of reported work injuries. This type of information can carry major uncertainties because the various sources do not use the same classification of sectors.

Sources of this type of data can be:

- Work injury statistics from the National Labour Inspection
- The company's annual reports/own statistics
- National statistics
- Sector associations

See Rasmussen (1997) for a more detailed review.

22.5 Modelling of the product system and aggregation of data

- If the reference product is very simple and contains only one or very few materials, the delimited product system is modelled in a linear structure as in a simple flow diagram.
- If the reference product is complex, the delimited product system is modelled in a hierarchical parts list structure.
- Model the product system in accordance with the scope definition. If the scope definition was done in the parts list structure, this model is, as described above, already set up.
- Model the product system using unit processes. Specify all processes in the product system as a number of key units of a process.
- Establish parts lists for the product system to and including product manufacture and for its use and disposal. These should be modelled in a complete hierarchy, where the complete product system is the top level.
- For each process describe:
  - geographic location,
  - product life stage,
  - special circumstances for the actual component/process,
  - any site- and region-specific circumstances (see Section 22.5.2).

22.5.1 Allocation during modelling

- Designate all joint processes and joint environmental exchanges when the complete product system is modelled, both those which are a consequence of co-products and those which are a consequence of recycling or reuse.
- Establish an overview of the products among which each of the joint exchanges is to be partitioned, and decide on where allocation can be avoided by expansion of the system boundaries. Extend the system boundaries at these points by modelling the actual alterations in the adjoining product systems.
- Find partitioning keys for the remaining joint exchanges.
- Enter allocation factors for the actual processes and exchanges.

22.5.1.1 Allocation on recycling

For allocation on recycling, all flows entering into the recycling must be identified. The various disposal routes must be determined, and losses of material grade ascribed to the quantities of materials consumed on this basis (cf. the example with corrugated cardboard in Examples 21.2, 21.4 and 21.5).

- Determine the disposal routes and the fraction of the products following each route.
- Divide the joint processes involved in consumption of materials into two fractions: one which corresponds in volume to the fraction sent for recycling, and one which corresponds in volume to the fraction subject to final disposal (cf. the examples for corrugated cardboard).
- Identify the loss of material grade applying to each of these fractions. Enter the relative grade losses as allocation factors for joint exchanges in the model.

22.5.2 Site- and region-specific aspects

- Identify the processes involving site-specific aspects and enter site factors.
- Identify the processes where region-specific normalization or weighting is required and define the actual reference areas for these normalizations and weightings in the processes in question.

Until the data are available, site-specific aspects such as those noted above should be defined with care, and only cases where obviously wrong environmental interpretations of the emissions are hereby avoided, but in such cases the exercise may be carried out. See also the description in Chapter 10.

22.5.3 Termination and aggregation

The modelled product system is now ready for computation. The modelling itself specifies the top level of the product system, namely the main stream from the materials to disposal.

Under this level there is a significantly larger part of the product system, partly the materials' product system back to raw material extraction and partly all of the "side streams", i.e. the product systems for energy and ancillary substances in all processes.

As described in Chapter 9, the inventory is concerned with terminating all processes, i.e. following all environmental exchanges back to the terminal input and forward to the terminal emissions to the recipient. Each of the exchanges is then aggregated. Aggregation is carried out *before* the calculations in the LCA if site-specific factors or region-specific normalizations and weightings *have not* been defined and *after* if these *have* been defined.

■ Compute the exchanges per functional unit. Use the following formula:

$$Q_i = T \cdot \sum_{up} Q_{i,up} + \frac{T}{L} \cdot \sum_{p} Q_{i,p}$$

where

Q_i = the sum of terminal exchanges (i) computed per functional unit,

T = duration of the functional unit (years),

L = lifetime of the product (years),

$Q_{i,p}$ = the terminal exchange (i) from the process (p) computed for the number of key units in the process entering into the product system; (p) designates all processes other than use and the non-terminal exchanges of the use process,

$Q_{i,up}$ = the terminal exhange (i) per annum from use (up) including the processes specified by all of the non-terminal exchanges deriving from the use process.

As also described previously, the exchanges can in principle be followed back infinitely. New non-terminal exchanges will continue to occur, however far one goes back. The process must therefore be stopped at some level. The EDIP method's PC program stops automatically after regression to an appropriate level.

22.5.4 Statistical data processing

■ Calculate a coefficient of variation for the total result when the LCA is to support strategically and economically very significant decisions in which it is important to know the degree of uncertainty in the conclusions of the LCA.

The statistical data processing in the modelling can, as described in Chapter 9, be done via simulation, i.e. by computing a number of times, and for each computation generating a value for each individual exchange on the basis of its statistical distribution as specified by the mean, the coefficient of variation and the distribution function. Mean values, standard deviations and confidence intervals are then calculated on the resulting aggregate figures.

This process is a very simple and suitable method of estimating an aggregate variability, and the only practicable method in an LCA. But it requires a very large number of calculations and can in practice only be done on a computer.

For modelling of one product the procedure is:

- Generate random numbers for each individual exchange x_i and aggregate separately for each individual exchange.
- Repeat n times.
- Calculate mean, standard deviation and confidence intervals for the resulting population of each individual exchange.

When comparing two products/solutions the procedure is:

- For each product/solution, model that part of the product system which separates the two solutions.
- Generate random numbers for each individual exchange, x_i, within the specified distribution, and aggregate for each individual exchange, x_i, separately.
- Then subtract the two systems from each other for each individual exchange separately, i.e. obtain a value of Δx_i for each exchange.
- Repeat n times, i.e. n results of Δx_i.
- Calculate mean, standard deviation and confidence intervals for the resulting population of Δx_i's for each individual exchange.

22.5.5 PC program

- Use a PC program to model the product and the calculations in the LCA.

Without a PC program, the calculations and especially the statistical calculations are extremely time-consuming. It is also a requirement for simulations of various changes in the product and the product system that modelling and calculation can be done within a reasonable time.

22.6 Presentation of the computations

- Present the computations in a table as in Table 9.8 in Chapter 9.
- If possible, include confidence intervals for the figures as calculated in the above simulations.
- If feasible, mark the exchanges which do not enter into the environmental interpretations in the tool's impact assessment section to inform the reader what the output of the impact assessment does not include.

23. Procedure for impact assessment

In the impact assessment, the inventory is translated into potential contributions to various impacts within the main groups environment, resources and working environment. This leads to a strong reduction in the size of the body of data, especially for emissions to the environment. At the same time there is a weighting of the individual interchanges according to the level of their potential impact.

The assessment also helps to focus work on any subsequent qualification of the data of the inventory. It shows which of the interchanges have the largest potential impacts and it should thus be performed with the highest possible degree of precision and certainty. The impact assessment thus qualifies the inventory as a basis for decisions in comparisons between products, and it can also be used to focus the further collection of data.

The LCA method's impact assessment phase is subject to the same general requirements with respect to transparency, reproducibility and scientific foundations as the other phases in the LCA. It is in the nature of the concept that an assessment can never be totally objective, but in the EDIP method's assessment phase, the unavoidable subjective choices are as far as possible postponed to the last element of the assessment.

The impact assessment progresses through three steps:

1. Calculation of potential contributions to various categories of impact.

In the first step, the types of environmental impacts to which the interchange may contribute are assessed. For each individual emission to the environment, a calculation is then made of the magnitude of the contributions to various impact categories. These contributions are called the emission's environmental impact potentials. The procedure for calculation of environmental impact potentials is given in Sections 23.1-23.9 of the Toolbox.

2. Comparison of impact potentials and resource consumptions with a common reference to show which are large and which small.

In many cases, which alternative will have the least impact cannot be determined on the basis of the summarized resource consumption and potential impacts on the working environment and the calculated environmental impact potentials. One alternative will often have least impact in some areas, while another alternative will have least impact in others. In such situations it is important to be able to assess which of the potential impacts and resource consumptions are large and which are small. This assessment can be difficult to perform on the basis of the figures alone. Potential impacts and resource consumptions are therefore compared with a common reference.

3. Weighting of the normalized impact potentials and resource consumptions to determine which are most significant.

Before the normalized impact potentials or resource consumptions are directly comparable, account must be taken of the seriousness of each individual impact in relation to the others. Scientific, political and normative considerations are involved, expressed in a weighting factor for each of the impact categories and resource consumptions.

The normalization and weighting elements are interdependent in the EDIP method and are therefore presented together in Section 23.10.

The elements of the impact assessment phase are shown in Figure 23.1 together with examples of results of the individual steps.

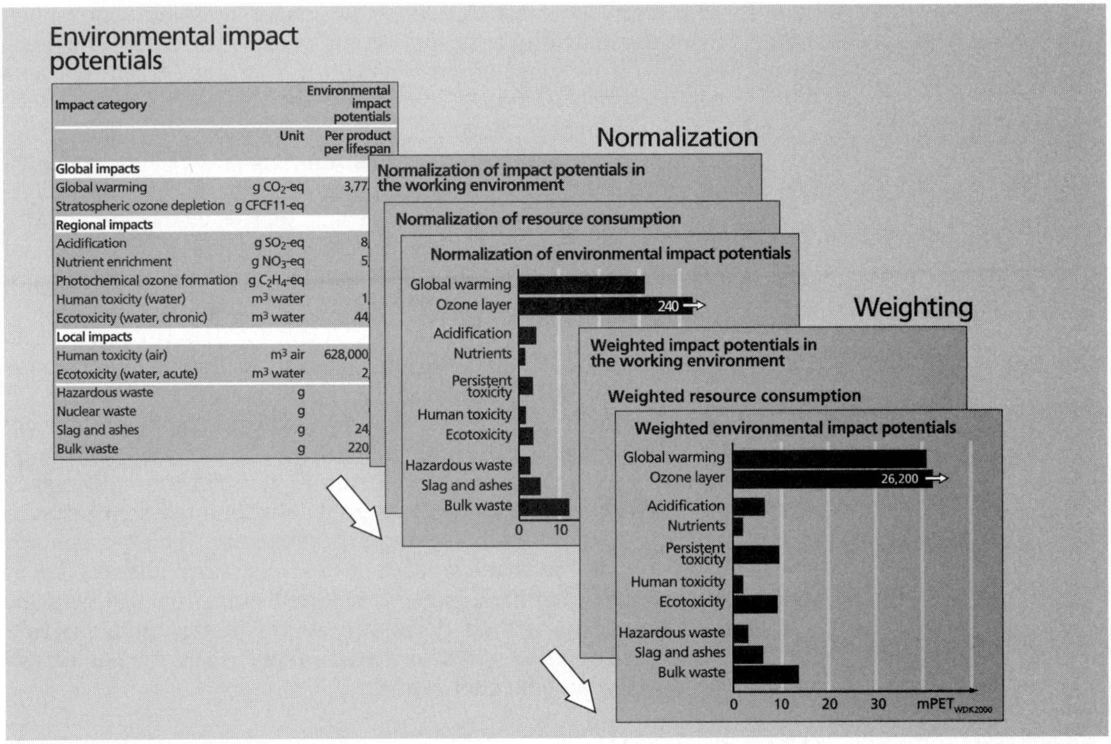

Figure 23.1. Steps in the impact assessment method

For the environmental impact categories "ecotoxicity" and "human toxicity", the quantitative assessment involves a considerable amount of work. A **qualitative assessment method** for these impact categories has therefore also been developed as a part of the EDIP method, based on the information presented in the chemicals' hazard classification and labelling.

The qualitative assessment method for ecotoxicity and human toxicity is given in Section 23.11.

Instructions for carrying out the sensitivity analysis are given in Chapter 24.

23.1 Assessment of environmental impact potentials

Chapter 1 gives the status of the environment, including a review of the various environmental impacts known today. The method of calculation of potential impacts for the categories of environmental impacts included in the EDIP method's assessment tool is given in Sections 23.2-23.9 of the Toolbox.

Each category of environmental impact is reviewed separately. Criteria are established to determine whether a substance contributes to a certain environmental impact category. In practice this can be decided for several impact categories by checking whether the substance is included in a subsequent table of equivalency factors for substances which contribute to the environmental impact in question. An equivalency factor expresses the potential environmental impact for a substance as the quantity of a reference substance which will make the same contribution to the environmental impact as one gram of the substance. For each environmental impact category, the reference substance is chosen as a typical or important contributor.

For non-global impact categories, it can be relevant to consider use of site factors in the calculation of potential impacts. Site factors and their application are discussed in the box on page 110, and Table 10.4 illustrates examples of situations where inclusion of site factors will qualify the calculated impact potentials. The use of site factors is, however, not yet generally implemented in the EDIP method, and it is therefore discussed in the Toolbox only under a single impact category.

There are thousands of different substances which can contribute to the impacts "ecotoxicity" and "human toxicity". Equivalency factors have, however, only been calculated for a few hundred substances. A detailed introduction is therefore given to how equivalency factors are calculated. Many users of the EDIP method will not possess the necessary chemical or ecotoxicological background to calculate equivalency factors for these two categories themselves. The work must then be given to an external consultant with the requisite expertise. This part of the task depends only on the characteristics of the substances, and as soon as these substances are identified, it can be carried out in parallel with the rest of the LCA. The method developed serves in this situation as a common reference method for assessment of potentials for ecotoxicity and human toxicity in connection with LCAs.

23.2 Global warming

The earth's atmosphere absorbs part of the energy emitted as infrared radiation from earth towards space, and is thereby heated. This natural greenhouse effect leading to a warming of the atmosphere has with certainty been increased over the past few centuries by human activities leading to accumulation of such compounds as CO_2, N_2O, CH_4 and halocarbons in the atmosphere. The most import human contribution to the global warming impact is attributed to the combustion of fossil fuels such as coal, oil and natural gas.

The consequences of the man-made global warming may include increased global average temperatures and sudden regional climatic changes.

The worldwide network of climate researchers and atmospheric chemists, the IPCC (Intergovernmental Panel on Climate Change), is following the latest development in our knowledge of global warming and issuing regular status reports. These status reports comprise the basis of the EDIP method's assessment tool for global warming.

Global warming is an impact affecting the environment on a global scale.

23.2.1 Determine which substances contribute to global warming

- Check whether the substance is found in Table 23.1, equivalency factors for global warming, in the next section.

For a substance to be regarded as contributing to global warming, it must be a gas at normal atmospheric temperatures and:

- be able to absorb infrared radiation and be stable in the atmosphere with a residence time of years to centuries,

or

- be of fossil origin and converted to CO_2 on degradation in the atmosphere.

The EDIP method's criteria for which substances contribute to global warming generally follow the IPCC's recommendation of excluding indirect contributions to the greenhouse effect, i.e. contributions attributable to a gas affecting the atmospheric lives of other greenhouse gases already present. The indirect effects are difficult to model, and the IPCC is therefore refraining for the time being from quantifying indirect contributions with the exception of contributions from the gas methane (Albritton *et al.*, 1995a). The EDIP method, however, goes further than the IPCC's recommendation in including the contribution from organic compounds which is due to their inevitable degradation to CO_2 in the atmosphere.

For every emission of CO_2 it is important to check whether it is a nett addition of CO_2 to the atmosphere or simply a manipulation of part of the natural biogeochemical carbon cycle. If the source of carbon is fossil (coal, oil, natural gas), conversion to CO_2 will mean a nett addition. If, however, there is a question of combustion or breakdown of material which does not derive from fossil carbon sources, but for example from biomass, there will normally be no nett addition because the material in question was generated recently by fixing of CO_2 from the atmosphere, and it would in any event inevitably be broken down to CO_2 again (see Hauschild & Wenzel, 1997d, for a more detailed discussion).

The list of substances considered to contribute to global warming is short and it can be regarded as exhaustive. It is thus not necessary in practice to check whether a substance fulfils the criteria in order to decide whether it is to be regarded as contributing to global warming. It is sufficient to consult the list of global warming equivalency factors in Table 23.1.

23.2.2 Calculate the global warming potential

- Decide the time scale for which the global warming potential is to be calculated. Unless specific reasons indicate otherwise, select 100 years.
- Find the substance's equivalency factor for the chosen time scale in Table 23.1.
- Calculate the global warming potential by multiplying the magnitude of the emission with the equivalency factor found: $EP(gw) = Q \cdot EF(gw)$.

The IPCC presents global warming potentials (GWPs), which for each individual greenhouse gas express the potential contribution to global warming from a given quantity of the gas relative to the contribution from a corresponding quantity of CO_2. Calculation of the GWP values is based on model simulations of the gases' expected behaviour in the atmosphere over a large number of years.

The present version of the EDIP method employs the GWPs presented by the IPCC in their 1994 status report (Albritton *et al.*, 1995a) for calculation of the global warming potential.

Equivalency factor

$$EF(gw)_i = GWP_i = \frac{\text{Contribution to global warming from gas (i) over T years}}{\text{Contribution from } CO_2 \text{ to global warming over T years}}$$

The IPCC does not include indirect contributions from gases other than methane. The EDIP method nevertheless offers the option of including that part of the indirect contribution from volatile organic compounds (VOCs) and carbon monoxide (CO) attributable to their predictable conversion to CO_2. This applies only if the gases originate from fossil resources.

As is clear from Table 23.1, the choice of time scale T plays a large role in the magnitude of the equivalency factor. For those gases with atmospheric lives significantly shorter than that of the reference gas CO_2, the equivalency factor decreases with an increase in T. The opposite is the case for those gases with significantly longer lives than CO_2. In accordance with general LCA practice, the EDIP method recommends using a time scale of 100 years, but equivalency factors for 20 and 500 years are also given in the table in order to show the significance of T and to provide the option of alternative choices.

Substance	Formula	GWP g CO_2/g substance 20 years	100 years	500 years
Carbon dioxide	CO_2	1	1	1
Methane	CH_4	62	25	8
Nitrous oxide	N_2O	290	320	180
CFC11	$CFCl_3$	5000	4000	1400
CFC12	CF_2Cl_2	7900	8500	4200
CFC113	$CF_2ClCFCl_2$	5000	5000	2300
CFC114	CF_2ClCF_2Cl	6900	9300	8300
CFC115	CF_2ClCF_3	6200	9300	13000
Tetrachloromethane	CCl_4	2000	1400	500
HCFC22	CHF_2Cl	4300	1700	520
HFC123	CF_3CHCl_2	300	93	29
HFC124	$CF_3CHCFCl$	1500	480	150
HCFC141b	$CFCl_2CH_3$	1800	630	200
HCFC142b	CF_2ClCH_3	4200	2000	630
HCFC225ca	$CF_3CF_2CHCl_2$	550	170	52
HCFC225cb	CF_2ClCF_2CHFCl	1700	530	170
1,1,1-trichloroethane	CH_3CCl_3	360	110	35
Chloroform	$CHCl_3$	15	5	1
Methylene chloride	CH_2Cl_2	28	9	3
HFC134a	CH_2FCF_3	3300	1300	420
HFC152a	CHF_2CH_3	460	140	44
Halon 1301	CF_3Br	6200	5600	2200
Carbon monoxide *	CO	2	2	2
Hydrocarbons (NMHC) *	various	3	3	3
Partly oxidized hydrocarbons *	various	2	2	2
Partly halogenated hydrocarbons *	various	1	1	1

* Contributes indirectly due to conversion into CO_2. Only compounds of petrochemical origin.

Table 23.1. Equivalency factors for global warming for time scales of 20, 100 and 500 years (Albritton et al., 1995a; Hauschild & Wenzel, 1997d)

23.3 Stratospheric ozone depletion

Stratospheric ozone is broken down as a consequence of man-made emissions of halocarbons, i.e. CFCs, HCFCs, halons and other long-lived gases containing chlorine and bromine. The ozone content of the stratosphere is therefore decreasing, and since 1985 a dramatic temporary thinning of the ozone layer, often referred to as the "ozone hole", has been observed each year, over the South Pole. In the last few years the depletion of ozone has also accelerated over the northern hemisphere. As a consequence of this the intensity of hazardous ultraviolet radiation at the earth's surface has increased over parts of the southern and northern hemispheres. This can have dangerous consequences in the form of increased frequency of skin cancer in humans and damage to the plants which form the basis of all ecosystems.

The stratospheric depletion of ozone is an impact which affects the environment on a global scale.

23.3.1 Determine which substances contribute to ozone depletion

- Check whether the substance is found in Table 23.2, equivalency factors for the ozone depletion impact category, in the next section.

For a substance to be considered as contributing to stratospheric ozone depletion, it must:

- be a gas at normal atmospheric temperatures,
- contain chlorine or bromine,
- be stable with a lifetime in the atmosphere of several years to centuries, to allow for its transportation up into the stratosphere.

The man-made substances contributing to the stratospheric breakdown of ozone are simple gaseous organic compounds with a substantial content of chlorine or bromine. The most important groups of ozone-depleting halocarbons are the CFCs, the HCFCs, the halons and methyl bromide. In contrast to these, the HFCs are a group of halocarbons which contain neither chlorine nor bromine but only fluorine, and which are therefore not regarded as contributors to stratospheric depletion of ozone.

CFCs and HCFCs are used mainly as foaming agents in foam plastic, as refrigerants and as solvents. Halons are used as fire-extinguishing agents in fire-fighting equipment. Methyl bromide is used in the disinfection of buildings and of soil in market gardens.

The production of halocarbons is regulated under the Montreal Protocol by the UN. Under this protocol, production of CFCs and halons ceased in the industrialized world on 1 January 1996. Consumption of methyl bromide was frozen in 1995, and consumption of HCFCs is to be decreased gradually. The deadlines for phasing out have been brought forward in a number of countries. CFCs and halons can, however, continue to be produced in Third World countries until 2010 (UNEP, 1993), and they will therefore also occur in future in inventories of product systems.

As for the greenhouse gases, the list of compounds considered as contributing to the stratospheric depletion of ozone is short and can be regarded as exhaustive. In practice it will therefore not be necessary to check a substance under the above criteria to decide whether it contributes to ozone depletion. It will be sufficient to consult the list of equivalency factors in the following section.

23.3.2 Calculate the ozone depletion potential

- Decide the time scale for which the ozone depletion potential is to be calculated. Unless specific reasons indicate otherwise, select time scale ∞ (infinite).
- Find the substance's equivalency factor for the chosen time scale in Table 23.2.
- Calculate the ozone depletion potential by multiplying the magnitude of the emission by the equivalency factor:
  $EP(od) = Q \cdot EF(od)$.

Together with the UN's environment programme UNEP and a number of other organizations, the World Meteorological Organization (WMO) organizes the "Global Ozone Research and Monitoring Project", a research network of experts in atmospheric chemistry. The network reviews international developments in scientific knowledge of stratospheric ozone depletion and every few years issues status reports summarizing the latest findings. The status reports present the Ozone Depletion Potentials (ODPs), which for individual gases express the ozone depletion potential as an equivalent emission of a reference substance CFC11 ($CFCl_3$). These ODP values are used as equivalency factors in the calculation of the ozone depletion potential. The equivalency factor is thus defined as:

Equivalency factor

$$EF(od)_i = ODP_i = \frac{\text{Contribution to stratospheric ozone depletion from gas (i)}}{\text{Contribution to stratospheric ozone depletion from CFC11}}$$

The focus in this definition is on the gas's total contribution to stratospheric ozone depletion throughout its life in the atmosphere. If the plans for phasing out halocarbons in the Montreal Protocol are followed, the breakdown of ozone in the stratopshere is predicted to peak within the next 10 years, and the problem will then gradually decline. On this basis the contribution to ozone depletion realized within the next few years can be viewed as more serious than contributions which will only be realized in the distant future. It can therefore be more relevant to view the gases' contributions in a significantly shorter time perspective. For the most short-lived of the gases, especially the HCFCs, this will result in some markedly larger equivalency factors.

In accordance with general LCA practice, the EDIP method does, however, recommend use, for equivalency factors, of ODP values representing the gases' full contributions, but the table also gives equivalency factors for 5, 20 and 100 years for some of the gases.

		ODP g CFC11/g substance			
Substance	**Formula**	**5 years**	**20 years**	**100 years**	∞
CFC11	$CFCl_3$	1	1	1	1
CFC12	CF_2Cl_2				0.82
CFC113	$CF_2ClCFCl_2$	0.55	0.59	0.78	0.90
CFC114	CF_2ClCF_2Cl				0.85
CFC115	CF_2ClCF_3				0.40
Tetrachloromethane	CCl_4	1.26	1.23	1.14	1.20
HCFC22	CHF_2Cl	0.19	0.14	0.07	0.04
HCFC123	CF_3CHCl_2				0.014
HCFC124	CF_3CHFCl				0.03
HCFC141b	$CFCl_2CH_3$	0.54	0.33	0.13	0.10
HCFC142b	CF_2ClCH_3	0.17	0.14	0.08	0.05
HCFC225ca	$CF_3CF_2CHCl_2$				0.02
HCFC225cb	CF_2ClCF_2CHFCl				0.02
1,1,1-trichloroethane	CH_3CCl_3	1.03	0.45	0.15	0.12
Methyl chloride	CH_3Cl				0.02
Halon 1301	CF_3Br	10.3	10.5	11.5	12
Halon 1211	CF_2ClBr	11.3	9.0	4.9	5.1
Methyl bromide	CH_3Br	15.3	2.3	0.69	0.64

Table 23.2. Equivalency factors for stratospheric ozone depletion for time scales of 5, 20 and 100 years and infinity (Solomon & Albritton, 1992; Solomon & Wuebbles, 1995)

23.4 Photochemical ozone formation

When solvents and other volatile organic compounds are released to the atmosphere, they are often degraded within a few days. The reaction involved is an oxidation, which occurs under the influence of light from the sun. In the presence of oxides of nitrogen (NO_x), ozone can be formed. The oxides of nitrogen are not consumed during ozone formation, but have a catalyst-like function. This process is termed photochemical ozone formation.

The volatile organic compounds or VOCs are broken down especially in the troposphere, the lowest region of the atmosphere, to which they are emitted. The most significant man-made sources of VOCs are road transport with its emission of unburnt petrol and diesel and the use of organic solvents, e.g. in paints.

Ozone is an unstable gas with a half-life in the troposphere of a few weeks, and ozone formed in the troposphere thus cannot rise to the stratosphere and there counteract the depletion of the ozone layer. Despite ozone's short life, the ozone concentration in the troposphere rises by about 1% a year over most of the northern hemisphere, where most VOC and NO_x emissions occur. In the southern hemisphere, however, the ozone concentration in the troposphere is practically constant.

Ozone attacks organic compounds in plants and animals or materials exposed to air. This leads to an increased frequency of problems of the respiratory tract in humans during periods of photochemical smog in cities. For agriculture it causes a reduction in yield which for Denmark is conservatively estimated to be about 10% of total production.

Photochemical ozone formation is an impact which affects the environment on both local and regional scales.

23.4.1 Determine which substances contribute to photochemical ozone formation

- Check whether the substance is an organic compound with a boiling point below 250°C and
  1) contains hydrogen or
  2) contains double bonds between carbon atoms

  and if so consider the substance as a contributor to photochemical ozone formation.
- Consider carbon monoxide (CO) as a further contributor to photochemical ozone formation.

Crude oil contains hundreds of compounds which fulfil these criteria. The use of crude oil products can therefore lead to emission of very many different substances with a potential for photochemical ozone formation. The list of equivalency factors in Table 23.3 is not exhaustive and therefore cannot be used alone as a basis for deciding which compounds contribute to photochemical ozone formation.

The presence of oxides of nitrogen can be equally as important a man-made factor in photochemical ozone formation as emission of VOCs. Despite this, a contribution from oxides of nitrogen to photochemical ozone formation cannot be calculated because the equivalency factor system used for calculation of ozone formation potentials does not facilitate calculation of an equivalency factor for NO_x.

The significance of NO_x for ozone formation is, however, reflected in the fact that two sets of equivalency factors are used - one for emissions of VOCs occurring in areas with a low background concentration of NO_x and one for emissions occurring in areas with a high background concentration of NO_x.

23.4.2 Calculate the photochemical ozone formation potential

- Determine whether the emission is occurring in an area with high or low background concentration of NO_x. The border between the two classes lies at a mean annual concentration of NO_x of 10 ppbv or 0.02 mg/m^3 over rural areas.
- For individual substances, find the substance's equivalency factor in Table 23.3.
- For composite VOC emissions, find the emission's equivalency factor in Table 23.4.
- Calculate the photochemical ozone formation potential by multiplying the magnitude of the emission by the equivalency factor found: $EP(po) = Q \cdot EF(po)$.

UNECE, the United Nations Economic Council for Europe, has established a committee to investigate transnational air pollution. For use in mapping the most significant sources of the rising ozone concentration in the troposphere and the increasingly more frequent episodes of photochemical smog, this committee applies factors which, for each individual compound, express the potential contribution to photochemical ozone formation. These factors are called Photochemical Ozone Creation Potentials (POCPs). In the same way as the GWP values for global warming and the ODP values for stratospheric ozone depletion, the POCP values express the ozone formation potential as an equivalent emission of a chosen reference substance. For photochemical ozone formation the reference substance is the gas ethylene (C_2H_4).

Equivalency factor

$$EF(po)_i = POCP_i = \frac{\text{Contribution to ozone formation from gas (i)}}{\text{Contribution to ozone formation from } C_2H_4}$$

There is no international panel of experts for the environmental impact of photochemical ozone formation such as there is for global environmental impacts. Agreement among participating countries in the UNECE on using the POCP factor system is therefore the closest approximation to international recognition of any equivalency factor system for photochemical ozone formation.

The POCP values are calculated with the aid of atmospheric chemical models and a series of assumptions must be made on climatic conditions and the magnitude of the simultaneous emissions of a number of other VOCs and of NO_x. The assumptions underlying the POCP values in Table 23.3 correspond to typical situations in areas with low and high background concentrations of NO_x. The assumptions are discussed in the references presenting these POCP values (Andersson-Sköld *et al.*, 1992; Derwent & Jenkin, 1990).

A time scale must also be chosen for ozone formation in model calculations of POCP values. A POCP value for a short time scale of 24 hours describes the photochemical ozone formation immediately surrounding the place of emission corresponding to potential contribution to episodes of photochemical smog. For a longer time scale of e.g. a week, most of the VOCs will be broken down and the POCP value provides a better expression of the total ozone formation potential. Table 23.3 presents POCP factors calculated for an intermediate time scale of 4-5 days.

Table 23.3. Equivalency factors for photochemical ozone formation, calculated for a time scale of 4-5 days for areas with high and low background concentrations of NO_x (above and below 10 ppbv or 0.02 mg/m³ over rural areas) (Andersson-Sköld et al., 1992; Derwent & Jenkin, 1990)

Substance	Formula	EF (po) (low NO_x) g C_2H_4/g gas	(high NO_x) g C_2H_4/g gas
Alkanes		0.4±0.1	0.4±0.1
Methane	CH_4	0.007	0.007
Ethane	C_2H_6	0.1	0.1
Propane	C_3H_8	0.5	0.4
n-butane	C_4H_{10}	0.5	0.4
Isobutane	$CH(CH_3)_3$	0.4	0.3
n-pentane	C_5H_{12}	0.3	0.4
Isopentane	$CH_2CH(CH_3)C_2H_5$	0.3	0.3
n-hexane	C_6H_{14}	0.5	0.4
2-methylpentane	$CH_2CH(CH_3)C_3H_7$	0.5	0.5
3-methylpentane	$C_2H_5CH(CH_3)CH_3$	0.4	0.4
2,2-dimethylbutane	$CH_3C(CH_3)_2C_2H_5$	0.3	0.3
2,3-dimethylbutane	$CH_3CH(CH_3)CH(CH_3)CH_3$	0.4	0.4
n-heptane	C_7H_{16}	0.5	0.5
2-methylhexane	$CH_3CH(CH_3)C_4H_{10}$	0.5	0.5
3-methylhexane	$C_2H_5CH(CH_3)C_3H_7$	0.5	0.5
n-octane	C_8H_{18}	0.5	0.5
2-methylheptane	$CH_3CH(CH_3)C_5H_{12}$	0.5	0.5
n-nonane	C_9H_{20}	0.4	0.5
2-methyloctane	$CH_3CH(CH_3)C_6H_{14}$	0.5	0.5
n-decane	$C_{10}H_{22}$	0.4	0.5
2-methylnonane	$CH_3CH(CH_3)C_7H_{16}$	0.4	0.4
n-undecane	$C_{11}H_{24}$	0.4	0.4
n-dodecane	$C_{12}H_{26}$	0.3	0.4
methylcyclohexane	C_7H_{14}	0.5	0.6
Alkanes		0.5±0.2	0.9±0.1
Ethylene	C_2H_4	1.0	1.0
Propylene	C_3H_6	0.6	1.0
1-butene	$CH_2CHC_2H_5$	0.5	1.0
2-butene (trans)	$C_2H_5CHCH_2$	0.4	1.0
Isobutene	$CH_3CCH_3CH_2$	0.6	0.6
1-pentene	$CH_2CHC_3H_7$	0.4	1.1
2-pentene (trans)	$C_2H_5CHCHCH_3$	0.4	0.9
2-methylbut-1-ene	$CH_2C(CH_3)C_2H_5$	0.2	0.8
2-methylbut-2-ene	$CH_3C(CH_3)CHCH_3$	0.5	0.8
3-methylbut-1-ene	$CH_2CHCH(CH_3)CH_3$	0.5	0.9
Isoprene	C_5H_8	0.6	0.8
Alkynes			
Acetylene	C_2H_2	0.4	0.2

Substance	Formula	EF (po) (low NO_x) g C_2H_4/g gas	(high NO_x) g C_2H_4/g gas
Aromatics		0.4±0.1	0.8±0.3
Benzene	C_6H_6	0.4	0.2
Toluene	$(C_6H_6)CH_3$	0.5	0.6
o-xylene	$(C_6H_6)(CH_3)_2$	0.2	0.7
m-xylene	$(C_6H_6)(CH_3)_2$	0.5	1.0
p-xylene	$(C_6H_6)(CH_3)_2$	0.5	0.9
Ethylbenzene	$(C_6H_6)C_2H_5$	0.5	0.6
1,2,3-trimethylbenzene	$(C_6H_6)(CH_3)_3$	0.3	1.2
1,2,4-trimethylbenzene	$(C_6H_6)(CH_3)_3$	0.3	1.2
1,2,5-trimethylbenzene	$(C_6H_6)(CH_3)_3$	0.3	1.1
o-ethyltoluene	$(C_6H_6)(CH_3)(C_2H_5)$	0.4	0.7
m-ethyltoluene	$(C_6H_6)(CH_3)(C_2H_5)$	0.4	0.8
p-ethyltoluene	$(C_6H_6)(CH_3)(C_2H_5)$	0.4	0.7
n-propylbenzene	$(C_6H_6)C_3H_7$	0.5	0.5
Isopropylbenzene	$(C_6H_6)CH(CH_3)_2$	0.5	0.6
Aldehydes		0.3±0.2	0.5±0.4
Formaldehyde	HCHO	0.3	0.4
Acetaldehyde	CH_3CHO	0.2	0.5
Propionaldehyde	C_2H_5CHO	0.2	0.6
Butyraldehyde	C_3H_7CHO	0.2	0.6
Isobutyraldehyde	$CH(CH_3)_2CHO$	0.3	0.6
Valeraldehyde	C_4H_9CHO	0.3	0.7
Acrolein	(CH_2CHCHO)	0.8	0.8
Ketones		0.2±0.1	0.4±0.2
Acetone	$(CH_3)_2CO$	0.1	0.2
Methyl ethyl ketone	$CH_3COC_2H_5$	0.2	0.4
Methyl isobutyl ketone	$CH_3COC(CH_3)_3$	0.3	0.6
Alcohols		0.2±0.02	0.3±0.1
Methanol	CH_3OH	0.2	0.1
Ethanol	C_2H_5OH	0.2	0.3
Isopropanol	$CH_3CHOHCH_3$	0.2	0.2
Butanol	C_4H_9OH	0.2	0.4
Isobutanol	$COH(CH_3)_3$	0.3	0.3
Butan-2-diol	$CH_3C(OH)_2C_2H_5$	0.3	0.3
Ethers		0.4±0.1	0.4±0.2
Dimethyl ether	CH_3OCH_3	0.3	0.3
Propylene glycol methyl ether	$CH_3CHOCH_2OCH_3$	0.5	0.5
Esters		0.2±0.1	0.2±0.1
Methyl acetate	CH_3COOCH_3	0.1	0.03
Ethyl acetate	$CH_3COOC_2H_5$	0.3	0.2
Isopropyl acetate	$CH_3COOCH(CH_3)_2$	0.2	0.2
n-Butyl acetate	$CH_3COOC_4H_9$	0.3	0.3
Isobutyl acetate	$CH_3COOC(CH_3)_3$	0.4	0.3
Propylene glycol methyl ether acetate	$CH_3CHOCH_3CH_2OOCCH_3$	0.2	0.1
Chloro-alkanes		0.01±0.01	0.004±0.004
Dichloromethane	CH_2Cl_2	0.02	0.01
Chloroform	$CHCl_3$	0.004	0.003
Methyl chloroform	CH_3CCl_3	0.002	0.001
Chloro-alkanes		0.2±0.3	0.3±0.4
Trichloroethylene	$C_2H_3Cl_3$	0.1	0.1
Tetrachloroethylene	$C_2H_2Cl_4$	0.01	0.01
Allyl chloride	CH_2CHCH_2Cl	0.5	0.7
Inorganic compounds			
Carbon monoxide	CO	0.04	0.03

In the references cited, POCP values have been calculated only for the individual VOCs of greatest significance for total photochemical ozone formation in Europe. But these are not necessarily the compounds of greatest significance for the ozone formation potential in an LCA. For example, styrene will give a significant contribution to the ozone formation potential in the LCA of the polymer polystyrene, but none of the references gives a POCP value for styrene. It can therefore be an advantage to be able to make an estimation of missing POCP values. Hauschild & Wenzel (1997a) describe various methods of estimating POCP values. As an example, a POCP value for styrene is estimated.

Emissions of VOCs will often figure in the inventory for a product system, without an indication of which individual compounds they are composed of. The composition can vary greatly for different sources, but if the *source* of the VOC emission is known (e.g. "exhaust from petrol-engined cars"), it may be possible in Table 23.4 to find an average POCP value which is representative for the VOC mixture in question. The average POCP values are calculated as a weighted average for the VOC mixture which is characteristic of the type of source in question.

Weighted average POCP values can be calculated for other types of sources if the average composition of the source's VOC emission is known. For example, a weighted POCP for the aggregated VOC "source" "polystyrene production from oil to final polystyrene polymers" is calculated in Hauschild & Wenzel (1997a).

Table 23.4. Average equivalency factors for photochemical ozone formation for VOC mixtures from various sources (Hauschild & Wenzel, 1997a)

VOC source	EF (po) (low NO_x) g C_2H_4/g VOC	EF (po) (high NO_x) g C_2H_4/g VOC
Petrol powered car, exhaust	0.5	0.6
Petrol powered car, vapour	0.4	0.5
Diesel powered car, exhaust	0.5	0.6
Power plants	0.4	0.5
Combustion of wood or twigs	0.6	0.6
Food industry	0.4	0.4
Surface coating	0.5	0.5
Chemical cleaning of clothes	0.3	0.3
Refining and distribution of oil	0.4	0.5
Natural gas leakage	0.02	0.02
Coal mining	0.007	0.007
Farming	0.4	0.4
Controlled landfilling of household waste	0.007	0.007

23.5 Acidification

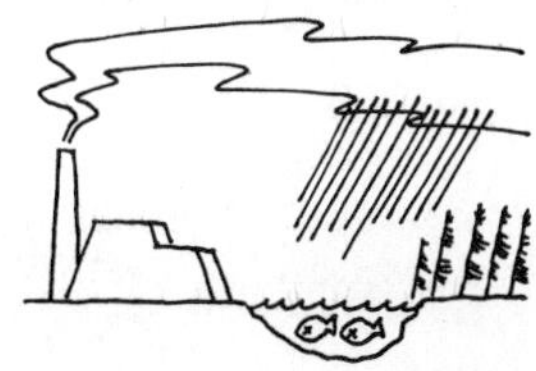

When acids and compounds which can be converted to acids are emitted to the atmosphere and deposited in water and soil, the addition of hydrogen ions may eventually result in a decrease in pH, i.e. an increase in acidity. This has consequences in the form of a widespread decline of coniferous forests in many places in Europe and the USA, and increased fish mortality in mountain lakes in Scandinavia and central Europe. Corrosion damage to metals and disintegration of surface coatings and mineral building materials are also caused by acidification on exposure to wind and weather.

The most significant man-made sources of acidification are combustion processes in electricity and heating production, and transport. The contribution to acidification is greatest when the fuels used contain sulphur.

Acidification is an impact which mainly affects the environment on a regional scale.

23.5.1 Determine which substances contribute to acidification

- Check whether the substance is found in Table 23.5, equivalency factors for acidification.

For a substance to be considered a contributor to acidification:

- it must cause introduction or release of hydrogen ions in the environment, and
- the anions which accompany the hydrogen ions must be leached or washed out from the system.

The addition of hydrogen ions occurs either when the substance itself is an acid or is converted to an acid, or when hydrogen ions are released as the substance is converted in the environment.

Acidification can be caused by emissions to air, water and soil. The number of substances which should be considered contributors to acidification is not large, and in practice the list of equivalency factors in Table 23.5 can be used to decide whether a substance contributes to acidification. Note that emission of organic acids is not regarded as a contribution to acidification.

23.5.2 Calculate the acidification potential

- Find the substance's equivalency factor in Table 23.5.
- Calculate the acidification potential by multiplying the magnitude of the emission by the equivalency factor found:
  $EP(ac) = Q \cdot EF(ac)$.

There is no internationally accepted system of equivalency factors for acidifying substances. In contrast to the global environmental impacts and photochemical ozone formation, it has therefore been necessary in the EDIP programme to develop equivalency factors for acidification.

Substance	Formula	EF (ac) g SO_2-eq/g substance
Sulphur dioxide	SO_2	1
Sulphur trioxide	SO_3	0.80
Nitrogen dioxide	NO_2	0.70
Nitrogen oxides	NO_x *	0.70
Nitric oxide	NO	1.07
Hydrochloric acid	HCl	0.88
Nitric acid	HNO_3	0.51
Sulphuric acid	H_2SO_4	0.65
Phosphoric acid	H_3PO_4	0.98
Hydrofluoric acid	HF	1.60
Hydrogen sulphide	H_2S	1.88
Ammonia	NH_3	1.88

* For NO_x the x is regarded as 2.

Table 23.5. Equivalency factors for acidification (Hauschild & Wenzel, 1997b)

Calculation of the equivalency factor for a substance is based on the number of hydrogen ions which can theoretically be released from the substance directly or after any conversions in the environment.

Whether or not the acidification potential is realized depends, as noted, on the accompanying anion release from the ecosystem which receives the emission. For some substances, the proportion of anions released can vary from ecosystem to ecosystem. To reflect this, Hauschild & Wenzel (1997b) propose a site factor, the value of which can vary between 0 and 1.

As for the potentials for the other types of environmental impacts, the acidification potential is expressed as an equivalent quantity of a reference substance. Sulphur dioxide (SO_2) is used as the reference substance for acidification. Should the inventory include compounds causing acidification other than those shown in Table 23.5, it is easy to calculate equivalency factors for them. The method is given in Hauschild & Wenzel (1997b).

23.6 Nutrient enrichment

Nutrient enrichment is an impact on ecosystems from substances containing nitrogen (N) or phosphorus (P). As a rule, the availability of one of these nutrients will be a limiting factor for growth in the ecosystem, and if this nutrient is added, the growth of algae or plants will be increased. In aquatic ecosystems this can cause the occurrence of situations without oxygen in the bottom strata due to the increased algal growth and subsequent breakdown of algae at the bottom. On land, ecosystems poor in nutrients such as raised bogs, commons and heathlands are gradually disappearing as a result of addition of nitrogen.

Society's emissions of nitrogen to the aquatic environment are blamed especially on the agricultural use of fertilizers, but oxides of nitrogen from combustion processes are also of significance for both aquatic and terrestrial ecosystems. The most significant sources of emission of phosphorus are sewage treatment plants for urban and industrial effluents and leaching from agricultural land.

The nutrient enrichment is an impact which affects the environment on both local and regional scales.

23.6.1 Determine which compounds contribute to nutrient enrichment

- Check whether the substance contains biologically available nitrogen (N) or phosphorus (P) and if so, regard it as contributing to nutrient enrichment. Free nitrogen (N_2) should not be so regarded.

For a compound to be regarded as contributing to nutrient enrichment, it must contain nitrogen or phosphorus in a form which is biologically available.

Free nitrogen is not regarded as a contributor to nutrient enrichment, even if it is available for certain bacteria and algae. This is because emission of N_2 has no additional fertilizing effect, inasmuch as the greater part of the atmosphere already consists of free nitrogen.

Nutrient enrichment can be caused by emissions to air, water and soil. The list of potentially nutrient enriching compounds is very long, but in practice most compounds occurring in the inventory of the product system will be in the list of equivalency factors in Table 23.6.

23.6.2 Calculate the nutrient enrichment potential

- Decide whether the nutrient enrichment potentials are to be calculated separately for nitrogen and phosphorus, or whether a combined potential is preferable.
- Find the substance's equivalency factor(s) in Table 23.6.
- Calculate potential(s) for nutrient enrichment by multiplying the magnitude of the emission by the equivalency factor(s) found, EF(N) for separate nitrogen potential, EF(P) for separate phosphorus potential and EF(ne) for combined potential for nutrient enrichment:

$$EP(N) = Q \cdot EF(N)$$
$$EP(P) = Q \cdot EF(P)$$
$$EP(ne) = Q \cdot EF(ne)$$

There is no internationally accepted system of equivalency factors for nutrient enrichment which can be used in calculation of its potential. As the environmental impact as here defined is attributed to addition of N- or P-containing nutrients, it is, however, easy to develop a system of equivalency factors, as the equivalency factor must simply show the compound's content of N or P.

It is possible to operate with two potentials for nutrient enrichment, one for each of the two nutrients N and P. It will be an advantage to treat N and P separately if an assessment is subsequently to be made of whether the contributions to nutrient enrichment are to be accorded major or little weight on the basis of site-specific considerations. The weight accorded nutrient enrichment depends on whether the nutrient which is added is the limiting one in the ecosystem in question. This can only be decided if the contributions from compounds containing N and P are treated separately in the calculation of their potentials.

It is also possible for the sake of simplicity to operate with only one potential for nutrient enrichment. The two nutrient enrichment potentials must then be combined, which requires an assumption of a fixed relative strength of the contributions from N and P which must be independent of the type and sensitivity of the ecosystem which is exposed. As there are on average about 16 times as many nitrogen atoms as phosphorus atoms in aquatic organisms, a strength of 16 is used for P relative to N when their impact potentials are aggregated into one. Table 23.6 shows the factors corresponding to both possibilities. As for several of the other potential environmental impacts, the potentials for nutrient enrichment are expressed as an equivalent quantity of a reference substance. If the two nutrients are treated separately, the reference substances are simply

Substance	Formula	EF (N) g N/g subst.	EF (P) g P/g subst.	EF (ne) g NO_3^-/g subst.
N-compounds				
Nitrate	NO_3^-	0.23	0	1
Nitrogen dioxide	NO_2	0.30	0	1.35
Nitrite	NO_2^-	0.30	0	1.35
Nitrogen oxides	NO_x *	0.30	0	1.35
Nitrous oxide	N_2O	0.64	0	2.82
Nitric oxide	NO	0.47	0	2.07
Ammonia	NH_3	0.82	0	3.64
Cyanide	CN^-	0.54	0	2.38
Total-N	N	1	0	4.43
P-compounds				
Phosphate	PO_4^{3-}	0	0.33	10.45
Pyrophosphate	$P_2O_7^{2-}$	0	0.35	11.41
Total P	P	0	1	32.03

* For NO_x the x is regarded as 2.

Table 23.6. Equivalency factors for nutrient enrichment (Hauschild & Wenzel, 1997c)

nitrogen and phosphorus. If they are combined in one impact potential, the latter is expressed as an equivalent quantity of nitrate, NO_3^-.

As noted, all substances containing N or P are potential contributors to nutrient loads. Compounds other than those for which equivalency factors are given here may therefore occur in the inventory. If this is the case, it is easy to calculate equivalency factors for them. The method is given in Hauschild & Wenzel (1997c).

23.7 Ecotoxicity

Chemicals emitted as a consequence of human activities contribute to ecotoxicity if they affect the function and structure of the ecosystems by exerting toxic effects on the organisms which live in them. If the concentrations of environmentally hazardous substances caused by the emission are high enough, the toxic effects can occur as soon as the substances are released. This form of toxic effect is called **acute ecotoxicity**. It often results in the death of organisms exposed.

Toxic effects which are not acutely lethal and which first appear after repeated or long-term exposure to the substance are called **chronic ecotoxicity**. Chronic ecotoxicity is often caused by substances which have a low degradability in the environment and which can therefore remain for a long time after their emission (persistent substances). Some substances also have a tendency to accumulate in living organisms, so that tissues and organs can be exposed to concentrations of the substance which are far higher than the concentrations in the surrounding environment. The chronic ecotoxicity of a compound is thus determined by its toxicity, its biodegradability and its ability to accumulate in living organisms. The result of a chronic ecotoxic impact can, for example, be reduced reproductive capacity, which means that the species' chances of survival in the longer term are reduced.

Ecotoxicity is an impact which predominantly affects the environment on local and regional scales. It can be a global impact for some toxic substances of very low biodegradability with a strong tendency to accumulate in living organisms.

Ecotoxic impacts can involve many different mechanisms, with the common feature that they all result in direct toxic impacts on ecosystems. Compared with the other environmental impact categories, ecotoxicity thus has the character of a composite category which includes all substances from the inventory which can have a direct effect on the health of the ecosystems. The list of substances classified as contributing to ecotoxicity will therefore be much more comprehensive than the corresponding lists of the other environmental impacts, and it will include many different types of substances with widely differing chemical characteristics.

The procedure for determining whether a compound should be regarded as contributing to ecotoxicity, and for calculating the ecotoxicity potentials, is given below.

An attempt has been made to make the method simple to use, and to fix the framework as far as possible. Nevertheless, some chemical and ecotoxicological background knowledge is required to use the method.

23.7.1 Determine which substances contribute to ecotoxicity

- Use the chemical hazard screening method for ecotoxicity presented in Section 23.11.
- Regard all substances scoring a value of 8 or more in the total ecotoxicological impact score as contributing to ecotoxicity.
- Scrutinize the inventory to assess whether there are other substances for which ecotoxicity potentials should be calculated.

For a substance to be regarded as contributing to ecotoxicity, it must be toxic to organisms in the environment or to the way in which they interact in the ecosystem, i.e. to the ecosystem's structure and function. Determination of which substances to include cannot be based solely on the substances' toxicity. The chronic toxicity which can be evident after long-term exposure to the substance is also influenced by the persistence and the ability of the substance to accumulate in living organisms.

The chemical hazard screening method in Section 23.11 therefore focuses on the following characteristics of a substance:

- The substance's ecotoxicity, which is determined experimentally in ecotoxicity tests.
- The substance's persistence, which is determined experimentally in tests of its biodegradability.
- The substance's potential for bioconcentration, which is determined directly in experiments or estimated on the basis of its octanol-water partitioning ratio, P_{ow}, which is used as an indicator of the substance's tendency to accumulate in adipose tissue.

23.7.2 Calculate the ecotoxicity potentials

- Find the substance's equivalency factors for the relevant emission route in Table 23.7.
- In case no equivalency factors are found for the substance in Table 23.7, determine equivalency factors as described in Sections 23.7.2.1-23.7.2.6
- Calculate the ecotoxicity potentials for the relevant compartments by multiplying the magnitude of the emission with the equivalency factors found: EP(et) = Q · EF(et).

As for nutrient enrichment and acidification, there is no internationally accepted set of equivalency factors expressing substances' potential toxicity in the environment in a way which can be used in the calculation of potentials for ecotoxicity. There are various proposals for how ecotoxicity can be handled in the LCA, but in the authors' view, their scientific foundation is too weak, or their demands on data are too high to make them operational and usable in practice. It has therefore been necessary in the EDIP program to develop a method for calculation of the ecotoxicity potentials of emissions. The EDIP method is inspired by the EU Commission's technical guidelines for risk assessment of chemicals in the environment (European Commission, 1996). It has been presented several times at conferences and workshops in SETAC (Wenzel & Hauschild, 1993; Hauschild *et al.,* 1993; Hauschild, 1994), and has entered into the methodological discussions in SETAC.

Ecotoxicity potentials

Up to four ecotoxicity potentials are calculated for each substance:

EP(etwa) = Q · EF(etwa) for acute ecotoxicity in water
EP(etwc) = Q · EF(etwc) for chronic ecotoxicity in water
EP(etsc) = Q · EF(etsc) for chronic ecotoxicity in soil
EP(etp) = Q_p · EF(etp) for ecotoxicity to microorganisms in sewage treatment plants

EP(et) is the ecotoxicity potential in one of the environmental compartments, water, soil and sewage treatment plant, determined as the product of the quantity of substance Q emitted and the equivalency factor EF for the emission regarding the compartment in question. The ecotoxicity potential is measured in cubic metres (m^3) of the compartment. It corresponds to the volume of the compartment to which the emission should be diluted in order to obtain a concentration of substance so low that no ecotoxic effects would be expected from the emission.

Substance	CAS no.	Emissions to air		Emissions to water			Emissions to soil		Emissions to WWTP
		EF(etwc) m^3/g	EF(etsc) m^3/g	EF(etwc) m^3/g	EF(etwa) m^3/g	EF(etsc) m^3/g	EF(etwc) m^3/g	EF(etsc) m^3/g	EF(etp) m^3/g
1,2-Propylene oxide	75-56-9	1.2	11	5.9	0.59	0	0	14	0.03
1,2-Dichlorobenzene	95-50-1	10	0.49	10	10	0.49	10	0.49	500
1,2-Dichloroethane	107-06-2	20	61	100	10	0	0	77	500
1-Butanol	71-36-3	0.01	0.09	0.07	0.04	0	0	0.11	0.08
2,3,7,8-Tetrachlorodibenzo-p-dioxin	1746-01-6	$5.6 \cdot 10^8$	$1.2 \cdot 10^4$	$2.8 \cdot 10^9$	$2.8 \cdot 10^8$	0	0	$1.5 \cdot 10^4$	$1.4 \cdot 10^8$
2,4-Dinitrotoluene	121-14-2	150	190	770	77	0	0	240	$3.8 \cdot 10^3$
2-Chlorotoluene	95-49-8	200	10	200	100	10	200	10	$5.0 \cdot 10^3$
2-Ethyl hexanol	104-76-7	0	0	2.7	1.3	0	0	0.16	67
2-Propanol	67-63-0	0.05	0.46	0.25	0.13	0	0	0.58	0.05
3-Chlorotoluene	108-41-8	200	14	200	100	14	200	14	$5.0 \cdot 10^3$
4-Chlorotoluene	106-43-4	200	12	200	100	12	200	12	$5.0 \cdot 10^3$
Acetic acid	64-19-7	0.08	0.79	0.4	0.2	0	0	0.99	0.02
Acetone	67-64-1	$4.0 \cdot 10^3$	$3.8 \cdot 10^4$	$2.0 \cdot 10^4$	10	0	0	$4.7 \cdot 10^4$	$5.0 \cdot 10^5$
Anionic detergent (worst case)	Not available	4.0	33	20	10	0	0	41	50
Anthracene	120-12-7	0	0	$5.0 \cdot 10^4$	$1.0 \cdot 10^4$	0	0	59	$5.0 \cdot 10^5$
Arsenic	7440-38-2	380	0.27	$1.9 \cdot 10^3$	190	0	0	0.33	2.5
Atrazine	1912-24-9	0	0	$6.7 \cdot 10^3$	670	0	0	530	$3.3 \cdot 10^4$
Benzene	71-43-2	4.0	3.6	4.0	10	3.6	4.0	3.6	500
Benzotriazole	95-14-7	4.0	13	20	2.0	0	0	16	1.0
Biphenyl	92-52-4	200	2.8	$1.0 \cdot 10^3$	100	0	0	3.5	$5.0 \cdot 10^3$
Cadmium	7440-46-9	$2.4 \cdot 10^4$	1.8	$1.2 \cdot 10^5$	$1.2 \cdot 10^4$	0	0	2.2	$1.5 \cdot 10^4$
Chlorobenzene	108-90-7	200	38	200	100	38	200	38	$5.0 \cdot 10^3$
Chloroform	67-66-3	20	25	20	10	25	20	25	500
Chromium (VI)	7440-47-3	130	0.01	670	67	0	0	0.01	1.0

Table 23.7. Equivalency factors for ecotoxicity in water, in soil and in waste water treatment plant (wwtp) for substances emitted to air, water or soil (for 'a' a global default value of 0.2 is used). The values are based on a rather simple fate modelling and for some substances the underlying database is weak. They should therefore not be used for purposes other than the impact assessment of LCA.

Substance	CAS no.	Emissions to air		Emissions to water			Emissions to soil		Emissions to WWTP
		EF(etwc) m^3/g	EF(etsc) m^3/g	EF(etwc) m^3/g	EF(etwa) m^3/g	EF(etsc) m^3/g	EF(etwc) m^3/g	EF(etsc) m^3/g	EF(etp) m^3/g
Cobalt	7440-48-4	400	9.1	$2.0 \cdot 10^3$	200	0	0	11	120
Copper	7440-50-8	$2.5 \cdot 10^3$	0.02	$1.3 \cdot 10^4$	$1.3 \cdot 10^3$	0	0	0.02	$2.5 \cdot 10^4$
Dibutyltinoxide	818-08-6	$2.0 \cdot 10^4$	530	$1.0 \cdot 10^5$	$1.0 \cdot 10^4$	0	0	670	670
Diethanolamine	111-42-2	0	0	0.91	0.45	0	0	2.2	0.005
Diethylaminoethanol	100-37-8	0	0	13	1.3	0	0	29	0.003
Diethylene glycol	111-46-6	0	0	0.03	0.02	0	0	0.07	0.006
Diethylene glycol mono-n-butyl ether	112-34-5	0	0	0.2	0.2	0	0	0.4	0.20
Ethanol	64-17-5	0.001	0.01	0.005	0.002	0	0	0.01	0.0004
Ethyl acetate	141-78-6	0.08	0.59	0.41	0.21	0	0	0.73	0.52
Ethylene glycol	107-21-1	0.001	0.01	0.005	0.002	0	0	0.01	0.001
Ethylenediamine tetraacetic acid	60-00-4	0	0	1.8	0.2	0	0	4.5	0.48
Ethylenediamine. 1.2-ethanediamine	107-15-3	0	0	0.87	0.43	0	0	2.1	0.5
Formaldehyde	50-00-00	24	200	120	60	0	0	250	3.6
Hexane	110-54-3	150	2.5	150	74	2.5	150	2.5	$3.7 \cdot 10^3$
Hydrogen cyanide	74-90-8	800	$7.6 \cdot 10^3$	800	$2.0 \cdot 10^3$	$7.6 \cdot 10^3$	800	$7.6 \cdot 10^3$	$1.0 \cdot 10^5$
Hydrogen sulphide	7783-06-4	0	0	$6.7 \cdot 10^3$	$3.3 \cdot 10^3$	0	0	0	$1.7 \cdot 10^5$
Iron	7439-89-6	20	0.53	100	10	0	0	0.66	6.7
Isopropylbenzene. cumene	98-82-8	2.9	0.08	2.9	7.1	0.08	2.9	0.08	1.9
Lead	7439-92-1	400	0.01	$2.0 \cdot 10^3$	200	0	0	0.01	$5.0 \cdot 10^3$
Manganese	7439-96-5	71	1.9	360	36	0	0	2.4	130
Mercury	7439-97-6	$4.0 \cdot 10^3$	5.3	$4.0 \cdot 10^3$	$2.0 \cdot 10^3$	5.3	$4.0 \cdot 10^3$	5.3	100
Methanol	67-56-1	0.01	0.09	0.05	0.025	0	0	0.12	0.008
Methyl methacrylate	80-62-6	0	0	0.54	0.27	0	0	0.48	0.5
Molybdenum	7439-98-7	400	3.9	$2.0 \cdot 10^3$	200	0	0	4.8	100

Substance	CAS no.	Emissions to air		Emissions to water			Emissions to soil		Emissions to WWTP
		EF(etwc) m^3/g	EF(etsc) m^3/g	EF(etwc) m^3/g	EF(etwa) m^3/g	EF(etsc) m^3/g	EF(etwc) m^3/g	EF(etsc) m^3/g	EF(etp) m^3/g
Monoethanolamine	141-43-5	0	0	27	13	0	0	66	0.008
n-Butyl acetate	123-86-4	0.56	1.0	2.8	0.56	0	0	1.3	0.43
Nickel	7440-02-0	130	0.05	670	67	0	0	0.07	69
Nitrilotriacetate	139-13-9	0	0	0.15	0.08	0	0	0.38	0.09
Nitrobenzenesulphonic acid	127-68-4	0.09	0.84	0.09	0.04	0.84	0.09	0.84	2.1
Phenol	108-95-2	0	0	44	22	0	0	110	0.78
Propylene glycol. 1.2-propanediol	57-55-6	0	0	0.02	0.01	0	0	0.05	0.5
Selenium	7782-49-2	$4.0 \cdot 10^3$	106	$2.0 \cdot 10^4$	$1.4 \cdot 10^3$	0	0	133	$1.0 \cdot 10^4$
Sodium benzoate	532-32-10	0.63	6.2	3.2	1.6	0	0	7.8	0.09
Sodium hypochlorite	7681-52-9	0	0	270	27	0	0	610	$1.3 \cdot 10^3$
Strontium (Sr)	7440-24-6	$2.0 \cdot 10^3$	53	$1.0 \cdot 10^4$	$1.0 \cdot 10^3$	0	0	67	$5.0 \cdot 10^3$
Styrene	100-42-5	0	0	0	40	0	0	0	0.69
Sulphamic acid	5329-14-6	2.8	28	14	7.0	0	0	35	0.35
Tetrachlorethylene	127-18-4	20	1.1	20	10	1.1	20	1.1	500
Thallium	7440-28-0	670	17.7	$3.3 \cdot 10^3$	330	0	0	22	$1.7 \cdot 10^3$
Thorium	7440-29-1	330	8.9	$1.7 \cdot 10^3$	$1.7 \cdot 10^3$	0	0	11	830
Titanium	7440-32-6	27.4	0.73	140	14	0	0	0.91	6.8
Toluene	108-88-3	4.0	0.97	4.0	10	0.97	4.0	0.97	500
Triethanolamine	102-71-6	0	0	5.6	1.1	0	0	14	5.0
Triethylamine	121-44-8	0	0	100	10	0	0	80	0.04
Vanadium	7440-62-2	40	0.34	200	20	0	0	0.43	290
Xylenes. mixed	1330-20-7	4.0	0.40	4.0	10	0.40	4.0	0.40	500
Zinc	7440-66-6	200	0.005	$1.0 \cdot 10^3$	100	0	0	0.007	44

Table 23.7. (continued)

Table 23.8 presents equivalency factors EF for around 70 compounds emitted to air, water or soil. If the table gives no equivalency factors for a substance, these must be calculated. The equivalency factors are determined as the product of three factors which represent the substance's dispersion in the environment, its ecotoxicological characteristics and its biodegradability. To calculate the equivalency factors it is therefore first necessary to:

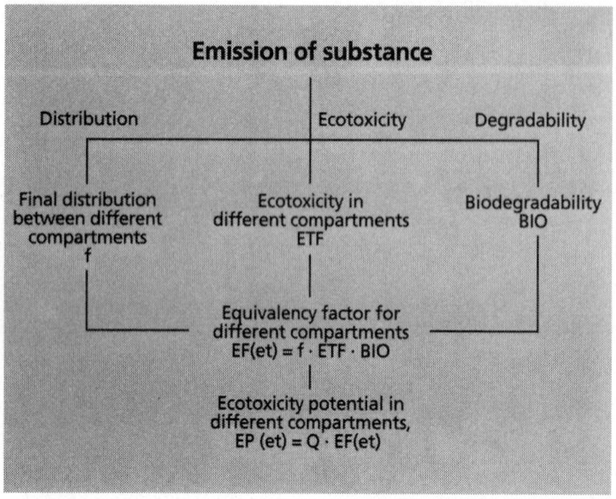

Figure 23.2. Calculation of ecotoxicity potentials involves investigation of the characteristics and the fate of the substance

- Determine the fraction (f) of the emission which reaches the different environmental compartments after dispersion.
- Calculate ecotoxicity factors (ETF) representing the substance's potential ecotoxicity in the three compartments.
- Determine the biodegradability factor (BIO) for the substance.

The equivalency factors for ecotoxicity depend exclusively on the characteristics of the substance. These will be the same independently of the context in which the emission occurs. It is therefore necessary to calculate them only once for each substance. The factors calculated can be reused every time the substance appears in an inventory to be assessed.

23.7.2.1 Determine the fraction 'f' of the emission which ends in the different compartments

The inventory may include emissions to water, air and soil. Even if a substance is emitted to one compartment, e.g. air, it may well contribute primarily to the ecotoxicity potential in another compartment. Redistribution calculations decide:

- For which parts of the environment ecotoxicity potentials are to be calculated.
- The quantity of substance contributing to the ecotoxicity potential in each of the compartments.

For each individual emission, the redistribution calculations result in distribution factors f_s and f_w, which specify for how large a fraction of the emission ecotoxicity potentials should be calculated in the compartments soil and water.

The redistribution calculations in the EDIP tool are simple and modest in their data requirements. The method was designed to ensure that the redistribution calculations can be done for all substances in the inventory.

23.7.2.1.1 Substances emitted to the air compartment

- Set f_{wa} =0.
- Check whether the substance's half-life in the atmosphere is more than 1 day. Atmospheric half-lives can be found in e.g. Howard *et al.* (1991) or in the database Hazardous Substance Data Bank (HSDB), or they can be estimated using software like AOPWIN (SRC, 1996).
- If this is the case, the substance will deposit to surface water and soil, so f_{wc} = a and f_{sc} = 1-a.
  Determine the relevant relationship between area of water surface and area of land surface, and fix 'a' accordingly (see explanation below). If the geographic location of the emission is unknown, use a global default value for 'a' of 0.2.
- If the atmospheric half-life does not exceed 1 day, no ecotoxicity potentials should be calculated for the substance, i.e. $f_{wa} = f_{wc} = f_{sc} = 0$.

A substance emitted to air can be deposited to the water and soil compartments if it is sufficiently stable, i.e. if it is not broken down in the atmosphere first. As a rule of thumb, the EDIP method assumes that deposition occurs only if the substance's half-life in the atmosphere exceeds 1 day.

If the substance is thus deposited, it can contribute to chronic ecotoxicity in water or soil, but it is unlikely that substances deposited from air can cause acute ecotoxicity in water; f_{wa} is therefore set to 0.

On deposition, the substance is distributed between the water and soil compartments in the ratio a:(1-a). The magnitude of 'a' should be fixed as that part of the area of deposition of the emission which is water; 'a' can be fixed for conditions representative of the region in question.

A substance deposited from air is assumed not to re-evaporate from the water or soil compartment, even if it is volatile. The distribution factors thus describe the final distribution of the substance.

As no ecotoxicity potentials are calculated for ecosystems in the air compartment, the substance does not contribute to any ecotoxicity potentials if it is emitted to air and has a half-life of less than 1 day.

Figure 23.3. Fixing of distribution factors for substances emitted to the air compartment

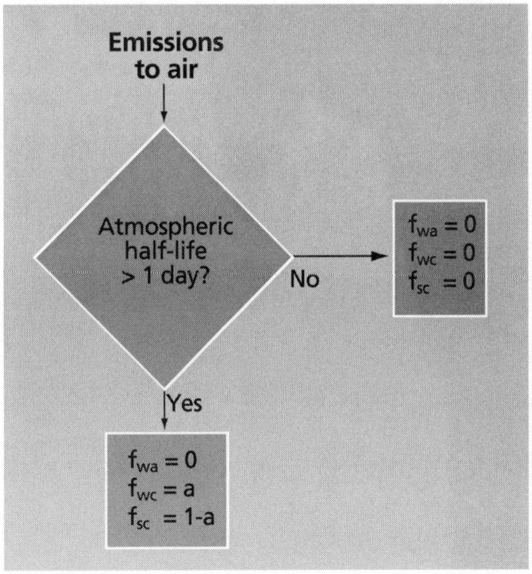

23.7.2.1.2 Substances emitted to the water compartment

- Set $f_{wa} = 1$.
- Check whether the Henry's law constant for the substance is greater than 10^{-3} atm · m^3/mol. The Henry's law constant for a substance can, for example, be found in Howard (1990) or in the database HSDB, or it can be estimated using software like HENRY (SRC, 1995).
- If the constant exceeds this limit, find the substance's half-life in the atmosphere. If the half-life is less than 1 day, set $f_{wc} = f_{sc} = 0$.
  If the half-life is more than 1 day, set $f_{wc} = a$ and $f_{sc} = 1-a$.
  Determine the relevant relationship between water and land areas and then fix 'a'. If the geographic location of the emission is unknown, use a global default value for 'a' of 0.2.
- If the Henry's law constant is not greater than 10^{-3} atm · m^3/mol, it is assumed that the entire quantity emitted remains in the water, i.e. $f_{wc} = 1$ and $f_{sc} = 0$.

Even if a substance is emitted with waste water, it may well evaporate to the air compartment. Even quite volatile substances may, however, remain in the recipient long enough to cause acute ecotoxicity; f_{wa} should therefore be set to 1 in all cases of emissions to the water compartment.

The substance's volatility from water is described by its Henry's law constant H, and as a rule of thumb it is assumed that if $H > 10^{-3}$ atm · m^3/mol, the substance will evaporate from water. In this case it is assumed that the entire quantity emitted will evaporate.

From the air compartment, the substance may deposit to water and soil, as described for substances emitted to the air compartment, provided that the substance is sufficiently stable not to be broken down first.

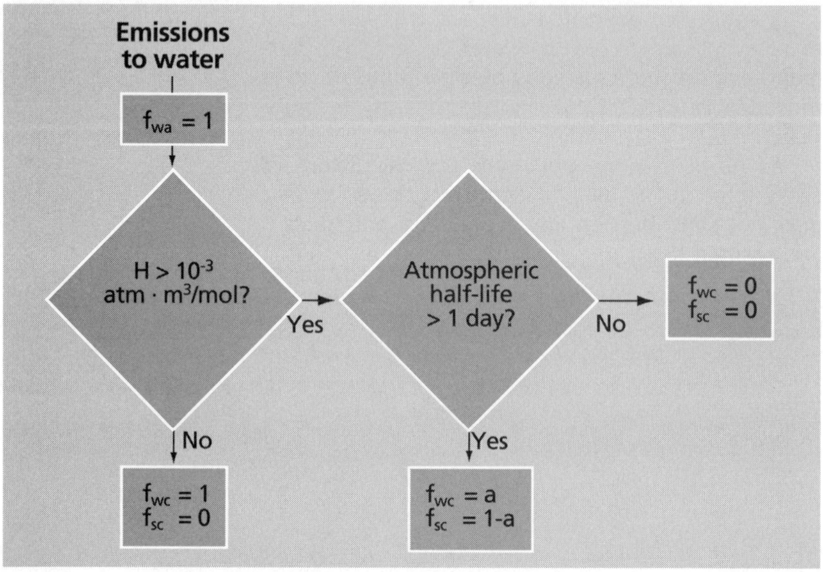

Figure 23.4. Fixing of distribution factors for substances emitted to the water compartment

23.7.2.1.3 Substances emitted to the soil compartment

- $f_{wa} = 0$.
- Check whether the substance's Henry's law constant is greater than 10^{-3} atm · m^3/mol.
- If this is the case, find the substance's half-life in the atmosphere. If the half-life is less than 1 day, set $f_{wc} = f_{sc} = 0$.

  If the half-life is more than 1 day, set $f_{wc} = a$ and $f_{sc} = 1\text{-}a$.

  Determine the relevant relationship between water and land areas and then fix 'a'. If the geographic location of the emission is unknown, use a global default value for 'a' of 0.2.
- If the Henry's law constant is not greater than 10^{-3} atm · m^3/mol, it is assumed that the entire quantity emitted remains in the soil, i.e., $f_{wc} = 0$ and $f_{sc} = 1$.

A substance can be emitted to soil either directly, e.g. by spraying of cultivated land with pesticides, or indirectly, e.g. through the spreading of sewage sludge. If the substance is volatile it can evaporate and possibly deposit again to the soil and water compartments in the same way as described for substances emitted to the water compartment.

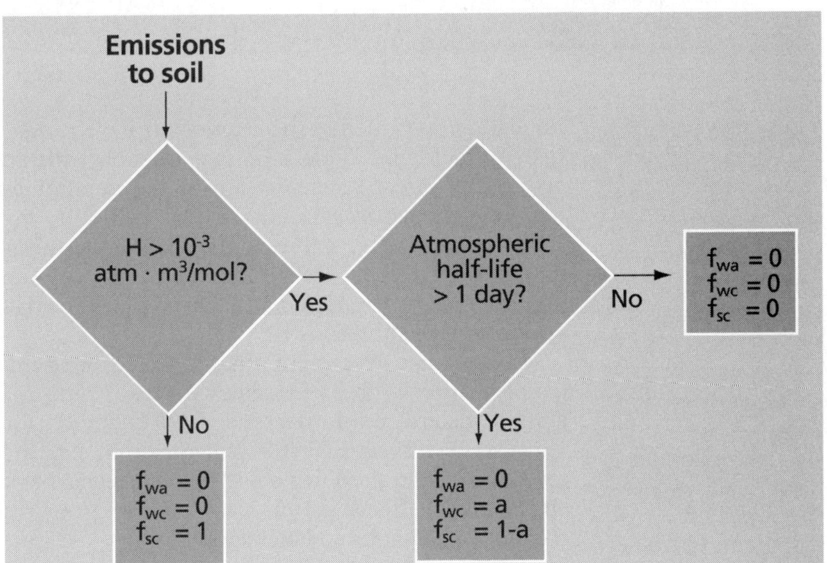

Figure 23.5. Fixing of distribution factors for substances emitted to the soil compartment

The procedure for the redistribution calculation is illustrated by the following example:

1.5 g of the substance n-butyl acetate is emitted in effluent from a sewage treatment plant to the water compartment.

n-butyl acetate has a Henry's law constant of $3.2 \cdot 10^{-4}$ atm · m³/mol. It is therefore assumed that the substance will not evaporate from water, so that ecotoxicity potentials in water are calculated for the entire quantity emitted. Thus

$f_{wa} = 1$

$f_{wc} = 1$

$f_{sc} = 0$

Example 23.1. Determination of distribution factors for n-butyl acetate.

23.7.2.2 Calculate ecotoxicity factors, ETF

- Decide on the basis of the redistribution calculations which ecotoxicity factors are to be calculated. If the distribution factor 'f' for a compartment is 0, no ecotoxicity potentials are to be calculated, and thus no ecotoxicity factor for this compartment.
- If the emission has passed through a biological sewage treatment plant, then calculate an ecotoxicity factor for the treatment plant.
- Calculate the relevant ecotoxicity factors with the expressions in Table 23.8.

Table 23.8. Ecotoxicity factors for the three compartments water, soil and sewage treatment plant

Final compartment	Acute/ chronic	ETF
Water	acute	$\frac{1}{PNEC_{wa}}$
	chronic	$\frac{1}{PNEC_{wc}}$
Soil	chronic	$\frac{1}{PNEC_{sc}}$
Waste water treatment plant	acute	$\frac{50}{LOEC_{aerobic\ heteorotroph}}$

For all compartments, the expression for the ecotoxicity factor in Table 23.8 is a fraction with an ecotoxicity effect concentration in the denominator. For the compartments water and soil, the substance's PNEC value is used. PNEC stands for Predicted No Effect Concentration, i.e. the concentration predicted to have no ecotoxic effects in the compartment.

For sewage treatment plants, the ecotoxicity concentration is the substance's LOEC value for aerobic heterotrophic bacteria. LOEC stands for Lowest Observed Effect Concentration, i.e. the lowest concentration which in an ecotoxicity test shows toxic effects in this case on the group of bacteria called aerobic heterotrophs.

23.7.2.2.1 Estimate $PNEC_{wc}$

- Find ecotoxicological test data for the substance in water (e.g. in the Aquatic Information Retrieval Toxicity Database AQUIRE).
- Determine the assessment factor for the test results used on the basis of Table 23.9.
- Estimate $PNEC_{wc}$ by dividing the lowest of the test results used by the assessment factor.

It will normally only be possible to find *laboratory* data for a substance's ecotoxicity. In the laboratory tests, selected test organisms have been exposed to the substance under standardized conditions for a shorter or longer time.

In a short-term laboratory test (acute ecotoxicity test), that concentration of the substance which kills e.g. 50% of the test organisms (LC_{50}, Lethal Concentration 50%) is determined.

For ecotoxicity tests of longer duration (chronic ecotoxicity tests), the highest concentration of substance which produces no observed effects on the test organisms (NOEC, No Observed Effect Concentration) or the lowest concentration which has resulted in observed effects (LOEC) is most often reported.

For LC_{50}, NOEC and LOEC values, the higher the value, the less toxic the substance.

PNEC, the concentration of substance found to have no toxic effect in the environment, is determined on the basis of the results of laboratory tests, i.e. LC_{50}, NOEC and LOEC values. The quality and the relevance of the available toxicity tests is assessed, and on this basis an assessment factor is established which, on division into the lowest of the ecotoxicity concentrations used, gives the PNEC value for the substance. The criteria in Table 23.9 should be regarded as a general guideline. Even if an attempt has been made to standardize the procedure for determination of PNEC values and to simplify it relative to a traditional hazard assessment, expert judgment will often be required. Some background in ecotoxicology is generally needed to determine the assessment factors.

This point is illustrated in the following example.

Table 23.9. Assessment factors for estimation of $PNEC_{wc}$ on the basis of test data for acute and chronic aquatic ecotoxicity

Assessment factor	Criterion
1000	Some data for acute ecotoxicity (LC_{50}) available, but data for species from one or more of the classes fish, Crustacea and algae are missing.
100	Data available for acute ecotoxicity (LC_{50}) for at least one species from each of the classes fish, Crustacea and algae.
20	Data available for chronic ecotoxicity (LOEC) for at least one species from each of the classes fish, Crustacea and algae.
10	Data available for chronic ecotoxicity (NOEC) for at least one species from each of the classes fish, Crustacea and algae.

The following laboratory test data for n-butyl acetate have been found in the aquatic ecotoxicity database AQUIRE.

Algae: The lowest observed effect concentration (LOEC) for growth inhibition in a chronic test with the green alga *Scenedesmus quadricauda* was 21 mg/l.

Crustacea: An LC_{50} of 32 mg/l was found in an acute ecotoxicity test on the brine shrimp *Artemia salina.* An LOEC value of 8.8 mg/l for growth inhibition was found in a chronic test on the water flea *Daphnia magna*.

Fish: An LC_{50} of 18 mg/l was found in an acute toxicity test on the fish *Pimephales promelas*.

Data are only available for chronic ecotoxicity for two of the three groups, *namely* Crustacea and algae. Whether the chronic ecotoxicity to fish would be the greatest of the three is thus unknown. From the two tests for the acute ecotoxicity of n-butyl acetate, it is seen that the acute ecotoxicity is greater for fish (the LC_{50} is lower). It cannot, therefore, be excluded that the chronic ecotoxicity is also greater for fish. Since the lack of data for chronic ecotoxicity for the most sensitive group of organisms cannot be excluded, the estimation of $PNEC_{wc}$ should not be based on the incomplete data for chronic ecotoxicity.

Instead, the estimation of $PNEC_{wc}$ is based on the data found for acute ecotoxicity, and a larger assessment factor is accordingly applied. Data for acute ecotoxicity for algae are indeed lacking, but this is probably not a problem: acute toxicity always occurs at higher concentrations than chronic toxicity, and the value for acute ecotoxicity for algae will therefore be higher than the 21 mg/l found for chronic effects on algae. It will thus also be higher than the lowest of the LC_{50} values found for fish and Crustacea, which was 18 mg/l. Even if data had been available for the substance's acute ecotoxicity to algae, it is therefore unlikely that estimation of the $PNEC_{wc}$, which is based on the lowest of the values found, would have been affected.

On the basis of these considerations the assessment factor is set at 100, and $PNEC_{wc}$ estimated as the lowest of the LC_{50} values found, 18 mg/l for the fish *Pimephales promelas*, divided by the assessment factor.

This gives an estimated $PNEC_{wc}$ of 0.18 mg/l.

Example 23.2. Estimation of $PNEC_{wc}$ for n-butyl acetate

23.7.2.2.1.1 Correction for the substance's capacity for bio-concentration

- Check whether the substance's octanol-water partitioning coefficient P_{ow} is greater than 1000, and whether $PNEC_{wc}$ was estimated on the basis of data for acute ecotoxicity. P_{ow} can, for example, be sought in the database HSDB.
- If this is the case, check Hauschild *et al.* (1997a) to see whether the $PNEC_{wc}$ value found has to be corrected for the substance's bioconcentration potential, and if so how this is done.

Any capacity of the substance for accumulation in living organisms will be able to increase its toxicity in chronic laboratory tests where the exposure lasts long enough for the bioconcentration to have time to occur in the test organisms. If the estimation of $PNEC_{wc}$ is based on the results of chronic ecotoxicity tests, the value found will therefore also reflect the compound's bioconcentration potential.

Among organic compounds, it is generally only the highly fat-soluble substances which bioconcentrate. As a rule of thumb, the substance's P_{ow} should be over 1000.

If $PNEC_{wc}$ for a substance which can bioconcentrate is estimated on the basis of data for acute ecotoxicity, the bioconcentration potential will not be represented in the resulting PNEC value. In this case the $PNEC_{wc}$ value found may require correction for the substance's bioconcentration potential. Hauschild *et al.* (1997a) discuss how a decision is made on whether this is the case and how the PNEC value should then be corrected.

In the example with n-butyl acetate the procedure is as follows:

In the database HSDB, it is found that n-butyl acetate has a P_{ow} value of 66, and it is therefore not suspected of bioconcentrating. Although $PNEC_{wc}$ for the substance was previously estimated on the basis of the data for acute ecotoxicity in Example 23.2, there is therefore no need to correct the value found to take account of n-butyl acetate's bioconcentration.

Example 23.3. Correction for bioconcentration in estimation of $PNEC_{wc}$ for n-butyl acetate

23.7.2.2.2 Estimate $PNEC_{wa}$

- Find test data for acute ecotoxicological effects of the substance in water.
- Determine the assessment factor for the test results used from Table 23.10.
- Estimate $PNEC_{wa}$ by dividing the lowest of the test results used by the assessment factor.

Assessment factor	Criterion
100	Some data for acute ecotoxicity (LC_{50}) available, but data for species from one or more of the classes fish, Crustacea and algae are missing.
10	Data available for acute ecotoxicity (LC_{50}) for at least one species from each of the classes fish, Crustacea and algae.

Table 23.9. Assessment factors for estimation of $PNEC_{wa}$ on the basis of test data for acute aquatic ecotoxicity

For acute ecotoxicity, the procedure for determination of the concentration of substance which is not expected to have any effects in the environment is the same as described for chronic ecotoxicity.

As illustrated in Table 23.10, estimation of $PNEC_{wa}$ is based on data for acute toxicity, and the assessment factors are lower by a factor of 10.

The following example illustrates the procedure for determination of $PNEC_{wa}$:

With the ecotoxicological data for n-butyl acetate from Example 23.2, it can be confirmed that

- data for acute ecotoxicity are available for both fish and Crustacea, and
- data are available for chronic ecotoxicity for algae (LOEC of 21 mg/l) which show that the acute ecotoxicity concentration for the species of alga in question will be higher than the acute ecotoxicity concentration (18 mg/l) for the species of fish tested.

It is thus appropriate to proceed as if there were data for acute ecotoxicity for organisms representing all three groups, and on this basis to fix the assessment factor at 10.

$PNEC_{wa}$ is estimated as the lower of the LC_{50} values at 18 mg/l divided by the assessment factor. This gives an estimated $PNEC_{wa}$ value of 1.8 mg/l.

Example 23.4. Estimation of $PNEC_{wa}$ for n-butyl acetate

23.7.2.2.3 Estimate $PNEC_{sc}$

- Estimate $PNEC_{wc}$ as described above.
- Find the coefficient of adsorption K_d for the substance in soil.

  For metals, find K_d in the table in Hauschild *et al.* (1997a).

  For other substances, K_d may be found in databases such as HSDB or Howard (1990) or it may be estimated using software like PCKOC (SRC, 1994).

  For organic compounds, missing K_d values can be estimated as $K_d = 0.02 \cdot P_{ow}$. K_d must be expressed in l/kg.
- Determine $PNEC_{sc}$ from $PNEC_{wc}$ by multiplying by (K_d + 0.27) and convert the value found from "g substance/kg dry weight" to "g substance/l soil" by multiplying by the density of the dry soil of 1.5 g/l. The complete expression is thus:
  $PNEC_{sc} = PNEC_{wc} \cdot (K_d + 0.27) \cdot 1.5$.

In principle, determination of PNEC for chronic terrestrial ecotoxicity follows the method presented for chronic aquatic ecotoxicity. It is, however, a problem that studies of ecotoxicity on soil-dwelling organisms have been undertaken for very few compounds only. In practice it is therefore almost never possible to base the calculation of a PNEC value for soil on data for soil-dwelling organisms. Data on the substance's ecotoxicity to aquatic organisms is instead used as an approximation. A correction is made for the difference in the substance's availability to organisms in water and soil. This is a generally used procedure in the environmental hazard assessment of chemicals in soil.

The correction for the reduced bioavailability is made by multiplying $PNEC_{wc}$ by (K_d + 0.27). K_d is the substance's coefficient of adsorption in soil, i.e. the ratio of the concentration of the substance on soil particles to the concentration in the soil's liquid phase. The value 0.27 is the average relative water content of soil (l water/kg dry matter soil).

If it is not possible to find some other value for K_d, it can be estimated for non-ionic organic compounds as $K_d \sim 0.02 \cdot P_{ow}$.

By multiplying by the soil's dry density of 1.5 kg/l, $PNEC_{sc}$ is expressed as a volumetric concentration similarly to the PNEC values for the other environmental compartments.

The procedure for determination of $PNEC_{sc}$ on the basis of ecotoxicity data for aquatic organisms can be illustrated by continuing the example of n-butyl acetate:

Even though f_{sc} was previously determined as zero for n-butyl acetate and in principle no equivalency factor for the soil compartment is thus to be calculated for an aqueous emission of this substance, a PNEC value for n-butyl acetate in soil is estimated for the sake of illustration:

$PNEC_{sc}$ is estimated on the basis of the value for $PNEC_{wc}$ previously found.

No K_d value has been found for n-butyl acetate, but P_{ow} is given as 66 in the database HSDB, and on this basis K_d is estimated as:

$K_d = 0.02 \cdot 66 = 1.32$.

$PNEC_{sc}$ can now be estimated as:

$PNEC_{sc} = 0.18 \text{ mg/l} \cdot (1.32 + 0.27) \cdot 1.5 = 0.43 \text{ mg/l}$.

Example 23.5. Estimation of $PNEC_{sc}$ for n-butyl acetate

23.7.2.2.4 Determine $LOEC_{aerobic\ heterotroph}$

- Find information on the lowest observed effect concentration, LOEC, of the substance to aerobic heterotrophic bacteria (bacteria which require oxygen) (e.g. in Verschueren, 1996; Nikunen *et al.*, 1991; Blum & Speece, 1991).

The procedure is illustrated by Example 23.6.

The lowest bacterial LOEC value of 115 mg/l found for n-butyl acetate is for population growth in the bacterium *Pseudomonas putrida. P. putrida* is an aerobic heterotroph and the ecotoxicity factor for n-butyl acetate in sewage treatment plants can therefore be calculated on the basis of this value.

Example 23.6. Determination of $LOEC_{aerobic\ heterotroph}$ for n-butyl acetate

23.7.2.3 Determine the biodegradability factor, BIO

- Find studies of the substance's biodegradability in tests for "Ready biodegradability" (OECD 301, 1992a) or "Inherent biodegradability" (OECD 302, 1992b) (e.g. in Howard, 1990) and determine the biodegradability factor BIO on the basis of the criteria in Table 23.11.

Table 23.11. Criteria for use in determination of the biodegradability factor BIO

Class	Criterion	Biodegradability factor BIO
Readily biodegradable	≥ 60% degraded in a 28 day Ready biodegradability test (OECD 301 A-F, 1992a, or 92/69/EEC C.4 A-F)	0.2
Potentially biodegradable	≥ 70% degraded in a 28 day Inherent biodegradability test (OECD 302 B-C, 1992b, or 87/302/EEC)	0.5
Non-biodegradable	< 70% degraded in a 28 day Inherent biodegradability test (OECD 302 B-C, 1992b, or 87/302/EEC)	1

- If no results of OECD or EU biodegradability tests or similar are available, assess the substance's biodegradability relative to the criteria in these tests on the basis of any other information which may be available.
- If no information which would permit an assessment of the substance's biodegradability is available, regard it as non-biodegradable and set BIO = 1.

The biodegradability factor represents the substance's biodegradability: the less the biodegradability, the greater the factor. Determination of BIO requires some knowledge of the biodegradability tests noted in Table 23.11 and of the tests which have been carried out on the substance. This is illustrated by the n-butyl acetate example:

Example 23.7. Determination of BIO for n-butyl acetate

Results for any of the suggested OECD tests have not been found for n-butyl acetate, but the following information was found in the database HSDB:

In an unnamed biodegradability test with an activated sludge culture in fresh water, 58% of the substance was broken down after 5 days and 83% was broken down after 20 days (HSDB, 1996).

For the substance to be easily biodegradable in the OECD 301 tests, 60% of the substance must be broken down within 10 days of commencement of the process, i.e. after 10% has been broken down.

It seems probable that this requirement is fulfilled in the n-butyl acetate test. The substance is thus assessed as "easily biodegradable" and assigned the BIO factor 0.2.

23.7.2.4 Calculate equivalency factors

■ Calculate the equivalency factors as:

$EF(etwa) = f_{wa} \cdot ETF_{wa}$

$EF(etwc) = f_{wc} \cdot ETF_{wc} \cdot BIO$

$EF(etsc) = f_{sc} \cdot ETF_{wc} \cdot BIO$

$EF(etp) = ETF_{p}$

The factors in the expressions are described above, and the equivalency factors can now be calculated for the ecotoxicity potential in the different environmental compartments. The procedure is illustrated by the n-butyl acetate example.

On the basis of information in the above examples, the following equivalency factors can be calculated for n-butyl acetate in the individual environmental compartments (as the emission has passed through a sewage treatment plant, it is also relevant to calculate an equivalency factor for this compartment).

Water

EF(etwa) $= f_{wa} \cdot 1/PNEC_{wa} = 1 \cdot 1/(1.8\ mg/l) =$ **0.56 m^3/g**

EF(etwc) $= f_{wc} \cdot BIO/PNEC_{wc} = 1 \cdot 0.2/(0.18\ mg/l) =$ **1.11 m^3/g**

Soil

EF(etsc) $= f_{sc} \cdot BIO/PNEC_{sc} = 0 \cdot 0.2/(0.43\ mg/l) =$ **0 m^3/g**

Sewage treatment plant

EF(etp) $= 50/LOEC = 50/(115\ mg/l) =$ **0.43 m^3/g**

Example 23.8. Calculation of equivalency factors for ecotoxicity of n-butyl acetate

23.7.2.5 Calculate ecotoxicity potentials

- Calculate the ecotoxicity potentials as:

$$EP(etwa) = Q \cdot EF(etwa) = Q \cdot f_{wa} \cdot ETF_{wa}$$
$$EP(etwc) = Q \cdot EF(etwc) = Q \cdot f_{wc} \cdot ETF_{wc} \cdot BIO$$
$$EP(etsc) = Q \cdot EF(etsc) = Q \cdot f_{sc} \cdot ETF_{sc} \cdot BIO$$
$$EP(etp) = Q_p \cdot EF(etp) = Q_p \cdot ETF_p$$

When the equivalency factors have been calculated for the different compartments, the ecotoxicity potentials can be determined as the product of the quantity of substance emitted, Q, and the associated equivalency factors. If the emission has passed through a sewage treatment plant, an ecotoxicity potential is calculated for the plant for the quantity which was discharged to it.

For the example given with n-butyl acetate, the following results are obtained:

Discharge of 1.5 g n-butyl acetate in the effluent from a sewage treatment plant has the following ecotoxicity potentials:

$EP(etwa) = Q \cdot EF(etwa) = 1.5\ g \cdot 0.56\ m^3/g = 0.84\ m^3$ water

$EP(etwc) = Q \cdot EF(etwc) = 1.5\ g \cdot 1.11\ m^3/g = 1.7\ m^3$ water

$EP(etsc) = Q \cdot EF(etsc) = 1.5\ g \cdot 0 = 0\ m^3$ soil

7.1 g n-butyl acetate originally entered the sewage treatment plant. 5.6 g were removed in the plant. The 7.1 g have an ecotoxicity potential in the plant of:

$MP(etp) = Q_p \cdot EF(etp) = 7.1\ g \cdot 0.43\ m^3/g = 3.1\ m^3$ waste water

Example 23.9. Calculation of ecotoxicity potentials for n-butyl acetate

23.7.2.6 Sources of data

Much of the work of the redistribution calculations and the calculation of equivalency factors consists of finding the necessary data for the substance's chemical and ecotoxicological characteristics.

Calculation of the **redistribution factors** requires a knowledge of the substance's

- Henry's law constant,
- atmospheric half-life,
- P_{ow}.

Calculation of **ecotoxicity factors** requires information on its

- Ecotoxicity,
- Possibly P_{ow} and K_d.

Calculation of the **biodegradability factor** requires a knowledge of the substance's

- biodegradability.

These data can be found in the literature or more readily in various databases in books or electronic data media. Table 23.11 recommends some databases for this information:

Ecotoxicity		**Henry's law constant, atmospheric half life**	**Biodegradability**
Water:	**Waste water treatment plants**	P_{ow}, K_d	
AQUIRE HSDB Verschueren, 1996 Nikunen *et al.*, 1991 IRIS IUCLID	Blum and Speece, 1991 Verschueren, 1996 Nikunen *et al.*, 1991	Howard, 1990 Howard *et al.*, 1991 HSDB SRC software	Howard, 1990 Howard *et al.*, 1991 HSDB

Table 23.11. Databases with the various substance data required in the calculation of equivalency factors for ecotoxicity

23.7.2.6.1 Brief description of the databases

AQUIRE:
Aquatic Information Retrieval Toxicity Database. Database with ecotoxicological substance characteristics developed for the US EPA. The latest version contains more than 100,000 test results for 5,600 chemicals collected from over 7,000 scientific publications. All test data are assessed and classified by the US EPA. Available in PC version.

Blum & Speece, 1991:
Study of toxicity of 52 different organic chemicals to nitrogen-fixing and heterotrophic bacteria.

Howard, 1990:
Database of various environmental chemical data relevant to assessment of the fates of organic substances in the environment. Data for 151 organic compounds in volumes 1 and 2.

Howard *et al.*, 1991:
Database of rates of degradation in the environment and in a sewage treatment plant. Data for 336 organic compounds.

HSDB:
Hazardous Substance Data Bank. Information on human toxicological and ecotoxicological characteristics and information relevant to assessment of a substance's fate in the environment. Contains comprehensive review of ca. 4,500 chemicals. Published by the National Library of Medicine, USA. Information included in HSDB is assessed and approved by a scientific review group. Available on CD-ROM.

IRIS:
Integrated Risk Information System. Database with information on human toxicity, ecotoxicity and fate in the environment. Prepared by the US EPA for over 300 environmentally toxic chemicals. Contains evaluated data. Available on CD-ROM.

IUCLID:
International Uniform Chemical Information Database. Database presenting the environmental data supplied to the European Chemicals Bureau (ECB) on more than 1,400 chemicals that are produced or marketed within the EU in annual quantities larger than 1,000 tonnes. The information is not evaluated. Available on CD-ROM.

SRC software:
Software developed by Syracuse Research Corporation for estimating properties like P_{ow}, atmospheric half-lives and Henry's law constant based on the molecular structure of the substance.

Verschueren, 1996:
Contains information on environmental characteristics for more than 2,300 organic compounds. The information is not evaluated.

Nikunen *et al.*, 1991:
Contains information on environmental characteristics for more than 1,700 chemicals. The information is not evaluated.

23.8 Human toxicity via exposure through the environment

Chemicals emitted as a consequence of human activities can contribute to human toxicity via exposure through the environment. This is the case if the substances are poisonous and if their characteristics in combination with their manner of emission result in humans being exposed to them. The most important routes of exposure are via the air breathed in or via other materials ingested orally, e.g. as food.

If the concentrations of hazardous substances caused by the emission are high enough, the toxic effects can occur immediately after human exposure. This is called acute toxicity.

Toxic effects which appear only on repeated or long-term exposure to the substance are called chronic toxicity. Chronic toxicity is often caused by substances which have only a low degradability in the environment and which therefore persist for a long time after their emission (persistent substances). Some substances also have a tendency to accumulate in the living organisms which function as food for humans. Our food can in this way contain concentrations of substances which are far higher than the concentrations in the surrounding environment. The chronic toxicity of a compound is thus determined both by its toxicity and its biodegradability and ability to undergo bioconcentration.

Toxicity is an impact which predominantly affects humans in the environment on local and regional scales. For a limited number of substances which are broken down with great difficulty, easily transported and poisonous, toxicity can be a global impact.

As with ecotoxicity, toxicity can be attributed to many different types of poisonous impacts, and a list of substances which can cause human toxicity in the environment may include thousands of entries.

The procedure for determining whether a compound should be regarded as contributing to human toxicity is given below, together with an explanation of how toxicity potentials are calculated.

Although an attempt was made to make the method simple to use and as systematic as possible, it nevertheless requires some chemical and toxicological background knowledge to use the method to calculate human toxicity potentials.

23.8.1 Determine which substances contribute to toxicity

- Use the chemical hazard screening method for toxicity presented in Section 23.11.
- Regard all substances scoring a total toxicological impact score of eight or more as contributing to human toxicity in the environment.
- Scrutinize the inventory to assess whether there are other substances for which toxicity potentials should be calculated.

For a substance to be regarded as contributing to toxicity, it must of course be poisonous to humans, but determination of which substances should be so regarded is not based solely on the substance's toxicity.

The substance's behaviour in the environment and its tendency to accumulate in human food determine the extent to which humans will be exposed to it. The chronic toxicity which can be evident after long-term exposure to the substance also depends on the persistence and ability of the substance to accumulate in human tissues.

The chemical hazard screening method in Section 23.11 therefore focuses on the following characteristics of a substance:

- The substance's toxicity to humans as determined in laboratory tests with experimental animals or by observation of humans who have been exposed to the substance.
- The substance's biodegradability, which is determined experimentally in biodegradability tests.
- The substance's potential for bioconcentration, which is determined directly in experiments or estimated on the basis of the substance's octanol-water partitioning ratio, P_{ow}.

23.8.2 Calculate the toxicity potentials

- Find the substance's equivalency factors for the relevant emission route in Table 23.13.
- In case no equivalency factors are found for the substance in Table 23.13, determine equivalency factors as described in Sections 23.8.2.1-23.8.2.7.
- Calculate the human toxicity potentials for the relevant exposure routes by multiplying the magnitude of the emission with the equivalency factors found: EP(ht) = Q · EF(ht).

As for ecotoxicity, it has been necessary in the EDIP method to develop a way by which toxicity potentials can be calculated for exposure of humans in the environment.

The EDIP method of calculation of human toxicity potentials was also inspired by the EU Commission's technical guidelines for risk assessment of chemicals in the environment (European Commission, 1996).

Toxicity potentials

Up to four toxicity potentials are calculated for each substance:

EP(hta) = Q · EF(hta) for toxicity to humans via air

EP(htw) = Q · EF(htw) for toxicity to humans via surface water

EP(hts) = Q · EF(hts) for toxicity to humans via soil

EP(htg) = Q · EF(htg) for toxicity to humans via groundwater

EP(ht) is the toxicity potential for humans by exposure to the substance via one of the four compartments air, water, soil or groundwater. The toxicity potential is determined as the product of the quantity of substance Q emitted and the substance's equivalency factor EF for exposure through the compartment in question.

The toxicity potential is expressed in m^3 of the compartment and corresponds to the volume of the compartment into which the emission should be diluted for its concentration to be so low that no toxicological effects could be expected from the emission.

Table 23.13 presents equivalency factors EF for around 90 compounds emitted to air, water or soil. If the table gives no equivalency factors for a substance, these must be calculated. The equivalency factors are determined as the product of five factors which represent the substance's dispersion in the environment, the efficiency of intake for the actual exposure route, the substance's toxic characteristics and its biodegradability. To calculate the equivalency factors it is therefore first necessary to:

- Determine the fraction (f) which reaches the different environmental compartments upon dispersion.
- Calculate the transfer factor (T) for the substance via the actual exposure route.
- Determine the intake factor (I) for the exposure route in question.
- Calculate toxicity factors (HTF) representing the substance's potential toxicity by exposure through the four compartments.
- Determine the biodegradability factor (BIO) for the substance.

Substance	CAS no.	Emissions to air			Emissions to water			Emissions to soil		
		EF(hta) m^3/g	EF(htw) m^3/g	EF(hts) m^3/g	EF(hta) m^3/g	EF(htw) m^3/g	EF(hts) m^3/g	EF(hta) m^3/g	EF(htw) m^3/g	EF(hts) m^3/g
1,1,1-trichloroethane	71-55-6	$9.2\cdot10^{2}$	$9.9\cdot10^{-4}$	$2.0\cdot10^{-3}$	$9.2\cdot10^{2}$	$9.9\cdot10^{-4}$	$2.0\cdot10^{-3}$	$9.2\cdot10^{2}$	$9.9\cdot10^{-4}$	$2.0\cdot10^{-3}$
1,2-Benzoisothiazolin-3-one	2634-33-5	$2.8\cdot10^{4}$	0	0	0	$1.3\cdot10^{-4}$	0	0	0	0.32
1,2-Dichlorobenzene	95-50-1	$8.3\cdot10^{3}$	0.37	$7.0\cdot10^{-3}$	$8.3\cdot10^{3}$	0.37	$7.0\cdot10^{-3}$	$8.3\cdot10^{3}$	0.37	$7.0\cdot10^{-3}$
1,2-Dichloroethane	107-06-2	$5.0\cdot10^{4}$	$3.9\cdot10^{-3}$	$7.5\cdot10^{-2}$	0	$2.0\cdot10^{-2}$	0	0	0	$9.4\cdot10^{-2}$
1,2-Propylene oxide	75-56-9	$3.3\cdot10^{4}$	$2.9\cdot10^{-6}$	$1.1\cdot10^{-3}$	0	$1.5\cdot10^{-5}$	0	0	0	$1.4\cdot10^{-3}$
1-Butanol	71-36-3	$1.3\cdot10^{4}$	$1.4\cdot10^{-3}$	0.14	0	$7.1\cdot10^{-3}$	0	0	0	0.18
2,3,7,8-Tetrachlorodibenzo-p-dioxin	1746-01-6	$2.9\cdot10^{10}$	$2.2\cdot10^{8}$	$1.4\cdot10^{4}$	0	$1.1\cdot10^{9}$	0	0	0	$1.8\cdot10^{4}$
2,4-Dinitrotoluene	121-14-2	$1.1\cdot10^{2}$	$5.8\cdot10^{-3}$	$9.6\cdot10^{-4}$	0	$2.9\cdot10^{-2}$	0	0	0	$1.2\cdot10^{-3}$
2-Chlorotoluene	95-49-8	$2.2\cdot10^{3}$	0.98	$1.9\cdot10^{-2}$	$2.2\cdot10^{3}$	0.98	$1.9\cdot10^{-2}$	$2.2\cdot10^{3}$	0.98	$1.9\cdot10^{-2}$
2-Ethyl hexanol	104-76-7	$1.8\cdot10^{3}$	0	0	0	$2.8\cdot10^{-2}$	0	0	0	$1.5\cdot10^{-3}$
2-Ethylhexyl acetate	103-09-3	$9.5\cdot10^{3}$	0	0	$9.5\cdot10^{3}$	0	0	$9.5\cdot10^{3}$	0	0
2-Propanol	67-63-0	$1.2\cdot10^{2}$	$7.5\cdot10^{-6}$	$2.8\cdot10^{-3}$	0	$3.7\cdot10^{-5}$	0	0	0	$3.5\cdot10^{-3}$
3-Chlorotoluene	108-41-8	$2.2\cdot10^{3}$	0.71	$2.4\cdot10^{-2}$	$2.2\cdot10^{3}$	0.71	$2.4\cdot10^{-2}$	$2.2\cdot10^{3}$	0.71	$2.4\cdot10^{-2}$
4-Chlorotoluene	106-43-4	$2.2\cdot10^{3}$	0.79	$2.2\cdot10^{-2}$	$2.2\cdot10^{3}$	0.79	$2.2\cdot10^{-2}$	$2.2\cdot10^{3}$	0.79	$2.2\cdot10^{-2}$
Acetaldehyde	75-07-0	$3.7\cdot10^{3}$	0	0	0	$7.1\cdot10^{-6}$	0	0	0	$9.2\cdot10^{-4}$
Acetic acid	64-19-7	$1.0\cdot10^{4}$	$3.3\cdot10^{-6}$	$1.6\cdot10^{-3}$	0	$1.6\cdot10^{-5}$	0	0	0	$2.0\cdot10^{-3}$
Acetone	67-64-1	$3.2\cdot10^{4}$	$8.5\cdot10^{-6}$	$4.1\cdot10^{-3}$	0	$4.3\cdot10^{-5}$	0	0	0	$5.2\cdot10^{-3}$
Acrylic acid	79-10-7	$6.7\cdot10^{5}$	$6.3\cdot10^{-5}$	$1.6\cdot10^{-2}$	0	$3.1\cdot10^{-4}$	0	0	0	$2.0\cdot10^{-2}$
Acrylic acid. 2-hydroxyethyl ester	818-61-1	$2.0\cdot10^{2}$	0	0	0	$6.4\cdot10^{-4}$	0	0	0	$7.6\cdot10^{-2}$
Anthracene	120-12-7	$9.5\cdot10^{2}$	0	0	0	11	0	0	0	$1.1\cdot10^{-4}$
Antimony	7440-36-0	$2.0\cdot10^{4}$	64	17	0	$3.2\cdot10^{2}$	0	0	0	21
Arsenic	7440-38-2	$9.5\cdot10^{6}$	7.4	$1.0\cdot10^{2}$	0	37	0	0	0	$1.3\cdot10^{2}$
Atrazine	1912-24-9	$1.4\cdot10^{5}$	0	0	0	1.1	0	0	0	$4.2\cdot10^{-2}$
Benzene	71-43-2	$1.0\cdot10^{7}$	2.3	14	$1.0\cdot10^{7}$	2.3	14	$1.0\cdot10^{7}$	2.3	14
Benzo(a)pyrene	50-32-8	$5.0\cdot10^{7}$	0	0	0	$3.2\cdot10^{2}$	0	0	0	$1.8\cdot10^{-3}$
Benzotriazole	95-14-7	$1.3\cdot10^{3}$	$9.3\cdot10^{-4}$	$2.0\cdot10^{-2}$	0	$4.6\cdot10^{-3}$	0	0	0	$2.5\cdot10^{-2}$
Biphenyl	92-52-4	$2.3\cdot10^{5}$	1.4	$2.9\cdot10^{-3}$	0	7.1	0	0	0	$3.6\cdot10^{-3}$
Butyl diglycol acetate	124-17-4	$1.3\cdot10^{4}$	0	0	0	$3.3\cdot10^{-2}$	0	0	0	0.27
Cadmium	7440-46-9	$1.1\cdot10^{8}$	$5.6\cdot10^{2}$	4.5	0	$2.8\cdot10^{3}$	0	0	0	5.6
Carbon monoxide	630-08-0	$8.3\cdot10^{2}$	0	0	$8.3\cdot10^{2}$	0	0	$8.3\cdot10^{2}$	0	0
Chlorine	7782-50-5	$3.4\cdot10^{4}$	0	0	$3.4\cdot10^{4}$	0	0	$3.4\cdot10^{4}$	0	$6.5\cdot10^{-2}$
Chlorobenzene	108-90-7	$2.2\cdot10^{5}$	0.27	$4.6\cdot10^{-2}$	$2.2\cdot10^{5}$	0.27	$4.6\cdot10^{-2}$	$2.2\cdot10^{5}$	0.27	$4.6\cdot10^{-2}$
Chloroform	67-66-3	$1.0\cdot10^{5}$	$5.4\cdot10^{-2}$	0.20	$1.0\cdot10^{5}$	$5.4\cdot10^{-2}$	0.20	$1.0\cdot10^{5}$	$5.4\cdot10^{-2}$	0.20
Chromium	7440-47-3	$1.0\cdot10^{6}$	3.6	1.1	0	18	0	0	0	1.4

Table 23.13. Equivalency factors for human toxicity via air, surface water and soil for substances emitted to air, water or soil (for 'a' a global default value of 0.2 is used). The values are based on a rather simple fate modelling and for some substances the underlying database is weak. They should therefore not be used for purposes other than the impact assessment of LCA.

Substance	CAS no.	Emissions to air EF(hta) m^3/g	Emissions to air EF(htw) m^3/g	Emissions to air EF(hts) m^3/g	Emissions to water EF(hta) m^3/g	Emissions to water EF(htw) m^3/g	Emissions to water EF(hts) m^3/g	Emissions to soil EF(hta) m^3/g	Emissions to soil EF(htw) m^3/g	Emissions to soil EF(hts) m^3/g
Cobalt	7440-48-4	$9.5\cdot10^{3}$	$2.5\cdot10^{-3}$	0.17	0	$1.2\cdot10^{-2}$	0	0	0	0.21
Copper	7440-50-8	$5.7\cdot10^{2}$	3.4	$4.0\cdot10^{-3}$	0	17	0	0	0	$5.0\cdot10^{-3}$
Dibutyltinoxide	818-08-6	$1.4\cdot10^{5}$	$3.7\cdot10^{-3}$	$4.2\cdot10^{-3}$	0	$1.9\cdot10^{-2}$	0	0	0	$5.3\cdot10^{-6}$
Diethanolamine	111-42-2	$4.0\cdot10^{4}$	0	0	0	$3.9\cdot10^{-5}$	0	0	0	$5.9\cdot10^{-3}$
Diethylaminoethanol	100-37-8	$2.7\cdot10^{4}$	0	0	0	$3.2\cdot10^{-3}$	0	0	0	0.30
Diethylene glycol	111-46-6	$2.5\cdot10^{5}$	0	0	0	$3.1\cdot10^{-6}$	0	0	0	$4.7\cdot10^{-4}$
Diethylene glycol mono-n-butyl ether	112-34-5	$2.0\cdot10^{6}$	0	0	0	$3.4\cdot10^{-3}$	0	0	0	0.16
Ethanol	64-17-5	$1.1\cdot10^{2}$	$2.9\cdot10^{-7}$	$1.5\cdot10^{-4}$	0	$1.5\cdot10^{-6}$	0	0	0	$1.8\cdot10^{-4}$
Ethyl acetate	141-78-6	$6.9\cdot10^{2}$	$8.9\cdot10^{-6}$	$1.2\cdot10^{-3}$	0	$4.4\cdot10^{-5}$	0	0	0	$1.5\cdot10^{-3}$
Ethylene glycol	107-21-1	$8.3\cdot10^{5}$	$1.4\cdot10^{-3}$	$2.0\cdot10^{-5}$	0	$7.0\cdot10^{-3}$	0	0	0	$2.5\cdot10^{-5}$
Ethylene glycol acetate	111-15-9	$3.7\cdot10^{3}$	0	0	0	$1.5\cdot10^{-3}$	0	0	0	$6.6\cdot10^{-2}$
Ethylene glycol mono-n-butyl ether	111-76-2	$2.1\cdot10^{4}$	0	0	0	$8.4\cdot10^{-5}$	0	0	0	$3.5\cdot10^{-3}$
Ethylenediamine tetraacetic acid. EDTA	60-00-4	$3.7\cdot10^{2}$	0	0	0	$6.7\cdot10^{-9}$	0	0	0	$2.5\cdot10^{-6}$
Ethylenediamine. 1.2-ethanediamine	107-15-3	$2.0\cdot10^{4}$	0	0	0	$1.4\cdot10^{-5}$	0	0	0	$1.5\cdot10^{-3}$
Fluoride	16984-48-8	$9.5\cdot10^{4}$	0	0	0	$1.2\cdot10^{-2}$	0	0	0	$6.3\cdot10^{-4}$
Formaldehyde	50-00-00	$1.3\cdot10^{7}$	$2.2\cdot10^{-5}$	$5.8\cdot10^{-3}$	0	$1.1\cdot10^{-4}$	0	0	0	$7.2\cdot10^{-3}$
Glycerol	56-81-5	70	0	0	0	$1.3\cdot10^{-6}$	0	0	0	$1.7\cdot10^{-4}$
Hexamethylene diisocyanate. HDI	822-06-0	$7.1\cdot10^{5}$	12	0.56	0	61	0	0	0	0.70
Hexane	110-54-3	$1.6\cdot10^{3}$	0.34	$9.7\cdot10^{-4}$	$1.6\cdot10^{3}$	0.34	$9.7\cdot10^{-4}$	$1.6\cdot10^{3}$	0.34	$9.7\cdot10^{-4}$
Hydrogen cyanide	74-90-8	$1.4\cdot10^{5}$	$1.5\cdot10^{-3}$	0.71	$1.4\cdot10^{5}$	$1.5\cdot10^{-3}$	0.71	$1.4\cdot10^{5}$	$1.5\cdot10^{-3}$	0.71
Hydrogen sulphide	7783-06-4	$1.1\cdot10^{6}$	$8.1\cdot10^{-4}$	0.26	0	$4.1\cdot10^{-3}$	0	0	0	0.33
Iron	7439-89-6	$3.7\cdot10^{4}$	$9.6\cdot10^{-3}$	0.77	0	$4.8\cdot10^{-2}$	0	0	0	0.96
Isobutanol	78-83-1	13	$2.9\cdot10^{-6}$	$3.8\cdot10^{-4}$	0	$1.5\cdot10^{-5}$	0	0	0	$4.7\cdot10^{-4}$
Isopropylbenzene. cumene	98-82-8	$1.0\cdot10^{4}$	0.21	$2.1\cdot10^{-2}$	$1.0\cdot10^{4}$	0.21	$2.1\cdot10^{-2}$	$1.0\cdot10^{4}$	0.21	$2.1\cdot10^{-2}$
Lead	7439-92-1	$1.0\cdot10^{8}$	53	$8.3\cdot10^{-2}$	0	$2.6\cdot10^{2}$	0	0	0	0.10
Maleic acid. dibutyl ester	105-76-0	$7.7\cdot10^{3}$	0	0	0	14	0	0	0	$3.4\cdot10^{-3}$
Manganese	7439-96-5	$2.5\cdot10^{6}$	$5.3\cdot10^{-3}$	0.42	0	$2.7\cdot10^{-2}$	0	0	0	0.53
Mercury	7439-97-6	$6.7\cdot10^{6}$	$1.1\cdot10^{5}$	81	$6.7\cdot10^{6}$	$1.1\cdot10^{5}$	81	$6.7\cdot10^{6}$	$1.1\cdot10^{5}$	81
Methacrylic acid	79-41-4	$4.5\cdot10^{4}$	0	0	0	$6.0\cdot10^{-3}$	0	0	0	0.22
Methanol	67-56-1	$2.5\cdot10^{3}$	$3.0\cdot10^{-4}$	$3.1\cdot10^{-4}$	0	$1.5\cdot10^{-3}$	0	0	0	$3.9\cdot10^{-4}$
Methyl isobutyl ketone	108-10-1	$3.3\cdot10^{3}$	$3.6\cdot10^{-3}$	0.12	0	$1.8\cdot10^{-2}$	0	0	0	0.15
Methyl methacrylate	80-62-6	$1.0\cdot10^{7}$	0	0	0	$4.9\cdot10^{-3}$	0	0	0	$3.2\cdot10^{-2}$

Substance	CAS no.	Emissions to air			Emissions to water			Emissions to soil		
		EF(hta) m^3/g	EF(htw) m^3/g	EF(hts) m^3/g	EF(hta) m^3/g	EF(htw) m^3/g	EF(hts) m^3/g	EF(hta) m^3/g	EF(htw) m^3/g	EF(hts) m^3/g
Methylenebis(4-phenylisocyanate). MDI	101-68-8	$5.0 \cdot 10^7$	0	0	0	$2.8 \cdot 10^2$	0	0	0	$4.0 \cdot 10^{-4}$
Molybdenum	7439-98-7	$1.0 \cdot 10^5$	$5.3 \cdot 10^{-2}$	1.5	0	0.27	0	0	0	1.9
Monoethanolamine	141-43-5	$2.7 \cdot 10^4$	0	0	0	$3.5 \cdot 10^{-5}$	0	0	0	$5.4 \cdot 10^{-3}$
Morpholine	110-91-8	$1.3 \cdot 10^4$	0	0	0	$1.0 \cdot 10^{-4}$	0	0	0	$1.6 \cdot 10^{-2}$
n-Butyl acetate	123-86-4	$1.1 \cdot 10^3$	$7.0 \cdot 10^{-3}$	$5.0 \cdot 10^{-2}$	0	$3.5 \cdot 10^{-2}$	0	0	0	$6.2 \cdot 10^{-2}$
Nickel	7440-02-0	$6.7 \cdot 10^4$	$3.7 \cdot 10^{-3}$	0.12	0	$1.9 \cdot 10^{-2}$	0	0	0	0.15
Nitrilotriacetate	139-13-9	$3.8 \cdot 10^4$	0	0	0	$8.2 \cdot 10^{-14}$	0	0	0	$5.1 \cdot 10^{-5}$
Nitrobenzenesulphonic acid.sodium salt	127-68-4	$2.6 \cdot 10^3$	$1.7 \cdot 10^{-7}$	$3.9 \cdot 10^{-5}$	$2.6 \cdot 10^3$	$1.7 \cdot 10^{-7}$	$3.9 \cdot 10^{-5}$	$2.6 \cdot 10^3$	$1.7 \cdot 10^{-7}$	$3.9 \cdot 10^{-5}$
Nitrogen dioxide and other NO_X	10102-44-0	$8.6 \cdot 10^3$	0	0	0	$3.7 \cdot 10^{-5}$	0	0	0	$3.7 \cdot 10^{-3}$
Nitrous oxide	10024-97-2	$2.0 \cdot 10^3$	0	0	$2.0 \cdot 10^3$	0	0	$2.0 \cdot 10^3$	0	0
Ozone	10028-15-6	$5.0 \cdot 10^4$	0	0	$5.0 \cdot 10^4$	0	0	$5.0 \cdot 10^4$	0	0
Phenol	108-95-2	$1.4 \cdot 10^6$	0	0	0	$3.4 \cdot 10^{-2}$	0	0	0	$6.4 \cdot 10^{-5}$
Phosgene	75-44-5	$2.0 \cdot 10^6$	0	0	$2.0 \cdot 10^6$	0	0	$2.0 \cdot 10^6$	0	0
Propylene glycol	57-55-6	$1.5 \cdot 10^3$	0	0	0	$4.8 \cdot 10^{-6}$	0	0	0	$7.7 \cdot 10^{-4}$
Selenium	7782-49-2	$1.5 \cdot 10^6$	28	$4.4 \cdot 10^{-2}$	0	$1.4 \cdot 10^2$	0	0	0	$5.5 \cdot 10^{-2}$
Silver	7440-22-4	$2.0 \cdot 10^5$	$5.3 \cdot 10^{-2}$	4.2	0	0.27	0	0	0	5.3
Sodium benzoate	532-32-10	$1.4 \cdot 10^4$	$4.0 \cdot 10^{-7}$	$1.4 \cdot 10^{-4}$	0	$2.0 \cdot 10^{-6}$	0	0	0	$1.7 \cdot 10^{-4}$
Sodium Hypochlorite	7681-52-9	$2.0 \cdot 10^3$	0	0	0	$2.6 \cdot 10^{-4}$	0	0	0	$2.5 \cdot 10^{-2}$
Styrene	100-42-5	$1.0 \cdot 10^3$	0	0	$1.0 \cdot 10^3$	0	0	$1.0 \cdot 10^3$	0	0
Sulphamic acid	5329-14-6	$9.0 \cdot 10^3$	$2.1 \cdot 10^{-9}$	$9.7 \cdot 10^{-6}$	0	$1.1 \cdot 10^{-8}$	0	0	0	$1.2 \cdot 10^{-5}$
Sulphur dioxide	7446-09-5	$1.3 \cdot 10^3$	0	0	$1.3 \cdot 10^3$	0	0	$1.3 \cdot 10^3$	0	0
Tetrachlorethylene	127-18-4	$2.9 \cdot 10^4$	0.36	$4.0 \cdot 10^{-2}$	$2.9 \cdot 10^4$	0.36	$4.0 \cdot 10^{-2}$	$2.9 \cdot 10^4$	0.36	$4.0 \cdot 10^{-2}$
Thallium	7440-28-0	$5.0 \cdot 10^5$	$1.3 \cdot 10^4$	10	0	$6.5 \cdot 10^4$	0	0	0	13
Titanium	7440-32-6	$1.8 \cdot 10^4$	$4.7 \cdot 10^{-3}$	0.38	0	$2.3 \cdot 10^{-2}$	0	0	0	0.47
Toluene	108-88-3	$2.5 \cdot 10^3$	$4.0 \cdot 10^{-3}$	$1.0 \cdot 10^{-3}$	$2.5 \cdot 10^3$	$4.0 \cdot 10^{-3}$	$1.0 \cdot 10^{-3}$	$2.5 \cdot 10^3$	$4.0 \cdot 10^{-3}$	$1.0 \cdot 10^{-3}$
Toluene diisocyanate 2.4/2.6 mixture	26471-62-5	$7.1 \cdot 10^5$	2.1	$1.2 \cdot 10^{-2}$	0	10	0	0	0	$1.5 \cdot 10^{-2}$
Toluene-2.4-diamine	95-80-7	$1.4 \cdot 10^3$	0	0	0	$1.3 \cdot 10^{-4}$	0	0	0	$1.1 \cdot 10^{-2}$
Trichloroethylene	79-01-6	$1.9 \cdot 10^4$	$9.1 \cdot 10^{-4}$	$6.9 \cdot 10^{-4}$	$1.9 \cdot 10^4$	$9.1 \cdot 10^{-4}$	$6.9 \cdot 10^{-4}$	$1.9 \cdot 10^4$	$9.1 \cdot 10^{-4}$	$6.9 \cdot 10^{-4}$
Triethanolamine	102-71-6	$1.3 \cdot 10^4$	0	0	0	$8.4 \cdot 10^{-5}$	0	0	0	$1.4 \cdot 10^{-2}$
Triethylamine	121-44-8	$1.4 \cdot 10^5$	0	0	0	0.23	0	0	0	1.2
Vanadium	7440-62-2	$1.4 \cdot 10^5$	$3.7 \cdot 10^{-2}$	0.96	0	0.19	0	0	0	1.2
Vinylchloride	75-01-4	$3.9 \cdot 10^5$	0.40	4.0	$3.9 \cdot 10^5$	0.40	4.0	$3.9 \cdot 10^5$	0.40	4.0
Xylenes. mixed	1330-20-7	$6.7 \cdot 10^3$	$1.1 \cdot 10^{-3}$	$6.7 \cdot 10^{-5}$	$6.7 \cdot 10^3$	$1.1 \cdot 10^{-3}$	$6.7 \cdot 10^{-5}$	$6.7 \cdot 10^3$	$1.1 \cdot 10^{-3}$	$6.7 \cdot 10^{-5}$
Zinc (as dust)	7440-66-6	$8.1 \cdot 10^4$	4.1	$1.3 \cdot 10^{-2}$	0	21	0	0	0	$1.6 \cdot 10^{-2}$

Table 23.13. (continued)

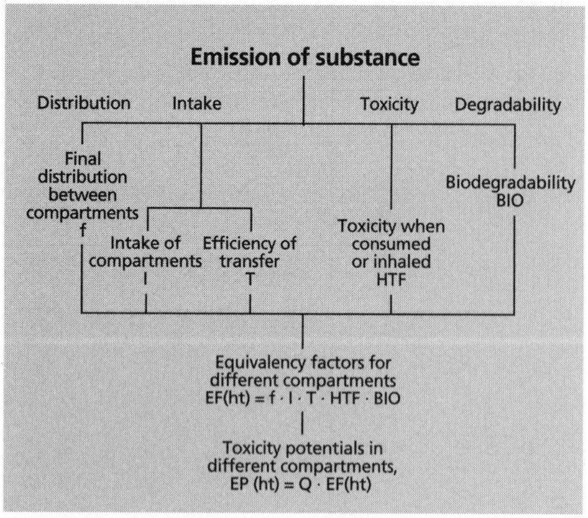

Figure 23.6. The calculation of human toxicity potentials involves investigation of the characteristics and the fate of the substance

The equivalency factors for toxicity depend exclusively on the characteristics of the substance. They will be the same independently of the context in which the emission occurs. It is therefore necessary to calculate them only once for each substance. The factors calculated can be reused every time the substance appears in an inventory to be assessed.

23.8.2.1 Determine the fraction 'f' of the emission ending in the different compartments

The inventory can include emissions to water, air and soil. Even if a substance is emitted to one compartment, e.g. air, it may well make its primary contribution to the toxicity potential in another compartment. The following are decided in the redistribution calculations:

- For which parts of the environment toxicity potentials for the emission should be calculated.
- The quantities of the substance which contribute to the toxicity potential in each of the compartments.

For each individual emission, the redistribution calculations result in distribution factors f_a, f_W and f_s, which specify the proportions of the emission for which toxicity potentials are calculated in the air, water and soil compartments. The calculation is discussed in more detail in Section 23.7.2.1.

23.8.2.1.1 Substances emitted to the air compartment

- Set $f_a = 1$.
- Check whether the substance's half-life in air is more than 1 day. Atmospheric half-lives can be found in Howard *et al.* (1991) or the database Hazardous Substances Data Bank (HSDB), or they can be estimated using software like AOPWIN (SRC, 1996).
- If this is the case, the substance will deposit to surface water or soil, so set $f_W = a$ and $f_s = 1\text{-}a$.

Determine the relevant ratio of water to land and fix 'a' accordingly.

See explanation below. If the geographical location of the emission is unknown, use a global default value for 'a' of 0.2.

- If the atmospheric half-life is not more than 1 day, toxicity potentials should only be calculated for the substance in the air compartment, i.e. $f_W = f_s = 0$.

A substance emitted to air can find its way into the water and soil compartments if it is stable enough not to be broken down in the atmosphere before it is deposited. If the substance is thus deposited from the atmosphere it can contribute to toxicity via water or soil.

On leaving the atmosphere the substance is distributed between the water and soil compartments in the ratio a:(1-a), where 'a' is determined as that part of the deposition area which is water. Under Danish conditions 'a' is ca. 0.5, i.e. the substance is distributed equally between the water and soil compartments. The value assigned to 'a' should reflect conditions at the place to which the emission is deposited.

A substance which is deposited from the atmosphere is assumed not to evaporate again from the water or the soil compartment, even if it is volatile. The distribution factors are thus regarded as the final distribution for the substance.

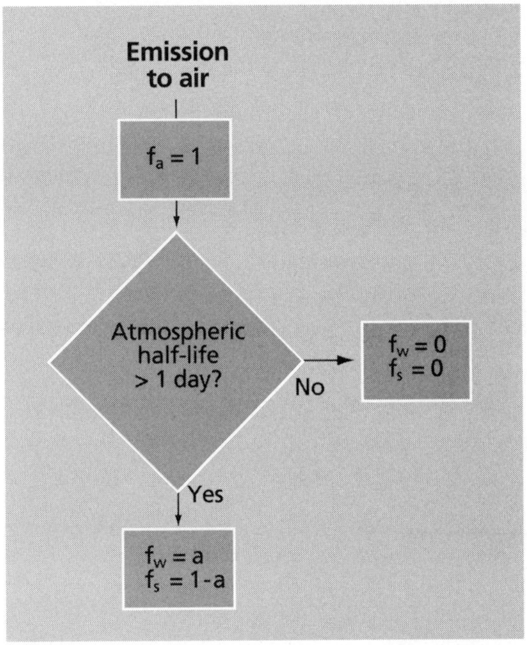

Figure 23.7. Determination of distribution factors for substances emitted to the air compartment

23.8.2.1.2 Substances emitted to the water compartment

- Check whether the substance's Henry's law constant is greater than 10^{-3} atm · m³/mol (the Henry's law constant for a substance can be found in Howard, 1990, or in the database HSDB), or it can be estimated using software like HENRY (SRC, 1995)).
- If this is the case, set $f_a = 1$ and find the substance's half-life in the atmosphere. If the half-life is less than 1 day, set $f_w = f_s = 0$.

If the half-life is greater than 1 day, set $f_w = a$ and $f_s = 1-a$.

Determine the relevant ratio of water to land and then fix 'a'. If the geographical location of the emission is unknown, use a global default value for 'a' of 0.2.

- If the Henry's law constant does not exceed 10^{-3} atm · m³/mol, assume that all of the substance emitted remains in the water, i.e. $f_w = 1$, $f_a = 0$ and $f_s = 0$.

Even if a substance is emitted with waste water, it may well evaporate to the air compartment. The substance's volatility from water is described by its Henry's law constant H, and as a rule of thumb it is assumed that if $H > 10^{-3}$ atm · m³/mol, the substance will evaporate from water. In this case it is assumed that the entire quantity emitted will evaporate.

From the air compartment, the substance may deposit to water and soil, as described for substances emitted to the air compartment, provided that the substance will not be broken down first.

Figure 23.8. Determination of distribution factors for substances emitted to the water compartment

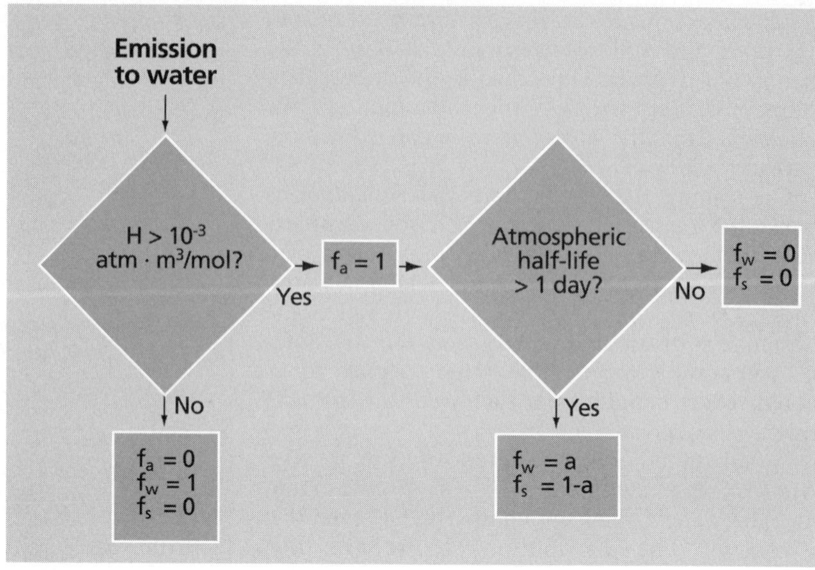

23.8.2.1.3 Substances emitted to the soil compartment

- Check whether the substance's Henry's law constant is greater than 10^{-3} atm $\cdot$ m^3/mol.
- If this is the case, set $f_a = 1$ and find the substance's half-life in the atmosphere. If the half-life is less than 1 day, set $f_w = f_s = 0$.
  If the half-life is greater than 1 day, set $f_w = a$ and $f_s = 1\text{-}a$.
- Determine the relevant ratio of water to land and then fix 'a'. If the geographical location of the emission is unknown, use a global default value for 'a' of 0.2.
- If the Henry's law constant does not exceed 10^{-3} atm $\cdot$ m^3/mol, assume that all of the substance emitted remains in the soil, i.e. $f_w = 0$, $f_a = 0$ and $f_s = 1$.

A substance can be emitted to soil either directly or indirectly with sewage sludge applied to the soil. If the substance is volatile it can evaporate and it may find its way to the soil and water compartments as described for substances emitted to the water compartment.

23.8.2.1.4 Distribution factor for the groundwater compartment

Irrespective of the compartment to which a substance is initially emitted, it can end in the groundwater if it has the right characteristics. Until recently it was assumed that percolation down to the groundwater of substances *diffusely* distributed in the environment was insignificant in the great majority of cases. The finding of numerous pesticides in an increasing number of the groundwater samples examined in recent years has shown that the transport of diffusely distributed substances from the surface of the earth to the groundwater zone must be re-evaluated, and

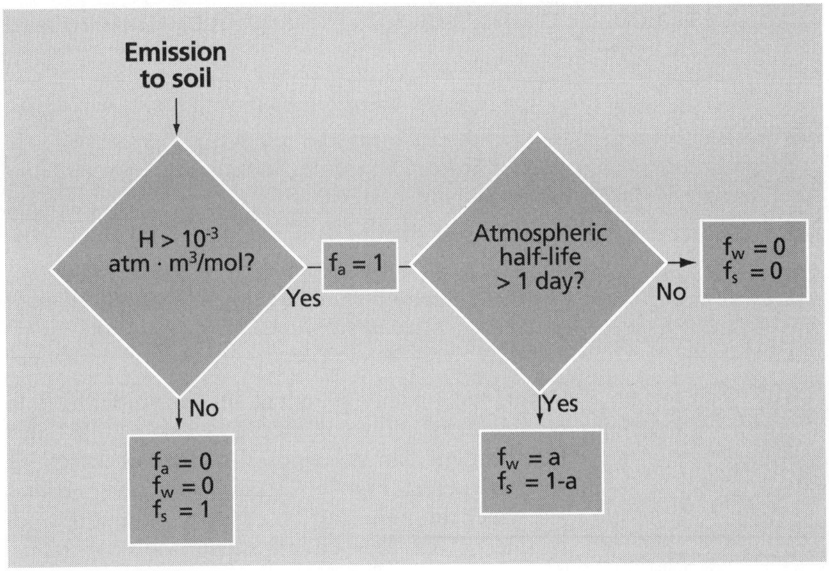

Figure 23.9. Determination of distribution factors for substances emitted to the soil compartment

transport in macropores taken into account. Since the transport of diffusely distributed substances to the groundwater is not at present satisfactorily modelled, the EDIP method does not yet offer a redistribution model for calculation of f_g.

The procedure for the redistribution calculation is illustrated by the following example for an aquatic emission of vinyl chloride:

Example 23.10. Determination of partitioning factors for vinyl chloride

3.2 g of vinyl chloride are emitted in waste water to the water compartment.

Vinyl chloride has a Henry's law constant of $5.6 \cdot 10^{-2}$ atm · m³/mol. It is therefore assumed that the substance will evaporate from the water, and the toxicity potential for the air compartment is therefore calculated.

Vinyl chloride has a half-life in the atmosphere of 1.5 days. In accordance with the rule of thumb, it is therefore assumed that the vinyl chloride which has evaporated will return to soil and water.

Under Danish conditions the quantity of substance emitted will be distributed equally between the water and soil compartments, i.e. $f_a = 1$, $f_w = 0.5$ and $f_s = 0.5$.Because of vinyl chloride's solubility in water and persistence in the soil, it may be assumed that it can percolate down to the groundwater, but for the time being this redistribution route has not been implemented in the EDIP method. No value for f_g is therefore determined.

23.8.2.2 Calculate transfer factors 'T' and determine intake factors 'I'

- Decide on the basis of the redistribution calculations which of the distribution factors differ from 0. Transfer factors are required for the corresponding compartments.
- Calculate the relevant transfer factors as described in the following section.
- Find intake factors for the relevant exposure routes in Table 23.14.

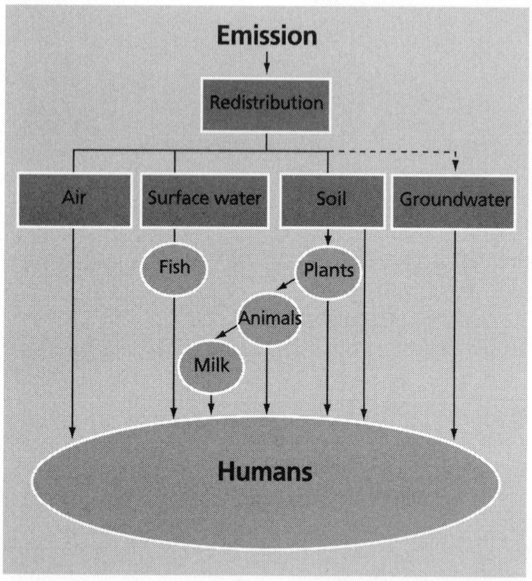

Figure 23.10. The method considers seven different exposure routes through which humans can be exposed to the substance from one of the four environmental compartments air, water, soil and groundwater.

When the toxicity potential for an emission is to be calculated, account should be taken of the fact that humans can be exposed to the substance via several routes. Figure 23.10 shows seven routes via which humans can be exposed to a substance in one of the four environmental compartments. The strength of the exposure depends on the exposure route. If there is no direct intake of material from the compartments air, water or soil, but there is intake via a medium such as fish or plants which have been exposed to the substance in the compartment, there may be a dilution or an increased concentration of the substance relative to the concentration in the compartment. This is expressed in a transfer factor 'T', defined as:

$$T = \frac{\text{Concentration of the substance in the medium inhaled or ingested (mg substance/kg medium)}}{\text{Concentration of the substance in the compartment (mg substance/kg compartment)}}$$

T is expressed in kg compartment/kg medium.

Apart from the availability of the substance to humans via various exposure routes, exposure is of course also affected by the quantities of the environmental compartment or medium in question inhaled or ingested by humans. This is expressed by the intake factor, defined as:

$$I = \frac{\text{Intake of compartment or medium (kg/day)}}{\text{Average body weight for humans (kg body weight)}}$$

The intake factors I for the various exposure routes are shown in Table 23.14. 'I' is expressed in kg compartment/kg body weight/day or kg medium/kg body weight/day.

Compartment	Exposure route	I	Unit
Air	direct	1	-
Water	fish	$3.71 \cdot 10^{-4}$	kg fish/kg bw/d
Soil	direct	$2.86 \cdot 10^{-6}$	kg soil/kg bw/d
	plants	$9.30 \cdot 10^{-3}$	kg plant/kg bw/d
	meat	$1.53 \cdot 10^{-3}$	kg meat/kg bw/d
	dairy products	$1.32 \cdot 10^{-2}$	kg dairy products/kg bw/d
Groundwater	direct	$2.86 \cdot 10^{-2}$	kg water/kg bw/d

Table 23.14. Intake factors for the exposure routes included in the method. The unit is kg compartment/kg body weight/day or kg medium/kg body weight/day

23.8.2.2.1 Air compartment

- Set the transfer factor to $T_a = 1$.
- Set the intake factor to $I_a = 1$.

On exposure to the air compartment, the substance is inhaled directly with the air and the human toxicity factor is based on a concentration of the substance. The intake factor and the transfer factor are therefore both 1.

23.8.2.2.2 Water compartment

- Find data for the substance's bioconcentration factor BCF. The BCF can be found in one of the databases AQUIRE and HSDB. BCF data for a number of metals are found in Hauschild *et al.* (1997b). If no BCF data are found in these sources, the BCF for organic substances can be estimated from the substance's octanol-water ratio or its solubility in water - see Hauschild *et al.* (1997b).
- Set the transfer factor T_w = BCF in the unit kg water/kg fish.
- Set the intake factor $I_w = 3.71 \cdot 10^{-4}$ kg fish/kg body weight/day.

For surface water, only the exposure route via intake of fish and shellfish is considered. The substance's ability to accumulate in living organisms is of major significance here.

23.8.2.2.3 Groundwater compartment

- Set the transfer factor to $T_g = 1$.
- Set the intake factor:
  $I_g = 2.86 \cdot 10^{-2}$ kg groundwater/kg body weight/day.

For the groundwater compartment, the substance is ingested directly in dissolved form with drinking water. The transfer factor is therefore 1.

23.8.2.2.4 Soil compartment

- Calculate the transfer factor T_s for each of the four exposure routes: direct ingestion of soil, ingestion of plants, ingestion of meat and ingestion of milk, as described in Sections 23.8.2.2.4.1-23.8.2.2.4.4.
- Find the intake factor I_s for each of these four exposure routes in Table 23.12.
- Calculate $I_s \cdot T_s$ for each of the exposure routes and select the exposure route with the greatest value.
- Divide T_s for the selected exposure route by the soil's dry density of 1.5 kg dry soil/l soil.

For the soil compartment, the assessment method includes four possible exposure routes for man:

- exposure via direct ingestion of soil,
- exposure via ingestion of plants which have grown in the soil,
- exposure via ingestion of meat from cattle which have grazed on the soil, and
- exposure via ingestion of milk products prepared from milk from cattle which have grazed on the soil.

To calculate the human toxicity potential via soil, one of the four exposure routes is selected which gives the greatest exposure determined as the product of the intake factor I and the transfer factor T.

23.8.2.2.4.1 Direct ingestion of soil

- Set the transfer factor to $T_{s,d} = 1$.
- Set the intake factor $I_{s,d} = 2.86 \cdot 10^{-6}$ kg soil/kg body weight/day.

Direct ingestion of soil occurs both when children intentionally eat it and when it is unintentionally ingested from hands or with vegetables. When soil is eaten, the substance in the soil is ingested directly, and the transfer factor is therefore 1.

23.8.2.2.4.2 Ingestion of plants

- Find data for the substance's stem concentration factor, SCF, and for its coefficient of adsorption in soil, K_d. SCF data for metals are available in Hauschild *et al.* (1997b). The SCF for non-ionic organic substances can be estimated from the substance's octanol-water partitioning ratio P_{ow} - see Hauschild *et al.* (1997b) - and K_d is estimated as $0.02 \cdot P_{ow}$.

- Set the transfer factor $T_{s,p} = \dfrac{SCF}{K_d + \dfrac{0.4}{1.5}}$
- Set the intake factor $I_{s,p} = 9.30 \cdot 10^{-3}$ kg plants/kg body weight day

For ingestion of the substance via plants, the transfer factor describes how easily the substance is taken up from soil by the plants growing on it.

23.8.2.2.4.3 Ingestion of meat

- Find data for the substance's meat (beef) transfer factor B_b, its stem concentration factor, SCF, and its coefficient of adsorption in soil, K_d. B_b values for metals are available in Hauschild *et al.* (1997b). B_b for non-ionic organic substances can be estimated from the substance's octanol-water partitioning ratio P_{ow} with the expression $\log B_b = \log P_{ow} - 7.6$.

- Set the transfer factor

$$T_{s,b} = \frac{16{,}9 \cdot B_b \cdot SCF}{K_d + \dfrac{0.4}{1.5}} + 0.41 \cdot B_b$$

- Set the intake factor $I_{s,b} = 1.53 \cdot 10^{-3}$ kg meat/kg body weight/day

For ingestion of the substance via meat from animals which have grazed on the soil, the transfer factor describes how easily the substance is taken up from soil into the plants growing on it, and thence transferred to the meat of the cattle grazing on the plants.

23.8.2.2.4.4 Ingestion of milk

- Find data for the substance's milk transfer factor, B_m, its stem concentration factor, SCF, and its coefficient of adsorption in soil, K_d. For metals, assume that B_m is zero. B_m for non-ionic organic substances can be estimated from the substance's octanol-water partitioning ratio P_{ow} with the expression $\log B_m = \log P_{ow} - 8.1$.

- Set the transfer factor

$$T_{s,m} = \frac{16.9 \cdot B_m \cdot SCF}{K_d + \frac{0.4}{1.5}} + 0.41 \cdot B_m$$

- Set the intake factor $I_{s,m} = 1.32 \cdot 10^{-2}$ kg milk/kg body weight/day

The transfer factor for ingestion of the substance via milk products describes how easily the substance is taken up from soil into plants, and thence transferred to the milk of those cattle grazing on the plants. T for all four exposure routes from soil has the unit kg dry soil/kg medium. To obtain the human toxicity potential for soil as a volume of soil, the selected transfer factor should be divided by the density of dry soil of 1.5 kg dry soil/l soil.

The expressions for the transfer factors for the various compartments and exposure routes are summarized in Table 23.15.

Table 23.15. Transfer factors for the various compartments and exposure routes

Compartment	Via	T	Unit
Air	direct inhalation	1	-
Water	consumption of fish	BCF	kg water/kg fish
Ground water	direct consumption	1	-
Soil	direct consumption	1	-
	consumption of plants	$\frac{SCF}{K_d + \frac{0.4}{1.5}}$	kg dry matter soil/kg plant
	consumption of meat	$\frac{16.9 \cdot B_b \cdot SCF}{K_d + \frac{0.4}{1.5}} + 0.41 \cdot B_b$	kg dry matter soil/kg meat
	consumption of dairy products	$\frac{16.9 \cdot B_m \cdot SCF}{K_d + \frac{0.4}{1.5}} + 0.41 \cdot B_m$	kg dry matter soil/kg dairy products

Example 23.11 illustrates the calculation of transfer factors for vinyl chloride.

Air

On exposure to vinyl chloride via air, $T_a = 1$ and $I_a = 1$.

Water

For exposure via water on ingestion of fish,

$T_w = 7$ l water/kg fish, as the BCF for vinyl chloride is estimated at 7 (HSDB). $I_w = 3.71 \cdot 10^{-4}$ kg fish/kg body weight/day.

Soil

Direct: in direct ingestion of soil,
$T_{s,d} = 1$ and $I_{s,d} = 2.86 \cdot 10^{-6}$ kg soil/kg body weight/day.
$I_{s,d} \cdot T_{s,d} = 2.86 \cdot 10^{-6}$ kg soil/kg body weight/day.

Via plants: based on a P_{ow} value for vinyl chloride of 4, SCF is estimated at 0.36 (Hauschild *et al.*, 1997b) and K_d is estimated at 0.08.

On this basis, $T_{s,p}$ is calculated as 1.0 kg soil/kg plant.

Since $I_{s,p} = 9.30 \cdot 10^{-3}$ kg plant/kg body weight/day,
$T_{s,p} \cdot I_{s,p} = 9.30 \cdot 10^{-3}$ kg soil/kg body weight/day.

Via meat: based on a P_{ow} value of 4, B_b is estimated at $1.0 \cdot 10^{-7}$, and BCF and K_d are estimated as described above.

$T_{s,b}$ is calculated as $1.8 \cdot 10^{-6}$ kg soil/kg meat.

Since $I_{s,b} = 1.53 \cdot 10^{-3}$ kg meat/kg body weight/day,
$T_{s,b} \cdot I_{s,b} = 2.75 \cdot 10^{-9}$ kg soil/kg body weight/day.

Via milk: based on a P_{ow} value of 4, B_m is estimated at $3.2 \cdot 10^{-8}$, and BCF and K_d are estimated as described above.

$T_{s,m}$ is calculated as $5.7 \cdot 10^{-7}$ kg soil/kg milk products.

With $I_{s,m} = 1.32 \cdot 10^{-2}$ kg milk products/kg body weight/day,
$T_{s,m} \cdot I_{s,m} = 7.5 \cdot 10^{-9}$ kg soil/kg body weight/day.

Exposure via plants with a value of $9.3 \cdot 10^{-3}$ kg soil/kg body weight/day gives the greatest potential transfer of vinyl chloride from soil to humans. The toxicity potential for the soil compartment is therefore based on the exposure route "ingestion of plants".

$T_s = 1.0$ kg soil/kg plant, corresponding to 0.69 l soil/kg plant, and $I_s = 9.30 \cdot 10^{-3}$ kg plant/kg body weight/day.

Example 23.11. Determination of intake factors and transfer factors for vinyl chloride from the four environmental compartments

23.8.2.3 Calculate human toxicity factors HTF

- Decide on the basis of the redistribution calculations whether human toxicity factors for the substance are to be calculated for inhalation (exposure via air) as well as for ingestion (exposure via water, soil or groundwater).
- Calculate the relevant human toxicity factors with the expressions in the table below.

Compartment	HTF
Air (inhalation)	$\frac{1}{HRC}$
Water, soil or groundwater (ingestion)	$\frac{1}{HRD}$

Table 23.16. Human toxicity factors for the four compartments air, water, soil and groundwater

The expression for the human toxicity factor is the reciprocal of a measurement of the substance's toxicity. For the air compartment, the relevant measure is a toxicity concentration, namely the substance's HRC value. HRC stands for Human Reference Concentration. It is the highest concentration of the substance in the inhaled air expected not to give any toxic effect on humans on life-long inhalation under standard conditions.

For the environmental compartments water, soil and groundwater, exposure occurs via ingestion, and the human toxicity factor for these compartments is therefore based on a toxicity dose of the substance, the substance's HRD value. HRD stands for Human Reference Dose. It is the largest quantity of the substance which is not expected to have any effect on humans on daily ingestion under standard conditions throughout an entire lifetime.

23.8.2.3.1 Estimate HRC

- Find toxicological test data for inhalation of the substance (e.g. in the databases HSDB and the Registry of Toxic Effects of Chemical Substances, RTECS).
- Determine the assessment factor for the test results used on the basis of Table 23.17.
- Estimate HRC by dividing the lowest of the test data used by the assessment factor.

On the basis of a short-term exposure on test animals (acute toxicity tests), it is possible to determine that concentration of the substance in air which kills 50% of the test organisms (LC_{50}, Lethal Concentration 50%) or the lowest concentration which causes death among the test organisms (LC_{Lo}, Lethal Concentration Low) or other effects.

In animal experiments of longer duration, the substance's LOAEC or NOAEC value can be found. LOAEC is the lowest concentration at which undesirable effects are observed in the experiment (Lowest Observed Adverse Effect Concentration) and NOAEC is the highest concentration at which no undesirable effects are observed (No Observed Adverse Effect Concentration).

For LC_{50}, NOAEC and LOAEC values, the higher the value, the less toxic is the substance.

HRC is a measure of the concentration of the substance in air which is assessed as having no toxic effect on life-long inhalation. The HRC value is determined on the basis of the results of inhalation experiments with animals or observations on humans who have been exposed to the substance via the respiratory tract, i.e. HRC is determined on the basis of LC_{50}, NOAEC and LOAEC values. The quality and the relevance of the available toxicity data are assessed, and on this basis an assessment factor is determined from Table 23.17; the weaker the data, the larger the assessment factor. The HRC value is determined by dividing the lowest of the test data used by the assessment factor.

23.8.2.3.2 Estimate HRD

HRD, the human reference dose, is a measure of the quantity of substance which is assessed not to have any toxic effect on life-long daily ingestion. The HRD is estimated in the same way as described for the HRC, with the difference that it must be based on results of ingestion experiments and thus dose values such as LD_{50}, NOAEL and LOAEL from ingestion experiments rather than inhalation experiments. An investigation in which the substance is given orally can also provide TD_{Lo} values (Toxic Dose Low). If acute effects are in question and if the reported effect is death, TD_{Lo} is in fact an LD_{Lo}. For chronic effects, TD_{Lo} is the same as LOAEL.

The criteria in Table 23.17 should be regarded as a general guideline. Although an attempt has been made to standardize the procedure for determination of HRC values and to simplify it relative to a toxicological hazard assessment, an expert judgment will often be required. A certain toxicological background knowledge will normally be required to determine the assessment factors.

Assessment factor	Criterion
10^5	Extrapolation from LC_{50} or LD_{50} from tests on animals
5×10^4	Extrapolation from LC_{Lo} or LD_{Lo} from tests on animals
10^4	Extrapolation from LOAEL from sub-chronic (duration < 1 year) tests on animals
5×10^3	Extrapolation from LC_{Lo} or LD_{Lo} in observations of acute toxicity to humans
10^3	Extrapolation from NOAEL from sub-chronic tests *or* extrapolations from LOAEL from chronic tests for toxicity (duration ≥ 1 year)
10^2	Extrapolation from NOAEL from validated chronic tests on animals (duration > 1 year) *or* extrapolated from LOAEL studies on humans *or* extrapolated from lowest irritant concentration when inhaled by humans
10	Extrapolation from NOAEL found in validated experimental chronic studies on humans

Table 23.17. Assessment factors for estimation of HRC or HRD on the basis of available toxicity data

This is illustrated by the following example.

Example 23.12. Estimation of HRC and HRD for vinyl chloride

The lowest reliable toxicity concentrations for vinyl chloride in the databases Registry of Toxic Effects of Chemical Substances, RTECS, and Hazardous Substances Data Bank, HSDB, are as follows:

Inhalation: The lowest observed effect concentration (LOAEC) on inhalation of vinyl chloride is a TC_{Lo} of 1 ppm in a test in which rats were exposed for 4 hours a day for 52 weeks.

Ingestion: For exposure of rats to vinyl chloride in their food, hepatic tumours were found at doses down to 1.8 mg vinyl chloride/kg body weight/day. The NOAEL value in the experiment was 0.13 mg/kg body weight/day.

On the basis of additional information in the database, the above toxicity information was assessed as having derived from experiments which could be considered reliable.

The inhalation toxicity demonstrated for vinyl chloride was reported for a chronic investigation and an LOAEC value was reported. This gives an assessment factor of 10^3 for the estimation of HRC. As the LOAEC value of 1 ppm vinyl chloride corresponds to 2.56 mg/m^3, HRC is estimated as:

$2.56\ mg/m^3/10^3 = 2.56 \cdot 10^{-3}\ mg/m^3$, which gives a human toxicity factor for exposure by inhalation of $HTF = 1/(2.56 \cdot 10^{-3}\ mg/m^3)$ $= 3.9 \cdot 10^2\ m^3/mg$ or $\mathbf{3.9 \cdot 10^5\ m^3/g}$.

The toxicity for oral ingestion was determined as an NOAEL value in an experiment which was assessed on the basis of other information in the database as a validated chronic animal experiment. On this basis, the assessment factor was set at 10^2 and HRD estimated as 0.13 mg/kg body weight/day/10^2 = $1.3 \cdot 10^{-3}$ mg/kg body weight/day. This gives a human toxicity factor for oral exposure to vinyl chloride of $1/(1.3 \cdot 10^{-3}$ mg/kg body weight/day) = **$7.7 \cdot 10^2$ kg body weight · day/mg.**

23.8.2.4 Determine the biodegradability factor, BIO

■ Section 23.7.2.3 discusses determination of the biodegradability factor BIO on the basis of results of biodegradability tests.

The biodegradability factor is determined in the same way for the impact categories "ecotoxicity" and "human toxicity via exposure through the environment".

For the example with vinyl chloride, the biodegradability factor is determined:

Example 23.13. Determination of BIO for vinyl chloride

Vinyl chloride shows only a very modest biodegradability under aerobic conditions, although it is biodegradable under anaerobic conditions (HSDB). The biodegradability factor is thus set to 1.

23.8.2.5 Calculate equivalency factors

- Calculate the equivalency factors as:
  $EF(hta) = f_a \cdot T_a \cdot I_a \cdot HTF_a$
  $EF(htw) = f_w \cdot T_w \cdot I_w \cdot HTF_w \cdot BIO$
  $EF(hts) = f_s \cdot T_s \cdot I_s \cdot HTF_s \cdot BIO$
  $EF(htg) = f_g \cdot T_g \cdot I_g \cdot HTF_g$

The values entering into the expressions are described in the previous sections. Equivalency factors can now be calculated for the human toxicity potentials in the various environmental compartments.

The procedure for calculation of equivalency factors for human toxicity is illustrated by the example for vinyl chloride.

On the basis of information in the previous examples, the following equivalency factors for vinyl chloride can be calculated for the individual environmental compartments:

Air

EF(hta) = $f_a \cdot I_a \cdot T_a \cdot HTF_a = 1 \cdot 1 \cdot 1 \cdot 3.9 \times 10^5$ m³ air/g = **3.9 × 10⁵ m³ air/g**

Water

EF(htw) = $f_W \cdot I_W \cdot T_W \cdot HTF_W \cdot BIO = 0.5 \cdot 3.71 \times 10^{-4} \cdot 7 \cdot 7.7 \times 10^2 \cdot 1$ l water/mg = 1.0 l water/mg = **1.0 m³ water/g**

Soil

EF(hts) = $f_S \cdot I_S \cdot T_S \cdot HTF_S \cdot BIO = 0.5 \cdot 9.30 \times 10^{-3} \cdot 0.69 \cdot 7.7 \times 10^2 \cdot 1$ l soil/mg = 2.5 l soil/mg = **2.5 m³ soil/g**

Example 23.14. Calculation of equivalency factors for human toxicity of vinyl chloride

23.8.2.6 Calculate human toxicity potentials

- $EP(hta) = Q \cdot EF(hta) = Q \cdot f_a \cdot T_a \cdot I_a \cdot HTF_a$
  $EP(htw) = Q \cdot EF(htw) = Q \cdot f_w \cdot T_w \cdot I_w \cdot HTF_w \cdot BIO$
  $EP(hts) = Q \cdot EF(hts) = Q \cdot f_s \cdot T_s \cdot I_s \cdot HTF_s \cdot BIO$
  $EP(htg) = Q \cdot EF(htg) = Q \cdot f_g \cdot T_g \cdot I_g \cdot HTF_g$

When the substance's equivalency factors have been calculated for the various environmental compartments, the human toxicity potentials can be determined as the product of the quantity of substance emitted, Q, and the relevant equivalency factors.

The above example for vinyl chloride gives the following results:

Emission of 3.2 g vinyl chloride in effluent has the following toxicity potentials:

$$EP(hta) = Q \cdot EF(hta) = 3.2\ g \cdot 3.9 \times 10^5\ m^3\ air/g = 1.2 \times 10^6\ m^3\ air$$

$$EP(htw) = Q \cdot EF(htw) = 3.2\ g \cdot 1.0\ m^3\ water/g = 3.2\ m^3\ water$$

$$EP(hts) = Q \cdot EF(hts) = 3.2\ g \cdot 2.5\ m^3\ soil/g = 7.9\ m^3\ soil$$

Example 23.15. Calculation of human toxicity potentials for vinyl chloride

23.8.2.7 Sources of data

The section on sources of data for use in calculation of ecotoxicity potentials in Section 23.7.2.6 presents a number of databases which can also be used for collection of the data entering into the calculation of human toxicity potentials. Apart from these databases, there is a further one which is relevant only for the calculation of human toxicity potentials because it contains human toxicological data, *namely:*

RTECS: Registry of Toxic Effects of Chemical Substances. Information on toxicological characteristics. Published by the National Institute for Occupational Safety and Health in the USA. The information is not assessed. Available on CD-ROM.

23.9 Waste to be dumped or landfilled

- Divide the inventory's emissions of waste into the four categories:
  - Bulk waste
  - Hazardous waste
  - Radioactive waste
  - Slag and ashes
- Calculate the total weight of waste for each of the four categories.

The inventory can include a number of emissions of waste to be dumped in various ways. These are actually non-terminal emissions, but it is not currently possible to terminate them in an acceptable manner as there still are no satisfactory unit processes for computation of emissions from dumping. Waste for dumping is therefore classified into four categories according to the type of landfilling or dumping. For the time being these "waste impact categories" must represent the environmental impacts which can be expected from the occurrence of the various types of dumping in the product system:

- Bulk waste, i.e. household waste, construction waste and similar brought to a (controlled) municipal landfill. The waste is distinguished by not containing environmentally hazardous substances.

- Hazardous waste, i.e. waste brought to special treatment facilities for hazardous waste and thereafter dumped. The waste is characterized by containing environmentally hazardous substances which can be released to the environment after dumping.

- Radioactive waste, i.e. waste of low radiation intensity from nuclear power plants brought to special storage sites for radioactive waste.

- Slag and ashes from incineration at coal-fired power plants and waste incineration plants brought to special dumps for incineration slag and ashes.

23.10 Normalization and weighting

In normalization, the product's impact potentials and resource consumptions are set in relation to a common reference impact to enable an assessment of which potentials are large and which small, and thus to facilitate subsequent weighting.

In weighting, the normalized impact potentials and resource consumptions are multiplied by a weighting factor which, within each of the main groups environment, resources and working environment, expresses the seriousness of the individual impact categories or resource consumptions relative to one another.

23.10.1 Normalization

- Carry out normalization by dividing the product's impact potentials or resource consumptions by the corresponding normalization references from Tables 23.16a and 23.16b.

- Then aggregate the normalized potentials for ecotoxicity and human toxicity into three impact categories by calculating the following averages:

Persistent toxicity (regional impact category)
NEP(pt) = {NEP(etwc) + NEP(etsc) + NEP(htw) + NEP(hts)}/n

Ecotoxicity (local impact category)
NEP(et) = {NEP(etwa) + NEP(etp)}/n

Human toxicity (local impact category)
NEP(ht) = NEP(hta)

where 'n' is the number of impact potentials entering into the numerators for which values have been calculated.

In normalization, all potential impacts gain the same unit, and they can therefore be added. This enables a reduction in the number of ecotoxicity and human toxicity categories. Before normalization there are up to four potentials for ecotoxicity and four for human toxicity. The two general impact categories ecotoxicity and human toxicity are thus represented by more impact potentials than all of the other types of environmental impacts in the product's environmental profile together. The impact potentials for ecotoxicity and toxicity are therefore collected in normalization into the three general impact categories according to the geographic scale which the impacts can affect and the time scale over which they are active. The three aggregated toxicity impact categories are:

- Persistent toxicity (ecotoxicity and human toxicity on a regional scale).
- Ecotoxicity on a local scale.
- Human toxicity on a local scale.

By focusing on the geographic distribution of the toxicity and its duration, the three aggregate toxicity potentials reflect the approach of society to regulation of the environmental hazards of substances better than do the original eight. They thus provide a better basis for the weighting of the environmental impact categories relative to one another.

Impact potentials and resource consumptions are calculated for the product's average annual environmental exchanges, and on normalization they are therefore expressed in person-equivalents, PE_{WDK90}, fractions of the community's average impact per person in 1990.

Note that if the definition of the product's performance in the functional unit does not include a duration, the unit for the normalized impact potentials and resource consumptions will be $PE_{WDK90} \cdot$ years.

The review of the method in Chapter 10 presents an option for carrying out region-specific normalization and weighting. This option is not yet implemented in the form of normalization references and weighting factors for the environmental impact categories for different regions. Currently, only the EDIP method's default values are available, which are person-equivalents calculated on the basis of the global background impact for the global impact categories and resource consumptions and Danish background impacts for regional and local impact categories.

23.10.2 Weighting

- Carry out the weighting by multiplying the normalized impact potentials or resource consumptions for the product by the corresponding weighting factors in Tables 23.18a and 23.18b on the following pages.

In weighting, each of the normalized impact potentials and resource consumptions is multiplied by a weighting factor which expresses its importance relative to the other impact categories within each of the main groups environment, working environment and resource consumption.

It has not been possible for a number of metal resources to find computations of the resource's current consumption for use in the normalizing reference, and a separate normalization cannot, therefore, be carried out. Normalization and weighting for these metals must be carried out in one step by multiplying the consumption by 1/person-reserve, which is the same as the weighting factor divided by the normalizing reference. Table 23.18b gives the value for 1/person-reserve for all resources.

It is important to note that the weighting factors for resource consumption and impacts on the working environment do not cover all aspects of potential relevance. Some of the missing aspects are discussed in the sections on interpretation of the weighted working environment and resource profiles in Chapter 10.

The weighting factors are selected such that the normalized environmental impact potentials are expressed on weighting in the unit $PET_{WDK2000}$ (or $mPET_{WDK2000}$), political target person-equivalents for the year 2000, i.e. the environmental impact potentials are expressed relative to the politically set target emissions per person in the year 2000 or, in other words, as fractions of the environmental latitude according to politically set targets in the year 2000.

Resource consumptions are expressed in PR_{90} (or mPR_{90}), person-reserves computed in 1990, i.e. proportions of total known reserves per person for the resource in question in 1990.

The potential impacts on the working environment are expressed in RWI_{DK90} (or $mRWI_{DK90}$), reported work injuries in 1990, i.e. the number of work accidents to be expected for the impact category in question as a consequence of the product system, based on total Danish work environment statistics for the years around 1990.

Impact category	Normalization reference Unit		Weighting factor	Reference year	Reference region	Ref.
Environmental impacts	**ER_{90}**		**WF**			
Global						
Global warming	kg CO_2-eq/pers/year	8,700	1.3	1990	World	[1]
Ozone depletion	kg CFC11-eq/pers/år	0.202	23	1990	World	[2]
Regional and local						
Photochemical ozone formation	kg C_2H_4-eq/pers/year	20	1.2	1990	Denmark	[3]
Acidification	kg SO_2-eq/pers/year	124	1.3	1990	Denmark	[4]
Nutrient enrichment	kg NO_3^--eq/pers/year	298	1.2	1989-92	Denmark	[5]
- N-equivalents	kg N-eq/pers/year	58	1.3	1989-92	Denmark	[5]
- P-equivalents	kg P-eq/pers/year	1.2	1.0	1992	Denmark	[5]
Human toxicity			2.8	1988-92	Denmark	[6]
- via air	m^3 air/pers/year	$9.2 \cdot 10^9$	-	1992	Denmark	[6]
Ecotoxicity			2.3	1988-91	Denmark	[7]
- water, acute	m^3 water/pers/year	48,000	-	1988-91	Denmark	[7]
- waste water treatment plant	m^3 water/pers/year	500,000	-	1991	Denmark	[7]
Persistent toxicity		-	2.5	1988-92	Denmark	[6,7]
- human toxicity via water	m^3 water/pers/year	$5.9 \cdot 10^4$	-	1988-92	Denmark	[6]
- human toxicity via soil	m^3 soil/pers/year	310	-	1988-92	Denmark	[6]
- ecotoxicity water, chronic	m^3 water/pers/year	470,000	-	1988-91	Denmark	[7]
- ecotoxicity soil	m^3 soil/pers/year	30,000	-	1988-92	Denmark	[7]
Waste						
- bulk waste	kg/pers/year	1350	1.1	1991	Denmark	[8]
- hazardous waste	kg/pers/year	20.7	1.1	1991	Denmark	[9]
- slag and ashes	kg/pers/year	350	1.1	1991	Denmark	[10,11,12]
- nuclear waste	kg/pers/year	0.035	1.1	1989	Sweden	[13]
Impacts on the working environment	**WR_{90}**		**WF**			
Chemicals and						
- cancer	hours/pers/year	47	$84 \cdot 10^{-6}$	ca. 1990	Denmark	[14]
- damage to the reproductive system	hours/pers/year	85	$5.4 \cdot 10^{-6}$	ca. 1990	Denmark	[14]
- allergy	hours/pers/year	62	$219 \cdot 10^{-6}$	ca. 1990	Denmark	[14]
- damage to the nervous system	hours/pers/year	79	$398 \cdot 10^{-6}$	ca. 1990	Denmark	[14]
Hearing impairments	hours/pers/year	514	$959 \cdot 10^{-6}$	ca. 1990	Denmark	[14]
Monotonous repetitive work	hours/pers/year	118	$774 \cdot 10^{-6}$	ca. 1990	Denmark	[14]
Accidents	accidents/pers/year	0.019	$19231 \cdot 10^{-6}$	ca. 1990	Denmark	[14]

References

[1]: Hauschild and Wenzel, 1997d, [2]: Hauschild and Wenzel, 1997e, [3]: Hauschild and Wenzel, 1997a, [4]: Hauschild and Wenzel, 1997b, [5]: Hauschild and Wenzel, 1997c, [6]: Hauschild et al., 1997a, [7]: Hauschild et al., 1997b, [8]: Hansen and Nielsen, 1992; Habersatter, 1991, [9]: Statistics Denmark, 1994, [10]: Association of Danish Electrical Utilities, 1992, [11]: ISWA, 1991, [12]: Skaarup, 1995, [13]: Swedish Nuclear Fuel and Waste Management Company, 1994, [14]: Rasmussen, 1997

Table 23.18a. Normalization references and weighting factors for the environmental impact categories and working environment impact categories implemented in the EDIP method

Resource consumption	Normalization reference Unit		Weighting factor (year⁻¹)	1 / pers.-reserve (pers/kg)	Reference year	Reference region	Ref.
Renewable	**RR_{90}**		**WF**	**WF/RR_{90}**			
Wood	m^3/pers/year	0.65	0.0030	0.0046	1990	World	[1]
Non-renewable	**RR_{90}**		**WF**	**WF/RR_{90}**			
Oil	kg/pers/year	590	0.023	0.000039	1991	World	[2]
Coal	kg/pers/year	570	0.0058	0.000010	1991	World	[2]
Brown coal	kg/pers/year	250	0.0026	0.000010	1991	World	[2]
Natural gas	kg/pers/year	310	0.016	0.000052	1991	World	[2]
Aluminium	kg/pers/year	3.4	0.0051	0.0015	1990	World	[1,3]
Antimony	kg/pers/year			1.0	1990	World	[4]
Beryllium	kg/pers/year			26	1990	World	[4]
Lead	kg/pers/year	0.64	0.048	0.075	1990	World	[1]
Cadmium	kg/pers/year			4.4	1990	World	[4]
Cerium	kg/pers/year			0.17	1990	World	[5]
Gold	kg/pers/year			87	1990	World	[4]
Iron	kg/pers/year	100	0.0085	0.000085	1990	World	[1,3]
Copper	kg/pers/year	1.7	0.028	0.016	1990	World	[1]
Cobalt	kg/pers/year			0.98	1990	World	[4]
Mercury	kg/pers/year			9.1	1990	World	[4]
Lanthanum	kg/pers/year			0.31	1990	World	[5]
Manganese	kg/pers/year	1.8	0.012	0.0067	1990	World	[1,3]
Molybdenum	kg/pers/year			0.25	1990	World	[4]
Nickel	kg/pers/year	0.18	0.019	0.11	1990	World	[1]
Palladium	kg/pers/year			140	1990	World	[4]
Platinum	kg/pers/year			120	1990	World	[4]
Silver	kg/pers/year			6.9	1990	World	[4]
Tantalum	kg/pers/year			21	1990	World	[4]
Tin	kg/pers/year	0.04	0.037	0.93	1990	World	[1]
Zinc	kg/pers/year	1.4	0.050	0.036	1990	World	[1]

References:

[1]: World Resources Institute, 1992, [2]: British Petroleum Company, 1992, [3]: British Geological Survey, 1991, [4]: US Bureau of Mines, 1990, [5]: Estimated from Ullman Encyclopedia of Industrial Chemistry, 1993.

Table 23.18b. Normalization references and weighting factors for the resource consumptions implemented in the EDIP method

23.11 Qualitative assessment

The assessment of the potential contributions of emissions to environmental impacts can be both quantitative and qualitative. As a rule, a qualitative assessment will be the least demanding, but for most applications it cannot stand on its own.

The qualitative assessment was developed for use in a screening to be carried out in order to focus work in a quantitative assessment. The qualitative assessment can also supplement a quantitative assessment in the areas where it has not been possible to generate the necessary information for a quantitative assessment.

Sections 23.1-23.10 describe the EDIP method's quantitative assessment tool. The quantitative assessment of the product's environmental impact potentials is designed so that for most of the product's impacts it can be summarized in a table of factors by which the emissions are simply multiplied. When the quantitative assessment is so simple and so undemanding to carry out, a qualitative assessment method is not required.

As indicated in Sections 23.7 and 23.8 on "ecotoxicity" and "human toxicity", the quantitative assessment of impact potentials for these categories is far more demanding than that for the other categories. It can be a help to use qualitative assessment tools to designate the most significant emissions, and thus to focus the quantitative assessment on them. The qualitative assessment tools are less demanding in terms of both quality and scope of data for the individual emissions.

A qualitative method was therefore developed for assessment of substances' ecotoxic and human toxic characteristics. The method can be used as an aid in the decision on which substances from an inventory should be regarded as potential contributors to these impact categories. It can also be used in a simple ranking of alternative products on the basis of the ecotoxicity and the toxicity of the substances emitted during the products' life cycles.

The qualitative assessment method commences with the existing classification of chemicals' toxicological hazards. For both ecotoxicity and human toxicity, the method is based on separate assessments of

- Exposure: the possibility that an emission will occur and the extent of exposure.
- Toxicity: the substance's ecotoxicity or human toxicity.

The qualitative assessment of a chemical is expressed in an impact score, which is the product of an exposure score and a human toxicity or ecotoxicity score.

In the same way as in the quantitative assessment methods for the two impact categories, the qualitative assessment requires some ecotoxicological and toxicological background in the user, especially in those situations where the substance has not officially been examined for classification.

23.11.1 Ecotoxicity

- Find the substance's environmental hazard classification in the official EEC list of hazardous substances (EEC, 1994) or in Pedersen *et al.* (1994).
- If the substance has not been examined for the purpose of environmental classification, find data for the properties underlying classification and assess how the substance in question should be classified (see Hauschild *et al.*, 1997a).

In the environmental hazard classification, chemicals are assigned relevant risk phrases on the basis of their environmental characteristics.

The risk phrases are given in Table 23.19. They often appear in combinations (e.g. R50/R53), but they are used individually in the screening method.

The criteria for this environmental hazard classification of chemicals are not very old, and the official EEC list of hazardous substances (EEC, 1994) is thus very incomplete with respect to environmental hazard classification. As a supplement to this list, it is recommended that the results of a Scandinavian project be used. The project proposes environmental classification of more than 400 additional chemicals (Pedersen *et al.*, 1994). In addition, it may be necessary to undertake a separate investigation of the chemicals' characteristics in relation to the criteria behind the risk phrases shown in Table 23.19. Refer in this case to Hauschild *et al.* (1997a).

R	Risk phrases	Criteria
50	Very toxic to aquatic organisms	Lowest LC_{50} for algae, crustaceans or fish < 1 mg/l
51	Toxic to aquatic organisms	1 mg/l < lowest LC_{50} < 10 mg/l
52	Harmful to aquatic organisms	10 mg/l < lowest LC_{50} < 100 mg/l
53	May cause long-term adverse effects in the aquatic environment	Not readily biodegradable and/or potential for bioaccumulation (log P_{ow} > 3 or BCF > 100)
54	Toxic to flora	Not determined
55	Toxic to fauna	Not determined
56	Toxic to soil organisms	Not determined
57	Toxic to bees	Not determined
58	May cause long-term adverse effects in the environment	Not determined
59	Dangerous for the ozone layer	Not ecotoxicologically relevant

Table 23.19. Risk phrases for classification of chemicals on the basis of their environmental hazardousness

23.11.1.1 Exposure score

- Determine the exposure score as the sum of the scores for expected emission and for labelling with R53 or R58 on the basis of the criteria in Table 23.20.
- If both scores are zero, assign the exposure score the value 1.
- If lack of data prevents determination of the exposure score, assign it the maximum value of 8.

Expected emission?	Score
yes	4
no	0
R53 or R58?	**Score**
yes	4
no	0

Table 23.20. Criteria for assignment of exposure score

The criteria for the exposure score are given in Table 23.20.

23.11.1.2 Ecotoxicity score

- Determine the substance's ecotoxicity score as the sum of the scores for aquatic and terrestrial ecotoxicities determined from the criteria in Table 23.21.
- If lack of data prevents determination of the ecotoxicity score, assign it the maximum value of 8 unless the substance is well known and considered by experts within the area to be free of significant hazardous impacts.

Aquatic ecotoxicity	Score
R50	4
R51	2
R52	1
Terrestrial ecotoxicity	**Score**
R54	4
R55	4
R56	4
R57	4

Table 23.21. Criteria for assignment of ecotoxicity score

The criteria for assignment of the ecotoxicity score are given in Table 23.21.

The ecotoxicity score is the sum of the scores for aquatic and terrestrial ecotoxicity. Criteria have not yet been established by the authorities for the four categories within terrestrial ecotoxicity, and there are thus no chemicals today carrying these designations. Until further notice, all substances must therefore be assigned the terrestrial ecotoxicity score 4 unless assessed by experts within the area as being non-toxic to soil ecosystems.

23.11.1.3 Ecotoxicological impact score

- Determine the substance's ecotoxicological impact score as the product of exposure score and ecotoxicity score.

The ecotoxicological impact score ranks emissions from the inventory according to their potential to cause ecotoxicity. If the goal of the screening is to focus work in a quantitative assessment, all emissions scoring 8 or above should be selected as a rule of thumb.

If the screening is carried out with a view to selecting the least environmentally damaging among several alternative products, the ecotoxicological impact scores should not simply be summed indiscriminately for each alternative. Instead, emissions with ecotoxicological impact scores of 16 or more should be regarded as potentially critical, and the product alternatives should be classified into groups according to whether the inventory shows emissions with ecotoxicological impact scores of 64, 32 or 16. Within each of these groups, the comparison among product alternatives can then be based on the product's total ecotoxicological impact scores.

23.11.2 Human toxicity

The hazard classification of chemicals offers the possibility of assigning them the risk phrases in Table 23.22, depending on their toxicological characteristics.

The source of information on the hazard classification of chemicals is the official EEC list of hazardous substances (EEC, 1994), which is much more comprehensive for the substances' toxicities than for their ecotoxicities.

The qualitative assessment of the chemicals is expressed in a toxicological impact score, which, as for ecotoxicity, is the product of an exposure score and a toxicity score.

23.11.2.1 Exposure score

- Determine the substance's exposure score as described above under ecotoxicity.

As the observed toxicity to humans is due to exposure on human contact with the surrounding environment, the criteria for determination of the exposure score are the same as for ecotoxicity.

23.11.2.2 Toxicity score

- Determine the substance's toxicity score on the basis of the criteria in Table 23.23.
- If the substance is not included in the official list of hazardous substances (EEC, 1994), assign it a toxicity score of 4 unless it is well-known and regarded as being without significant hazardous impacts.

R	Risk phrase
20	Harmful by inhalation
21	Harmful in contact with skin
22	Harmful if swallowed
23	Toxic by inhalation
24	Toxic in contact with skin
25	Toxic if swallowed
26	Very toxic by inhalation
27	Very toxic in contact with skin
28	Very toxic if swallowed
33	Danger of cumulative effects
34	Causes burns
35	Causes severe burns
36	Irritating to eyes
37	Irritating to respiratory system
38	Irritating to skin
39	Danger of very serious irreversible effects
40	Possible risk of irreversible effects
41	Risk of serious damage to eyes
42	May cause sensitization by inhalation
43	May cause sensitization by skin contact
45	May cause cancer
46	May cause heritable genetic damage
48	Danger of serious damage to health by prolonged exposure
49	May cause cancer by inhaltion
60	May cause sterility
61	May cause teratogenicity
62	Possible risk of damage to capacity for reproduction
63	Possible risk of teratogenicity
64	May cause damage to infants during period of breast feeding

Table 23.22. Risk phrases for classification of chemicals on the basis of their toxic characteristics

	Score 1	Score 4	Score 8
General	Irritating and hazardous substances	Caustic or toxic substances and substances with allergenic effects or neurotoxic effects and substances causing irreversible damage	Substances with carcinogenic or mutagenic properties or which cause reproductive damage, highly toxic substances, and substances causing very serious organ damage
Acute toxicity	Xn; R 20-21-22	T; R 23-24-25	Tx; R 26-27-28
Irritation	Xi; R 36-37-38	C; R 34-35-41	
Allergenicity		Xi; R 43 or included in at least two of the list used in Thomsen, 1991	R 42 or stated to be allergenic by inhalation in Thomsen, 1991
Irreversible effects, organ damage (including damage to the nervous system)		Xn; R 40 or R 48 in combination with R 20-21-22 or T; R 39 in combination with R 23-24-25 or R 33	T; R 48 in combination with R 23-24-25 or Tx; R 39 in combination with R 26-27-28
Genotoxicity			T; R 46 or Xn; R 46 or R 40
Carcinogenicity			T; R 45 or R 49 or Xn; R 40 or included in Danish National Labour Inspection, 1994
Damage to the reproductive system or deformities			T; R 60 or R 61 or Xn; R 62 or R 63 or R 64

Table 23.23. Criteria for assignment of toxicity score

The criteria for assignment of toxicity score are given in Table 23.23. A substance can have several different toxic impacts, but the toxicity score is assigned after the impact giving the highest score in Table 23.23.

23.11.2.3 Toxicological impact score

■ Determine the substance's toxicological impact score as the product of exposure and toxicity scores.

As a rule of thumb, if the goal of the screening is to focus work in a quantitative assessment, all emissions having toxicological impact scores of 8 and above should, as for ecotoxicity, be selected.

If the screening is carried out with a view to selecting the least environmentally damaging among several product alternatives, the toxicological impact scores should not simply be summed indiscriminately for each alternative. Instead, emissions with toxicological impact scores of 16 or more should be considered as potentially critical and the product alternatives should thus be classified into groups according to whether the inventory shows emissions with toxicological impact scores of 64, 32 or 16. Within each of these groups, the comparison among product alternatives can then be based on the product's total toxicological impact scores.

24. Procedure for sensitivity analysis

After assessment of the exchange's resource consumption and potentials for impact on the environment and the working environment, a sensitivity analysis is carried out to identify which of the assumptions made is of greatest significance for the result of the LCA. Included are assumptions in the modelling of the product system, in the scope definition of the system, for individual processes and for the method's impact assessment phase. The sensitivity analysis investigates the significance of the assumptions for the total result and for the conclusions to be reached on that basis. The assumptions or choices of major significance are reconsidered in detail and documented in the reporting of the LCA.

Review the product system stage by stage and:

- Identify the assessment's key figures, i.e. the most important modelling assumptions, processes and environmental exchanges
- Identify the most important impact assessment factors.

The sensitivity analysis includes a review of the scope definition and the inventory one further time in the light of the insights provided by the assessment.

The sensitivity analysis is carried out after the impact assessment phase of the LCA, in order to utilize the knowledge of the environmental exchanges' potential impact contribution that is introduced here, for the assessment of which exchanges are the most significant and which are of no significance in the total picture.

Assumptions in the modelling

The most important assumptions concern the choice of which processes to include in the product system. No detailed guideline for the sensitivity analysis can be given for this choice, other than that the whole product system should be reviewed and the delimitations assessed. Other important modelling assumptions are often:

- The choice of distribution of the product among different export markets which can be of significance for all processes in the use and disposal stages.
- The choice of patterns of use and scenario for electricity production, especially for products which use electricity in the use stage.
- Distribution of the product among different disposal routes.

Data for processes

Uncertainties for processes are determined both for the computed magnitudes of environmental exchanges and for the extent to which the process enters in the modelling of the product system. If assessment of the data uncertainties is not performed for all processes, it should at least be carried out for those processes which the sensitivity analysis shows to have the greatest significance for the total result.

Impact assessment factors

The equivalency factors for all environmental impact categories are also encumbered by uncertainties. An actual uncertainty assessment of the equivalency factors is not implemented in the EDIP method, but for the ecotoxicity and toxicity equivalency factors calculated as a part of the assessment, the user can assess the uncertainty based on the quality of the underlying data.

■ Carry out the sensitivity analysis on a key figure by

1) determining the variation in the key figure which can occur when
   - assumptions and choices are changed to other relevant possibilities
   - process data and assessment data vary in agreement with the uncertainty determined
2) letting the key figure assume the maximum and minimum values found and investigating what this means for the total result of the LCA by recalculating the inventory and the assessment with the new values.

■ Investigate the result's sensitivity for the key figures among the modelling assumptions made.

■ Investigate the result's sensitivity for missing process data.

■ Investigate the result's sensitivity for the uncertainty in the key figures among the product's environmental exchanges and assess how representative these are for the processes in question. Also assess how representative the processes are to the objective of the modelling.

■ Investigate the result's sensitivity to the uncertainty in the most significant of the assessment factors, especially for impact types related to ecotoxicity and human toxicity, where the uncertainty in equivalency factors and normalization references can be large.

■ Systematize the sensitivity analysis by classifying the key figures after their relationship to Material selection, Energy, Chemicals or "Others" (the MECO principle).

The sensitivity analysis can be carried out by hand, but it often includes such large calculations that it is practical only when the product system is modelled by computer.

Part V. Case Histories

- *Which of the reference products' resource consumptions and impact potentials are significant?*
- *What can be changed in the reference product and what are the environmental consequences of such changes?*
- *What are the environmental specifications for new products?*
- *Which environmental improvements have been implemented in new products and how large are the gains?*

25. Introduction to the case histories

Part V contains five case histories of environmental assessment and design in product development at the Danish companies Gram, Bang & Olufsen, KEW Industri, Grundfos and Danfoss.

This part of the book presents the results of the LCAs and illustrates how they are used in product development and as a basis for a product-oriented environmental policy in the companies. Emphasis has been placed on documenting the methodology of the LCA and on showing the types of results which the LCA can yield. The case histories should therefore be seen as a library of examples for solution of the environmental tasks for which the environmental specialist and the designer are responsible in their work on environmental aspects in product development.

The most important message of Part V is that "environmental design is possible". The examples from the five Danish companies show that the LCAs can be carried out, and that the results are highly applicable in building environmental features into new products. In the experience of the companies, environmental considerations can enter into product development on an equal footing with all other aspects, and all companies have been successful in developing more environmentally friendly products.

Content of the case histories

The case histories comprise the following products: a refrigerator, a television, a high pressure cleaner, a pump and an electro-hydraulic activation unit for a hydraulic valve. Each of these products has been assessed, and the experience from the LCA has been used at the company in a variety of ways in product development and in the setting of goals, strategies and environmental policies.

Each case history is divided into four chapters:

1. **Environmental assessment of references:** indicates the product's most significant resource consumptions, environmental impact potentials and potential impacts on the working environment. Also designates the components and the stages of the product life which make the greatest contribution.

2. **Environmental diagnosis:** reviews the options for improving the product from an environmental point of view and indicates where in the product they reside. Assesses the environmental consequences of changes in the product.

3. **Environmental specification:** analyses the competitive framework in the market, i.e. customers' perceptions of the product in environmental terms and the environmental priorities of competitors; then shows the environmental specifications for the product type prompted by the market analysis and the environmental assessment and diagnosis.

4. **Environmental design of new products:** shows the environmental considerations which the company has built into new products, quantifies the gains and describes the company's experience with the environmental work involved in product development.

Figure 25.1 illustrates the structure of the individual case history.

Figure 25.1 Structure of each case history

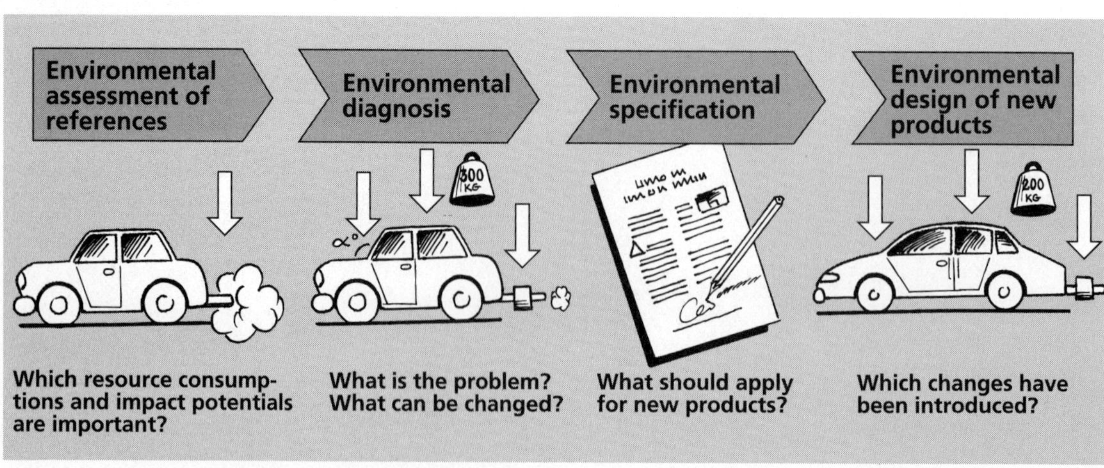

Guide to reading

Table 25.1 provides a summary of where the various items are reviewed in each of the five case histories.

If the reader has a special interest in a particular product type or in seeing use of the method in its entirety, one example can be read from end to end. If the reader has a special interest in a particular issue and would like to see experience from several products at the same time, this specific issue can be read in each of the examples.

The following comments may be of assistance as a guide to the reading of Part V: the LCA itself in Section 1 of each example is described more or less uniformly for the five case histories, as is the environmental specification in Section 3. The environmental diagnosis (Section 2), however, is described in greatest detail for the example for Gram, and the work on environmental design of products (Section 4) is described in greatest detail in the example for Bang & Olufsen. In documentation of the method as a whole, the highest degree of detail is given in the examples for Gram and Bang & Olufsen.

Table 25.1. Extended table of contents

Guide to use of the results

The object of the LCAs in the EDIP programme has been to create a basis of knowledge for use in the development of new products. The results in the various figures and tables therefore represent answers to the questions posed by the designer when the product's environmental aspects are to be clarified, e.g.:

- What are the product's resource consumptions and impact potentials?
- How large are they?
- Which impact potentials are the most significant?
- Where do they originate in the product system?
- Which sub-assemblies in the product produce the greatest impacts?
- Where in the product are the greatest potentials for improvements in environmental terms?
- How strong a parameter of competition is the product's environmental characteristics?
- Which parameters should be targeted and what is the environmental specifications for the product?

Emphasis is placed on presenting the results in a way which will make them best suited to reaching conclusions in product development. The aim of the presentation is partly to show *that* the relevant questions can be answered and partly *how* they can be answered. The results can therefore be used as examples of solutions to the LCA tasks arising in connection with product development.

Emphasis has *not* been placed on *documenting* the results by including all of the basic data in this presentation. This would have been too extensive a task for the purpose of this book. The database for each product is available at the individual company. The presentation of the LCAs is not regarded as external documentation of the product, and is thus not directed towards marketing and comparison with competing products in a sales situation. If this had been the objective, greater emphasis would have been placed on documentation of the database. As presented here, the results should therefore not be used in marketing and sales.

Database

The data are stored as so-called "unit processes" in the EDIP database (Frees, 1996), i.e. in a form which makes them available to and usable by third parties in the modelling and assessment of other industrial products. A large part of the data covers general processes for e.g. production of materials, energy production, transport and disposal, i.e. data which are of general interest in the LCA of products.

Method

The LCAs were carried out in accordance with the EDIP method as described in Parts II and IV. The method is therefore not further described in the case histories.

26. Gram: Refrigerators

Anne-Marie Mose, Gram
Henrik Wenzel, Institute for Product Development
Michael Hauschild, Institute for Product Development

Gram A/S manufactures domestic and commercial high quality white goods, refrigerators and freezers. The company has 2,000 employees and is situated in the town of Vojens in the southern part of Denmark, approx. 50 km from the German border

26.1. LCA of a refrigerator - the LER200

The LER200 is a low-energy refrigerator with an internal volume of 200 litres. The designation "LER" stands for "Low Energy Refrigerator"; the "200" indicates the volume.

Figure 26.1. The reference product, the LER200 refrigerator

Choice of reference product

Gram has selected an LER200 as reference product for the development of new products in order to set a standard at the highest possible level in environmental terms. It is the company's most optimized unit in terms of energy consumption, and among all refrigerators on the market it has one of the very lowest electricity consumptions. Under the EU's energy labelling system it is ranked in the best category, which includes only 10% of all refrigerators labelled under the system.

The functional unit

The unit's primary function is the refrigeration of foodstuffs. A measure of this function is normally the volume and the unit's temperature class, i.e. the temperature range which the refrigerator can maintain internally at a given ambient temperature.

The LER200 can maintain a constant temperature within the range 2-8°C at an ambient temperature of 16-32°C. As standard values, the internal temperature is set at 5°C and the ambient temperature is set at 25°C. The average life span of the LER200 is set at 13 years. The primary functional unit for an LER200 is thus defined as:

a volume of 200 litres refrigerated at 5°C for 13 years at an ambient temperature of 25°.

The refrigerator is further characterized by several secondary qualities:

It is self-defrosting, it has evaporation of the water from defrosting, and it has a door and a number of shelves, baskets and boxes as shown in the adjacent drawing. These functions have a durability which corresponds to the life span.

When comparing alternatives, care must be taken to ensure that the comparison is made for the same primary functional unit and the same level for the secondary qualities.

Scope definition of the product system

The exploded view in Figure 26.2 shows an LER200's components. They are specified with item numbers and described in Table 26.1 in the product's parts list on the following pages. The information includes component weights and material content as well as the manufacturing processes and ancillary substances used in manufacture. The parts list in Table 26.1 is a simplified version of the original parts list for the product: for simplicity of presentation, some of the small items have been combined into groups of the same material, but nothing has been omitted from the parts list, i.e. it covers the whole product.

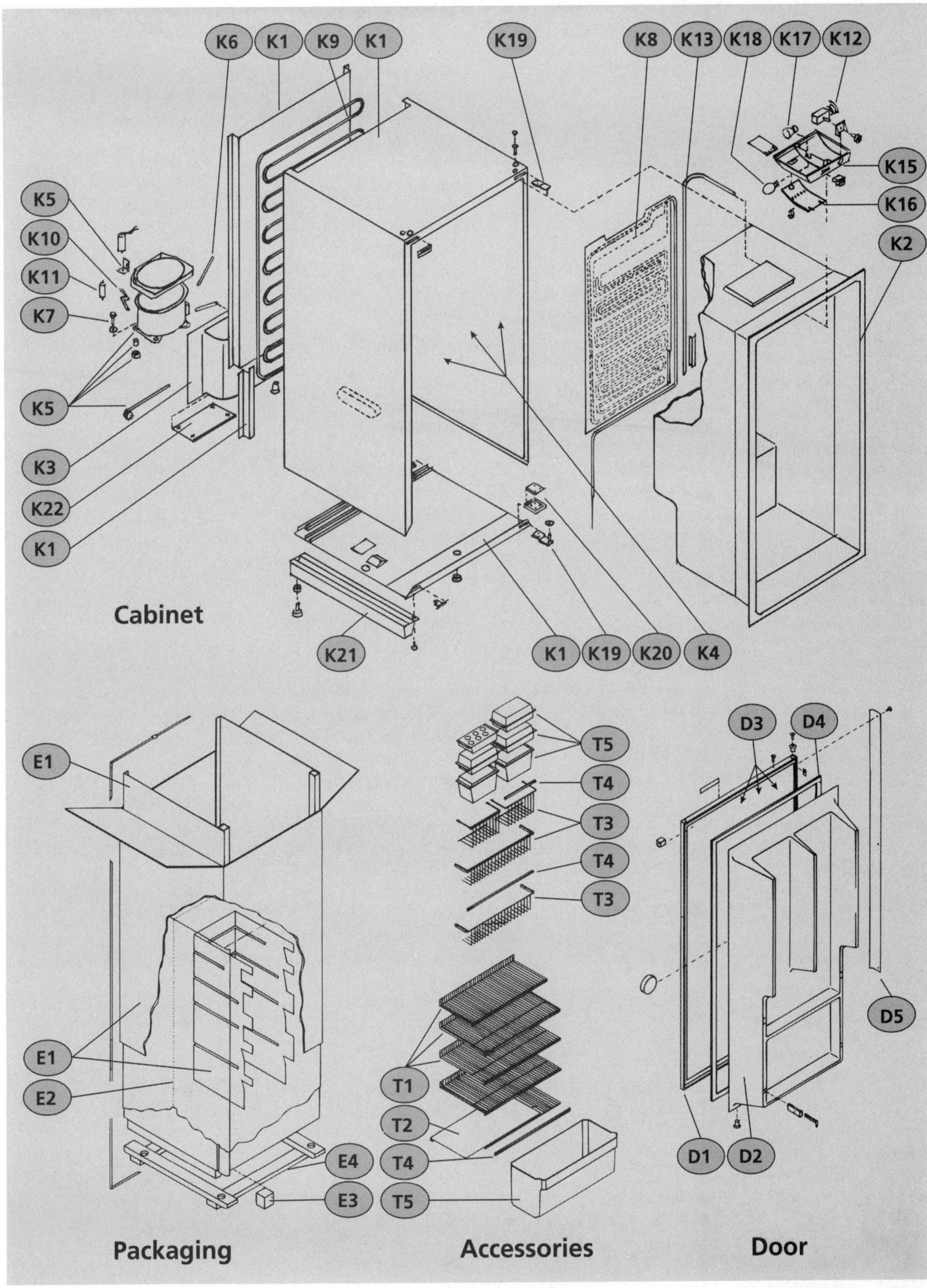

Figure 26.2. Exploded view of the LER200 unit

Pos.	Number	Item	Material	Total weight (g)	⏚	Manufacturing processes	*Incl. in model	Ancillary substances	Weight (g)	⏚
Cabinet										
K1	1	Steel cabinet	Steel sheet	17,587	✓	Punching operations,	✓✓✓			
						welding operations, etc.	✓✓✓	Electrodes		÷
						A total of 9 processes	✓✓✓	Water	50	✓
			Epoxy powder	494	÷	Surface treating	✓✓	Salt	18.5	÷
						A total of 2 processes		Detergent	38	÷
								Phosphoric acid	3	÷
								Water	20,000	✓
K2	1	Inner box	ABS sheet	3,320	✓	Vacuum forming, aftertreatment	✓✓			
K3	1	Compressor box	ABS sheet	419	✓	Vacuum forming, aftertreatment	✓✓			
K4	1	PUR foam	Isocyanate	2,790	÷	Foaming	✓	CFC 11	340	÷
			Polyalcohol	1,886	÷					
K5	1	Compressor with accessories	Steel, copper, PVC, brass, rubber, aluminium	6,825	✓[1]	Standard component[1]	✓[1]	[1]	[1]	[1]
K6	1	Pipes for compressor	Copper pipe	10	✓	Blow-through, cutting	✓✓	Nitrogen		÷
K7	1	Refrigerant			✓	Fillling		CFC 12	100	÷
K8	1	Evaporator	Aluminium	1,340	✓	Subcontractor	÷			
K9	1	Condenser	Steelpipe	363	✓	Winding and plugging, bending	✓✓			
K10	1	Pipe for condenser	Copperpipe	20	✓	Blowing, cutting, drifting of	✓✓	Nitrogen		÷
						holes, bending	✓✓			
K11	1	Drying filter	Copper/synth. zeolite	31	✓	Subcontractor	÷			
K12	1	Thermostat	Steel, plastics	85	✓[1]	Standard component[1]	✓[1]	[1]	[1]	[1]
K13	1	Sensor/thermostat	PVC pipe	25	✓	Cutting	✓			
K14	3	Wiring	Copper, PVC, steel, brass	281	✓	Subcontractor	÷			
K15	1	Lamp panel	ABS granulate	95	✓	Injection moulding	✓	Water	560	✓
K16	1	Lamp screen	Lurane	22	✓	Injection moulding	✓	Water	170	✓
K17	1	Lamp socket	Steel, "plastic"	18	✓	Standard component	÷			
K18	1	Light bulb	Glass, brass	13	✓	Standard component	÷			
K19	2	Hinge fittings	Stainless steel sheet	130	✓	Punching, abrasive tumbling, etc.	✓✓✓	Water	80	✓
						A total of 6 processes	✓✓✓	Abrasives	2	÷
K20	4	Threaded portion Hinge fittings	Steel sheet	204	✓	Punching, tumble deburring, etc.	✓✓			
						A total of 4 processes	✓✓	Abrasives	1	÷
K21	2	Mounting rail	Plated steel	1,220	✓	Subcontractor	÷			
						+ own processes (2)	✓✓	Water	2,500	✓
K22	1	Base plate	Galvanized steel	178	✓	Subcontractor	÷			
						+ own processes	✓✓			
						cutting, punching, etc. (5)	✓✓✓			
Door										
D1	1	Door casing	Steel sheet	4,250	✓	Punching operations	✓✓✓			
						welding operations, etc.	✓✓✓✓	Acetylene, oxygen		÷
						A total of 7 processes				
			Epoxy powder	157	÷	Surface treatment	✓✓	Salt	6	÷
						A total of 2 processes		Detergent	12	÷
								Phosphoric acid	0.2	÷
								Water	8,000	✓
D2	1	Door lining	ABS sheet	1,800	✓	Vacuum forming, deburring	✓✓			
D3	1	PUR foam	Isocyanate	860	÷	Foaming	✓	CFC 11	215	÷
			Polyalcohol	615	÷					
D4	1	Sealing frame	PVC	330	✓	Subcontractor	÷			
D5	1	Handle	Aluminium	260	✓	Extrusion	✓			

Table 26.1. Simplified parts list for the LER200 unit

Pos.	Number	Item	Material	Total weight (g)	⏚	Manufacturing processes	* Incl. in model	Ancillary substances	Weight (g)	⏚
Accessories										
T1	5	Thread shelves	Steel wire	2,400	✓	Cutting, spot welding etc.	✓✓✓			
						A total of 6 processes	✓✓✓			
			Epoxy powder	80	÷	Surface treatment	✓✓	Salt	3.75	÷
						A total of 2 processes		Detergent	6.30	÷
								Phosph. acid	0.65	÷
								Water	6,250	✓
T2	1	Coverplate	PVC	93.4	✓	Cutting, punching, bending	✓✓✓			
T3	4	Thread baskets	Steel wire	580	✓	Cutting, spot welding, etc.	✓✓✓			
			Steel sheet	140	✓	A total of 7 processes	✓✓✓✓	Abrasives	0.1	÷
			Epoxy powder	73.4	÷	Surface treatment	✓✓	Salt	0.8	÷
						A total of 2 processes		Detergent	1.35	÷
								Phosph. acid	0.14	÷
								Water	2,400	✓
T4	7	Front rim	PVC/aluminium	137	✓	Subcontractor	÷			
T5	9	Various plastic boxes	Lurane	2,213	✓	Injection moulding	✓			
Packing										
E1	-	Packaging	Corrugated cardboard	3,650	✓					
E2	1	Plastic bag	Polyethylene	105		Subcontractor				
E3	1	Door support	Expanded PS	14	✓	Cutting	✓			
E4	1	Base frame	Wood	1,600	✓	Manual assembling	✓			
Miscellanous										
	1	Sealing compound	Thermogrip	45	÷	Jointing	✓			
			Contact mass	117	÷					
			White	100	÷					
	10	Other steel items	Steel	20	✓	Mostly standard components	÷			
	5	Rubber items	Rubber	15	✓	Mostly standard components	÷			
	5	Other ABS items	ABS regranulate	47	✓	Injection moulding	✓	Water	280	✓
	6	Nylon items	Polyamide	6	✓	Mostly standard components	÷	Water	40	✓
	7	Adhesive tape	Adhesive tape	14	÷	Standard components	÷			
	4	Lupolene items	Lupolene	4	✓	Injection moulding	✓	Water	30	✓
	4	Acetal items	Acetal	40	✓	Injection moulding	✓	Water	300	✓
	28	Other small items	Various	174	÷	Various	÷			
		Tin solder	Tin solder	ca. 10	÷	Soldering	÷			

Notes

1) Data for all processes from raw material extraction to and including manufacture of the item are received from the subcontractor and included.

⏚ The column indicates whether or not the product system of the material or the ancillary substance is included all the way back to the extraction of raw materials.

✓ Means that it is included.

÷ Means that it is not included.

* In the column "Incl. in model" is likewise indicated if the manufacturing processes itself is included or not. ✓ means that energy consumption and use of ancillary substances is measured, that emissions and waste are either measured or calculated and that impacts on the working invironment are measured or assessed. The number of ✓'s indicates how many processes are included.

Table 26.1. (continued)

The parts list illustrates the refrigerator's components. It also illustrates how much of the product system from extraction of raw materials to and including manufacture is included in the LCA. For each component, the symbols ✓ and ÷ have been used to show whether the material's and the ancillary substances' product systems are included back to extraction of raw materials. Whether or not the manufacturing processes are included is also shown. The scope definition of the product system has followed the MECO principle described in Parts II and IV, i.e. each of the areas "materials", "energy", "chemicals" and "others" has been analysed separately, and the extent of the refrigerator's product system to be included in the inventory has been decided on the basis of this analysis.

Materials
An attempt was made to include all materials in the inventory, and the attempt was substantially successful. Only very small items have not been included (see for example the group of 28 items under "miscellaneous" items in the parts list). The items concerned are principally small plastic items of one kind or another.

For those units which were not included, an assessment was made of whether they contain scarce resources which were not otherwise represented. This was found not to be the case.

Energy
The energy content in the materials is covered under the above scope definition. All significant manufacturing processes in terms of energy have also been included, and the use stage and disposal are included completely. In the first instance it was assumed that use and disposal will occur in Denmark. Section 26.2 then illustrates the consequences of their occurrence in other countries.

Chemicals
All manufacturing processes which include chemicals are included in the inventory, and all chemical emissions from the manufacturing processes are thus also included. An attempt was made to obtain information on the production of chemicals and to follow them all the way "back to nature", but, as is evident in the parts list, this attempt was not successful. An attempt was also made to obtain information on the additives which were contained in the materials, especially the plastic, but this did not prove possible either.

Others
The category "Others" covers especially the working environment, which for certain impacts is not linked to the above three categories. Impacts on the working environment are included for all manufacturing processes at Gram, and these are judged to cover over 80% of all manufacturing processes of the refrigerator. The working environment is also included for production of the steel, for electricity generation and for transport.

Production of materials

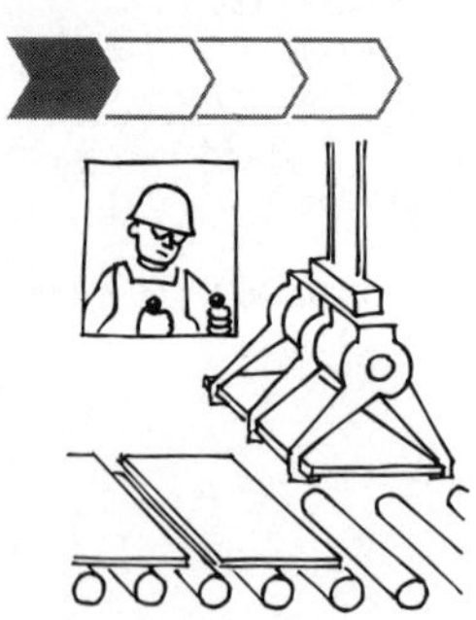

The refrigerator's product system begins in the ground and at the bottom of the sea, where the raw materials used in energy production and which enter into the materials are found. They are distributed over the whole world, and even though they are extracted as close as possible to the production localities in Europe, raw materials from all parts of the world enter into the refrigerator.

The use of raw materials for the energy which enters into the production of materials are determined by the energy systems in the countries in which the materials are produced. This is mainly Germany, France, Holland, Italy and Sweden. The most important sources of energy in these countries are oil, gas, coal, uranium ore and dammed water. Oil and gas furthermore serve as raw materials for the production of plastic and chemicals, and together with the metal ores, they are the most important resources for production of materials.

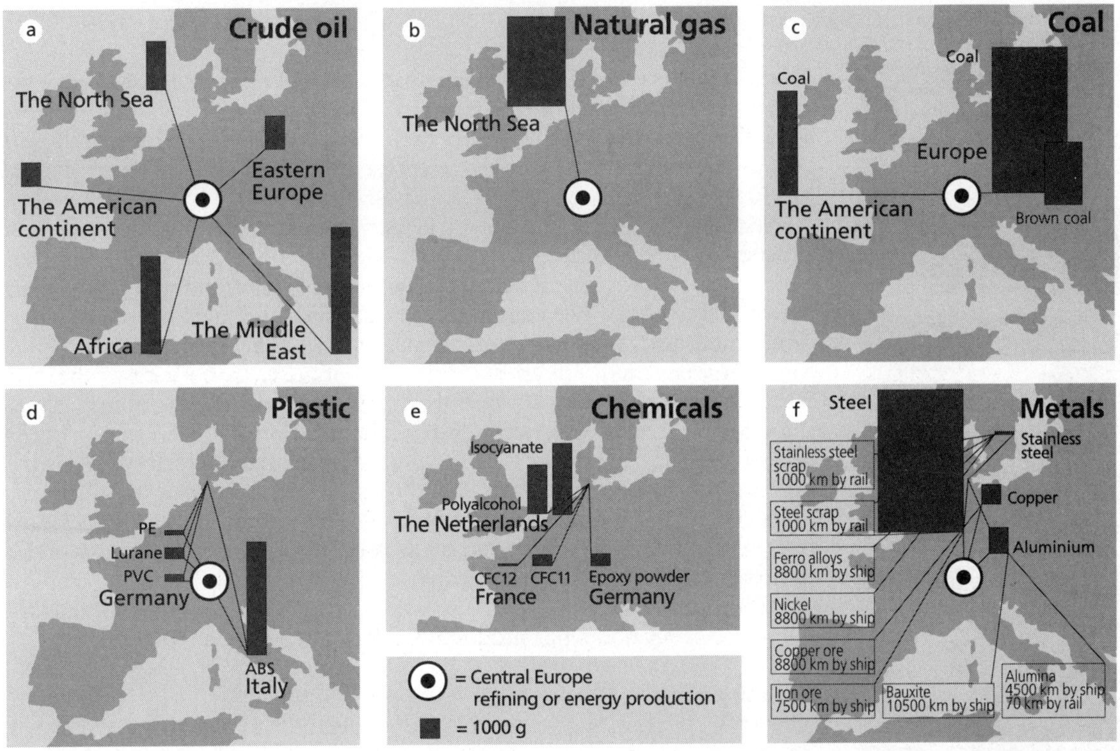

Figure 26.3. The product system for materials and chemicals including fossil fuels for their energy consumption

Figure 26.3 shows the quantities and extraction sites for the most important raw materials. The figure shows both raw materials for the materials themselves and for the energy consumed in the material production processes.

Crude oil, natural gas and coal are transported firstly to central Europe, where they are either refined or used in energy supply (see Figures 26.3a, b and c). From refining, they are transported via pipes, lorry or railway to the production sites for plastic and chemicals, and from there further to the manufacturing processes at Gram in Vojens, as shown in Figures 26.3d and e. Note that in Figure 26.3e, as is also evident in Table 26.1, it has not been possible to include the product system of chemicals prior to the manufacture. It was not possible to obtain information from the suppliers. The significance of this is discussed later under the sensitivity analysis.

The metal ores are transported from all parts of the world for processing in Europe, and from there on to Gram in Vojens or to subcontractors (cf. Figure 26.3f).

All transport is included for all components which weigh more than 100 g. More than 90% of the transport for materials is thus included.

Manufacture

The refrigerator is manufactured at the Gram factory in Vojens, with the exception of a few items made by subcontractors. All processes at the factory are included in the inventory, which means that almost all of the manufacturing stage is included.

The processes consist principally of processing of metal and plastic including forming, machining, assembly and surface treatment. Figure 26.4 shows some of the processes. Injection moulding and vacuum forming of plastic units are much used, as are trimming, bending, punching and welding processes in steel.

Figure 26.4. Manufacturing processes at Gram: vacuum forming, surface treatment, installation of the compressor

Most of the processes use electricity. Only certain assembly processes are entirely manual, and therefore not energy-consuming. Energy consumption was measured by both the regional electricity supply authority and Gram for the most energy-intensive processes, and by Gram for the residual processes.

Some of the processes use ancillary substances. An inventory of these substances was made on the basis of the parts list, and emissions were either measured or calculated for all processes using environmentally significant substances. Measurements for the most important process emissions to air and water and measurements on the air in the working environment were made by an authorized laboratory. In both cases the processes concerned were surface treatment and vacuum forming. Emissions for other processes were either calculated on the basis of consumption of ancillary substances, or the data were extrapolated from measurements of corresponding processes. Quantities of waste were weighed.

Noise in the working environment was measured at the process. All processes were also reviewed and assessed by an expert on the working environment from the Danish Employees' Health Service, and the employees were interviewed. Statistics on work accidents were included, and the number of man-hours per product was computed for each process.

Rejection rate

The rejection rate for items for the refrigerator is set at an average of 2%, i.e. all wastage, including wastage attributable to errors in the parts list, errors in purchasing, differences in stores and faulty production of various items. In other words, 1.02 refrigerators are produced on average for each refrigerator sold. The total inventories of both the manufacturing stage and the material stage are therefore multiplied by 1.02.

Overhead

The energy consumed in heating, lighting, central ventilation and other joint facilities at both production and administration facilities was measured, and a portion thereof was ascribed to the refrigerator. The portion was calculated by dividing the annual overhead energy consumption by the number of refrigerators produced per annum, and thus apportioning the factory's consumption per unit produced. This procedure is judged to be reasonable, given the existing degree of uniformity of the production.

Use

The LER200 is sold in many different countries, but the largest market is in a few countries close to Denmark. Figure 26.5 illustrates the distribution of sales. In the LCA, it is assumed in the first instance that the unit will be used in Denmark, and a simulation of use in other countries is then also given in Section 26.2.

In Denmark, the refrigerator is transported (in 1993) by rail to central stores (in 1995 by lorry), and thence to dealers by lorry. It is installed primarily in private households and used for an average of 13 years, after which it is discarded. While in use, the refrigerator contributes to the residence's heating. In the winter months this means that there is a corresponding saving in other heating, and the quantity of energy which the refrigerator thus replaces was therefore included in the calculations. In other words, the quantity of energy replaced was subtracted from the refrigerator's energy consumption. Account was taken of the fact that the degree of efficiency in electricity production is not as high as in district heating or in oil or gas fires, and compensation for this was made in the calculation.

It is assumed that electricity produced in Denmark is used for operation of the refrigerator, and in the first instance no account was taken of imports and exports of electricity between Denmark and neighbouring countries. Use of electricity produced in Germany and in the Netherlands is simulated in Section 26.2.

During the use stage, the refrigerator leaks CFCs, both refrigerant and blowing agent from the foam. A leakage of 1-2 g CFC12 per annum from the refrigeration system and ca. 14 g CFC11 per annum from the foam was estimated. The latter figure in particular is uncertain; it derives from a measurement of the residual CFC content in four refrigerators on disposal, but it is the best estimate available.

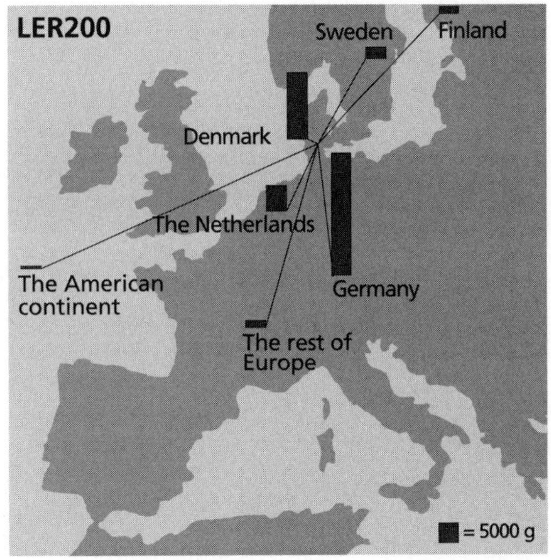

Figure 26.5. Distribution of sales for the LER200 refrigerator

Servicing

The servicing percentage for an LER200 was set at 4%, i.e. about 4% of refrigerators will have a service call during the 13 years of their life spans. In by far the greatest number of cases, repairs are made directly in the user's home. Average travelling of 50 km per service call was assumed, i.e. 2 km per unit in total. Leakages from the refrigeration system during servicing are included in the above estimates of CFC escape during use.

Life span

The life span of 13 years is a conservative estimate. It is determined not only by the technical functioning of the unit; about half of all refrigerators are discarded because of moving or kitchen modernization. For the remaining part of the refrigerators, problems with the compressor or leakages in the refrigeration system are the most common causes for the refrigerator to be replaced.

Disposal

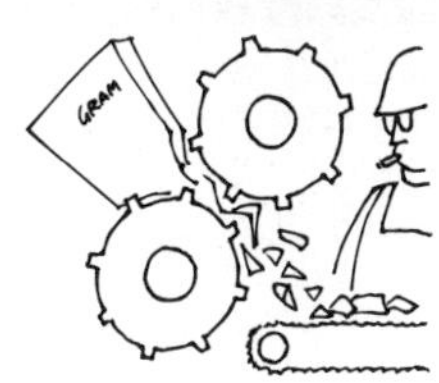

After use, the refrigerator follows various routes to a scrap dealer, where it ends its life in a shredder. Here it is crushed, and the materials separated. The plastic and the foam finish in a dump, and most of the CFCs are released to the atmosphere. The metals are sent for re-melting and then enter into new product systems. The various disposal routes are shown in Figure 26.6.

This disposal scenario generally applies to the most important markets, namely Denmark, Germany and the Netherlands. In the calculations, the situation as it was in Denmark in 1992 is assumed in the first instance (Thorsen, 1993). At that time there were four different ways for the user to dispose of a refrigerator, as shown in Figure 26.6. About 25 municipalities (of the total of 375 Danish municipalities) had arrangements for the draining of CFC12 from the refrigeration system, and only some of these were able to send the refrigerator to the CFC recovery firm in Aarhus, where CFC11 in the foam is also extracted and destroyed.

In all, refrigerant was drained from the refrigeration system in ca. 30% of refrigerators, and ca. 7% were sent to the CFC recovery firm. A new central CFC recovery facility was opened in Copenhagen in 1995.

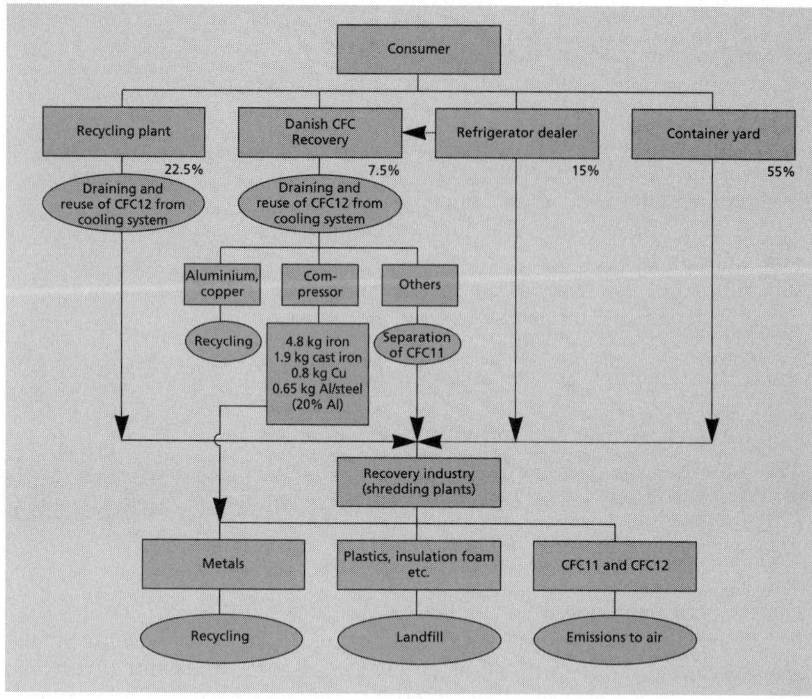

Figure 26.6. Distribution of disposal routes for refrigerators in 1992

This has increased the number of refrigerators from which the CFC has been removed from the foam to ca. 15% and the proportion from which the refrigerant is drained to a little over 30% (Grau, 1995). Once the plastic and the foam have been through the shredder they are polluted, and waste incineration plants do not want them. These material fractions are therefore currently dumped. They require space, and their energy content is not utilized.

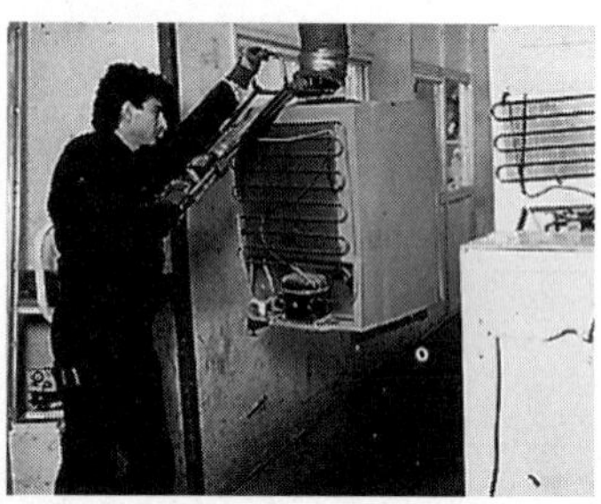

Figure 26.7. Disposal of a refrigerator at the Danish CFC recovery plant (Photo: Danish CFC Recovery)

It is assumed in the calculations that all iron, aluminium and copper are sent for recycling. This assumption is fulfilled in optimal operation of the shredder and the subsequent separation plant, but in practice it clearly does not apply completely. The next section therefore also provides a simulation of other disposal routes so that the consequences thereof can be assessed. The small stainless steel units probably follow the iron in the sorting after shredding and are therefore not sent for production of new stainless steel. The nickel content has no value in conventional recycled steel, and it is therefore assumed that the nickel in stainless steel is not recovered in the disposal stage.

Database

Collecting information on the many processes in the product system has been a major task and it has taken a long time. However, a database which also covers most of the product system for Gram's other products has now been established. Most of the processes in the product system are not specific to the LER200, but apply generally to extraction of raw materials, production of materials, transport, energy systems and disposal systems. Data for these processes comprise the main part of the database for the refrigerator's product system. The manufacturing processes measured at the factory in Vojens comprise over 50% of the factory's production. This means that 80-90% of the database is already available next time a new refrigerator or freezer is to be analysed. In future, the task will therefore consist mainly of maintenance of the database.

Data reference									
	Data specifity			Data source type					Comments
Product system & process type	Product specific	Site specific	General	1	2	3	4	5	
Extraction of raw materials									
Extraction of raw materials			x			x			Literature Trade organizations
Production of materials									
Steel		x		x					Contractor to Gram
Primary aluminium			x			x			Trade organizations
Plastic			x				x		Trade organizations
Other materials			x			x			Literature
Product manufacturing									
Shaping of plastic	x		x	x		x			Gram: measurements of energy consumption[a] and emissions[b] performed by an authorized laboratory. Assessments of the working environment have been performed by the Danish Industrial Health Service. Emissions of CFCs from foaming process estimated by Gram.
Steel processing	x			x					
Steel surface treatment	x			x					
Foaming	x						x		
Assembly	x	x		x		x			
Other processes	x	x		x	x	x			
Use									
Energy consumption	x			x					Physical Lab. III, TUD, continuous measurements on 120 refrigerators during one year.
Lifetime			x			x			Extrapolated from other refrigerators. Conservative estimate.
Emissions of CFC			x		x				Based on calculations from the content in four disposed refrigerators.
Disposal									
Routes of disposal		x				x			Danish average (Thorsen,1993)
Shredding		x	x	x					Largets scrap dealer in Denmark
Steel		x	x	x					Danish recovery steelworks
Aluminium		x	x	x					Danish recovery steelworks
Others			x			x			Literature
Transport									
Distances and means of transportation	x		x		x	x			Gram
Energy consumption and emissions			x			x			Common European transport database Copert 1990. Literature
Energy systems									
Energy systems			x			x			System Analysis Dept., Risø National Laboratory, literature

Notes

1) Measurements
2) Computations (from mass balance considerations and input data for the process in question)
3) Extrapolation of data from similar type of process or technology
4) Extrapolation of data from different type of process or technology
5) Estimate

a) Gram and the regional electricity corporation
b) Danish Environmental Center, Ltd

Product-specific data: concern processes specifically handling the LER200
Site-specific data: concern data from actual sites in the product system of the LER200, but the process data are not specific for the LER200.
General data: all others

Table 26.2. Data sources

Table 26.2 provides an overview of the specificity of the data which enter into the LCA and the type of source from which they are derived. The data are described in more detail in the EDIP database "EDIP unit process database" (Frees, 1996).

Exchanges with the environment

An inventory of the processes' exchange of substances with the surroundings and impacts on the working environment was made process by process. The most significant results in environmental terms were selected and are presented in Table 26.3.

Table 26.3. Exchanges with the environment in the refrigerator's product system

			Production of materials	Product manufacturing	Use	Disposal	Transport	Total
Resource consumption								
Crude oil		g	17,000	5,500	13,300	-900	1,600	36,500
Natural gas		g	14,000	900	8,700	-80	90	23,600
Coal		g	36,000	11,000	265,000	-90	10	311,900
Brown coal		g	6,400	-	-	-50	-	6,400
Uranium ore		g	0.45	-	-	-0.01	-	0.44
Water for hydro power		litre	12,000	550	12,000	-40	-	25,000
Aluminium	Al	g	1,900	-	-	-1,430	-	470
Iron	Fe	g	38,500	-500	-	-28,500	-	9,500
Copper	Cu	g	1,040	-	-	-940	-	100
Manganese	Mn	g	230	-2	-	-170	-	60
Nickel	Ni	g	14	-3	-	-	-	11
Zinc	Zn	g	17	-	-	-	-	17
Calcium carbonate	$CaCO_3$	g	6,600	-	-	-	-	6,600
Quartz	SiO_2	g	150	-	-	-	-	150
Sodium chloride	NaCl	g	970	20	-	-	-	990
Soft wood		g	9,880	40	940	-2,500	-	8,400
Groundwater		litre	2.1	51	-	-	-	53
Unspecified water		litre	265	1.2	2.0	-0.5	0.3	268
Unspec. fuel (oil-eq)		g	900					900
Unspec. resources		g	210					210
Materials and ancillary substances for which resource consumption is not included								
Epoxy powder		g	810	-	-	-	-	810
Isocyanate		g	3,650	-	-	-	-	3,650
Trichlorofluoromethane	CFC11	g	555	-	-	-	-	555
Dichlorodifluoromethane	CFC12	g	100	-	-	-	-	100
Polyalcohol		g	2,500	-	-	-	-	2,500
Sealing compound		g	220	-	-	-	-	220
Emissions to air								
Carbon dioxide	CO_2	g	168,000	50,300	779,000	-3,400	4,100	998,000
Carbon monoxide	CO	g	840	10	130	-20	60	1,020
Nitrogen oxides	NO_x	g	510	200	2,900	-40	60	3,630
Sulphur dioxide	SO_2	g	780	440	4,140	-10	10	5,360
Nitrous oxide	N_2O	g	6	2	50	0	0	58
Unspec. particles (dust)		g	120	20	380	-70	5	455
Trichlorofluoromethane	CFC11	g	-	30	180	320	-	530
Dichlorodifluoromethane	CFC12	g	-	0	20	60	-	80
Hydrocarbons	HC	g	550	260	5,990	0	-	6,800
Volatile org. compounds	VOC	g	20	5	15	0	15	55
Unspec. aldehyde		g	0.2	0.1	2.7	0	-	3.0
Dioxin		mg	0.0007	0	-	-	-	0.0007
Epichlorohydrin		mg	5	-	-	-	-	5
Arsenic	As	mg	4	0	7	0	-	11
Lead	Pb	mg	8	6	30	0	-	44
Cadmium	Cd	mg	0.6	0.2	0.7	0	-	1.5
Copper	Cu	mg	6	4	60	0	-	70
Mercury	Hg	mg	1	0	10	0	-	11
Manganese	Mn	mg	5,000	-	-	-	-	5,000
Vanadium	V	mg	230	280	550	0	-	1,060

			Production of materials	Product manufacturing	Use	Disposal	Transport	Total
Emissions to water								
Chem. oxygen demands	COD	g	40	1	0	0	0	41
Total nitrogen	tot-N	g	0.03	0.007	-	-	-	~ 0
Total phosphorus	tot-P	g	1	1	-	-	-	2
Hydrocarbons		g	3	0	0	0	0	3
Unspec. oil		g	1.4	0.5	0.3	0	0	2.2
Unspec. anionic detergent		g	-	0.5	-	-	-	0.5
Unspec. non-ionic detergent		g	-	1.1	-	-	-	1.1
Lead	Pb	mg	20	0	-	-	-	20
Cadmium	Cd	mg	10	0	-	-	-	10
Chromium	Cr	mg	650	-	-	-	-	650
Copper	Cu	mg	100	10	-	-	-	110
Nickel	Ni	mg	690	2	-	-	-	692
Zinc	Zn	mg	180	-	-	-	-	180
Waste								
Unspec. hazardous waste		g	125	3	-	-	-	128
Unspec. dust with heavy metals		g	5	-	-	-	-	5
Unspec. industrial waste		g	180	-	-	-	-	180
Unspec. nuclear waste		g	0.05	0	0	0	0	0.05
Unspec. slag and ashes		g	2,200	890	21,000	-20	0	24,000
Slag with chromium		g	50	-	-	-	-	50
Polyvinyl chloride	PVC	g	-	20	-	540	-	560
Epoxy powder		g	-	20	-	-	-	20
Unspec. plastic		g	-	-	-	9,280	-	9,280
Polyurethane foam	PUR	g	-	-	-	6,200	-	6,200
Steel		g	-	-	-	8,500	-	8,500
Aluminium		g	-	-	-	440	-	440
Copper		g	-	-	-	100	-	100
Unspec. bulk waste		g	20,700	6,400	154,000	-50	0	181,000
Working environment								
Monotonous repetitive work		hours	0.001	0.080	0.003	-	0.006	0.091
Hearing impairment inducing noise		hours	0.050	0.790	0.427	-	0.009	1.276
Substances causing damage to the nervous system		hours	0.004	0.051	0.014	-	-	0.069
Allergic substances		hours	0.008	0.413	0.071	-	0.007	0.492
Substances causing damage to the reproductive system		hours	0.002	0.050	-	-	-	0.052
Carcinogenic substances		hours	0.005	0.052	0.051	-	-	0.108
Accidents		number	$5.6 \cdot 10^{-6}$	$113 \cdot 10^{-6}$	$17.2 \cdot 10^{-6}$	-	$0.25 \cdot 10^{-6}$	$136 \cdot 10^{-6}$

Notes

- \- means that no information is available for the life cycle stage in question
- 0 means that the figure is very small compared to the other stages of the life cycle

Table 26.3. (continued)

The table shows which substances are exchanged with the surroundings by the processes in the refrigerator's product system, and it provides an overview of where the large quantities occur. The exchanges were summarized in resource consumption and emissions to air and water, as waste, and in impacts on the working environment, which were quantified as the time per product for which the impact was assessed to last. It is not necessarily the case that the exchanges result in impacts on the environment or the working environment. Large quantities are not necessarily the same as large environmental impacts because the substances possess different hazard properties. An impact assessment of the individual exchanges is therefore required in order to understand where in the product system the most significant environmental impacts can occur. This is presented in the next section, but the inventory presented in the table is discussed first. The overview provided in Table 26.3 represents a large quantity of underlying data, available at Gram.

Resource consumption

The refrigerator weighs over 54 kg plus 6 kg of packaging. But as shown in the table, much larger quantities of resources are extracted in the material stage in order to produce these 60 kg. Resource consumption is shown as the pure metal content in the ores and not the ore itself. If resource consumptions for the production of materials are added, it is seen that more than 140 kg of pure resources are used to produce the 60 kg product, i.e. 80 kg do not enter into the product. About 50 kg of the resources which do not enter into the refrigerator are fuel for the energy consumed in the production of materials, and 30 kg or ca. 20% of the resource consumption in the material stage is thus waste.

The manufacturing stage uses ca. 17 kg fuel, while the use stage consumes almost 300 kg throughout the refrigerator's life span.

It is noteworthy how few resources are used by the transport. Transport data are included in data for material production from extraction of raw materials up to production of materials, but all transport from the production of materials to manufacture in Denmark and further in the rest of the product system to and including disposal are included in the transport stage shown in the table. This transport only uses 1.7 kg of fuel. This shows both how rationalized the transport is and how little it means in the total picture.

There are negative figures in Table 26.3 for both the disposal stage and the manufacturing stage. This is attributable to the fact that metals and corrugated cardboard are sent for recycling, and part of the resource consumption and the emissions from the extraction of raw materials therefore follows them to the next product system. In the total accounts, a portion of the resources is therefore subtracted again in the stage where recycling occurs. The procedure involved is given in Wenzel (1997).

Emissions to air

The emissions to air derive mainly from the energy systems. Note also that CFC11 and CFC12 are emitted during manufacture, use and disposal. As shown later, these are of decisive environmental significance.

Emissions to water

The emission of organic substances is very small, as shown by the total chemical oxygen demand (COD), as are the emissions of nitrogen and phosphorus. In the manufacturing stage, aqueous washing processes of oil-coated steel occur, but there is an effective treatment including skimming of oil and chemical precipitation of the effluent, and the emission of detergents, oil and heavy metals is very low. The sludge from this waste water treatment is included as "unspecified hazardous waste" during manufacture. The heavy metal emissions are significantly greater in the material stage, where production of steel makes the biggest contribution.

Waste

The largest quantity of waste arises in the use stage, namely 154 kg. This derives from extraction of the coal used in producing the electricity used to run the refrigerator. The coal is refined when mined, and the waste from refining is accumulated in mounds which over the years are processed into a "natural part" of the landscape. But this takes many years, and it is therefore reasonable to include the quantities concerned as waste.

Disposal of the refrigerator itself is responsible for a portion of the waste which ends at the dump.

The working environment

The measurements in the working environment in the material stage cover only the steel production, and the data are not from the specific supplier to Gram, but are extrapolated from data for another steelworks. Steel comprises over half of the refrigerator's weight. In the manufacturing stage, the working environment has been studied very thoroughly, and the stage has been covered virtually completely. Data for the use stage are an average for Danish power plants, and for the transport stage the data are an average for goods transport by lorry in Denmark.

Energy profiles

The energy consumption through the product system accounts for a significant proportion of resource consumption and emissions, waste, and impacts on the working environment. It is therefore important that it be studied in more detail. A profile of the energy consumption divided by stages of the product life shows some simple points which it is worth highlighting.

Figure 26.8a shows firstly the direct energy consumption, i.e. the energy which enters directly in the processes of the product system. This is the energy consumption on which the designer has direct influence in his choice of solutions for new products. The figure for direct energy consumption does not, however, include the efficiency by which fuel is converted to electricity or heat in the various energy systems which underlie the processes. Neither does it include the calorific value in the materials. Figure 26.8b illustrates the energy consumption after inclusion of the efficiency and the calorific value. This is also called consumption of primary energy, i.e. the energy content in the resulting consumption of fuel in the energy systems and in the material itself as calorific value. Note that Figure 26.8b includes the energy content in polyalcohol and isocyanate on the basis of an estimate from literature.

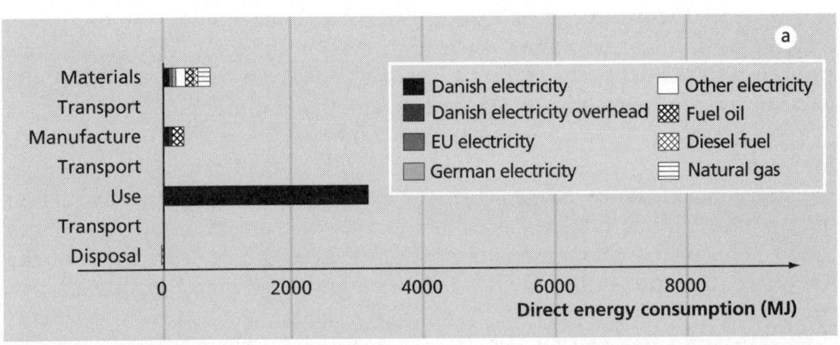

Figure 26.8a. Energy profile. Direct energy consumption

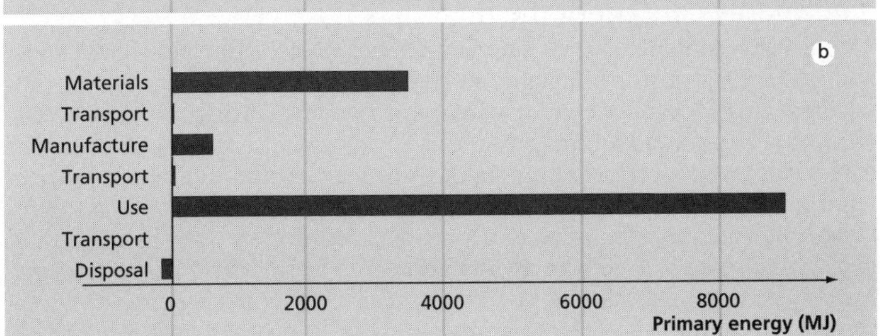

Figure 26.8b. Energy profile. Primary energy consumption

The designer can indirectly influence the consumption of primary energy by taking it into consideration when ranking the individual processes, components and product life stages relative to one another. It is therefore important to present consumption of primary energy, and it is this which enters into the LCA.

Materials vs. product manufacturing
As is evident in Figure 26.8a, the direct energy consumption of the processes in the material production is twice as large as in the product manufacturing. If the efficiency of the energy production and the calorific value of the materials is included, the situation shown in Figure 26.8b is reached, where consumption of primary energy for the materials is over five times larger than consumption in the manufacturing stage. That the difference between consumption of primary energy for the materials and for manufacturing is so large is important for a ranking between savings in materials and material recycling vs. process optimization in the manufacture. The processes' impact on the materials' life is therefore important, e.g. the significance of the surface treatment for the materials' recycling properties.

Overhead
The above recognition is reinforced by the fact that for the manufacturing stage, electricity consumption for overhead and fuel oil consumption, which is also an overhead, comprise over half of the energy consumption, i.e. the manufacturing processes' own consumption of energy is less than 10% of the total energy consumption for the materials.

The use stage
The use stage is still very dominant even for the LER200, which is one of the most energy-efficient refrigerators on the market. Refrigerators normally have an energy consumption 2-4 times higher than the LER200, so for other refrigerators, the use stage is normally even more dominant than is shown in Figure 26.8b. Consumption of primary energy of the foam, including the calorific value, is ca. 600 MJ or only ca. 7% of the consumption of primary energy in the use stage, so even for the LER200, the insulation is far from having reached the energy-optimal thickness.

Disposal
On recycling of the metals, part of the energy consumption for extraction of raw materials is credited, but in the total picture this does not mean much. The plastic and the foam are dumped, so the calorific value of the materials after use is utilized only for wood and corrugated cardboard.

The energy content of the materials

The energy content of the materials is thus of considerable significance, and it is therefore worth looking more closely at how it is distributed. Figure 26.9 provides an overview, where it is distributed first to the individual materials in the quantities present in the refrigerator itself, and then distributed to the individual components. On incineration in a waste incineration plant, part of the calorific value (ca. 75%) is utilized, and on recycling, both the calorific value and part of the process energy is utilized. It is clear that over 80% of the energy content is to be found in the ten largest components, so from an energy point of view, any recycling can concentrate on these.

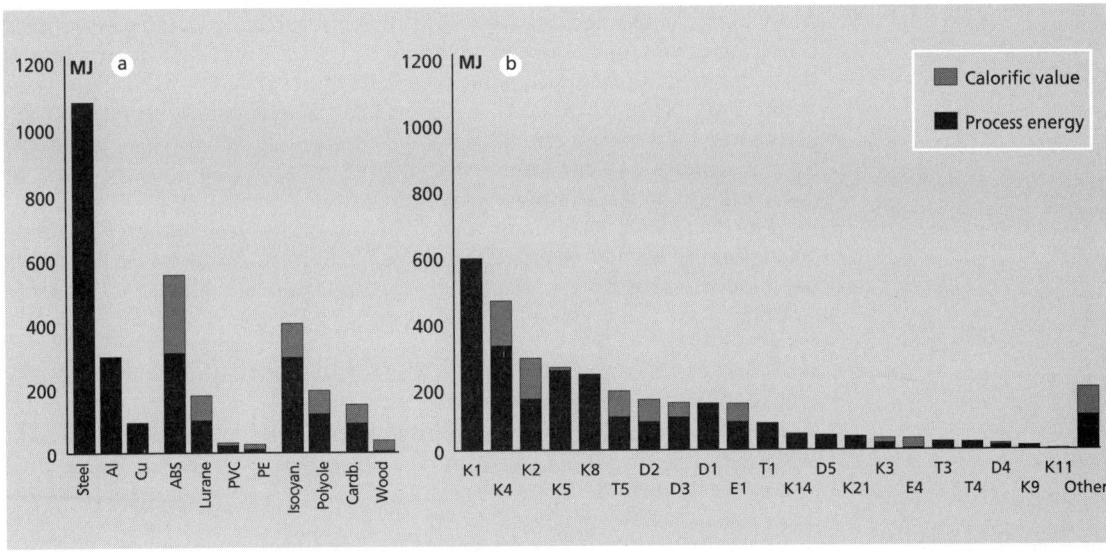

Figure 26.9. The content of primary energy in the materials found in the finished refrigerator (a) and in the materials in components (b). Numbers refer to the parts list in Table 26.1

The calorific value comprises only 20% of the energy content in the refrigerator, so it is not a type of product which should be disposed of by incineration of the whole product. Separation and recycling of the materials are preferable in terms of energy.

Impact assessment

The exchanges with the environment from the various processes in the product system were assessed, and the most significant resource consumptions and potential impacts on the environment and the working environment were designated on the basis of the EDIP method's weighting criteria. At the same time, the most important *sources* of resource consumption and potential impacts were found, and the locations in the product system where the greatest contributions arise were designated, together with the sub-assemblies of the product which contribute the most.

Environmental impact potentials

The environmental impacts which can arise from the product throughout its product life will derive from the emissions to air and water and the dumping of waste. Whether *actual* effects will arise cannot be determined, but an assessment can be made of how large the *potential* for environmental impacts is. A significant element of the LCA is the conversion of data for the emissions to impact-equivalents, which are an expression of how large their potential environmental impacts are.

The result of this conversion of the emissions in the product system for an LER200 is shown in Table 26.4. Note that although all computed emissions are not given in Table 26.3 for reasons of clarity, they *are* included in the environmental impact potentials in Table 26.4 if they are known to contribute to any of the impact categories.

Table 26.4. Conversion of the environmental exchanges to environmental impact potentials

Impact category	**Potential impact**	
	Unit	**Per product per lifetime**
Global impacts		
Global warming	g CO_2-eq	3,722,000
Ozone depletion	g CFC11-eq	600
Regional impacts		
Acidification	g SO_2-eq	8,000
Nutrient enrichment	g NO_3-eq	5,200
Photochemical ozone formation	g C_2H_4-eq	60
Human toxicity (water)	m^3 water	1,600
Ecotoxicity (water, acute)	m^3 water	44,000
Local impacts		
Human toxicity (air)	m^3 air	628,000,000
Ecotoxicity (water, acute)	m^3 water	2,200
Hazardous waste	g	750
Nuclear waste	g	0.05
Slag and ashes	g	24,300
Bulk waste	g	220,000

Most of the substances in aerial and aqueous emissions make a contribution to toxicity, i.e. toxicity to man and to ecosystems. The quantifying of these contributions is done after the method described in Hauschild *et al.* (1997), and data for toxicity are found for all relevant substances in the emissions. However, it should be noted that the uncertainty in such toxicity assessments is very large, so that interpretations of toxicity must be made with care. An example is given below.

The category "hazardous waste" in this example includes PVC with plasticizers. This is not an official classification of PVC, but it is used in this case to focus on the plasticizers during disposal, where they finish in landfill.

Resource consumption and potentials for impacts on the working environment

Resource consumption expressed in grams and the impacts on the working environment in hours or numbers from the inventory in Table 26.3 enter directly into the further assessment.

The magnitude of the product's contribution

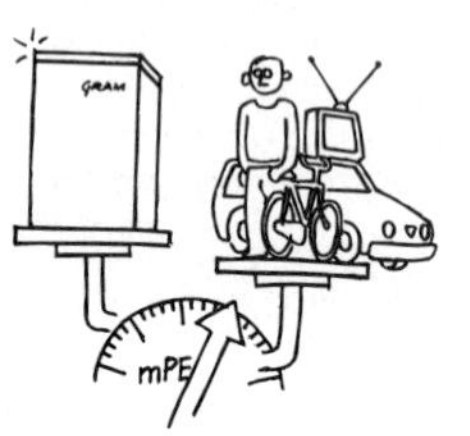

The refrigerator's contribution to the various impact categories is seen in relation to a person's average contribution in Figure 26.10. The magnitude of the refrigerator's contribution is thus expressed in person-equivalents, or more correctly milli-person-equivalents (mPE), which are the product's contribution in thousandths of an average person's contribution. The magnitude is thus made immediately comprehensible, and it is possible to make a mutual comparison of the contributions to the various impact categories. This is called *normalization.*

Figure 26.10 shows a normalized profile for an LER200: a resource profile, an environmental profile and a working environment profile. As is clear, resource consumption and environmental impact potentials are of the same magnitude, the level for most impact categories being about 5-10 mPE, corresponding to 0.5-1% of the contributions for an average person in 1990. The potentials for impacts on the working environment are lower than for resource consumption and the environmental impact potentials. This is attributable to the fact that the use stage does not contribute any noteworthy impact on the working environment. The normalization over the life span therefore means a "dilution" of the impacts on the working

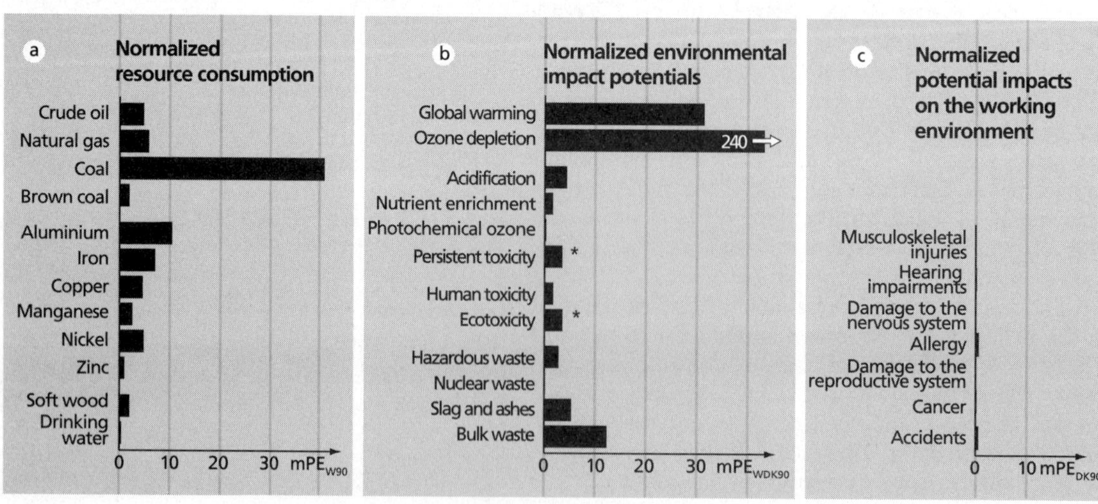

*Figure 26.10. Normalization of resource consumption and impact potentials for the refrigerator. * See explanation of persistent toxicity and ecotoxicity in the text of Figure 26.12*

environment seen per annum relative to the environmental impacts and the resource consumption, where the use stage is significant. Another part of the explanation is that the inventory for the working environment does not cover the whole product system, but only about 3/4 of it.

The resource profile shows that an LER200 used in Denmark is characterized by a relatively high consumption of coal relative to the other resources. The LER200 accounts for over 4% of an average person's consumption of coal worldwide in 1990, the reference year. For aluminium and iron, the refrigerator accounts for about 1% of normal consumption, and for the other metals 0.5% and below, and similar magnitudes are seen for crude oil and natural gas. The profile shows what is characteristic for an LER200 used in Denmark, but it does not show what is most critical. This requires a relative weighting of the various resource types. See the next section.

The environmental profile shows that the LER200 is distinguished by a particularly large contribution to ozone depletion, with the CFC content being as it was in the reference product when it was chosen in 1992. It should be noted that the units are now produced without CFCs. A CFC-containing LER200 contributes ca. 240 mPE or ca. 24% of the contribution to ozone depletion from an average person in the world in 1990. It may appear surprising to most people that it is not more, as there are normally not many other CFC-containing products in the household apart from the refrigerator and the freezer. If it is assumed that these products are normally shared by two persons, the refrigerator's contribution is only ca. 12% per person. But the person-equivalent also includes commercial consumption of CFCs, and this is a significantly greater contribution. At the same time, the consumption in CFC-based air conditioning systems is also included, because the reference for the person-equivalent is global and thus includes all CFC applications in the world. For global warming the contribution is over 3%, and most of this also derives from the CFC compounds. The contribution to bulk waste is also significantly larger than to the other impact categories because of the large consumption of coal for electricity, as the bulk waste, as noted above, derives predominantly from the operation of coal mines. The other impact categories lie around 0.5% and below.

The working environment profile shows special characteristics for allergy and work accidents, where one refrigerator contributes more than 0.05% of the average impact on workers in Denmark. The LER200 is also distinguished by potentials for hearing impairment and cancer of the order of 0.02% of the impact on an average Danish worker.

The most significant impact potentials

The various impact categories are, however, not equally critical, and they must therefore be mutually weighted before the most significant impact potentials can be designated. This weighting follows the EDIP method as explained in Part II. It is shown in Figure 26.11.

The resource consumption in Figure 26.11a is weighted relative to the size of known reserves. This weighting ranks the resource consumption on the basis of a "scarce resources" criterion, i.e. the resources which are most threatened by depletion. On the basis of this criterion, coal is still the highest priority, followed by copper, crude oil, natural gas and nickel. For the metals, copper and nickel are the most significant because of the very limited reserves known to be remaining in 1990. At the present rate of consumption world-wide, they will probably be exhausted in only about 35 years.

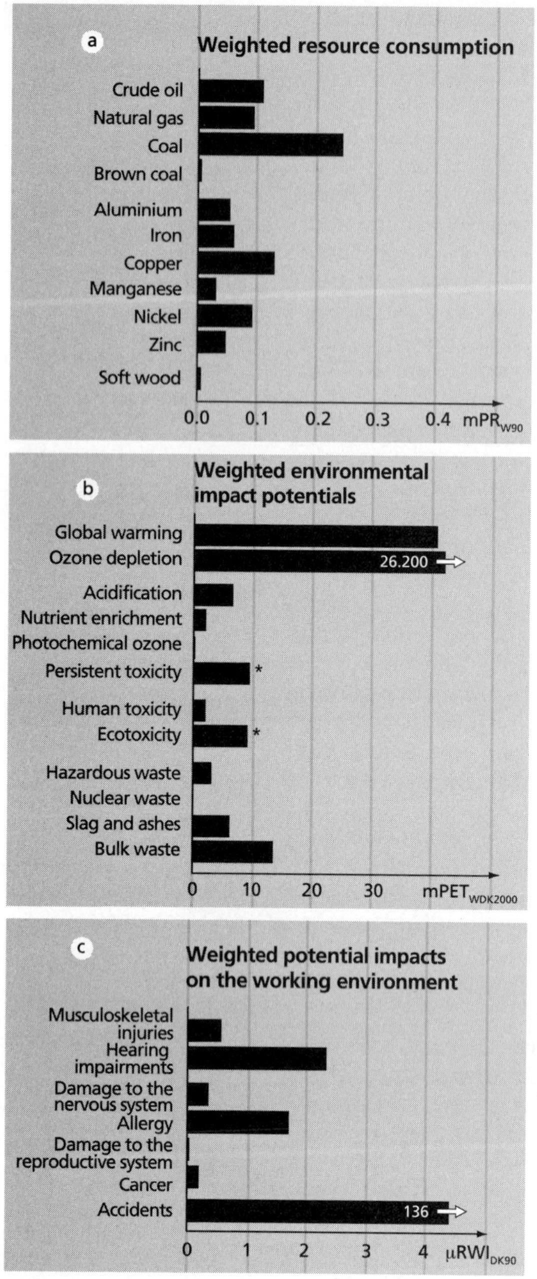

Figure 26.11. Weighted resource consumptions and impact potentials for the refrigerator

Aluminium and iron are ranked lower in the weighting because both occur in the earth's crust, and the supply horizon based on known reserves is significantly greater. The figures express how large a part of the known remaining reserves per person the refrigerator claims. As is evident, the LER200 uses 0.24 mPR (**m**illi-**P**erson-**R**eserves) of coal, i.e. 0.25‰ of the quantity of coal which was known to remain in 1990 for one person and all of his or her successors for all time. On closer consideration, this is quite a lot. Note that in this weighting, no account is taken either of the value of the various resources or of whether the consumption is irreversible as in combustion of fossil fuels, or reversible in principle as for landfilling of metals.

The environmental profile in Figure 26.11b is weighted with the international political reduction targets for the global impact categories and the national reduction targets for the regional and local impact categories. On the basis of this criterion, the contribution to ozone depletion, not surprisingly, has the highest priority. This is followed by global warming and then by bulk waste. It was surprising that persistent toxicity and ecotoxicity were found to be high priorities. This was unexpected for the types of emissions in the product system of the refrigerator. The cause was investigated further and the result is described in the next section. The figures express the product's contribution relative to the target contribution per person in the year 2000. It is evident, for example, that a CFC-containing LER200 is responsible for ca. 2600% of the target contribution to ozone depletion for a person in the year 2000, and 4% of the target global warming potential.

The working environment profile is weighted by the number of reported work injuries per impact-hour, and the figure thus expresses the number of expected incidents per LER200 (cf. Figure 26.11c). On the basis of this weighting, accidents have top priority, followed by hearing impairment and allergy, which are also weighted highly.

This weighting does not include any consideration of the seriousness of the reported effects. The weighting is, however, the best basis for assessing the seriousness. In a comparison between alternatives, it is possible on the basis of Figure 26.11c to ask: what is weighted most highly, 'x' fewer work accidents or 'y' fewer cases of cancer?

The figures in Figure 26.11 are mutually weighted only among resource types, environmental impact categories and working environment impact categories. At this level the figures are therefore not comparable across resources, environment and working environment, as they are in principle at the normalization step in Figure 26.10. Where the normalization gives a total common order of magnitude relative to the "normal", the weighting supports the mutual ranking within each of the three main groups of impact categories.

The most significant sources of impact potentials

For the designer to see from where the contributions to the various impact categories come, they are grouped into contributions from "materials" and "processes", again after the MECO principle described in Part II, and classified as shown in Table 26.5.

The classification gives a total of six types of "sources" of potential impacts, and their contributions are shown in Figure 26.12 as weighted resource consumptions and potential environmental impacts. How much the choice and handling of material during disposal means for the individual impact categories, and how much the processes' energy consumption and consumption of ancillary substances means, is clear from the classification. At an aggregate level, the designer can thus see the consequences of the chosen solutions in the product.

Materials	
Material	- consumption - disposal
Processes	
Energy	- thermal - electric
Chemicals	- ancillary substances
Others	- overhead

Table 26.5. Sources of resource consumption and impact potentials classified after the MECO principle

By nature, consumption of materials is decisive for resource consumption. For oil and natural gas, electricity consumption is, however, just as important, and for coal, electricity consumption accounts for 90%.

For the metals, the recycling in the disposal stage appears significant, as a large portion of the consumption is credited again on disposal. Even if 100% recycling of the refrigerator's aluminium, iron, copper and manganese content is assumed, not as much is credited on disposal as enters in consumption of materials. This is attributable to two things. Firstly, there is wastage during production of materials which does not enter into the refrigerator, and which is therefore not recycled on disposal. Secondly, there is a loss of material grade when the materials are converted to recycling materials, and a resource consumption corresponding to the loss of material grade is therefore ascribed even to recycled materials, as described by the allocation procedure in Parts II and IV.

Figure 26.12 shows that the consumption of ancillary substances in the processes is very decisive for ozone depletion and global warming, of course caused by the CFCs.

Disposal is also decisive for ozone depletion, and it is again the CFCs in the foam and the refrigeration system which make the contribution. In all, CFCs account for 100% of the contribution to ozone depletion and 75% of the contribution to global warming. The small contribution to

Figure 26.12. Weighted resource consumptions and environmental impact potentials classified by type of source

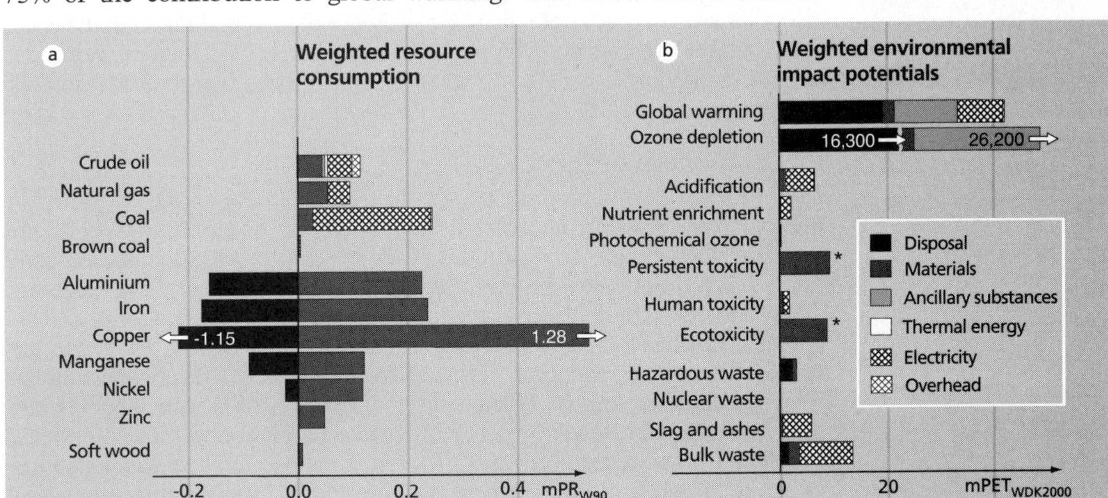

ozone depletion from the materials derives from the CFC content in the thermostat. This is bought in as a standard component and is therefore grouped under materials. Disposal is also significant for hazardous waste, where the dumping of PVC with a content of plasticizers is occurring.

The consumption of thermal energy is very low. Not many processes in the product system use fuel directly, and the highly energy-intensive drying and hardening processes are, for example, also electricity-based. Only the consumption of oil in the transport processes is categorized as thermal energy, and the only place where this contribution is seen is in the resource consumption of crude oil.

The consumption of electrical energy provides the most significant contribution to many of the impact categories. This is true of acidification, nutrient enrichment, slag and ashes and bulk waste. For global warming, the electricity consumption contributes 25%.

Persistent toxicity and ecotoxicity

It is clear from Figure 26.2 that the potentials for persistent toxicity and ecotoxicity derive from the consumption of materials. As noted, this was unexpected, and it was therefore investigated further. The computation showed that it derived from the steel production. By going a step further back in the computation it was clear that the two toxicity contributions derived almost 100% from the iron content in the emission from the steelworks in question. Data from the steelworks simply showed the actual quantity of "iron" in the emission, not the form in which the iron existed. In the computations of toxicity contributions it was assumed that the iron would exist as ions, which are very toxic. But a subsequent check quickly showed that this was improbable given the actual quantity of iron in presumed ionic form, because iron is rapidly oxidized in water to iron oxides, which are much less toxic. An inquiry to the steelworks confirmed that "iron" in the effluent meant "iron oxides". The impact potentials demonstrated here for the two toxicity categories were thus expressions of an error, and the toxicity potentials are in fact significantly less. They are therefore marked with an asterisk in the figures.

The examples are included to show that toxicity assessments must be handled with care in the LCA. When a high priority arises, there should be a further analysis of the underlying factors. The toxicity assessments are encumbered with a greater uncertainty than the assessments of the other environmental impact potentials. But it is valuable to include them because they provide an introductory orientation on how the toxicity from the known emissions is weighted relative to the other environmental impact potentials.

Simplification of the LCA

The information revealed by the profiles classified by source as shown in Figure 26.12 provides an opportunity to simplify the presentation of the LCA in subsequent investigations and to avoid too much superfluous information. Firstly, the impact categories which have a very low weighting relative to the others are excluded. On the basis of this criterion, photochemical ozone, persistent toxicity, ecotoxicity and radioactive waste are excluded, as are brown coal, zinc and wood. Impact categories which can be represented by others, because they derive from the same sources, are then excluded. On this criterion, nutrient enrichment and human toxicity, which can be represented by acidification, are excluded because the sources are materials and electricity in the same ratio as for acidification, and slag and ashes, which can be represented by bulk waste. Manganese is also excluded as it enters into the steel and can be represented by iron. Nickel is excluded from the next presentations because it only enters into a single component, but the consumption of nickel in this component is shown again later as a focal point.

The most significant stages of the product life

The LCA in Figure 26.13 is divided into the various stages of the product life. This provides a picture of where in the product system the greatest resource consumptions and potential impacts arise.

Figure 26.13a shows the resource profile divided into stages. As is evident, the manufacturing stage and the transport stage are of very little significance for resource consumption. Only small quantities of fuel enter into these stages. The use stage is of considerable significance for consumption of coal because of the large consumption of Danish electricity.

An LER200 contains 1,000 g copper and has thus occupied 1.3‰ of the 1990 person-reserve, i.e. 1.3‰ of the "ration" of remaining known reserves in 1990 for one person and his or her successors for all time. It shows the importance of recycling copper in the disposal stage, as is presumed to be done here.

Figure 26.13b shows the weighted environmental profile divided by stages, where the potentials for global warming and ozone depletion are dominant. The disposal stage is the most significant, and the CFC emission accounts for the entire contribution here. In the use stage the potentials for global warming are equally great for the CFC emission and the

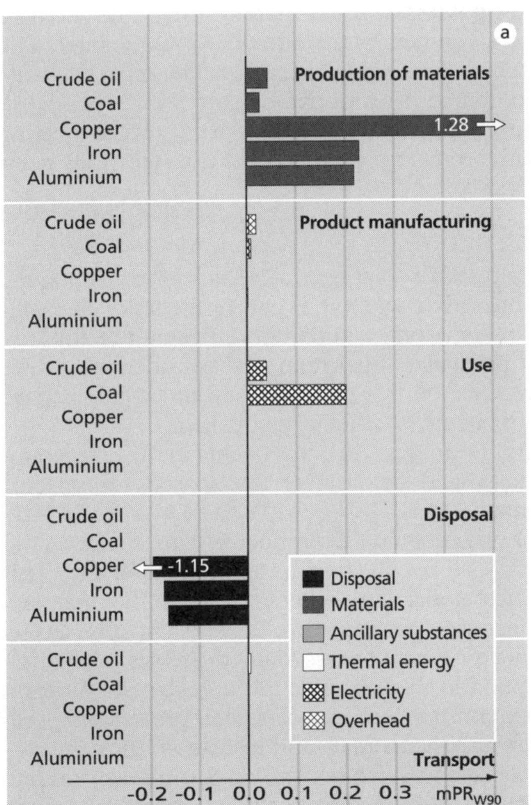

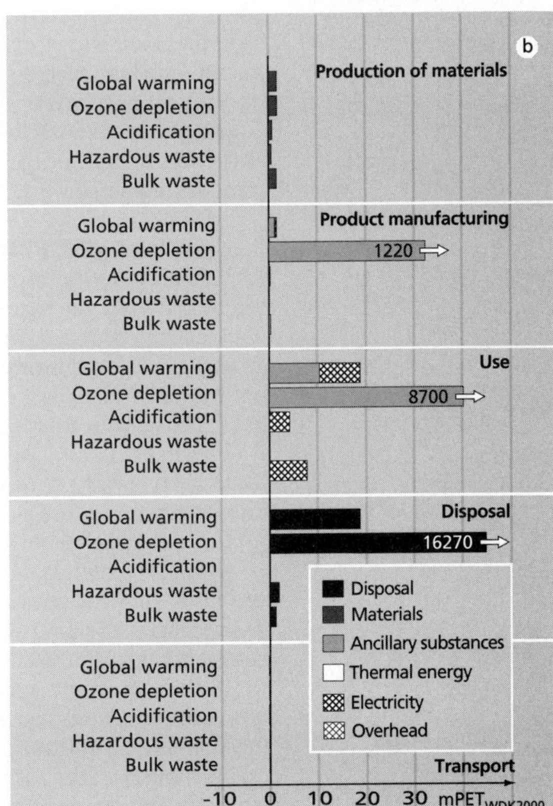

Figure 26.13a. Weighted resource consumption divided by stages of the product life

Figure 26.13b. Weighted potentials for environmental impacts divided by stages of the product life

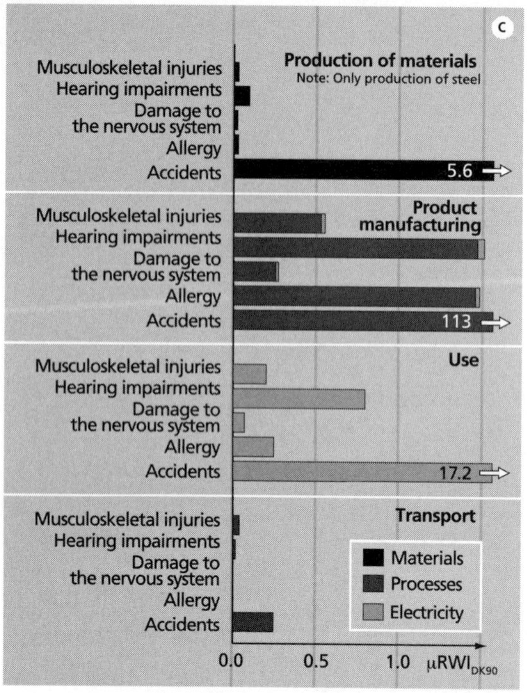

Figure 26.13c. Weighted potentials for impacts on the working environment divided by stages of the product life

electricity consumption. The figure shows that the use stage, due to its energy consumption, will be very dominant when alternatives to CFCs which can neither contribute to ozone depletion nor to global warming are used.

It is evident that the contribution to bulk waste is much greater in the use stage than in the disposal stage. This is attributable to the fact that the quantities of waste from coal mining significantly exceed the quantities of materials which end in landfill after disposal. The content of plasticizers in the PVC means that the PVC waste in the disposal stage will be dominant for hazardous waste.

Figure 26.13c shows the weighted working environment profile divided by stages. Only steel production is included for the material stage, but the steel comprises over 50% of the refrigerator's weight. The figure thus provides some impression of the material stage's contribution.

Note the small size of the contribution to impacts on the working environment from the steel production relative to that in the product manufacturing stage. This is attributable to the fact that time used per refrigerator for production of the steel is significantly less than for the manufacturing stage, because significantly greater tonnages per unit of time are produced during steel production than during the refining processes in the manufacture. The measure of impacts on the working environment is the duration of the impact. The manufacturing stage is therefore most significant when the impacts on the working environment are computed per product.

There are relatively many reports of, especially, work-related hearing impairment from co-generation power plants, and the use stage thus gains some significance.

The most significant components

An LER200 is one of many possible solutions to the task of providing the functional unit. Each component is a solution to part of the task. The solutions chosen are an expression of a long tradition in refrigerator technology in general and at Gram in particular. Environmental considerations have not hitherto had a large influence on this tradition, and the solutions have not previously been subject to an environmental assessment.

Figure 26.14 is an attempt to do this. All components in the refrigerator have been assessed separately, and the lists of top scorers for selected impact categories are presented.

For the great majority of processes in the product system, the exchanges with the environment can be ascribed to a single component. This applies to all processes in the material stage and almost all in the manufacturing stage. Certain manufacturing processes do, however, involve several components at the same time, especially the assembly processes.

But the energy consumption in the use stage in particular is a function of the interplay among many components, especially compressor, evaporator, condenser, control (relay and thermostat) and insulation. In many cases, exchanges in the disposal stage will also depend on the interplay among components, but for the LER200 with its presumed disposal in a shredder,

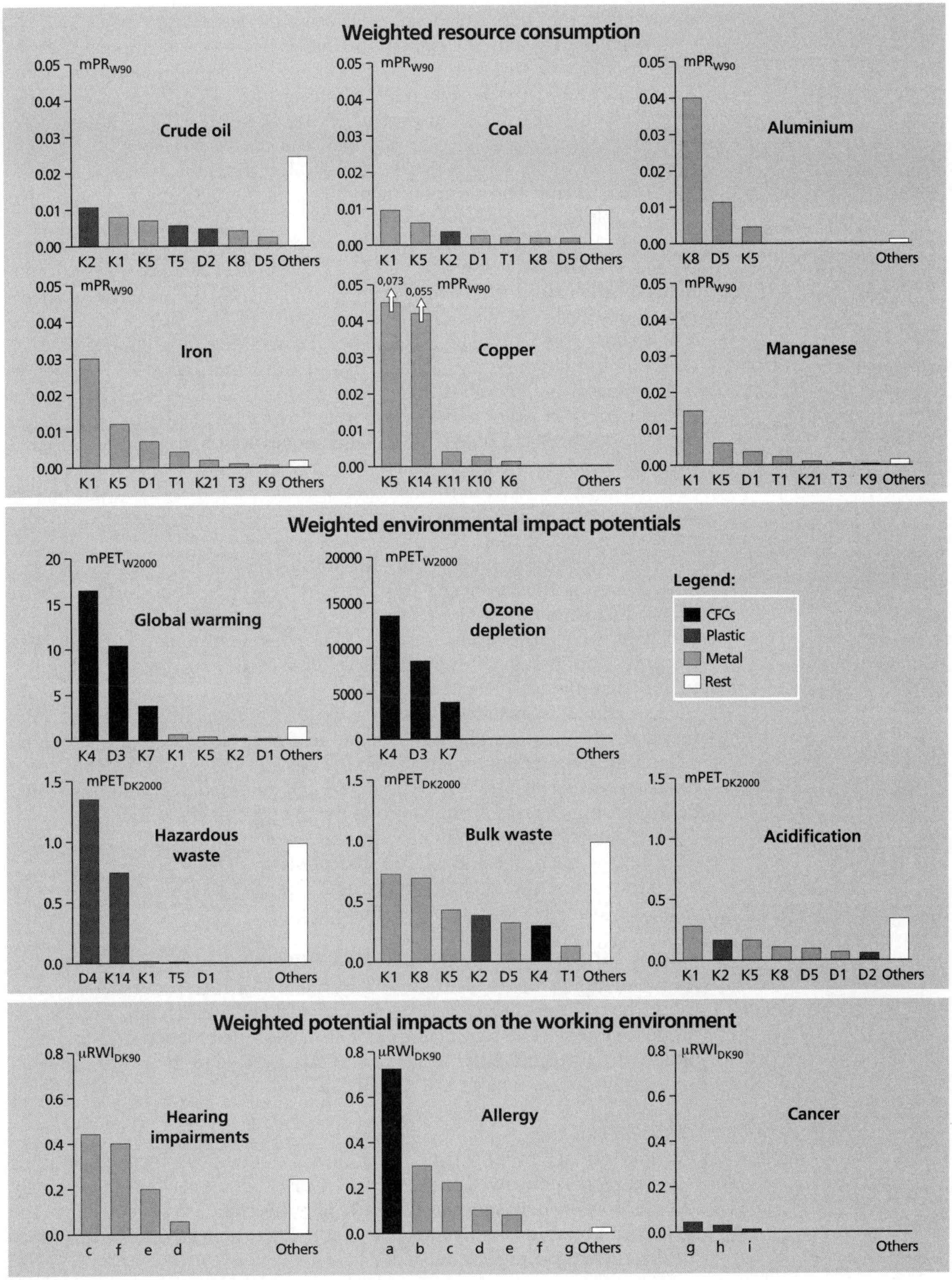

Figure 26.14. Top scorers among the refrigerator's components

they will in practice be able to be traced to a single component.

Figure 26.14 shows weighted potentials for various impacts for the entire product system with the exception of the energy consumption in the use stage. The material stage, manufacturing stage, CFC leakage in the use stage, disposal stage and transport stages are all included. The figure therefore provides a holistic picture of which choices for solutions at the component level in the LER200 have the greatest potential consequences for the environment. The component numbers in the figure refer to the parts list in Table 26.1.

Apart from showing which components make the biggest contributions, the figure also shows the residual contribution from all other components together. In this way the figure shows how widely the impact category in question is distributed in the product, and thus how many or how few components an initiative will affect if something must or can be done about the problem. Note that these lists of top scorers must not be seen as a designation of environmental points of focus. It may be that a given component is already the best solution in environmental terms, even if it is the top scorer. It is a matter of the alternatives which are available. Figure 26.14 is therefore merely a neutral overview and a good basis for a discussion between the environmental specialist and the designer. Alternatives are discussed, and points of focus designated, under environmental diagnosis in the next section.

The lists of top scorers show only an overview at the component level. The causes of environmental problems are also to be found in the choice of product structure and in the chosen concepts. But a designation of problems or "top scorers" at the level of structure and concept also requires a discussion of alternatives at these levels, and this is also given under environmental diagnosis in the next section.

As disposal is included, it is assumed that the metals shown are 100% recycled. The consumption of metal resources in the figure therefore reflects the consumption ascribed on the basis of the loss of material grade discussed earlier. If 100% recycling were not assumed, the columns would be higher, but the relationship among them would be the same.

The contributions to bulk waste are greater for certain of the metal components than for the plastic components. This may be surprising when they are sent entirely for recycling. The reason, however, is the same as before, namely loss of material grade.

Lack of information, uncertainty and sensitivity analyses

As noted under scope definition, information on the processes which it would have been desirable to include is lacking for certain parts of the product system. The significance of this lack of information and the magnitude of the uncertainties in general is discussed here in order to give an impression of how tenable the conclusions are.

The material stage

The greatest lack of information is in the product system for the chemicals from extraction of raw materials up to their production. This is especially true for polyalcohol, isocyanate, CFCs, joint filler and epoxy powder in the quantities noted in the inventory in Table 26.3, and in smaller quantities for detergents and phosphoric acid.

The energy consumption for production of these is estimated, and for polyalcohol and isocyanate, which enter into the foam, it amounts to ca. 600 MJ. As noted previously, this energy consumption is included in Figure 26.8b to emphasize the point of the figure. For the other chemicals, the total energy content including both process energy and calorific value is estimated to be less than 200 MJ. In the inventory in Table 26.3 and in the impact assessments which follow it, the material stage is thus lacking resource consumption and emissions from the energy production corresponding to ca. 800 MJ.

The emissions from the production processes for these chemicals are also lacking, but as noted previously, the emission of chemicals in the processes in the manufacturing stage where they are used is included. For CFCs and detergents, for example, the emissions in the manufacturing stage will probably be greater by a factor of at least ten, as the percentage loss in the chemical industry during production of these substances is estimated to be significantly lower than 10%. In general, the emission of chemicals during their production is insignificant relative to their use, but there may be exceptions which are not known here. For example, there may be special exposures in the working environment.

Impacts on the working environment in the material stage have been assessed only for the steel production. The number of man-hours per kilogram of material is probably greater for plastic and chemicals, but how much greater is not known. There may be exposures for allergenic substances (epoxy powder) and substances which cause damage to the reproductive system (bisphenol A) during production of epoxy powder which are not known. There may also be unknown exposures for carcinogenic substances (acrylonitrile and styrene) during production of ABS.

The uncertainties in the data for steel production are small. They derive from a single steelworks which is the sole supplier to Gram. The uncertainty is greater for the plastic types as the data are average data from the European union of plastics manufacturers. The choice of plastics supplier is, however, also more variable, and it is therefore appropriate to use average data.

The manufacturing stage

For the manufacturing stage, all processes have by and large been measured and/or assessed. The energy measurements are assessed as being highly reliable, and the uncertainty small. It was ensured for the measurements of emissions that operating conditions were normal, but they were measured only once over several hours. There may therefore be considerable uncertainty in the measurements of emissions, estimated at up to 50%. However, uncertainties of 50% for the measured manufacturing processes are of no significance in the total picture. The distribution of the overhead contribution per produced unit, as done in this case, could have been done differently. For example, an economic partitioning key could have been used, i.e. distribution of overhead consumption per dollar of refrigerators sold instead of per unit. The ascribing of overhead contribution to an LER200 will, however, vary at most by 10-20%, according to which partitioning key is used, and in the total picture this is of minor significance.

The use stage

Data for the use stage are highly reliable. Operation of the LER200 refrigerator was measured continuously in 120 units over a year (Pedersen & Lawætz, 1989), and the uncertainty in the determination of their energy consumption was assessed as being less than 10%. Data were also assessed as being highly reliable for measurements of the emissions for Danish power plants. In the partitioning of resource consumptions and emissions to electricity and heat for power/co-generation power plants, the electricity companies' own internal partitioning key was used. It was advised that 90% was ascribed under this key to the electricity. This partitioning could have been made under other keys, and it is of significance for both the manufacturing stage and the use stage. Figures 26.12 and 26.13 show how much the electricity consumption contributes to resource consumption and potential environmental impacts and impacts on the working environment, and the consequence of a changed partitioning key can be immediately seen here, as the electricity contribution in the two stages will be changed in proportion to the percentage ascribed to the electricity. The electricity contribution is assessed as being able to change by a maximum of 15% as a function of the partitioning key used between electricity and heat.

It is assumed in the assessment that the refrigerator is used in Denmark. A pattern of use for an average family is also assumed. The significance of the use stage does, however, depend on these assumptions, and both alternative sites and patterns of use are therefore simulated. This is done together with the other simulations in Section 26.2 (cf. Figures 26.15g and 26.15h).

As noted above, the uncertainty in the emission of CFCs from the foam in the use stage is large, as an estimate is available for only four refrigerators to support the data. This can be of significance for the assessment of the magnitude of the contribution from the use stage, but its significance in the total picture is small, as only 7.5% is collected in the disposal stage. CFCs which do not escape in the use stage escape during disposal, and vice versa.

An investigation of the life spans for refrigerators on the market in general has shown that it is probably about 17 years (Thorsen, 1993). As the LER200 was developed in 1988, there is no experience with this unit, and Gram has proposed a conservative estimate of the life span; 13 years was thus chosen. In purely technical terms the average life span is probably significantly longer. A longer life span means that the use stage gains a greater significance than the assessment has shown, as the significance of the other stages is reduced with increased life span.

The disposal stage

The individual processes in the disposal stage were assessed with a relatively high degree of certainty as the data are from specific disposers of refrigerators and of metals. The assumption that CFCs were drained from the refrigeration system in 30% of the units and from the foam in 7.5% is highly reliable, as these were the actual average figures for refrigerators in Denmark in 1992. There is a higher degree of uncertainty in the assumption that 100% of the metals are recycled. There is also a general uncertainty in the data deriving from the fact that the processes in the landfill do not enter into the calculations, and that environmental problems from the landfill are instead merely indicated by the actual quantity of waste.

Simulation of changes

Other routes of disposal, altered life spans, other electrical systems during use and a number of other changes in relation to those assumed for the reference product are simulated in the environmental diagnosis in Section 26.2. This shows the sensitivity of the most significant assumptions made for the product system of the LER200 refrigerator.

26.2. Environmental diagnosis for a refrigerator - the LER200

The goal of the environmental diagnosis is to designate environmental points of focus in the product, i.e. to designate which of the resource consumptions and potential environmental impacts shown by the LCA should be considered problematic, and to identify where they reside in the product.

As noted earlier, it is not necessarily the most highly weighted impact potentials which are the most problematic for the product, but to a large extent it is those which are greater than necessary. The goal of the environmental diagnosis is therefore to identify the potentials for improvement in the product so that, together with the environmental assessment of the reference product, these can form the basis for the designation of points of focus.

One part of the task is to simulate various changes in the product or its product system, both theoretical changes and concrete alternatives, and to assess these changes. The other part of the task is to generate information on known alternatives or technologies which are judged to be underway, as potentials for improvement can only be designated on the basis of information about alternatives.

Simulation of changes in the refrigerator or its product system

Simulation of changes in the product or its product system follows the MECO principle, i.e. changes in choice of materials and disposal of materials, changes in the processes' consumption of energy and chemicals (ancillary substances), and "other" changes such as changes in life span and patterns of use. The simulations are based on the reference product's actual composition and product system, and the environmental consequences of the simulated changes are compared with the reference (see Table 26.6).

Table 26.6. Simulations of changes in the product and its product system

M	E	C	O
Disposal	Consumption of electricity	Consumption of chemicals	Others
• Landfill 100% instead of the present scenario • Incinerate 100% instead of the present scenario • Reuse 100% instead of the present scenario	• Optimize energy efficiency of compressor to 100% • Optimize energy efficiency of evaporator to 100% • Optimize energy efficiency of cooling system to 100% • Optimize energy efficiency of refrigerator to 100% incl. theoretically optimal insulation • Increase the insulation with PUR foam by 50% for the R134a unit and the pentane/isobutane unit • Increase the insulation with PUR foam by 100% for the R134a unit and the pentane/isobutane unit	• Substitute CFC with R134a in foam and as refrigerant. • Substitute CFC with pentane in the foam and isobutane as refrigerant	• Reduce lifetime to 10 years • Increase lifetime to 20 years • Different amount of users (one simulation at a time) • Use of product in different regions or locations (one simulation at a time)

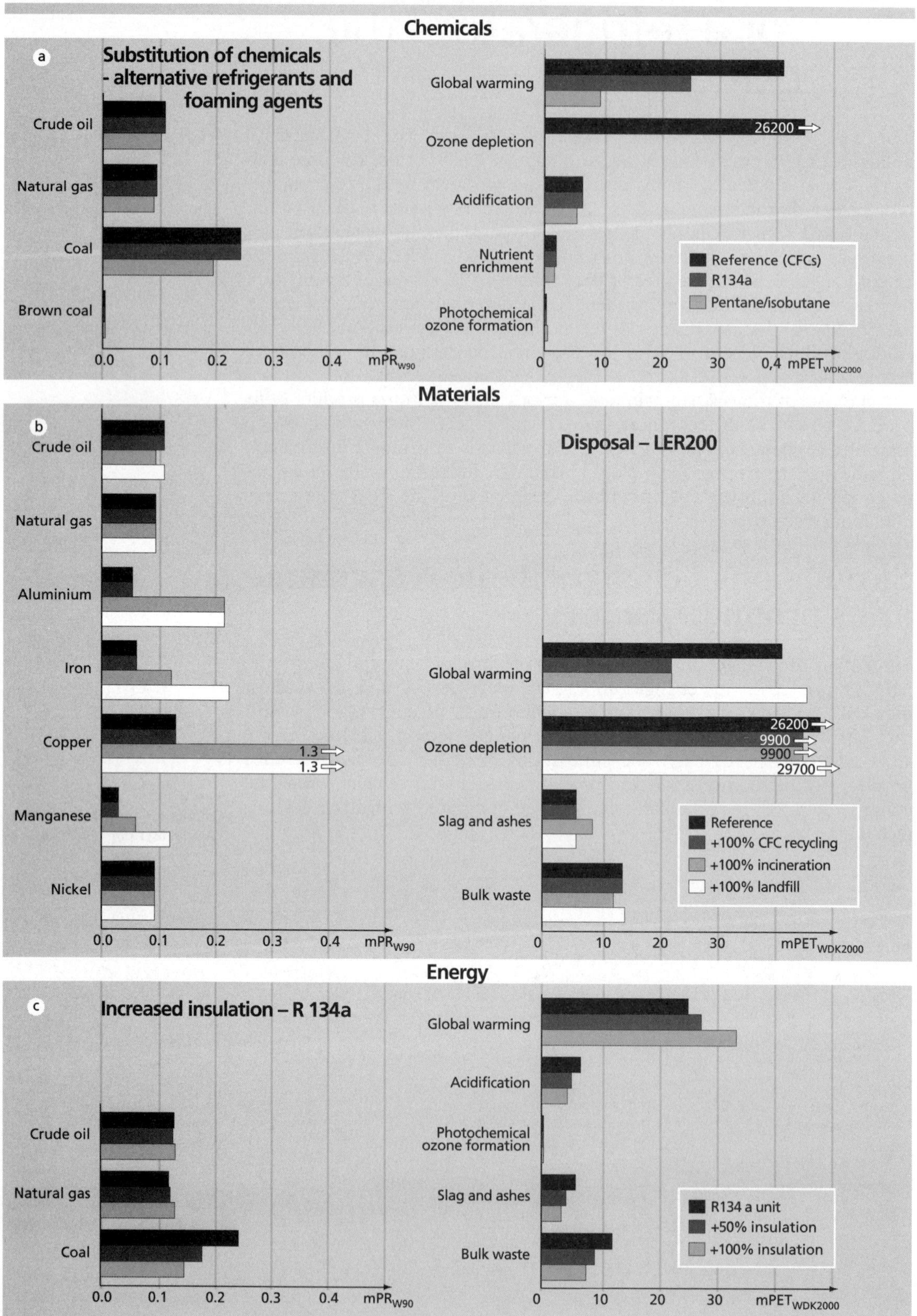
Chemicals
a
Substitution of chemicals - alternative refrigerants and foaming agents
Crude oil
Natural gas
Coal
Brown coal
0.0 0.1 0.2 0.3 0.4 mPR W90
Global warming
Ozone depletion
26200
Acidification
Nutrient enrichment
Photochemical ozone formation
Reference (CFCs)
R134a
Pentane/isobutane
0 10 20 30 0,4 mPET WDK2000
Materials
b
Crude oil
Natural gas
Aluminium
Iron
Copper
1.3
1.3
Manganese
Nickel
0.0 0.1 0.2 0.3 0.4 mPR W90
Disposal – LER200
Global warming
Ozone depletion
26200
9900
9900
29700
Slag and ashes
Bulk waste
Reference
+100% CFC recycling
+100% incineration
+100% landfill
0 10 20 30 mPET WDK2000
Energy
c
Increased insulation – R 134a
Crude oil
Natural gas
Coal
0.0 0.1 0.2 0.3 0.4 mPR W90
Global warming
Acidification
Photochemical ozone formation
Slag and ashes
Bulk waste
R134 a unit
+50% insulation
+100% insulation
0 10 20 30 mPET WDK2000

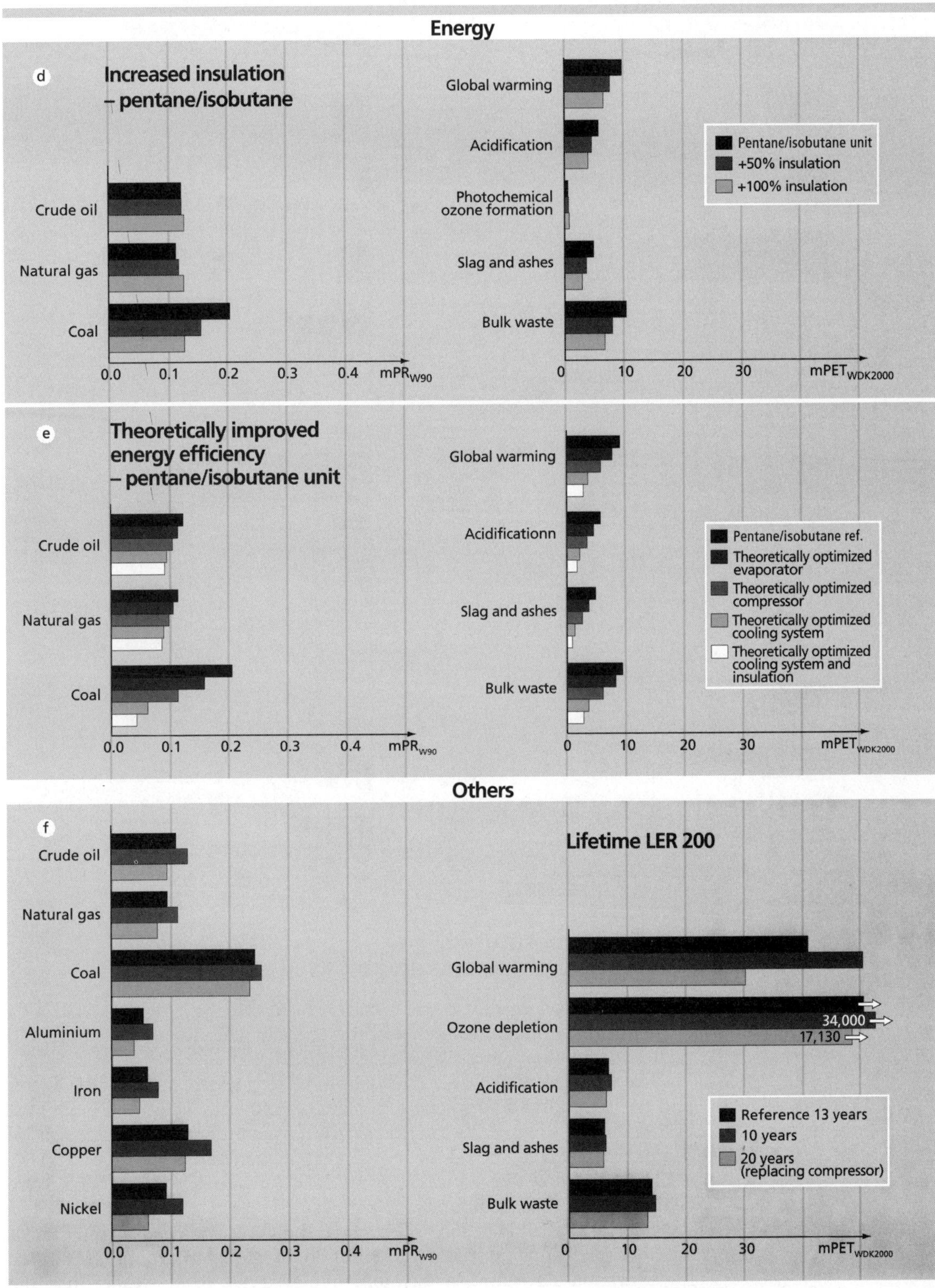

Figure 26.15. Simulation of the environmental consequences of changes in the product and its product system

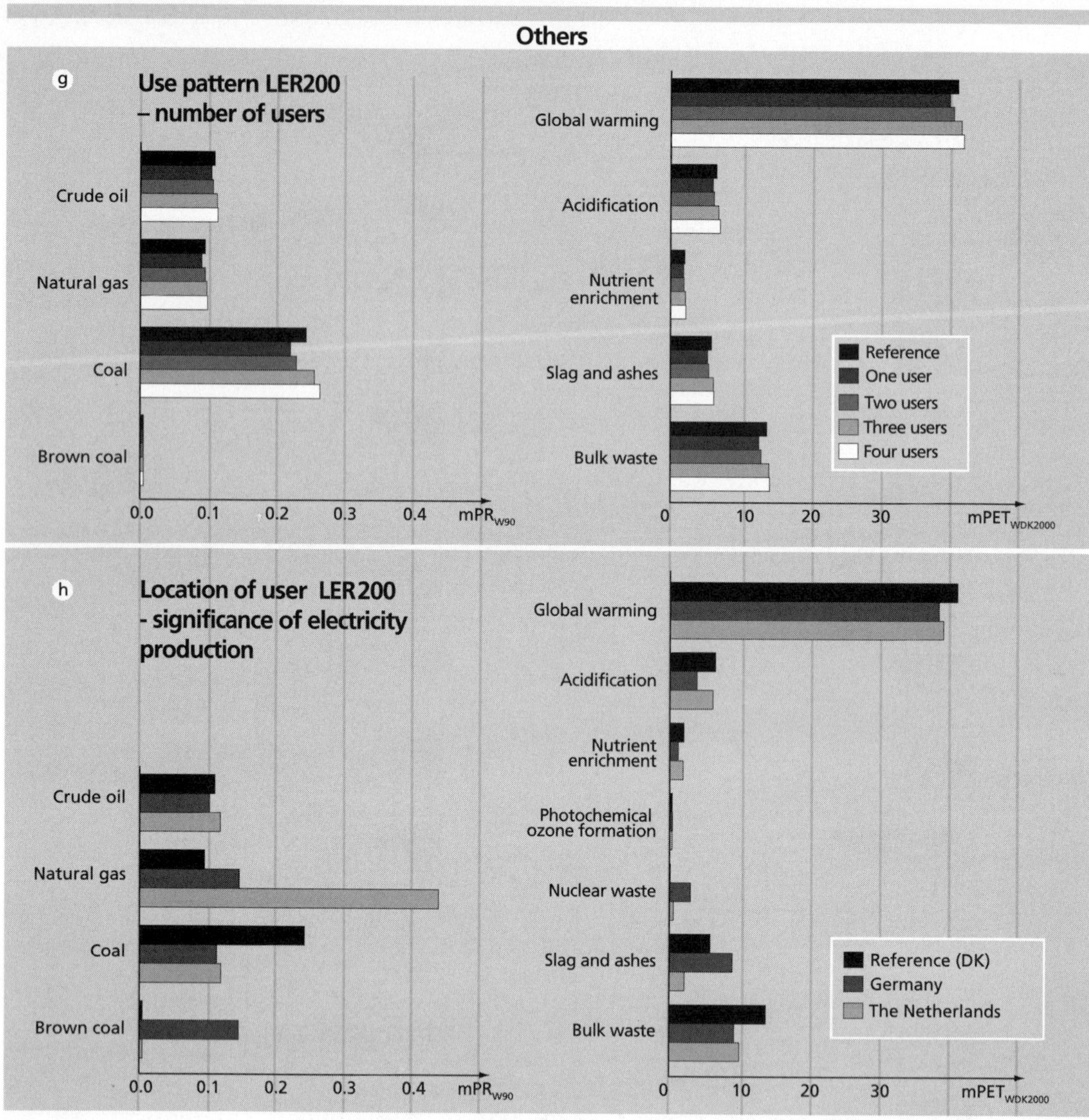

Figure 26.15. (continued)

Chemical substitution. Alternative refrigerants and blowing agents

Two alternatives to CFCs were analysed, viz. R134a, which is an HFC, i.e. a fluorine-containing hydrocarbon without chlorine or bromine, and pentane/isobutane, which are alkanes.

The first alternative uses R134a both as refrigerant and blowing agent, and the second alternative uses pentane as blowing agent and isobutane as refrigerant. Figure 26.15a shows the environmental consequences of the alternatives. It should be noted that in general, the figures for the R134a unit show the situation from 1994 onwards, where CFCs were replaced by R134a in all units.

The R134a substitution will remove the contribution to ozone depletion entirely, and halve the contribution to global warming. R134a is, however, still a greenhouse gas, although assumed to be weaker than the CFCs, so there will continue to be a contribution to global warming, and as will be evident, it will be equally as great as the contribution from the electricity consumption. The other impact potentials are unaffected by the substitution. The pentane/isobutane solution will also eliminate the contribution to ozone depletion, and it will reduce the contribution to global warming by 75%, so that only the contribution from the energy consumption remains. The pentane/isobutane refrigerator has also been assessed as being 20% more economical in terms of energy in the use stage, and this is reflected in a reduction for the energy-related impact categories in general. The potential for formation of photochemical ozone is increased by this alternative. The increase is 60% relative to the reference and the R134a refrigerator, but in the weighted picture the significance of this increase in not very great.

The simulation shows that in environmental terms, the pentane/isobutane solution is preferable to the R134a unit. This conclusion is further strengthened by Figures 26.15c and 26.15d, which show the options for energy savings by increasing the insulation in the two solutions.

Disposal of an LER200

The significance of the disposal route is very great, as shown in Figure 26.15b. Three alternative disposal scenarios are simulated for an LER200 refrigerator, viz.:

- that it is delivered 100% for CFC recovery from both foam and refrigeration system,
- that it is sent 100% for incineration,
- that it is sent 100% for landfill.

On 100% CFC recovery, a reduction of 50% is obtained in the contribution to global warming, namely that part accounted for by the CFC emissions during disposal. For the contribution to ozone depletion, there is a corresponding reduction of 60%. The remaining 40% derives from the manufacturing and use stages. There are no changes for other environmental impact categories and resource consumptions. It should be noted that the working environment was not assessed, and that there may be significant considerations involved in manual handling in connection with draining of the CFCs.

On 100% incineration, corresponding reductions are obtained for the contributions to global warming and ozone depletion. The incineration of plastic and foam emits CO_2, which contributes to global warming, but at the same time there is a saving in other fuel because it is assumed that the heat from incineration plants is utilized. In all, this results in a small additional nett contribution to global warming, because the nett energy efficiency for the incineration plant is a little smaller than for conventional heating plants, but the difference is too small to be seen in the figure. On the other hand, there is a visible effect in the reduction in use of other fuel. It is assumed here that this fuel is oil, and the effect is seen in the form of a reduction in oil consumption of ca. 10%. Incineration reduces the contribution to bulk waste, but it makes a relatively larger contribution to slag and ashes because of the substantial metal content. Incineration of the refrigerator without first removing the metal will have a very unfortunate effect on resource consumption, especially for copper, where it will lead to a consumption of 1.3‰ of the person-reserve. The consequences will be less for iron and manganese because after incineration the steel is separated magnetically from the slag in certain incineration plants, here assumed to be 50%.

One hundred percent dumping of the refrigerator has the same unfortunate effects as incineration, but is worse for iron and manganese and worse again for global warming and ozone depletion.

The simulations show that the disposal route is of great significance for both the refrigerator's resource consumption and its potential environmental impacts. The correct disposal route for an LER200 is:

- drain the CFCs from the refrigeration system, collect CFCs from the foam and destroy the CFCs,
- separate metals, plastic and foam,
- recycle the metals, including stainless steel,
- burn plastic and foam in an incineration plant with utilization of the energy, preferably a co-generation power plant,
- if possible, regranulate and recycle the large plastic items in ABS and lurane.

The optimal solution is to collect all CFC-based refrigerators in use and to replace them with non-CFC-containing units, and then to follow the above disposal route, also in order to avoid the significant emission of CFCs in the use stage.

Increased insulation. R134a and pentane/isobutane refrigerators

Figures 26.15c and 26.15d show the effect of increased insulation, i.e. 50% and 100% increased insulation for both the R134a and the pentane/isobutane refrigerators. An increase in the insulation on the outer side is assumed in order to maintain the volume of 200 litres as required by the functional unit.

For the R134a refrigerator the contributions are reduced to the impact categories which are energy-related only. For the potential for global warming, on the other hand, the contribution is increased with the thickness of the insulation because R134a is a greenhouse gas. As is clear, the increase is greatest for the last part of the insulation layer because the energy saving per quantity of R134a falls with the thickness of the layer. Consumption of oil and natural gas is increased with increasing thickness of insulation because they enter into the insulating material, but not to a significant degree into the Danish electricity production which enters into the use stage. For these resources, the optimum point for the insulation lies below a 50% increase.

For the pentane/isobutane refrigerator, the same trade-off problem is not seen as in the R134a with respect to the contribution to global warming. There is, however, an increase in the potential for formation of photochemical smog, but in the weighted picture the increase is again not as significant as the improvements in the other impact categories. Consumption of oil and gas is increased as a function of increased insulation for the same reason as before, but is more pronounced because the energy consumption for this model in the use stage is less than for the R134a model. The total resource depletion is reduced by increased insulation because of the improvement in the consumption of coal. It is evident that the optimum point for energy consumption (see the potential for global warming) is not reached with a 100% increase in the thickness of the insulation, but the degree of improvement decreases from 50 to 100%.

The consumption of additional materials, including extra steel to increase the insulation, is of course included in the computations illustrated in the figures.

Theoretical improvement in the energy efficiency

The energy efficiency of an LER200 refrigerator is 3-4%, and of all refrigerators it has one of the lowest energy consumptions in the world. The energy efficiency is defined as the energy required to cool down the contents and to compensate for door openings divided by the refrigerator's energy consumption. As noted, in practice the energy required to cool down the contents and to compensate for door openings is 25% of the total refrigeration, i.e. 1/4. The rest is lost via the insulation. At the same time, the efficiency of the refrigeration system, i.e. compressor, evaporator and condenser, is ca. 14%, i.e. ca. 1/7. The refrigerator's total energy efficiency is thus $1/4 \times 1/7 = 1/28$ or ca. 3.5% of the theoretical maximum for a fully optimized system (Jakobsen, 1995a).

This provides considerable scope for technological development in the years ahead, and significant improvements are already obtainable with known technology. Figure 26.18e shows the theoretically attainable improvements by:

- optimizing evaporator to 100% efficiency,
- optimizing compressor to 100% efficiency,
- optimizing the entire refrigeration system to 100% efficienc,
- optimizing the refrigeration system plus insulation to 100% efficiency.

The improvements are shown for the pentane/isobutane model so that the contributions from CFCs and R134a do not disturb the picture of the consequences of energy reductions.

A theoretically optimal evaporator can increase the efficiency of the refrigeration system by about 50%. An optimal compressor can similarly increase efficiency by ca. 100% and if the entire refrigeration system is optimized, energy consumption falls by the above factor of seven. If insulation is optimal, there is a further reduction by a factor of four. The energy consumption for the insulating materials is *not* taken into account here, but a theoretically optimal vacuum insulation of one kind or another is assumed.

Figure 26.15e reflects the consequences of these reductions in electricity consumption in the use stage. Apart from these improvements, an improvement in the efficiency of society's electrical system can yield further reductions in consumption of primary energy in the use stage by a factor of nearly two.

It is thus theoretically possible to reduce energy consumption very greatly relative to that of the reference product as it is today. This has been a recognized fact for many years (Nørgård, 1979; Guldbrandsen *et al.*, 1986), and technological developments over the last 20 years have led to the refrigerators' efficiency becoming better and better.

Relative to the reference product LER200, it is believed to be possible in practice to achieve an improvement in the energy efficiency by a factor of 5 within the next 5-10 years: a factor of 2.5 by a combination of optimized evaporator and compressor, and a factor of two by better insulation, e.g. with vacuum panels (Jakobsen, 1995b).

The life span of the LER200

The life span is of considerable significance for the relative weighting of all stages other than the use stage, i.e. the material, manufacture, transport and disposal stages. This is attributed to the LCA being expressed per unit of time, and the contributions therefore normalized with the life span.

Figure 26.15f shows the significance of two theoretical changes in the life span, namely:

- a life span of 10 years,
- a life span of 20 years.

It is clear that the significance of the life span is greatest for the contribution to global warming and ozone depletion, because they are most significant during disposal. To be conservative, a compressor replacement is included in the 20 year life span, which means that the copper consumption is not significantly reduced in the 20 year life span relative to that for 13 years.

Figure 26.15f shows that increased life span results in an environmental improvement. It is, however, not certain that this will be the case, because the technology is developing so rapidly. It is expected, for example, that refrigerators in the year 2000 will have much smaller environmental impacts than in 1996. As the use stage will probably also be decisive in future, shorter life spans for units with today's technology may be environmentally preferable.

The pattern of use for an LER200

It is of interest to know the significance of the user's habits for the refrigerator's energy consumption. It is estimated in advance that ca. 10% of the refrigeration is used in refrigerating goods, ca. 15% in compensating for opening of the door, and ca. 75% in compensating for entry of heat through the walls. The significance of the pattern of use thus lies within 25% of the use stage's contribution. The basic energy measurements on 120 units are grouped by number of users, and the classification provides a certain impression of the significance of the pattern of use (Pedersen & Lawætz, 1989). Figure 26.15g shows this. Four simulations were carried out:

- one user,
- two users,
- three users,
- four users.

The significance of the number of users is greatest for the purely energy-related effects, and the significance is large, namely ±10% relative to the reference, especially for coal consumption, which derives predominantly from the use stage's electricity consumption. In other words, there is a variation of 20% between one user and four users, which agrees well with the assessment made in advance. For the other impact categories, contributions from sources other than electricity in the use stage play a greater role, and the variation is therefore less for these. The purpose of the figure is to show the significance of the pattern of opening/closing and the quantity of goods. Of course it will not demonstrate that it is an advantage to have only one person per refrigerator. If the impact potentials are expressed *per person*, it is self-evident that it is an advantage for there to be several persons per refrigerator.

The place of use for an LER200

The LCA concerns a refrigerator used in Denmark with only Danish electricity used for operation. There is therefore, for example, no radioactive waste, even if nuclear power actually enters into the product system for an LER200. Figure 26.15h illustrates the consequences of using an LER200 in other countries, namely:

- Germany, or
- The Netherlands

which are the most important markets for the LER200 apart from Denmark. Nuclear power, brown coal and larger quantities of natural gas are used in Germany. This leads to radioactive waste and more slag and ashes (from the burning of brown coal), but smaller contributions to global warming, acidification, nutrient enrichment and bulk waste. Much larger quantities of natural gas are used in the Netherlands as an alternative to the Danish consumption of coal, and this leads to slightly lower contributions especially to global warming, slag and ashes, and bulk waste.

Uncertainties in the simulations shown

The purpose of Figure 26.15 is to show the differences between the alternatives. In an uncertainty assessment it is therefore also the uncertainty in the difference which is of greatest interest, and it is often less than the uncertainty in the impact assessment expressed in mPR of resource consumption and mPET of potential environmental impacts. The differences in the impact potentials shown reflect real differences in the environmental exchanges from the processes in the product systems being compared, e.g. energy consumption, CO_2 emission, quantities of waste, etc. The uncertainty in the impact assessment does not enter here.

The reason why the differences are not expressed directly in energy consumption etc., but nevertheless as impact potentials, is that for many of the alternatives there may be trade-off situations, i.e. improvements in some impact potentials and worsenings in others. In such trade-off situations between various impact categories the uncertainties are larger because the uncertainty of the impact assessment is involved.

It should therefore be noted that the uncertainty is not so great when all of the differences point in the same direction, but larger when trade-offs have to be judged.

Simulations - summary

The simulations show that the substitution of CFCs has the highest priority, and it is assessed that the pentane/isobutane solution is environmentally preferable because R134a is a greenhouse gas. The R134a solution makes it difficult to optimize more in the insulation, because further insulation results in increased global warming from the foaming agent, and at this point the solution thus stands in the way of further development.

The energy optimizations have the second highest priority, and both increased insulation and improved efficiency of the refrigeration system are significant.

It is also evident that a responsible disposal is very important, both for the CFC-related environmental impacts and for resource consumption for many of the metals.

Finally, the life span is significant, but not unambiguous. Increased life span can reduce certain resource consumptions and potential environmental impacts by 10-20%, but with the expected future technological developments for refrigerators, it can be argued that refrigerators with present levels of energy consumption should not have too long a life span, the use stage being as significant as it is.

The simulations supplement the LCA of the reference product, and together they form a basis for designating the environmental points of focus in the product, as described in the next section.

Where do potentials for improvement reside in the refrigerator?

The LCA has shown which impact potentials are weighted most highly, where in the product system they occur, and which components account for the greatest contributions. The simulation of changes in the product and its product system has shown which improvements can be obtained for various theoretical changes, and they were quantified relative to one another. The degree to which the improvements can actually be achieved depends, however, on which solutions can be found in practice.

The LCA and the simulations together show in which directions it can be of environmental interest to seek alternatives. Together with a certain advance knowledge of alternative solutions, they form a basis for desig-

Table 26.7 Examples of points of focus at the concept level

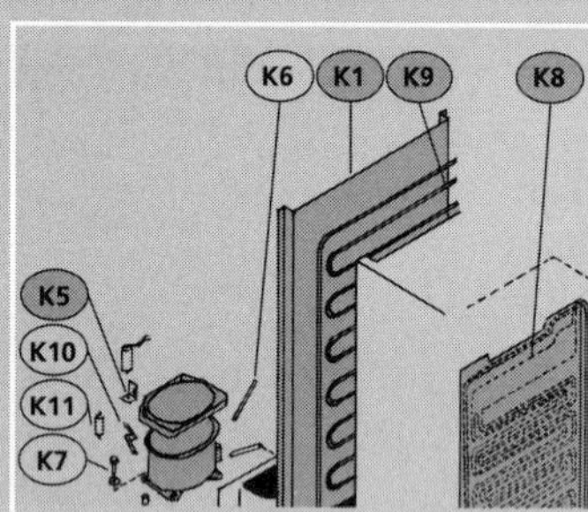

Refrigeration: **evaporator/condenser.** A very basic conceptual choice in the LER200 is the heat pump principle, i.e. evaporation/condensation of a refrigerant and use of a compressor, as known in virtually all refrigerators. As noted, the efficiency of this system is only 3-4% for the best refrigerators in 1997.
Alternatives: 1) When the outdoor temperature falls to a certain level, refrigeration can be provided by the outside air - thermostat-controlled and as a supplement to the existing system. The potential for savings in Denmark is about 50% of the energy consumption in the use stage.
2) Refrigeration can be provided geothermally. Soil temperature in Denmark seldom exceeds 8°C at a depth of 1 m, suggesting the possibility of using a shaft/goods elevator principle. The potential for savings is 100% of the energy consumption in the use stage, albeit at the cost of greater consumption of materials.

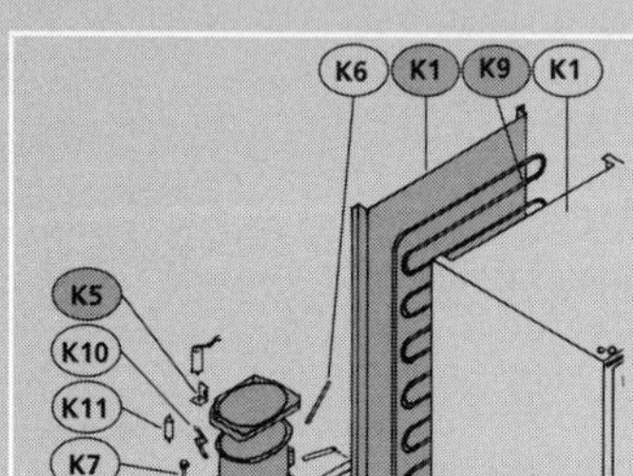

Heat exchange - compressor and condenser. Heat from the compressor and the condenser is removed by heat exchange with the air. The heat exchange usually occurs under less than optimum conditions because of poor air circulation about the refrigerator.
Alternative: the condenser and the compressor can be water-cooled and the water used as pre-heating for household water, increasing the efficiency of the refrigeration system and utilizing the waste heat in summer, when it is otherwise not used for indoor heating.

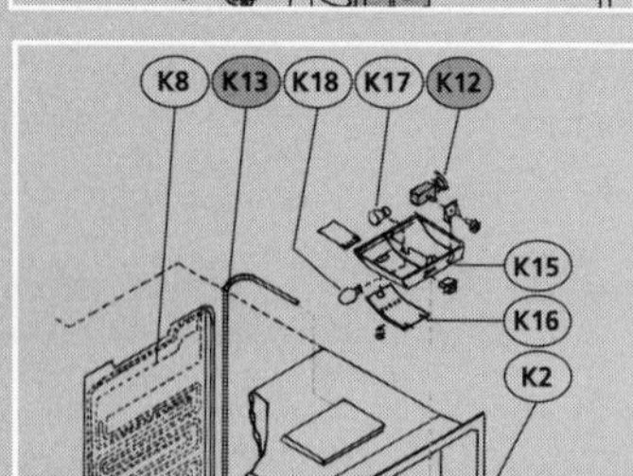

Control: relay control. The temperature in the LER200 is an on-off relay control of the compressor. This results in a large energy consumption each time the compressor starts. Energy consumption in the use stage can probably be reduced by 30% if the compressor runs continuously.
Alternative: the temperature can be regulated by continuous control of the compressor by a frequency converter, so that the compressor can run all the time.

Table 26.8 Examples of points of focus at the structure level

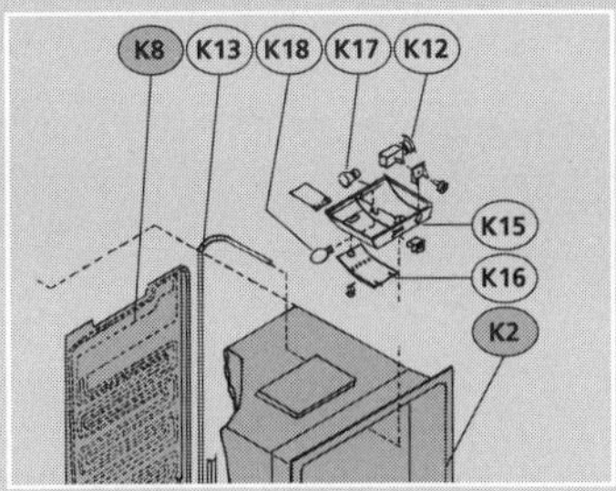

Evaporator: location and size. The evaporator in the LER200 is placed at the rear of the inner box to make cleaning easier. Its size is shown in the exploded view in Figure 26.2.

Alternatives: the evaporator can be placed on the inner side of the box as in many refrigerators, increasing the efficiency by at least 10%. A greater evaporator area will also be able to increase efficiency by at least 10%. An evaporator could be placed at the top of the refrigerator, which would also give a better temperature distribution inside the refrigerator: the refrigerated air will fall, resulting in some air circulation. It may be possible to increase the efficiency with a small fan (1 W) to circulate the air past the evaporator.

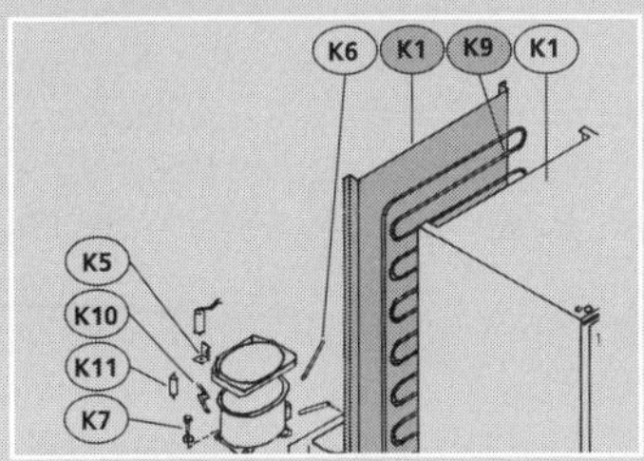

Condenser: installing on the rear panel. The condenser in the LER200 is installed on the rear panel with tape and thermokit to ensure heat conduction. The rear panel is relatively thick. Heat conduction is not optimal and the installation could be simpler.

Alternatives: Gram have worked on alternatives, but their results cannot be discussed for reasons of industrial secrecy.

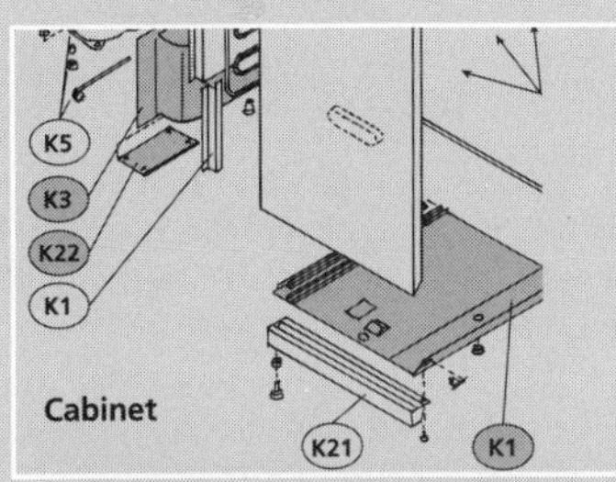

Compressor compartment and base panel. The base panel could be made in a way which will permit the compressor to be mounted directly on to it, possibly clipped into place. This would save some spot welding, and all of the galvanized steel base panel, which weighs 178 g, could be omitted. The result would be a saving in material, and the galvanizing process would be eliminated. Installation of the compressor would be easier, and removal and disposal would be simpler and faster because a crowbar could be used on the clips in place of a screw wrench. At the same time, the compressor compartment can be simplified so that there is more room for installation, and there will be savings in insulating materials without loss of insulating ability.

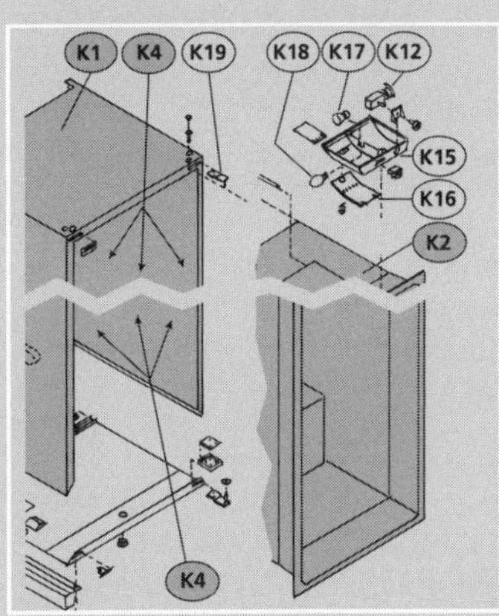

Sandwich construction. The strength and rigidity of the LER200 are based on a sandwich construction, where the inner box in ABS and the outer box in steel are "glued" together by the foam. This saves materials, but separation during disposal will be difficult, and if manual disassembly becomes an important disposal route in the future, the construction will be a problem.

Alternatives: none currently known.

Number of materials. An LER200 contains a relatively large number of materials. This also makes manual disassembly difficult, and it is possible to revise the construction and rationalize the choice of materials with a view to reducing their number.

Table 26.9 Examples of points of focus at the component level. Materials

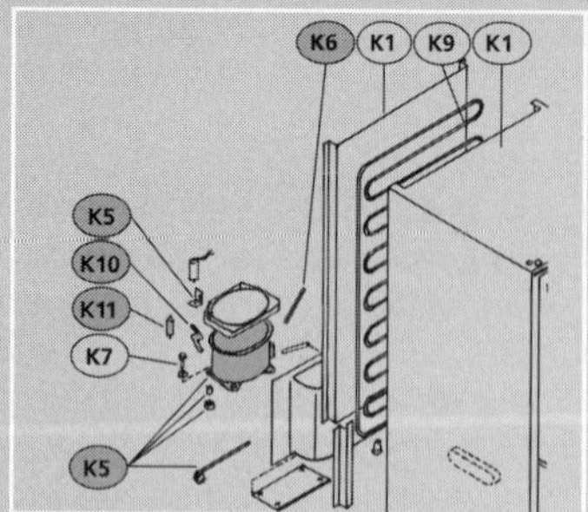

Copper. Copper is used in the compressor coils, in the leads, in various small pipes and in the drying filter.

Alternatives: the copper piping in the compressor could possibly be made of soft steel. New compressors which contain less copper are available. Recycled copper can be used, and it may be possible to use aluminium, which is not a scarce resource, in the leads.

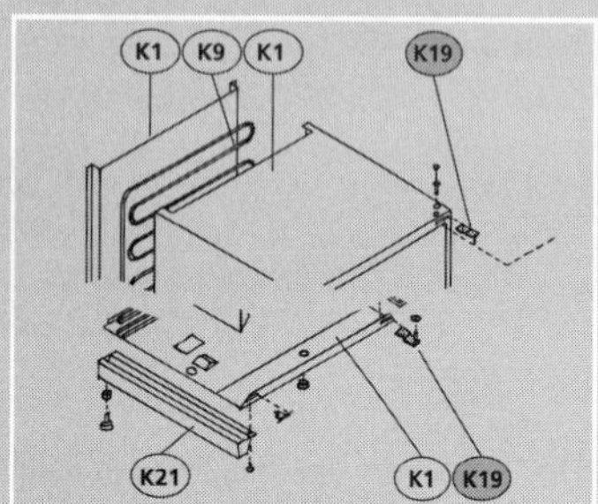

Stainless steel. The two door hinges are made of stainless steel. With current shredder technology, the hinges are removed magnetically in the disposal stage and sent to a recycling plant for ordinary steel. The nickel content is thus lost, and nickel is thus one of the most highly weighted resource consumptions in the refrigerator.

Alternatives: it is possible to avoid stainless steel. Painted hinges can be used instead.

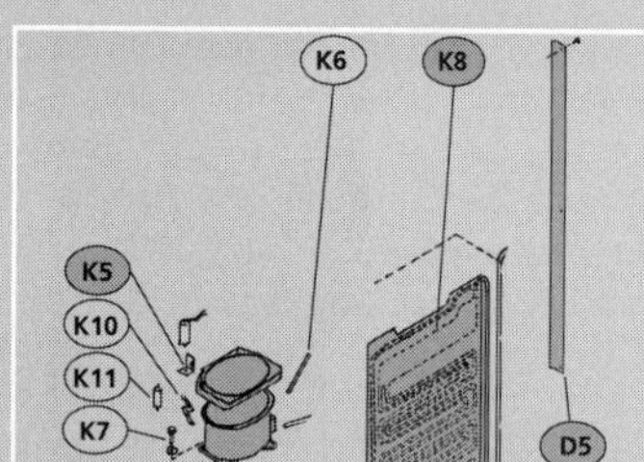

Primary aluminium. Primary aluminium is used in all three important aluminium components: evaporator, door handle and compressor. Production of primary aluminium consumes much more energy than recycling of aluminium. Energy consumption in the material stage would thus be reduced if it were possible to use recycled aluminium.

Alternatives: it may be possible to use a painted steel handle or a painted handle in recycled aluminium.

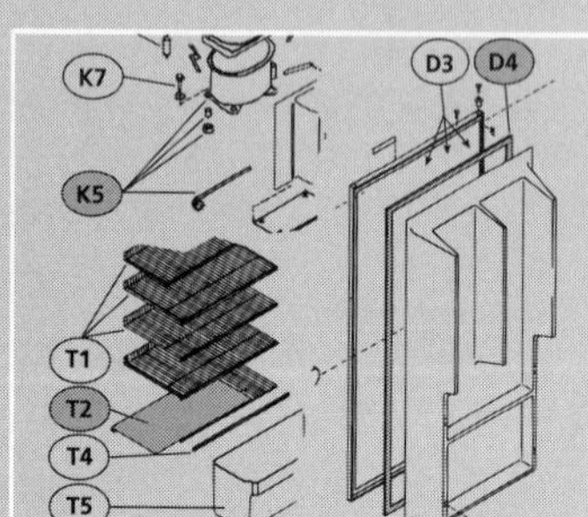

PVC. Use of plasticized PVC for the sealing frame and as insulation on the leads is an environmental problem because the plasticizers are believed to damage the reproductive system. As the PVC ends in landfills (1997), the content of plasticizers is a real problem.

Alternatives: no immediate alternatives to plasticized PVC for the sealing frame are obvious. The PVC cover panel for the bottom shelf can, however, be replaced immediately by a different type of plastic or by glass. No alternatives are currently known for the leads.

ABS. Production and thermal processing of ABS carries a risk of exposure of workers to styrene and acrylonitrile, both of which are suspected carcinogens.

Alternatives: it is difficult to see equally good alternatives. The temperature in the vacuum forming processes at Gram is kept in the range 180-200°C, and it is believed that carcinogenic substances are not released at these temperatures.

Table 26.10 Examples of points of focus at the component level. Processes

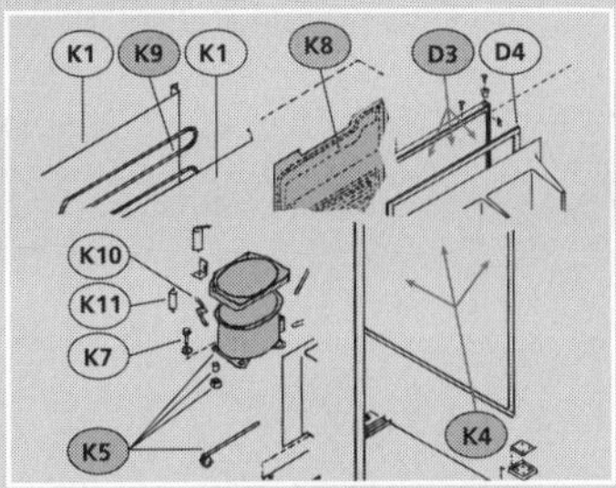

Refrigerant and foaming agent. The consumption of CFCs in an LER200 is the most significant point of focus. It should also be noted that CFCs are no longer used in Gram's refrigerators. The potentials for improvement by substituting CFCs are reviewed.
Alternatives: the primary focus is on two alternatives discussed previously: R134a and pentane/isobutane. The R134a solution has been implemented and is not leading to any problems in construction. For the pentane/isobutane solution, there has been discussion of whether the risk of explosion is significant in the manufacturing, transport, use and disposal stages. Tests and practical use have provided some experience. All indications are that the risk is very small.

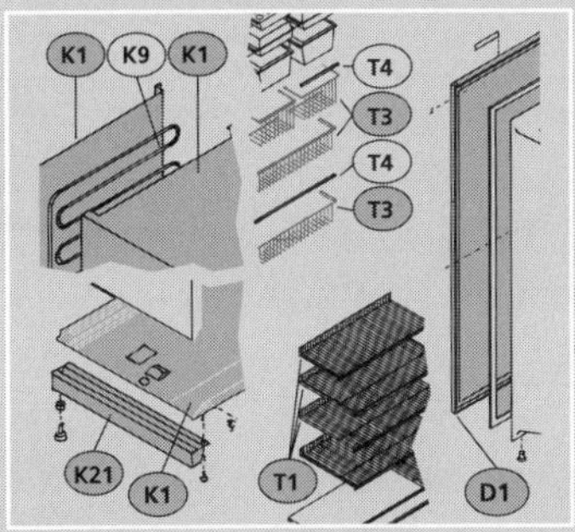

Epoxy powder. Gram have replaced all solvent-based painting with powder coating, and have thus removed solvents from the factory. Painting was previously done manually with solvent-based paints. Replacing these with automatic powder coating has been a major improvement for the environment and the working environment at the company. However, powder coating does have some impact, both in manufacturing and in the production of epoxy powder. Epoxy powder is allergenic, and workers are exposed to it. One of the ingredients of epoxy powder is bisphenol A, which has an oestrogenic effect and therewith the potential to damage the reproductive system.
Alternatives: no better alternatives are known at present (1997).

Noise and impacts on the musculoskeletal apparatus

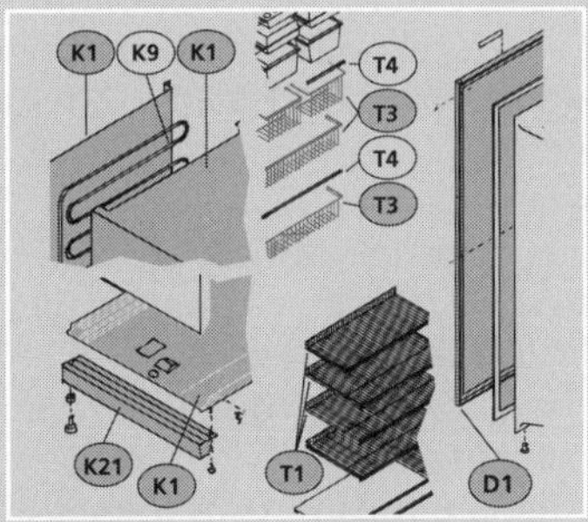

Punching, bending and drilling processes. The metals are the most significant components with respect to exposure of workers to noise. The noise is due primarily to the punching and bending processes. There is a very high noise level in the drilling processes for the door lining.
Alternatives: it is possible to sound-proof the processes or possibly to substitute a different type of material.

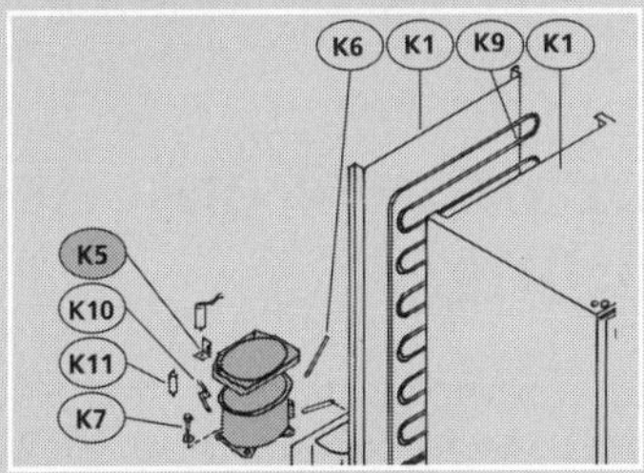

Manual installation: leads, condenser and compressor. The manual installation of these items is somewhat difficult and it results in an impact on the working environment in the form of monotonous repetitive work.
Alternatives: alternative ways of installing all three items have been found and implemented. See next section.

Table 26.11 Examples of points of focus at the component level. Others

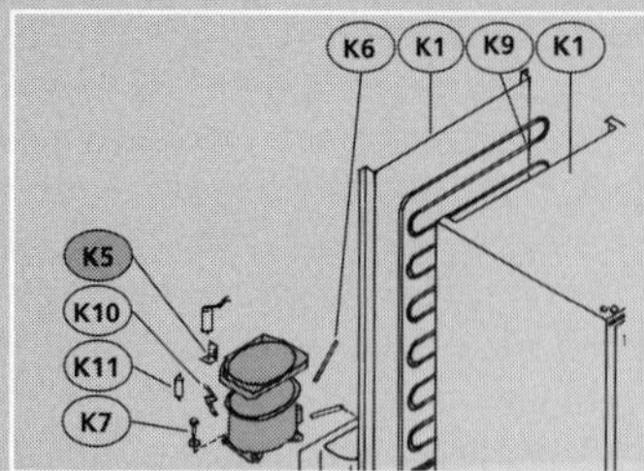

Compressor. The motor in the compressor used in the LER200 had an efficiency of ca. 0.6.
Alternatives: a compressor with a permanent magnet motor has an efficiency of ca. 0.8. Such a compressor is available. Potential savings of ca. 25% in energy consumption in the use stage are involved.

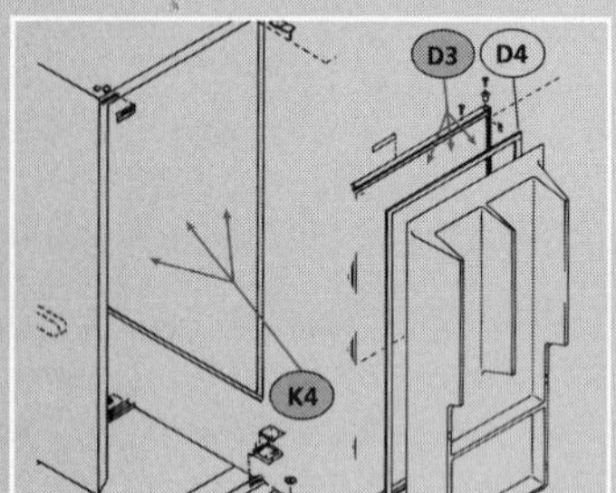

Insulation. The insulation in the LER200 is significantly better than that used in other refrigerators. This also means that the 200 litres internal volume cannot be maintained without the external dimensions becoming too large for building in a 55 × 55 cm kitchen refrigerator. This is one of the most significant impediments to sales of the LER200.
Alternatives: the insulation thickness can be increased, but this will lead to problems of building in while current kitchen architecture dictates the refrigerator's external dimensions. An alternative is to use vacuum panels, which have the potential to increase insulating ability by a factor of two.

nating environmental points of focus in the product, i.e. designating where in the product the options for environmental improvements lie.

Where the LCA of the LER200 refrigerator is an assessment of an individual solution to providing the functional unit, the environmental diagnosis admits the possibility of looking at entirely different concepts or other structural solutions than those represented by the LER200. Alternative solutions and points of focus are therefore discussed at the levels of concept, structure and component, as shown in Tables 26.7-26.11.

In the overview given in the tables, the options are not ranked and are not assessed in either technical or commercial terms. For some, the alternatives to the selected solution in the LER200 are well known, while for others they are less certain. The tables do not provide a complete picture, but they do include some of the most significant options. The alternatives are not actual design considerations, but they are meant as examples and as inspiration. Section 26.4 discusses some of the items which Gram has already implemented and which appear as future options.

Where can potentials for improvement be found in the surroundings?

The environmental problems and potentials for improvement in the product system of an LER200 do not, however, reside only in the LER200 itself. The surroundings, that is the systems into which the LER200 enters, carry part of the responsibility. The greatest impediments set by the surroundings to achieving improvements in the LER200 product system are reviewed in Table 26.12.

Table 26.12 Examples of points of focus in the surroundings

Shredding. As noted, CFC-containing refrigerators are still shredded. At present (1996), there are no plans to store there refrigerators until the CFC content can be collected and destroyed. Until there is a mandatory system of facilities for processing CFCs and until these are in full operation, CFC-containing refrigerators will therefore most probably continue to be shredded with consequent release of CFCs.

Alternatives: refrigerators could be stored. With an annual disposal of 200,000-250,000 refrigerators in Denmark, an area averaging 10 m × 10 m per annum per municipality will be required if the refrigerators are stacked in two layers. Such areas are available, and no additional expense would be involved.

The municipalities must in any case establish disposal facilities, and the capacity does not have to be greater to handle a backlog when the time comes. The facilities will merely have to be operated for a few more years, and when all CFC-containing refrigerators have been disposed of, the facilities will be removed. The total additional expense in collecting the refrigerators for storage is therefore only the additional years of operation required. Considering what industry invests in the search for alternatives and the investments otherwise made in environmental plans of action, this is an area where a major environmental return on investment can be achieved.

Shredding. Because the waste from shredders is too polluted to be accepted by waste incineration plants, plastic and foam are currently dumped after shredding together with their contents of both plasticizers and residual CFCs.

Alternatives: 1) Manual disassembly could sort out plastic and foam for incineration. Very large potentials for improvement are involved for both CFC emissions and quantities of waste. 2) Better planning of the shredding, e.g. by runs with refrigerators only, will probably result in sufficiently pure fractions for acceptance of the waste by incineration plants. The potentials for improvement will also be very large for both CFC emissions and quantities of waste. Manual disassembly will result in the greatest improvements as the CFC emissions in the shredder will be avoided.

The electrical system. Most electrical systems, including Danish systems, are based on a few large power plants. The plants are often some distance from populated areas, so that waste heat from electricity generation is difficult to utilize for household heating. The total energy efficiency of the electrical system is therefore relatively low, less than 60% in Denmark. At the same time, coal is used as fuel, and the contribution to global warming from coal is large, e.g. almost double that of natural gas per MJ. Large potentials for improvement are to be found in increasing this efficiency, about 70% reduction in primary energy consumption in the use stage, and the contribution to global warming from the use stage can be eliminated entirely by using renewable energy.

Alternatives: 1) Decentralized co-generation power plants located where the heat can be utilized. 2) Renewable energy sources as fuel, e.g. pyrolysis of chips with subsequent production of heating and power.

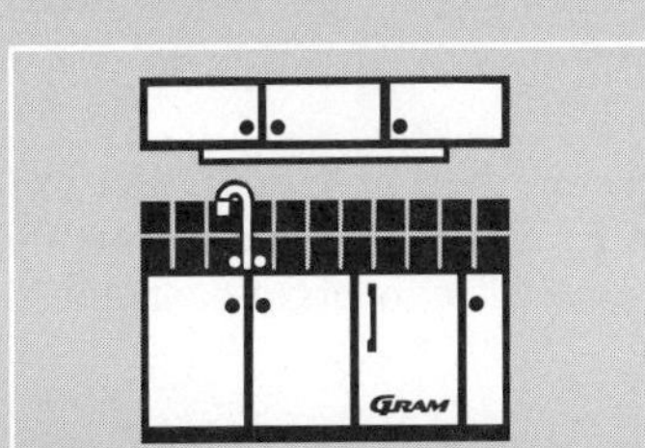

Kitchen architecture. Modular kitchens with standard dimensions of 55 cm × 55 cm for refrigerators dominate the market, and it is usual to build the refrigerator into a refrigerator module. This sets obvious limits to the thickness of the insulation.

Alternatives: it may be possible to find solutions where larger modules for refrigerators are adapted to the general architecture. International standards for these should be adopted.

26.3. Environmental specification for refrigerators

The LCA and the environmental diagnosis have provided part of the basis for being able to specify environmental objectives and targets for refrigerators and for product development in the longer term. This basis is now supplemented by a commercial assessment of the environmental points of focus which have been designated.

The competitive framework

An analysis was carried out of competing products' environmental characteristics and of customers' perception and priorities with respect to the environment. These factors have helped to set the requirements which must be met by the refrigerators of the future. Designation of the environmental points of focus for development of refrigerators occurs both on the basis of the environmental ranking provided by the LCA and the diagnosis, and on the basis of the competitive framework defined by the customers and the competitors.

The refrigerator's content of refrigerant and blowing agent was (1995) a very important parameter of competition, and the competition among manufacturers to find alternatives to CFCs has been very strong. As early as September 1993, Gram replaced CFCs in the foam with R134a, and from January 1994, R134a was also used as refrigerant. In the German market especially, customer focus on ozone depletion has been great, but global warming has also attracted the attention of German customers. The choice between R134a and other solutions such as pentane/isobutane has therefore become an important parameter of competition in Germany, and Gram's biggest competitor in Germany markets mainly pentane/isobutane models.

In Denmark, customer focus on CFCs is not nearly as great, but energy consumption has become a decisive parameter. From 1 January 1995, an energy labelling system was introduced in the EU, where refrigerators are classified into categories from A to G according to relative energy consumption (Danish Energy Agency, 1994). "A" represents the lowest energy consumption and "G" the highest. The LER200 is in class A, and for the LER200 this is one of the most decisive sales parameters.

It is difficult to designate noteworthy differences in the competing products for parameters other than those noted here. The next parameter to appear in the future is expected to be the option of disassembly when alternatives to shredder technology are generally established.

Long-term objectives

In general, Gram will work towards increasing the implementation of life cycle assessments in product development. As a consequence of the points of focus designated by LCAs to date, work will concentrate on the following:

- Energy efficiency: energy consumption is also expected to be an important parameter of competition in the future, and focus will be directed towards both the refrigeration system's performance and the insulation.

- Working environment: the impact of production processes on the working environment will continue to be in focus.
- Choice of materials and disposal: recognition of how much choice and disposal of materials means in terms of resources and environment relative to the manufacturing processes' own impact will be incorporated by the manufacturers, i.e. choice of materials and the material's fate in the product system will be central.
- Scarce resources: in the longer term, work will be directed towards reducing copper consumption and, if possible, replacing copper entirely with other materials. Alternatively, copper will be so placed in the construction that it is very simple to separate it from the other materials in the product. Attempts will also be made to avoid use of stainless steel, or alternatively to place and label it in a manner which ensures its recycling to a stainless steel works.
- Plastic: attempts will be made to replace softened PVC with other types of materials. The circumstances surrounding production of ABS will be investigated further, and potential problems will be identified.
- Chemicals: focus will be on the substitution of R134a because it is a greenhouse gas, and alternatives will be introduced when it is believed to be responsible to do so relative to other problems to which the alternatives may lead.

Product specifications for refrigerators

The experiences from the LCA, the environmental diagnosis and environmental considerations concerning competitors and the market's perception of the environment mean that the product specifications for a refrigerator today (1997) include environmental aspects as outlined in Table 26.13.

Table 26.13. Example of the environmental section of the product specifications for refrigerators in 1997

Environmental product specifications for refrigerators
Materials
Copper: choose a compressor containing the least amount of copper. If possible, avoid the use of copper in any of the other parts.
Stainless steel: should only be used if strictly necessary. If used one is to ensure that the stainless steel is reused by labelling and product design.
Primary aluminium: should be substituted with recycled aluminium or steel, if possible.
Softened PVC: avoid using softened PVC unless it is strictly necessary.
Energy
The product must apply to the demands of category A according to EUs energy labelling.
Chemicals
CFCs must not be used.

26.4. Environmental design of new refrigerators

Since the start of the EDIP project in 1991, Gram has worked with the products' environmental characteristics on the basis of the life cycle concept. The experiences presented in this example originated in this work over the years during which the project has continued. Many of the findings have been included in product development and are incorporated into new products which are underway, including new variants of the LER200.

Product development at Gram

New ideas for product development come from many different sources: customers, sales and marketing, designers, employees in production, and the environmental department. The ideas are analysed, and if they are judged to be viable, a specification is worked out. At this point this comprises a brief description of the confirmed market demand or desired product, with an interim assessment of the expected price level, anticipated sales, sales areas, etc. The most important requirements with respect to e.g. equipment, dimensions, capacity, operation, variants and necessary approvals are also established. If there are environmental requirements which are to be fulfilled, an indication of the possibility of compliance with such requirements is obtained at this point. There may be requirements concerning energy consumption, the use of particular materials, or requirements with respect to avoiding particular materials.

The design department then makes a product assessment. This covers brief proposals for solutions and contains information which is necessary to decide whether the developmental work is to be commenced. The information includes time and investments required, etc. On this basis, a decision is made on whether product development is to be commenced. During product development, the other departments in the organization, such as manufacture, tool technology, purchasing and environment, are involved. During development, the solutions which have been prepared are continually compared with specifications. Product development finishes with a final product documentation, both technical and commercial. Marketing information for the product can be prepared on the basis of the product documentation. The object of the marketing information is to inform employees concerned of the new product, and on this basis the internal and external information materials can be prepared. Everything is now ready for starting production of the new product, and operational plans, production plans, etc. are now prepared. The sequence can be sketched roughly as shown in Figure 26.16.

Figure 26.16. Product development at Gram

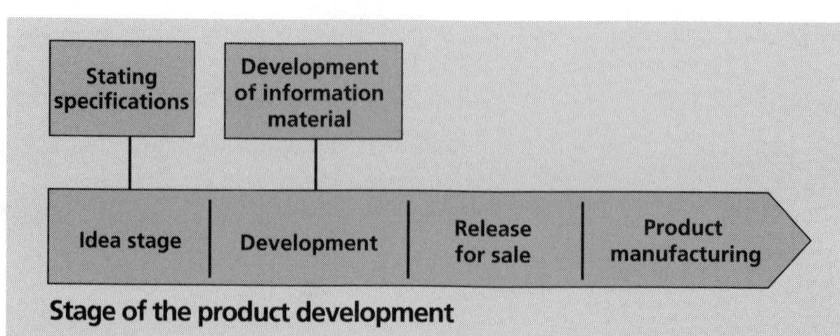

Examples of environmental considerations in new refrigerators

During the EDIP project, Gram carried out a large number of product development projects which have built environmental considerations into new refrigerators. The results were broadly implemented in Gram's product range. The projects have concerned all levels: conceptual changes, structural changes and changes in individual components, and they have had an influence on savings in materials, energy savings, chemical substitutions and impacts on the working environment. An icon beside the description of each design project shows the areas on which the project has had an influence, and for each project there is a brief quantification of the extent of savings in materials, energy, chemicals or others. To see how much these savings mean in environmental terms in the whole, they can be compared with the presentations in the environmental diagnosis in Figure 26.15.

Project 1. Control system

An electronic control, which has optimized the refrigeration system, has been introduced on all models with dimensions 60 cm × 60 cm. The control ensures more accurate maintenance of the temperature set by the user in the refrigerator. It also ensures a less frequent but better-controlled defrosting.

	M	E	C	O
Concept		●		
Structure				
Component				

Energy: This has led to total energy savings of ca. 5-10%.

Project 2. Placing of the evaporator

To reduce energy consumption, Gram has stopped producing refrigerators with evaporators enclosed in foam, i.e. with the evaporator placed at the rear side of the inner box in ABS, in a very large proportion of refrigerators. In these models the evaporator has been moved into the refrigerator itself.

	M	E	C	O
Concept				
Structure		●		
Component				

Energy: This has increased energy efficiency by ca. 10%.

Project 3. Condenser

A new construction of condenser and rear panel has been developed which has resulted in both savings in materials and increased heat conduction, and thus increased the efficiency of the condenser. The thickness of the rear panel has been reduced from 1.5 to 0.6 mm, and heat conduction has increased.

	M	E	C	O
Concept				
Structure	●	●		
Component				

Materials: Savings in materials of ca. 5 kg steel were achieved, corresponding to 15% of the steel in the refrigerator.

Energy: Better conduction and therewith efficiency were achieved, which have not yet been quantified.

Project 4. Compressor compartment

	M	E	C	O
Concept				
Structure	●			●
Component				

A new compressor compartment was developed and introduced. The compartment is now broader and has provided more room for installing the compressor and, at the same time, a significant reduction in the amount of insulating material without loss of insulation value.

Materials: Savings of 1 litre of foam corresponding to ca. 40 g for insulation of the compressor compartment. This corresponds to ca. 0.5% of the foam in the refrigerator.
Other: More room, and thus better ergonomic conditions during installation.

Project 5. Base plate

	M	E	C	O
Concept				
Structure	●	●	●	
Component				

A new base plate was constructed which was made in one piece instead of four. This resulted in a saving in materials and chemicals, namely the entire galvanized base plate in steel, a few pins, and energy (32 spot welds were avoided). At the same time, installation and removal of the compressor were made easier, as the compressor is no longer screwed, but clipped into position. Attachment with clips also makes it easier in the disposal stage, as disposal does not require work with different screwdrivers, but merely a crowbar.

Materials: Savings of 178 g steel, corresponding to ca. 0.5% of the steel in the refrigerator. Easier disassembly can result in a higher degree of recycling of materials, but this cannot be quantified.
Energy: Process energy corresponding to 32 spot welds was saved. This accounts for less than 1% of the refrigerator's total energy consumption in the product system.
Chemicals: Galvanizing and associated chemicals were avoided.

Project 6. Placing of the light panel

	M	E	C	O
Concept				
Structure	●			●
Component				

The positioning of the light panel was changed. In 60 cm × 60 cm refrigerators, the panel was integrated into the top of the refrigerator. This resulted in minor savings in materials and easier installation; the leads were ducted externally and not, as before, inside the foam. This means that it is now easier to access the leads in manual dismantling of the refrigerator. It is necessary only to remove the cover panel, providing free access to the leads. There is, however, still one lead inside the unit, namely the thermostat's sensor lead, but it is not enclosed in foam.

Materials: A minor saving, unquantified.
Others: Easier assembly and thus better ergonomic conditions.

Project 7. Alternatives to CFCs

	M	E	C	O
Concept				
Structure				
Component			●	

CFCs were replaced by R134a from 1 January 1994 as both refrigerant and blowing agent in all models.

Chemicals: CFCs were replaced 100%, thus removing the entire contribution to ozone depletion together with 75% of the potential for global warming.

Project 8. New foaming agent

Experiments were carried out with alternative foaming agents to replace the R134a foam. A decision was made to introduce alternatives when other problems which may be associated with these have been solved.

	M	E	C	O
Concept				
Structure				
Component			●	

Chemicals: A future 100% substitution of R134a will entirely eliminate the chemical contribution to the potential for global warming.

Project 9. Compressor

Energy-optimized compressors were introduced in many models, and at the same time the evaporators and the condensers were optimized. The new compressors have motors with better efficiency and have so-called semi-hermetic injection of refrigerant.

	M	E	C	O
Concept				
Structure				
Component		●		

Energy: Energy consumption was reduced by a total of 10-15%.

Project 10. Hinges

So-called "tabletops" were developed for most refrigerators. A tabletop is a frame in moulded plastic covered by a painted steel panel and mounted above the refrigerator. In this design, one stainless hinge was removed and replaced by a single pin in sintered steel. The last stainless hinge at the bottom will be replaced by a steel fitting painted in white.

	M	E	C	O
Concept				
Structure				
Component	●		●	

Materials: Stainless steel was eliminated, and with it, nickel consumption.

Chemicals: A small extra consumption of powder coating. The advantage of eliminating nickel is considered to be considerably greater.

It should be noted that with respect to the improvements listed in energy efficiency in the various projects, the percentage improvements are not simply additive.

Gram's experience with environmental considerations in product development

It is Gram's conclusion that it is possible to include environmental considerations during the design of refrigerators/freezers.

The individual designers do not necessarily require a broad environmental knowledge or a computer to support their conclusions in product development. On the contrary, Gram's experience is that it is sensible to have one person responsible for the LCAs, namely the environmental specialist. On the basis of LCAs undertaken by the environmental specialist, it is possible to single out some individual guidelines which help the designers to think and to construct in more environmentally friendly terms.

It has proved not to be more complicated to build environmental aspects into products than to build in the economic or other aspects with which Gram is used to working. The biggest task is to designate the correct environmental points of focus in the product, and for this purpose the life cycle assessment is an outstanding tool. When the points of focus have been designated, the designers can handle the environmental considerations in precisely the same way as all other considerations.

The EDIP LCA method provides a good and instructive way of presenting the product's environmental impacts. Throughout the project period, data and figures were presented for resources, environment and working environment for a large number of internal and external persons, and the impression is only positive: most people quickly understand the message.

An internal course was held for the safety organization, and internal information materials were prepared which are also used by the Dutch subsidiary. Environmental documentation for a refrigerator was prepared in connection with a tender to a housing association, and this was a contributing reason for an order for 2,300 refrigerators.

In a complaint about an environmental marketing campaign, the results of the EDIP project's LCAs were used as documentation for the consumer ombudsman, and as a result of the documentation, the case was decided in favour of Gram.

27. Bang & Olufsen: Televisions

Rikke Nedermark, Bang & Olufsen
Marianne Wesnæs, Institute for Product Development
Henrik Wenzel, Institute for Product Development

Bang & Olufsen manufactures high technology audio and video equipment with a high emphasis on design and quality. The company is situated in the western part of Denmark, only 45 km from the North Sea Coast. The company has approx. 2,500 employees

Page

27.1. LCA of a television - the Beovision LX 5500

Bang & Olufsen's choice of reference product

The Beovision LX 5500 is a 28″ colour television with two bass and two treble loudspeakers. The television has teletext and remote control, and up to 50 programmes can be selected.

Bang & Olufsen have selected the Beovision LX 5500 as reference product because:

- Bang & Olufsen assume that a TV has a greater environmental impact than any other Bang & Olufsen product.
- Among Bang & Olufsen's products, the greatest environmental focus is on televisions.
- The Beovision LX 5500 represents a series of products which have been Bang & Olufsen's flagship for many years, and it was therefore desired to obtain more detailed information on this particular product.
- The Beovision LX 5500 is a complex product, and by testing the LCA method on this product, it was expected that experiences gained with the method could be transferred to the development of all types of new products in the company.

Figure 27.1. The reference product, the Beovision LX 5500

The television's functional unit

The primary function of a television set is reception of TV programmes. This function can be defined both quantitatively and qualitatively, and by the period of time in which the service is provided.

The screen size of 28″ is a quantitative description of the Beovision LX 5500. The operating time for the television's total life was chosen as a measure of the time for which the service is provided. The operating time is summarized as the number of hours for which a television is operated on average in normal use. The operating time will often be greater than the "viewing time", as some people use the television as background entertainment, and others simply forget to turn it off. The operating time for the Beovision LX 5500 was computed on the basis of Bang & Olufsen's dimensioning criterion: 6 hours a day with a life of at least 10 years, giving a total of 21,900 hours of operation. The functional unit for the Beovision LX 5500 is therefore quantified as:

Reception of TV programmes for 6 hours a day with 18 hours standby for 10 years for a 28″ colour television.

When new solutions in product development are compared, apart from having the same quantitative functional unit, attention must also be paid to the television's quality and secondary functions such as picture quality (colours and sharpness), sound quality, number of channels, and whether, for example, there are remote control, teletext, decoders or the option of connecting external loudspeakers.

Scope definition of the product system

The television consists principally of the picture tube, electronics, front panel, casing, rear cover and loudspeakers. The stand and the remote control are not included.

A Beovision LX 5500 weighs a total of 42.7 kg. The picture tube alone weighs 24.5 kg and thus comprises more than half of the total weight. The materials can be broadly divided into:

- 24.5 kg picture tube, of which
  - 21.8 kg is glass
  - 1.5 kg is iron
  - 0.5 kg is copper
  - 0.7 kg is other materials
- 2.9 kg contrast screen, of which
  - 2.75 kg is glass
  - 0.15 kg is iron
- 0.96 kg loudspeakers, of which
  - 0.43 kg is iron
  - 0.53 kg is magnets
- 11.0 kg plastic, used especially for the casing, rear cover, front panel and loudspeaker boxes
- 0.23 kg aluminium used for edging
- 3.15 kg electronics, of which
  - 2.0 kg is plastic
  - 0.5 kg is iron
  - 0.42 kg is copper
  - 0.10 kg is aluminium
  - 0.12 kg is lead

Division of the television into components - parts list and exploded drawing

The LCA started with an inventory of the television's components. The product systems of the components were included from extraction of raw materials to the product manufacturing processes. Use and disposal of the television and transport were also included.

The components are given in the parts list in Table 27.1 and the exploded drawing in Figure 27.2. The components are given item numbers on the exploded drawing. The corresponding numbers are found in Table 27.1 on the following pages.

Table 27.1 is a simplified parts list for the Beovision LX 5500. The weight and material content of the components are shown together with the manufacturing processes and ancillary substances used during production of the components. A "✓" or a "÷" in Table 27.1 indicates which processes, materials and ancillary substances are included in the LCA and which are not included.

The LCA includes most parts of the television. For the great majority of components, the manufacturing processes for components and the materials' product system back to extraction of raw materials are included. It has not been possible to include ancillary substances such as paint, glue and soap back to extraction of raw materials.

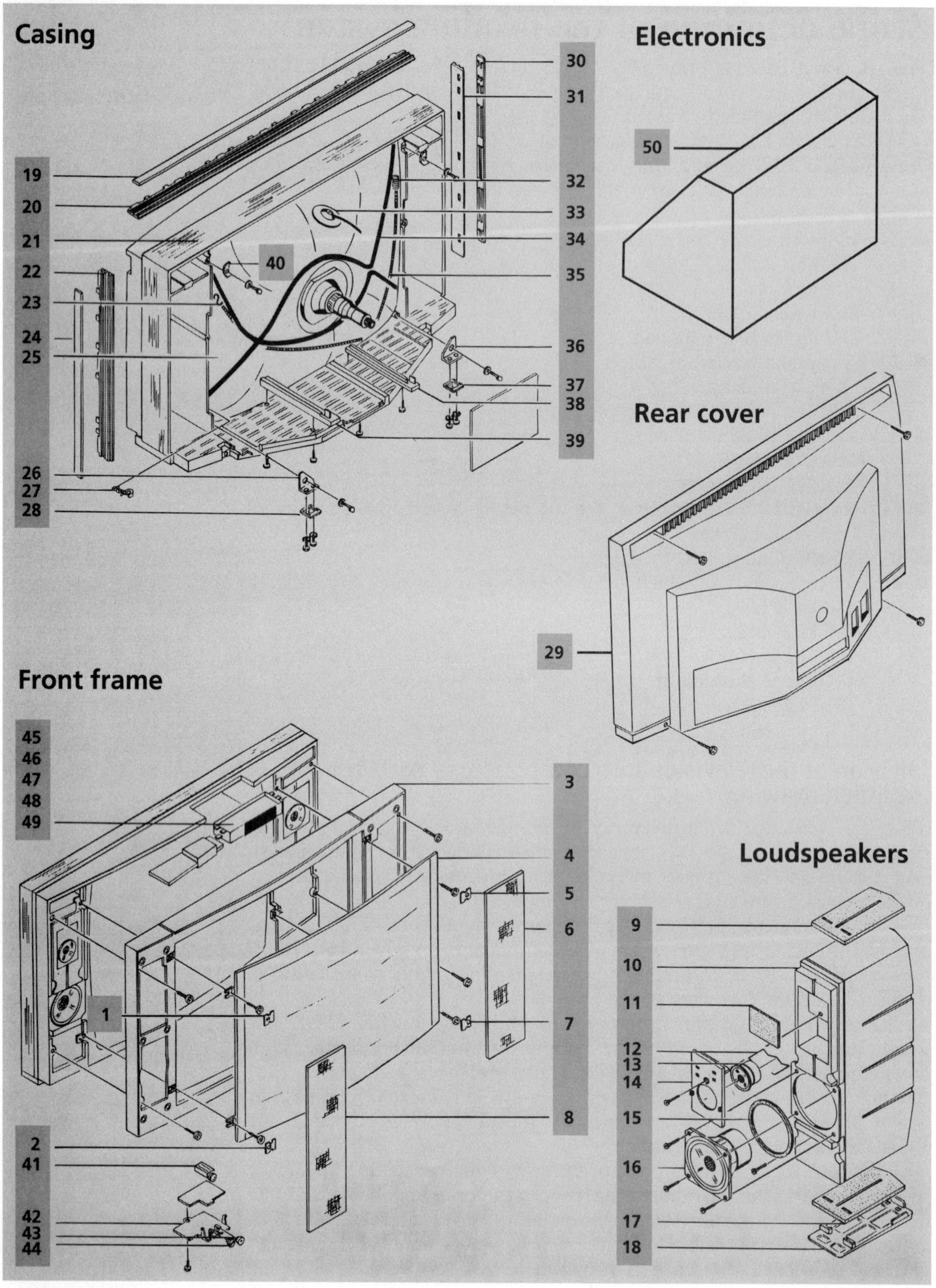

Figure 27.2. Exploded drawing of a Beovision LX 5500 television, showing front panel casing, rear cover, loudspeakers and electronics

Pos.	Num-ber	Item	Material	Total weight (g)	⏚	Manufacturing processes	*Incl. in model	Ancillary substances	⏚
List of components for casing									
19,24, 30	1+2	Top profile and 2 side profiles	Aluminium	170.32	✓	Cut, CNC cut, grind, degrease, polish, anodize, dry, placed, roll-coating, inspecting, placed, serigraphy, packing	✓✓✓ ✓✓ ✓✓ ✓✓ ✓✓ ✓✓	Oil, grinding paste, shop primer, paint thinner, cleanser thinner, dye	÷÷ ÷ ÷ ÷ ÷
20,22, 31	1+2	Holders for top profile and side profiles	ABS plastic	52 83.25	✓ ✓	Mould, clean, packing	✓✓✓		
21	1	Cabinet	PS plastic, foamed	5,260	✓	Mould, paint	✓✓		
23	1	Clamp	Phosphor bronze	1	÷	Subcontractor	÷		
25 a	1	Picture tube	Glass Iron	21,900 1,500	✓	Subcontractor	✓		
25 b		Deflecting coil	Copper	400					
26,36	1+1	Holder, picture tube	Band iron	58.64	✓	Punching, bending, pressing thread, alkaline degreasing	✓✓ ✓✓	Lubricating oil, ethanol, cutting oil	÷ ÷÷
27	8	Holder, degaussing coil	POM-plastic, dye	1.64	÷	Injection moulding Subcontractor	✓		
28,37	2	Spacer, picture tube	Zinc	33	✓	Zinc casting, Subcontr.	✓		
29	1	Rear cover	PS-plastic	2,100	✓	Mould, paint. Subcontr.	✓✓		
32	1	Spring	Spring steel	5.5	✓	Subcontractor	÷		
33,34	1+1	EHT cable and degaussing coil	Copper and PVC plastic	Not incl.	÷ ÷	Subcontractor	÷		
38,39	1+1	Holder, rail	PC plastic, 10% glass	33.52	✓	Mould	✓		
40	2	Spacer, picture tube	PPO/SB-plastic, 20% glass	7.1	✓	Mould	✓		
50	1	Electronics	Cu, Al, Fe, plastic, Ni, Pb, Sn, Ag, Au, Be, Cd, Cr, Pd, ceramics	2,646	✓✓✓✓✓ ÷÷÷÷÷ ÷÷÷÷	B&O's electronics factory in Skive, DK, and sub- contractors	✓ ÷		
List of components for front frame									
1,2, 5,7	4	Cap	ABS plastic	0.8	✓	Moulding	✓		
3	1	Front frame	ABS plastic	670	✓	Moulding, cleaning, painting Subcontractor	✓✓✓	Paint thinner, topcoat	÷ ÷
4	1	Rubber string	Silicon rubber	17	÷	Subcontractor	÷		
4	1	Glass plate 28″	Band iron Glass plate Label	76.9 2,756	✓ ✓ ✓	Cutting, punching, bending, deburring, cleaning, painting, inspecting, grinding Mounting glass plate, mounting tape, new placem. Printing	✓✓✓ ✓✓✓ ✓✓ ✓ ✓✓ ✓	Polyester lacquer, acrylic glue, activator, isopropanol	÷ ÷ ÷ ÷
6,8	2	Loudspeaker panel	PS plastic, impact strengthened Loudspeaker fabric	148	✓ ÷	PS moulding, cleaning, packing Mounting fabric	✓✓ ✓ ✓		
41	1	Jack plug		Not incl.	÷	Subcontractor	÷		
42	1	House for stand-by	ABS plastic, impact strengthened	21.3	✓	Moulding, inspecting, packing	✓✓✓		
43,44	1+1	Button for mains switch + button pstep	PC-ABS plastic mixture	1.04	✓	Moulding, inspecting, cutting, printing, cutting	✓✓ ✓✓✓	Dye	÷
45,46	1+1	Shield, top + shield bottom	Band iron	4.46 3.72	✓ ✓	Punching	✓	Lubricating oil	÷
47	1	Holder, diode	PBT plastic	1.44	÷	Moulding	✓		
48	1	Lens, IR	Acrylic,	1.44	✓	Moulding, cutting, inspecting	✓✓✓		
49	1	Housing for IR	PC plastic	8.73	✓	Moulding, cleaning, packing	✓✓✓		

Table 27.1. Simplified parts list for the LX 5500 television

Pos.	Number	Item	Material	Total weight (g)	⏚	Manufacturing processes	*Incl. in model	Ancillary substances	⏚
List of components for loudspeakers									
9,17	2	Pressure pad	Polyether foam	45	÷	Water jet cutting Subcontractor	÷		
10 a 10 b	2	Loudspeaker cabinet a: rear part b: front part	PS plastic, impact str. Polyester	1,894 106	✓ ✓ ✓	Moulding, cleaning, packing	✓✓✓	Slipping agent, detergent	÷ ÷
11	2	Pressure pad	PUR foam	1	÷	Subcontractor	÷		
12	2	Spring	Spring steel	1	✓	Subcontractor	÷		
13	2	Treble speaker			÷	Subcontractor	÷		
16	2+2	Bass speaker	Iron Magnet	215 265	✓ ÷	Subcontractor	÷		
14	2	Loudspeaker suspension	PS plastic, impact str.	20	✓	Moulding, subcontractor	÷		
15	2	Gasket	Not inventoried	0.26	÷	Subcontractor	÷		
18	2	Plastic foot	ABS plastic	16.5	✓	Moulding, subcontractor	✓		

Notes

⏚ The column indicates whether or not the product system of the material or the ancillary substance is included all the way back to the extraction of raw materials.

✓ Means that it is included.

÷ Means that it is not included.

* In the column "Incl. in model" is likewise indicated if the manufacturing processes itself is included or not. ✓ means that energy consumption and use of ancillary substances is measured, that emissions and waste are either measured or calculated and that impacts on the working environment are measured or assessed. The number of ✓'s indicates how many processes are included.

Table 27.1 (continued).

System boundaries for the television's product system - what is included?

The principles behind setting the system boundaries are given in more detail in Parts II and IV. A number of general system boundaries have been set for the Beovision LX 5500:

- With few exceptions, all components in the Beovision LX 5500 are included in the LCA as shown in Table 27.1.
- Product manufacturing processes occurring at Bang & Olufsen and subcontractors are included in the LCA for all items or components weighing more than 100 g. As indicated in Table 27.1, a number of manufacturing processes are also included for items weighing less than 100 g. In all, the manufacturing processes are thus included for more than 90% of the television's weight. This covers to a significant degree the environmental problems concerning energy-related impacts and non-scarce resources.
- For the remaining 10%, an estimate was made of where scarce resources are consumed, and inclusion of consumption of these was sought to the extent to which data were available.
- An estimate was also made of where there may be processes with significant consumption of chemicals with special environmental impacts, especially toxicity, and inclusion of these was attempted.

The working environment at Bang & Olufsen is described for the internal product manufacturing processes. A significant number of the subcontractors are foreign (cf. Figure 27.3), and it was therefore not possible to include the working environment for these. As it was not possible to include a significant part of the impact on the working environment in the manufacturing stage, the working environment was omitted in the LCA.

Apart from these general principles, the following specific considerations apply to the scope definition:

- The electronics were included as consumption of materials, i.e. as quantities of plastic, copper, aluminium, lead, iron and various other metals used. The manufacturing energy at Bang & Olufsen's printed circuit factory in Skive was also included. Subcontractors' manufacturing processes for electrical components and electronics were not included, as it was not possible to generate adequate information. However, production of materials was followed as far as possible all the way back to extraction of raw materials.
- Minor quantities of copper are used in the demagnetization coil and the other leads. It was not possible to work out these quantities of copper, but it was estimated that they comprise less than 10% of the total quantity of copper in the television.
- Most of the data used include overhead energy, i.e. the energy consumption used for lighting, heating, canteen, administration, etc. Overhead energy is thus included in most of the processes for production of materials, for manufacture at Bang & Olufsen and for all subcontractors except that for the picture tube. Overhead energy for Bang & Olufsen was distributed among products by weight based on one year's sales.
- Ancillary substances for the manufacturing processes, e.g. lubricants and cleaning agents, were included from Bang & Olufsen's own manufacturing as far as the process emissions of these substances are concerned. The production of ancillary substances themselves was not included, as it was not possible to obtain data from the subcontractors.
- For some of the materials used in the Beovision LX 5500 it was not possible to generate data for all of the preceding steps in their life. This was the case for e.g. lead, tin, silver and gold. The inventory of these materials is documented in Frees (1996).
- Service and repairs of the TV were not included.

Division of the television's life into stages

The television's life cycle is divided into stages:

- Production of materials: extraction of raw materials, processing of raw materials and production of materials.
- Manufacturing: processes at Bang & Olufsen and subcontractors.
- Use: the consumer's electricity consumption for the time during which the television is used.
- Disposal.

Transport is described under each stage.

Production of materials

Production of materials covers extraction of raw materials, their processing, production, and transport between these steps.

The geographic location of extraction of raw materials and production of materials is shown in detail in the LCA of the refrigerator (see the example from Gram); the same situation applies here.

All transport from processing of raw materials to and including production of materials is included as average transport for each material. The materials are bought on international markets, and their route back to the site of extraction is not possible to trace, and will also vary significantly from shipment to shipment. It was therefore judged correct to commence with average observations.

It was not possible to include transport of finished materials to Bang & Olufsen's subcontractors.

Manufacturing

Manufacturing at Bang & Olufsen and subcontractors was fully mapped, with information on consumption of raw materials, energy consumption, emissions to air and water, and quantities of waste.

Figure 27.3 shows the places of origin of the various components and semi-manufactured goods for the Beovision LX 5500. The figure should be seen as an expression of where supplies come from at the time of data collection, and the picture will change when Bang & Olufsen occasionally change suppliers. As is evident from the figure, most of the components come from Germany and France. All transport is by lorry, with the exception of the base units for the loudspeakers, which are transported by ship from Korea. The figure does not show all of the components included, but only the largest, where the most significant part of the transport occurs.

The country of origin has an influence on the distance over which the components have to be transported, and therewith on the environmental impacts to which this transport leads. Electricity production also depends on the country where it is produced, which is of significance for consumption of fuels (coal, crude oil, natural gas and uranium) for the generation of electricity.

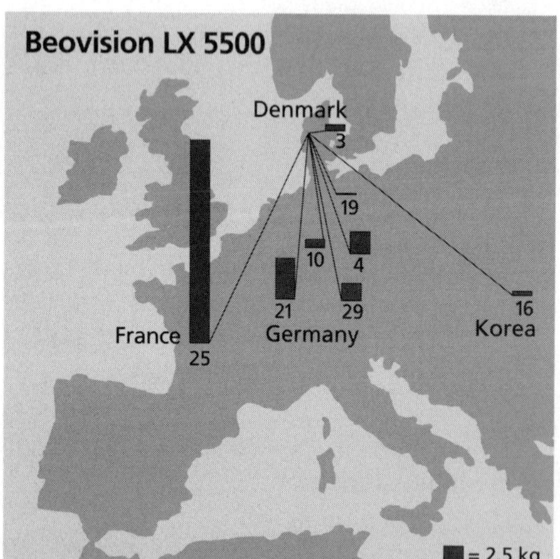

Figure 27.3. Components and semi-manufactured goods in the Beovision LX 5500 and their geographical origin at the time of inventory. The numbers refer to the component numbers in Table 27.1

The transport of materials, components and semi-manufactured goods from supplier to Bang & Olufsen was included for all items weighing more than 100 g. The transport was thus included for 90% of the television on a weight basis.

Figure 27.4 shows the manufacturing layout at Bang & Olufsen. The manufacturing itself occurs in factory 4, factory 5 and the factory in Skive. Factory 5 produces the items, the factory in Skive installs the electronics, i.e. the printed circuit boards, and the sets are finally assembled in factory 4.

The following were measured or calculated for all processes used in connection with the manufacture of televisions in factory 5: energy consumption, consumption of materials and

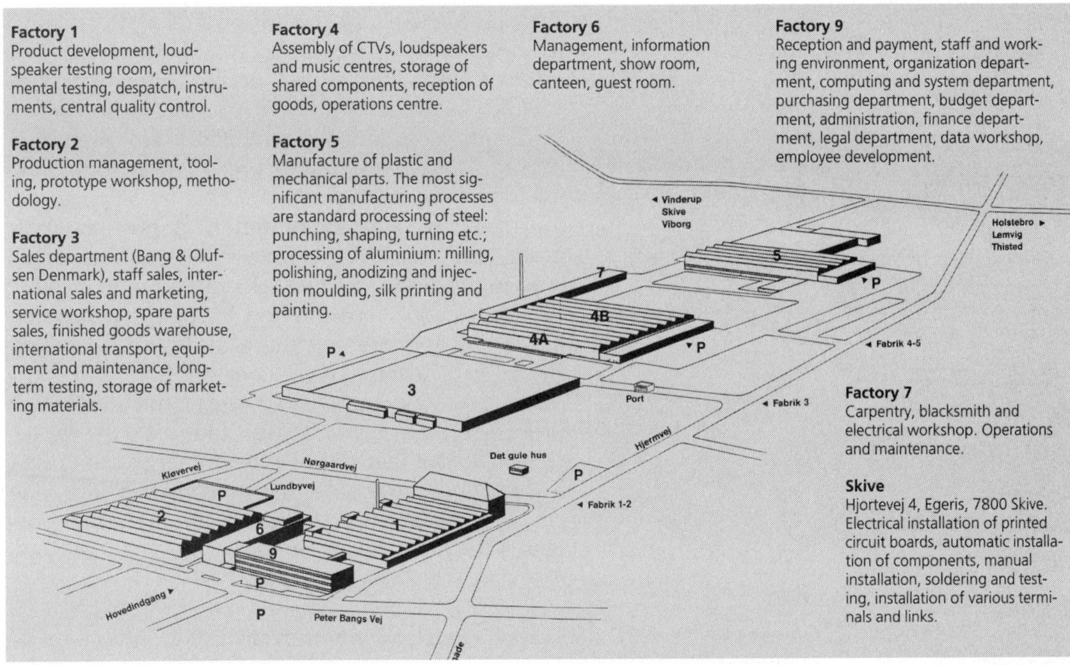

Figure 27.4. Plan of Bang & Olufsen's factories in Struer

ancillary substances, emissions to air and water, and waste. Aerosol measurements were included.

For assembly of the television in factory 4, emissions to air from various processes were computed on the basis of inputs and mass balance considerations.

The individual processes were not computed or measured for the factory in Skive where the electronics are produced. The printed circuit production is highly uniform - all printed circuits are subject to the same processes - and it is therefore sufficient to take the total consumption and emissions and the total waste from the factory and to set them in relation to the quantity of goods produced.

The effluent from Bang & Olufsen's waste water treatment plant is checked about nine times a year. The content of chromium, nickel, zinc, lead, copper, mercury, cobalt, oil, anionic detergents, nitrogen and phosphorus is measured. These emissions derive principally from the anodizing, and they are there-

Figure 27.5a. Assembly hall in factory 4

Figure 27.5b. Anodizing of aluminium units in factory 5

Figure 27.5c. Assembly of printed circuit boards in the factory in Skive

Figure 27.5d. Assembly of television in factory 4

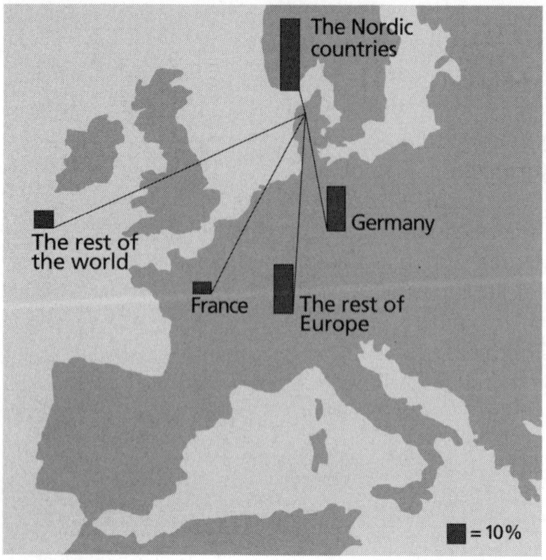

Figure 27.6. Worldwide distribution of Bang & Olufsen television sales

fore allocated to the Beovision LX 5500 relative to the area of surface anodized. The waste water effluent from the printed circuit factory in Skive was not included.

Faulty components and items are discarded during production. The degree of discarding varies from process to process and is included in the inventory, so that consumption of raw materials is greater than the quantity of raw materials entering into the finished television.

The energy consumption for lighting, heating, hot water and administration is found by taking Bang & Olufsen's total consumption and subtracting the quantity used in the manufacturing processes themselves. It was found at the factories in Struer that the overhead energy amounted to ca. 75% of the total energy consumption. The energy consumption for the manufacturing processes thus comprised only ca. 25%.

Use

The Beovision LX 5500 uses ca. 81 W when in operation and ca. 3 W in standby. Bang & Olufsen build televisions designed to be able to operate for at least 6 hours a day for 10 years.

As is evident from Figure 27.6, Bang & Olufsen export most of their televisions, principally to Europe. Most televisions are sold to Germany and Scandinavia.

The country in which the television is used is of significance for:

- Transport from Bang & Olufsen to the customer and the environmental impacts to which the transport contributes.
- The customers' patterns of use - an average Englishman watches more television than a Dane.
- The environmental impacts of the electricity consumption. There is a difference in how the electricity used by the television in the use stage is made - Danish coal-fired co-generation power plants have different environmental impacts from French nuclear power plants.
- Disposal of the television is not the same in the various countries, and the environmental impact of the disposal is thus also different.

Figure 27.7. The use stage - children in Hong Kong watching "Peanuts"

The energy consumption during operation and standby yields heat, which during winter in cold regions to some degree substitutes for other heating of the house. As Bang & Olufsen's television is also exported to warmer countries which do not heat their houses, but in some cases use air-conditioning for cooling, no account has been taken of this factor.

Transport from Bang & Olufsen to dealers is partly by large lorries (38 tonnes) to a central distribution point in Germany, and partly by small lorries (5 tonnes). Transport between TV dealer and customer is not included.

Disposal

A consumer typically disposes of a worn out television set to a reconditioning organization, a recycling station or a television dealer. The dealer receives the television in case there is a question of whether or not it will pay to repair the old set. It is assumed that on average, 70% of televisions are then dumped, while 30% are incinerated in waste incineration plants. Materials from the television are not believed to be recycled today, and all consumption of materials is thus lost.

Transport in connection with disposal of the television is assumed to occur with 5 tonne lorries. The distance to an incineration plant is assumed to be 30 km, with 15 km to a dump. The consumer's transport in connection with disposal is not included.

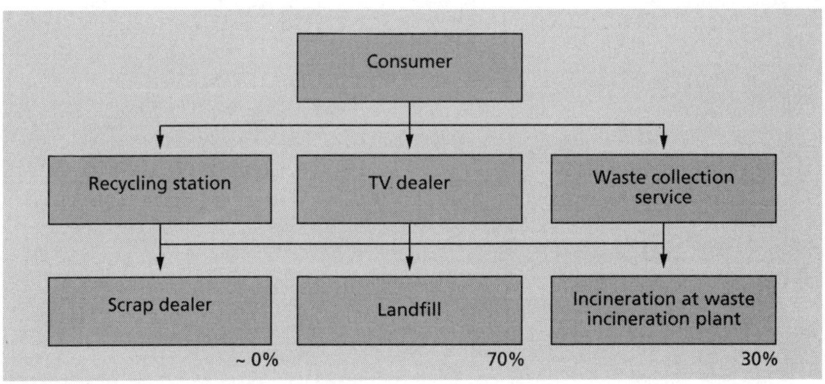

Figure 27.8. Disposal routes for the Beovision LX 5500

Database

Information on the product system for the Beovision LX 5500 was based on specific data for manufacturing of the Beovision LX 5500 to the extent to which this was possible. Table 27.2 shows the type of data used in the different stages in the television's product life.

Information on processes for manufacturing of components and assembly of the television was obtained from Bang & Olufsen's factories and subcontractors. This was a major task. Twelve subcontractors who contribute to the manufacturing of items for the Beovision LX 5500 were contacted. Bang & Olufsen and subcontractors have made inventories of ca. 45 different processes. The most significant of these appear in the parts list in Table 27.1. The inventory of processes at Bang & Olufsen comprises ca. 25 different types of resource consumption, ca. 80 types of emissions to air, ca. 40 different types of emissions to water and ca. 30 types of waste.

It is assumed that the television is disposed of in the same way as the average household waste in the EU. This is supported by information on disposal routes for televisions found by contacting dealers and recycling centres and through an EU report on disposal of waste in the EU (Commission for the European Communities, 1992).

Beovision LX 5500									
Data reference									
	Data specifity			Data source type					Comments
Product system & process type	Product specific	Site specific	General	1	2	3	4	5	
Extraction of raw materials									
Glass			x			x			Literature
Plastic			x			x	x		Trade organizations
Aluminium			x			x			Trade organizations
Steel			x			x			Literature and subcontractors
Copper			x			x			Literature
Product manufacturing at B&O and subcontractors									
Shaping of plastic			x			x			Measurements of energy consumption
Metal processing	x			x					performed by B&O
Metal surface treatment	x			x					Measurements of emissions performed
Assembly	x			x					
Other processes	x	x		x	x	x			by an authorized laboratory
Use									
Energy consumption	x			x					
Lifetime	x					x			B&O's criteria for dimensioning the Beovision LX 5500
Disposal									
Beovision LX 5500									
Means of disposal			x					x	TV dealers, recycling centres and EU reports
Incineration			x			x			Literature
Transport									
Distances and means of transportation	x		x	x	x	x			Raw materials and materials: literature Components for B&O: B&O From B&O to dealer: B&0 Disposal: estimate
Energy consumption and emissions			x			x			Common European transport database, Copert, 1990. Literature
Energy systems									
Energy systems			x			x			RISØ System Analysis Dept. Literature, see Frees, 1996. Presumed EU average everywhere

Notes

1) Measurements
2) Computations (from mass balance considerations and input data for the process in question)
3) Extrapolation of data from similar process type or technology
4) Extrapolation of data from different process type or technology
5) Estimate

Product-specific data: concerns processes specifically handling the Beovision LX 5500
Site-specific data: concerns data from actual sites in the product system of the Beovision LX 5500, but the process data are not specific for the Beovision LX 5500
General data: all others

Table 27.2. Sources of data for the LCA of the Beovision LX 5500

Exchanges with the environment

Table 27.3 shows a summary of the inventory of resource consumptions, emissions to air and water, and waste produced. The inventory is divided by stages of the product life, and all transport is included here to illustrate its total significance. It is important to remember that the stage "manufacturing" covers manufacturing at both Bang & Olufsen and subcontractors. The emissions given are thus summed figures for all of the companies.

The resources, emissions and quantities of waste presented in Table 27.3 were selected on the basis of an expert assessment of the environmental impact potentials to which they contribute. Only the most relevant contributions are therefore presented.

			Production of materials	Product manufacturing	Use	Disposal	Transport	Total
Resource consumption								
Crude oil		g	15,100	2,300	59,700	10	3,300	80,000
Natural gas		g	59,800	6,200	43,000	6	200	109,000
Coal		g	4,500	14,300	215,000	200	16	234,000
Brown coal		g	1,030	610	160,000	-	-	162,000
Uranium ore		g	0.05	0.06	22	-	-	22
Water for hydro power		litre	3,900	870	88,000	9	~ 0	92,800
Aluminium	Al	g	710	-23	11	~ 0	~ 0	700
Iron	Fe	g	3,200	-200	4	~ 0	~ 0	3,000
Copper	Cu	g	800	-	-	-	-	800
Manganese	Mn	g	23	-1	-	-	-	22
Nickel	Ni	g	5	-	-	-	-	5
Zinc	Zn	g	46	-	-	-	-	46
Barium carbonate	$BaCO_3$	g	3,100	-	-	-	-	3,100
Lead oxide	Pb_3O_4	g	1,700	-	-	-	-	1,700
Calcium carbonate	$CaCO_3$	g	1,950	-160	7	~ 0	~ 0	1,800
Quartz	SiO_2	g	15,400	-290	-	-	-	15,000
Sodium carbonate	Na_2CO_3	g	3,800	-	-	-	-	3,800
Sodium chloride	NaCl	g	615	12	18	~ 0	~ 0	650
Soft wood		g	530	80	10,200	1	~ 0	10,800
Groundwater		litre	-	44	-	3	~ 0	47
Surface water		litre	~ 0	1	~ 0	~ 0	~ 0	1
Unspecified water		litre	180	280	33	~ 0	1	500
Materials and ancillary substances for which resource consumption is not included								
Paint		g	-	319	-	-	-	319
Various materials*		g	-	2,160	-	-	-	2,160
Various ancillary substances*		ml	-	6,000	-	-	-	6,000
Emissions to air								
Carbon dioxide	CO_2	g	166,000	62,000	1,051,000	13,000	6,400	1,300,000
Carbon monoxide	CO	g	200	22	680	43	750	1,700
Nitrogen oxides	NO_x	g	1,000	210	4,200	17	170	5,600
Sulphur dioxide	SO_2	g	550	260	9,300	3	18	10,100
Dinitrogen oxide	N_2O	g	18.1	5.2	84.8	~ 0	0.2	108
Unspec. part. (dust)		g	76	87	500	~ 0	5	670
Hydrocarbons	HC	g	470	350	5,050	4	9	5,900
Volatile organic comp.	VOC	g	1.6	38	1.1	~ 0	0.4	41
Unspec. aldehydes		g	~ 0	0.16	3.75	~ 0	~ 0	3.9

Table 27.3. Exchanges with the environment in the television's product system

			Production of materials	Product manufacturing	Use	Dis-posal	Transport	Total
2-propanol, isopropanol		g	-	1.8	-	-	-	1.8
Benzene		g	0.032	-	-	-	-	0.032
Butyl acetate		g	-	10	-	-	-	10
Butyl diglycolacetate		g	-	15	-	-	-	15
Ethyl acetate		g	-	88	-	-	-	88
Phenol		g	-	0.25	-	-	-	0.25
Toluene		g	-	4	-	-	-	4
Trichlorethylene		g	-	1.1	-	-	-	1,1
Unspec. xylene		g	-	51	-	-	-	51
Dioxin		mg	-	1 E-06	-	128 E-06	~ 0	130 E-06
Epichlorhydrin		mg	37	-	-	-	-	37
Arsenic	As	mg	~ 0	1	99	~ 0	~ 0	100
Lead	Pb	mg	990	1	140	~ 0	1,330	2,460
Cadmium	Cd	mg	~ 0	~ 0	12	~ 0	~ 0	12
Emissions to air								
Copper	Cu	mg	1	4	99	140	~ 0	240
Mercury	Hg	mg	~ 0	1	18	~ 0	~ 0	19
Manganese	Mn	mg	410	260	-	-	-	670
Vanadium	V	mg	28	39	2,770	~ 0	~ 0	2,800
Emissions to water								
Chem. oxygen demand	COD	g	18	~ 0	1	~ 0	~ 0	19
Total nitrogen	tot-N	g	1	8	2	~ 0	-	11
Total phosphorus	tot-P	g	-	0.15	-	-	-	0.15
Hydrocarbons	HC	g	6.7	0.06	1.4	~ 0	0.06	8.2
Unspec. oil		g	5.9	0.3	2.4	~ 0	~ 0	8.6
Phenol		g	0.04	~ 0	0.035	~ 0	~ 0	0.08
Arsenic	As	mg	15	-	-	-	-	15
Lead	Pb	mg	2	-	-	-	-	2
Cadmium	Cd	mg	8	-	-	-	-	8
Chromium (VI)	Cr	mg	-	9	-	-	-	9
Copper	Cu	mg	55	3	-	-	-	58
Mercury	Hg	mg	0.003	0.033	-	-	-	0.036
Manganese	Mn	mg	-	0.2	-	-	-	0.2
Nickel	Ni	mg	52	28	-	-	-	80
Zinc	Zn	mg	48	35	-	-	-	83
Waste								
Unspec. hazardous waste		g	0.03	0.001	-	-	-	0.03
Unspec. dust containing heavy metals		g	-	2	-	-	-	2
Unspec. industrial waste		g	64	1	-	-	-	65
Unspec. radioactive waste		g	0.004	0.01	3	-	-	3
Unspec. slag and ashes		g	330	1,220	41,000	8,200	8	50,800
Unspec. bulk waste		g	5,600	33,600	125,000	30,000	9	194,000

Notes

The values from each stage of the product system and the totals are rounded. The uncertainties of the data has not been assessed and the rounded figures should not be seen as an expression of this.

- means that no information is available for the life cycle stage in question.

~ 0 means that the figure is very small compared to the other stages of the life cycle.

* Only volumes above 100 g included. B & O have made an inventory of the consumption of a large number of materials and ancillary substances. For reasons of transparency these have here been collected under the headings: " various materials" and "various ancillary substances".

The product manufacturing stage includes both the production at B & O and at the subcontractors. The stated figures for resource consumption and emissions are therefore added up from all the manufacturers.

Table 27.3. (continued)

The inventory in Table 27.3 provides an overview of which substances enter into the Beovision LX 5500's product system. The substances and quantities alone say nothing about their significance for the environment, but it is possible to gain an impression of the categories of substances and of where they occur in the product system.

Further details and documentation of the inventory are available at Bang & Olufsen.

The information in Table 27.3 is amplified below on the basis of a background knowledge of how the figures were obtained.

Consumption of energy resources

It is immediately evident that the consumption of energy resources is greatest in the use stage for crude oil, natural gas, coal, brown coal and uranium ore. The consumption of coal in this stage comprises 92% of the total consumption. For brown coal and uranium ore, the use stage accounts for more than 99% of the consumption in the entire product system of the television.

The consumption of uranium ore is attributed exclusively to production at nuclear power plants for the electricity used abroad. The electricity consumption abroad is attributed principally to the fact that a significant fraction of the televisions is exported, and therefore operated abroad.

The consumption of dammed water is ascribed to the production of electricity at hydroelectric power stations, which, as for nuclear power, can be the case for electricity consumption outside Denmark.

Crude oil and natural gas are used both as resources for the production of energy and for the production of plastic materials, but by far the greater part is used for energy production.

It is clear from the table that the total transport does not use particularly large quantities of energy resources relative to the television's total product system.

Consumption of resources for production of materials

When viewing the resources used for production of materials (aluminium, iron, copper, etc.), it is seen that 15.4 kg of quartz is used. This is used for production of the picture tube in the television, together with 1.7 kg of lead oxide and other substances. The lead oxide is added to screen out X-rays.

During manufacturing, negative values appear in the inventory for aluminium and iron and other substances. This is attributed to the fact that the losses associated with processing and cutting out of the materials are collected and reused for other products, and the material thus used in new products is therefore credited, i.e. registered as "negative consumption".

It is otherwise noteworthy that in its product life, the television uses ca. 630 kg of materials (dammed water for electricity production is not included), when the television itself weighs only 42.7 kg. The greatest quantities of resources enter into energy production.

Emissions to air

Among the emissions to air, carbon dioxide clearly comprises the largest quantity. In its entire product life, a Beovision LX 5500 contributes 1.3 tonnes of carbon dioxide emissions, of which 1 tonne or 81% derives from the use stage. The other typical energy-related emissions similarly derive to a high degree from use of the television, namely 75% for nitrogen oxides, 92% for sulphur dioxide, 79% for nitrous oxide and 40% for carbon monoxide. A large part of the emissions of heavy metals are also ascribed to energy production and they therefore occur in greatest quantities in the use stage. All emissions in the use stage derive from electricity production.

Various solvents in the manufacturing stage are emitted. This is ascribed especially to the surface treatment of materials (cleaning and painting) at Bang & Olufsen and subcontractors.

Note that the transport contributes significantly to emissions: carbon monoxide (44% of the total quantity) and lead (54% of the total quantity).

The incineration of waste during disposal is the most significant source of emission of dioxins. There are also small quantities of dioxins from the manufacturing stage, ascribed to the incineration of production waste.

Emissions to water

Especially noteworthy in the aqueous emissions are the heavy metals, which are emitted mainly during production of materials and extraction of raw materials.

Waste

The greatest quantities of waste are unspecified bulk waste and slag and ashes. Most of this is ascribed to electricity production in the use stage. The operation of coal mines produces bulk waste and combustion of coal produces slag and ashes. Some of the waste also derives from disposal of the television, where during incineration the picture tube contributes significantly to the formation of slag.

Energy profiles

As is clear from the discussion of Table 27.5, the energy consumption is of considerable significance for the total energy consumption, emissions to air and quantities of waste. It is therefore of interest to look more closely at energy consumption. Figure 27.8 shows the total energy consumption throughout the entire product life for the Beovision LX 5500. The energy consumption is specified as "primary energy", i.e. the quantity of energy extracted from the earth for the television's entire product system.

The largest part of the energy consumption in the television's product system is ascribed to the electricity used when the television is switched on and when it is in standby - ca. 78% of the total energy consumption is in the use stage. This is not surprising in view of the fact that the largest part of the energy resources was seen to be crude oil, gas and coal

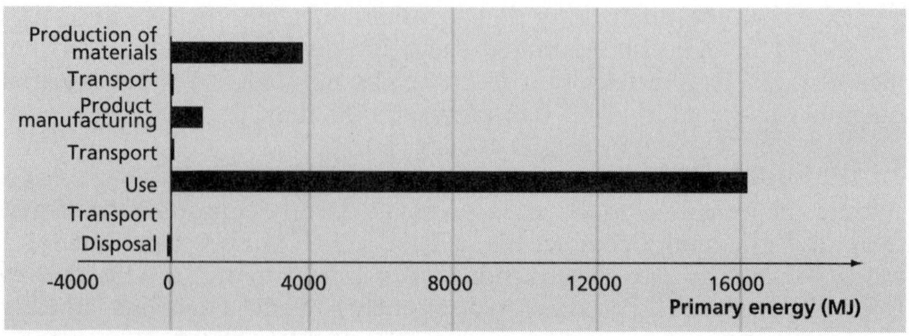

Figure 27.9. Energy consumption in the Beovision LX 5500's entire product life, expressed in primary energy

used in precisely this stage. Operation of the television accounts for ca. 70%, and standby ca. 8%, of the total energy consumption. As is clear from Figure 27.9, transport has practically no significance for total energy consumption.

During product development, the energy consumption is a suitable indicator of relevant resource consumption and potentials for environmental impacts for products which use energy in the use stage. An overview of the energy consumption gives the product developer a quick indication of the product's environmental impact and of the environmental impacts of new solutions. Figure 27.9 shows that one of the most important considerations in the development of new televisions is to construct a television which uses less power during operation and when in standby.

The energy content of the materials

The designer's options for saving energy in the choice of materials and for enabling recycling are clear in Figures 27.10a and 27.10b, which present the primary energy for the first stage in the television's life: extraction of raw materials and production of materials. These product life stages comprise a total of 22% of the total energy consumption in the television's entire product life.

The energy is divided into:

- process energy, i.e. the energy used for extraction and processing, and
- calorific value, i.e. the energy bound in the materials, and which can subsequently be used in incineration with utilization of the energy (e.g. in incineration of plastic, wood and paper).

As Figures 27.10a and 27.10b show, most of the primary energy is associated with the picture tube and the glass which the picture tube contains. But the plastic parts, the casing and the rear cover also contain a significant part of the energy. Not much energy is to be obtained by incinerating a Beovision LX 5500 - the calorific value is only 420 MJ, corresponding to 11% of the total energy used in extraction and material production. When the options for energy recovery in disposal are considered, it can therefore pay better to recover the materials than to incinerate the television.

Figure 27.10a. The materials' primary energy in the most used materials in a Beovision LX 5500

Figure 27.10b. The materials' primary energy in the components in a Beovision LX 5500. The numbers refer to the component numbers in Table 27.1

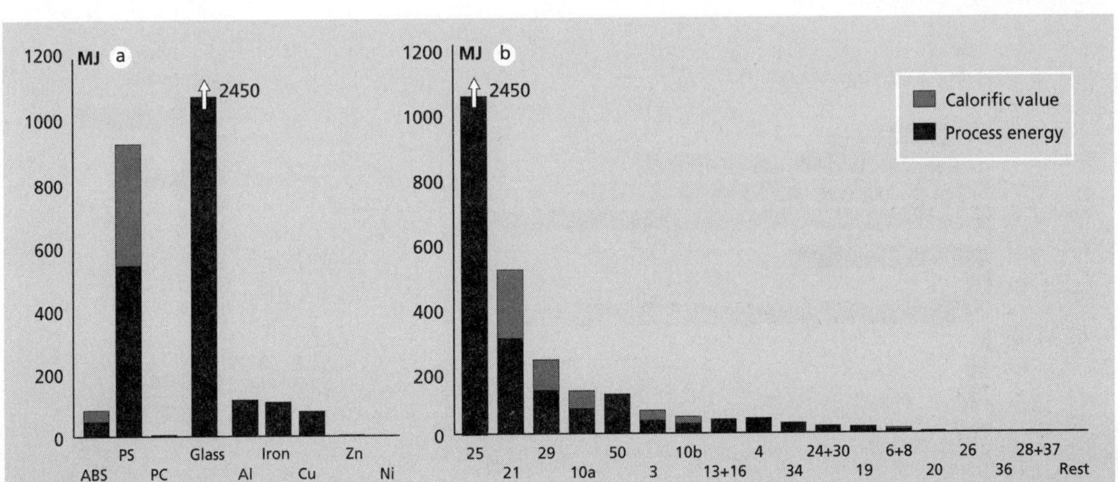

Impact assessment

The LCA includes resource consumption and environmental impacts.

Impact category		Potential impact
	Unit	Per product per lifetime
Global impacts		
Global warming	g CO_2-eq	1,340,000
Ozone depletion	g CFC11-eq	0
Regional impacts		
Acidification	g SO_2-eq	14,000
Nutrient enrichment	g NO_3-eq	7,900
Photochemical ozone formation	g C_2H_4-eq	190
Human toxicity (water)	m^3 water	8,300
Ecotoxocity (water, chronic)	m^3 water	3,700
Local impacts		
Human toxicity (air)	m^3 air	806,000,000
Ecotoxicity (water, acute)	m^3 water	200
Hazardous waste	g	2.8
Nuclear waste	g	3.0
Slag and ashes	g	50,800
Bulk waste	g	197,000

Table 27.4. Conversion of the environmental exchanges to environmental impact potentials

Environmental impact potentials

Emissions to air and water can give rise to environmental impacts. Whether these will really occur is not known, and the impacts are therefore assessed as potentials for environmental impacts. The contributions to the environmental impact potentials are given in Table 27.4.

The table shows emissions from the television's total product system converted to impact equivalents. The impacts are not ranked in relation to some order of magnitude, and the potentials for environmental impacts are not weighted according to how critical they are. The table therefore exclusively shows each of the individual contributions.

The waste in Table 27.4 is classified into four categories. The quantities in Table 27.4 are larger than in Table 27.3, as Table 27.3 comprises only an extract of the total amount of waste produced.

Waste is not converted to the impacts which they have on the environment (such as e.g. pollution of groundwater, pollution of drinking water or consumption of land), as it has not yet been possible to include the dumping of waste in landfills as a process and to predict the relationship between quantity of waste and environmental impact. The quantities of waste are therefore used as indicators of the environmental impact potentials to which they can lead. For resource consumptions, a further assessment is made directly from Table 27.3.

The magnitude of the product's contribution

To gain some perspective of whether the Beovision LX 5500 contributes a lot or a little to the environmental impact potentials, the contributions are ranked relative to a person's average annual contribution, normalized relative to the year 1990. Figures 27.11a and 27.11b show nor-

Figure 27.11. Normalization of resource consumption and environmental impact potentials for the television

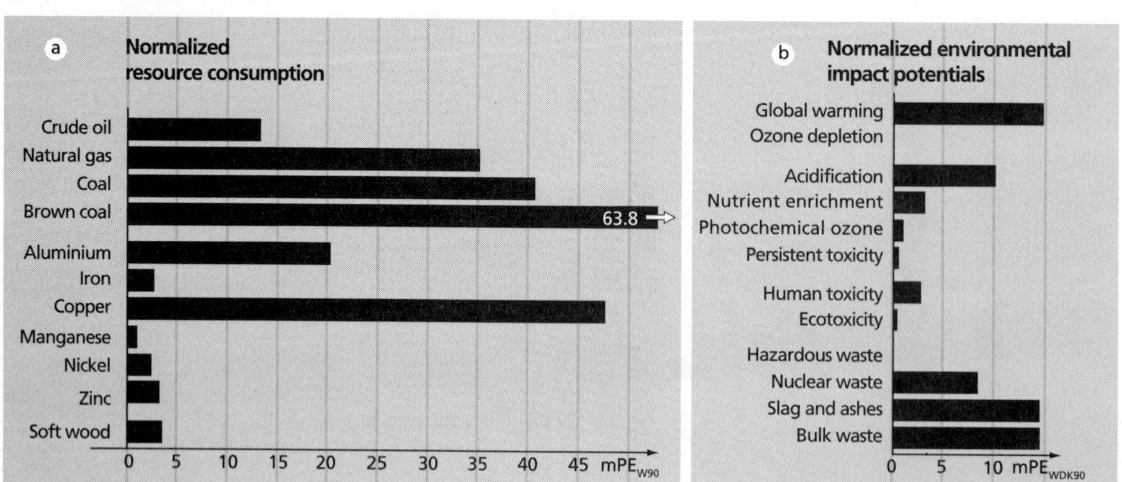

malized resource consumptions and environmental impact potentials. It is clear from the figures that consumption of energy resources (crude oil, natural gas, coal and brown coal) corresponds to between 10 and 65 milli-person-equivalents or 1-6.5% of a person's average annual consumption in 1990. Use of a television similarly corresponds to ca. 1.4% of a person's annual contribution to global warming in 1990 and 1.4% of the annual contribution to bulk waste and slag and ashes.

The most significant impact potentials

Not all environmental impacts are equally critical. To be able to assess which resource consumptions and environmental impact potentials in Figures 27.11a and 27.11b are judged to be the worst, it is necessary to apply weightings. The results of these weightings are shown in Figures 27.12a and 27.12b.

The resource consumption is weighted relative to the size of known reserves for each resource. This means that scarce resources are weighted highly, while abundant resources are given a low weighting. With these weightings, the most significant resource consumptions for a Beovision LX 5500 are consumption of copper and energy-related resources, especially natural gas. The consumption of copper in a Beovision LX 5500 corresponds to 1.3 milli-person-reserves. The television has thus consumed 1.3 parts per thousand of the entire quantity of copper available for one person and all of his or her successors for all time. It is obvious that recycling of copper will be of considerable significance for the television's complete resource profile.

The potentials for environmental impacts are weighted on the basis of Danish and international political reduction targets, i.e. according to how much the environmental impact is targeted for reduction within a fixed reference year. With this weighting, the most significant potentials for environmental impacts for a Beovision LX 5500 are global warming and acidification, and the waste categories radioactive waste, slag and ashes, and bulk waste.

The figures express the television's fraction of the acceptable contribution to an impact category for a person in the year 2000. It is evident that the most highly weighted impact categories comprise about 1.5-2% of the contribution targeted for a person in the year 2000.

Figure 27.12. Weighted resource consumptions and environmental impact potentials for the television

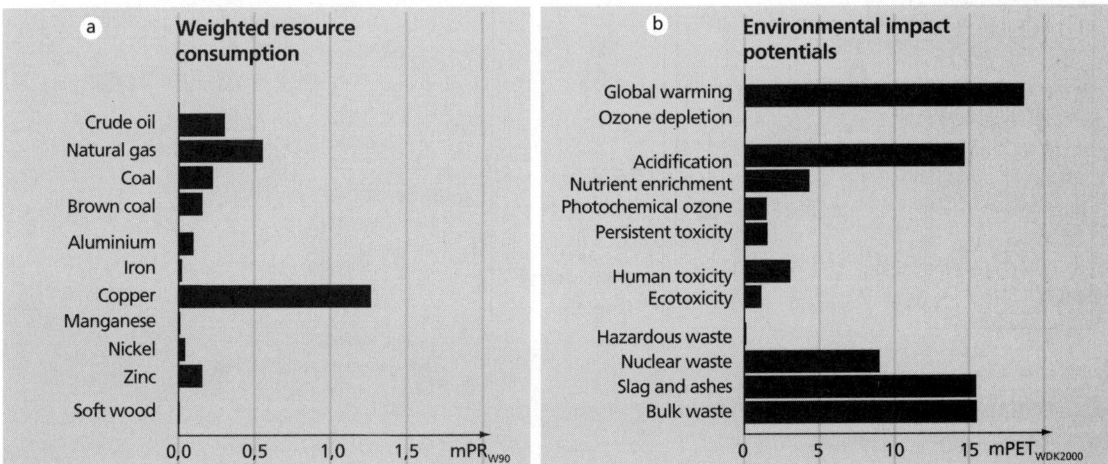

Materials		
Material	- consumption - disposal	
Processes		
Energy	- thermal - electric	
Chemicals	- ancillary substances	
Others	- overhead	

Table 27.5. Sources of resource consumption and environmental impact potentials

The most significant sources of resource consumption and impact potentials

In order to see where the causes of the resource consumptions and the environmental impact potentials are to be found, it is an advantage to make a classification into different categories:

- **Consumption of materials** and consequent resource consumptions, and environmental impact potentials from extraction, processing, transport and energy production for the energy consumption in this stage.
- **Disposal of materials** and the consequent environmental impact potentials from the disposal, and energy production for the energy consumption in this stage.
- **Consumption of electricity** during manufacturing and recycling. This covers resource consumption and environmental impacts for production of electricity, and extraction of raw materials for this.
- **Consumption of thermal energy** (oil, natural gas and coal) during manufacturing, use and transport. This covers both emissions from combustion of fossil fuels in the product manufacturing processes and from extraction of raw materials for the fossil fuels.
- **Consumption of ancillary substances and consequent emissions** from use of the ancillary substances in product manufacturing (e.g. evaporation of solvents and cleaning agents).
- **"Overheads"**, i.e. impacts consequent upon energy consumption for lighting, heating, canteen, administration, etc.

The designer's choices directly and indirectly influence materials and processes, including ancillary substances and energy consumption. By dividing up the resource and environmental profiles in this way, it is possible to see the consequences of the decisions which are made.

Figures 27.13a and 27.13b show the weighted resource consumptions and environmental impact potentials for a Beovision LX 5500 classified by source.

The resource consumption is ascribed mainly to the materials entering into the television and to the consumption of electricity.

Figure 27.13. Weighted resource consumptions and environmental impact potentials classified by source

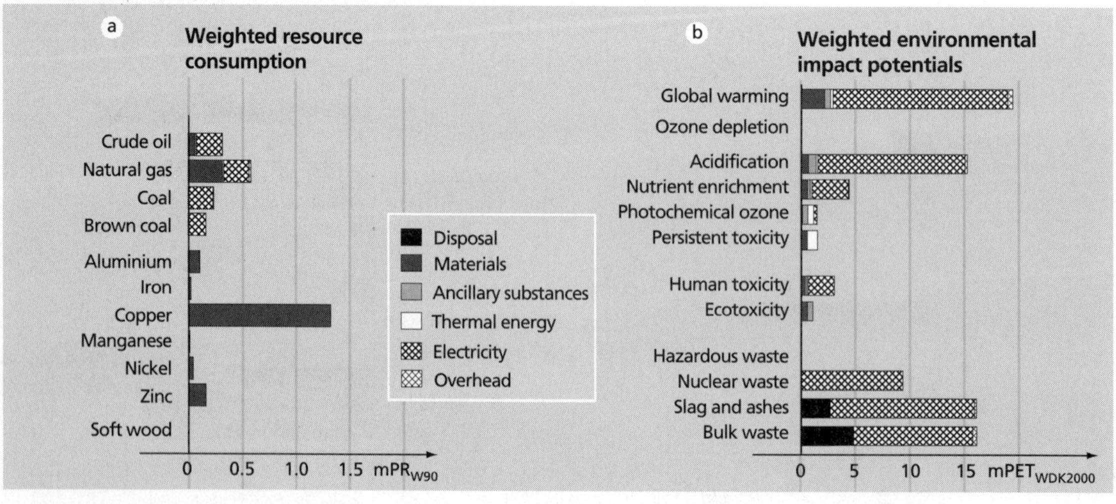

Natural gas and crude oil are used for production of both materials and electricity. About 20% of the oil consumption and more than half of the natural gas is used for production of materials. A large amount of natural gas is used in production of the picture tube and some of the plastic of which the television is made. The metals enter into the television as materials.

The consumption of electricity is responsible for a significant fraction of the contribution to many of the environmental impact potentials: global warming, acidification, nutrient enrichment, and toxic impacts on humans. The consumption of electricity accounts for 83% of the total contribution to global warming.

Some of the contributions to photochemical ozone formation, persistent toxicity and ecotoxicity are ascribed to chemicals used in manufacturing at Bang & Olufsen and subcontractors.

The disposal is of some significance for the quantities of slag and ashes and bulk waste, where, as noted, the glass in the picture tube in particular accounts for significant quantities of slag. The disposal of materials is, however, not dominant. Electricity production contributes 82% of the quantities of slag and ashes and 68% of the bulk waste.

The most significant stages of the product life

Resource consumption is classified into the stages in the Beovision LX 5500's product life in Figure 27.14a, which shows very clearly that the most significant consumption of the energy-related resources (oil and coal) is in the use stage, as by far the greatest fraction of the energy con-

Figures 27.14 Weighted resource consumptions and environmental impact potentials for a Beovision LX 5500 classified by stages of the television's product life

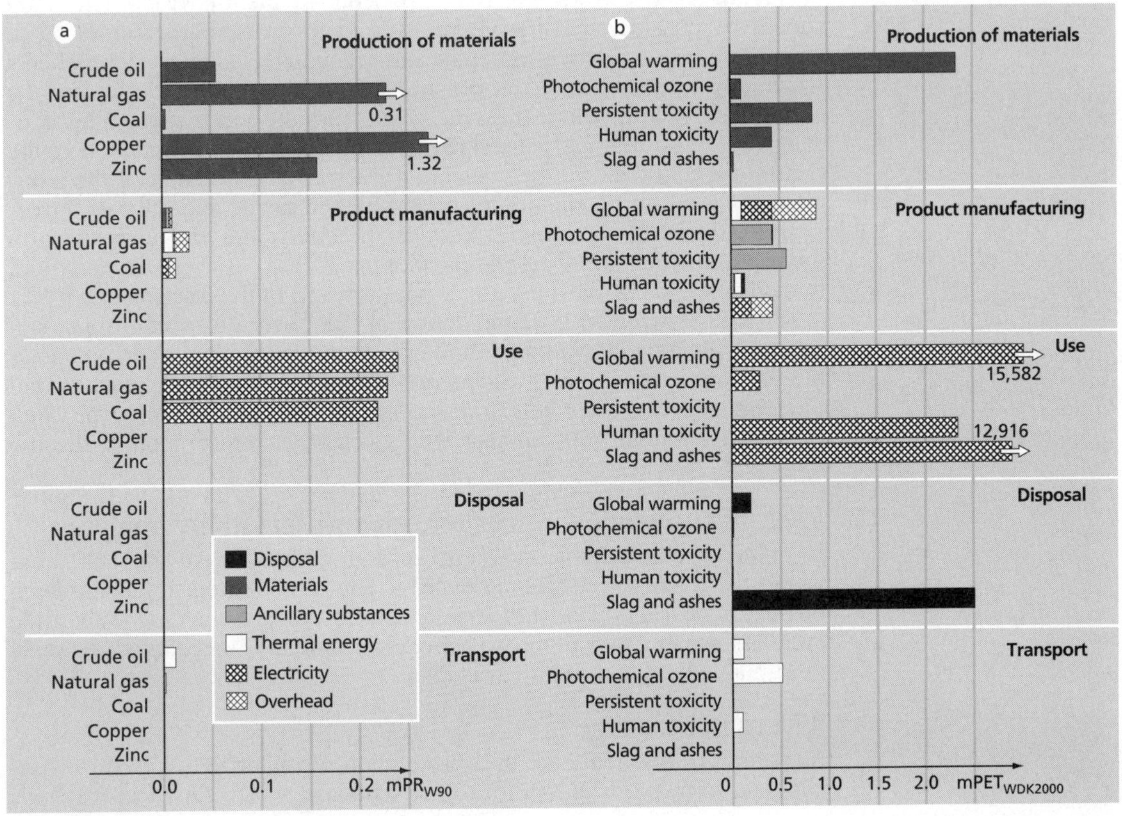

sumption occurs here. The energy-related resources (natural gas, crude oil, coal, brown coal and uranium) are represented by crude oil, natural gas and coal in Figure 27.14a.

Consumption of the material-related resources lies in the material production stage. This is shown in Figure 27.14a in the consumptions of copper, zinc, natural gas and oil.

A division of the environmental impact potentials into stages of the television's life shows that the use stage is extremely dominant (cf. Figure 27.14b). The use stage accounts for most of the contribution to global warming, human toxicity and slag and ashes, attributed to the fact that these are energy-related. These environmental impacts also represent acidification, nutrient enrichment and the other categories of waste, which are also strongly related to the consumption of electricity.

There are contributions to photochemical ozone formation and persistent toxicity during production of materials and product manufacturing. The contribution to photochemical ozone formation derives mainly from the production and use of thermal energy (38%), the production of electrical energy (22%), and a significant part (27%) also derives from ancillary substances, namely painting of the rear cover and the casing.

The most significant components

In order to gain an impression of which components contribute to selected resource consumptions and environmental impact potentials, the components which contribute most are given in Figures 27.15a and 27.15b. The figures represent only the stages extraction of raw materials, production of materials and product manufacturing.

Figure 27.15 shows which components in the Beovision LX 5500 should be considered if the intention is to focus on particular resource consumptions and environmental impact potentials. The components which contribute most to the consumption of crude oil and coal are the casing, the picture tube, the rear cover and the electronics. The consumption of natural gas is ascribed to a high degree to production of the picture tube, and as will be evident, a much smaller fraction of the natural gas is used for production of plastic for the casing and the rear cover.

Aluminium is consumed mainly in the electronics and in profiles on the top and sides of the television. Copper is used in the electronics and in deflector coils. Zinc is used in separators and in the electronics.

Processes involved in manufacture of the picture tube contribute significantly to persistent toxicity and human toxicity. The other potentials for environmental impacts are energy-related to a significant degree, and therefore reflect energy consumption during the manufacturing processes, which is significantly smaller than electricity consumption in the use stage.

Lack of knowledge, uncertainties and sensitivity analyses

In spite of considerable work in collecting data, there are still areas within the LCA where more knowledge could be desired. It has not been possible to collect all of the desired data at the present time. When life cycle assessments are more widespread in industry, the desired data will presumably be more readily available.

The material stage

Extraction of raw materials and production of materials is based on general data from the EDIP unit process database, which for most materials

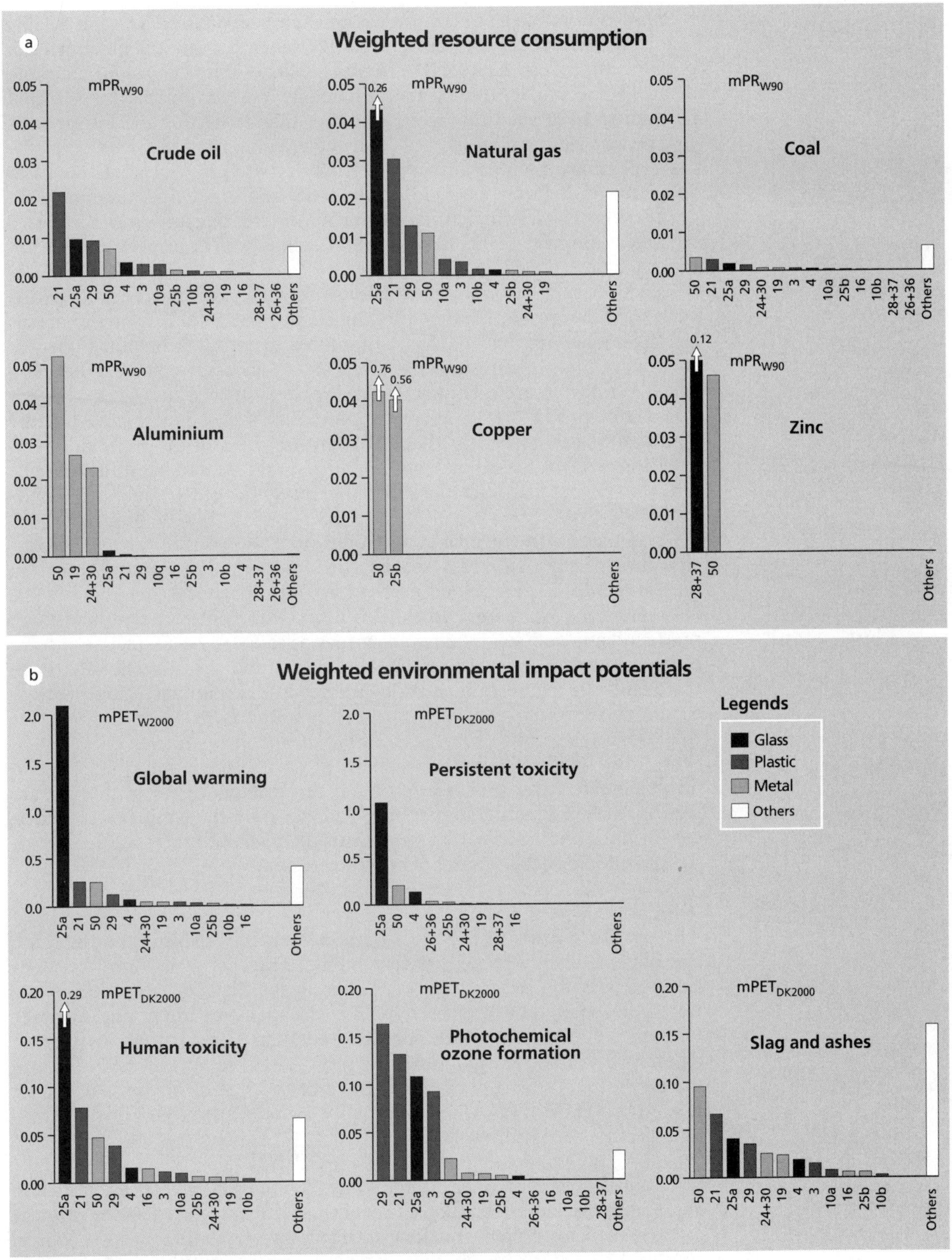

Figures 27.15. Top scorers among components in the Beovision LX 5500 which contribute most to resource consumption and environmental impact potentials. The numbers refer to the component numbers in Table 27.1

is based on data in the literature. Some of these data are of high quality (e.g. for steel and aluminium), while for other materials it has not been possible to collect data of the same quality. This is covered in more detail in Frees (1996). For most materials it is reasonable for Bang & Olufsen to have used average data rather than collecting the information on specific production sites. As the materials are often bought "on the market", their origin can vary from batch to batch.

It has not been possible to include considerations pertaining to destruction of the physical landscape or destruction of areas of special biological interest in the EDIP method. This will be especially relevant for extraction of raw materials.

The product system of ancillary substances has not been included back to their origins. Neither production nor extraction of raw materials has been mapped. The ancillary substances for product manufacturing at Bang & Olufsen and subcontractors are mainly solvents and cleaning agents, and they are included in the processes into which they enter, as noted. For production of these substances in the chemical industry, the loss is estimated to be less than 10%, and as most of the substances end as emissions when used at Bang & Olufsen and their subcontractors, this omission is not considered significant, although there could be exceptions.

The metal content in the electronics and the picture tube is of considerable significance for resource consumption. The metal content in the electronics is determined on the basis of a chemical analysis, but the manufacturing processes for the electronics were not included, and there may be significant environmental impacts here. Some of the metals are scarce resources (e.g. copper, tin, lead, silver and gold), while others are toxic and can therefore be a problem during the manufacturing processes and on disposal.

The manufacturing stage

The manufacturing processes at Bang & Olufsen and their subcontractors have been thoroughly mapped, and there are no significant deficiencies in the basic information here, with the exception of manufacture of the basic loudspeaker components from Korea.

The use stage

The use stage leads to the most significant energy consumption and the consequent energy-related environmental impacts and resource consumptions in the product system. This means that determination of the television's daily consumption and its life span are important. Bang & Olufsen's dimensioning criterion for operating time, namely 6 hours a day for 10 years, corresponding to 21,900 hours of operation, enters into the reference scenario. The significance of this choice is great, as will be clear from Table 27.6, where alternative choices are given. Two alternative scenarios are proposed:

- Danish average: since 1988, Bang & Olufsen have been installing meters in some televisions. The meter is read when the television is returned for repair, and Danish televisions have thus been shown to be switched on for an average of 4 hours a day. The study covered ca. 800 televisions from Denmark. To assess the television's life span, 20 randomly selected Danish Bang & Olufsen dealers were asked for their assessments of the life spans of different brands of television. The dealers often receive televisions for assessment of whether the set is worth repairing or whether it should be replaced by a new one. The study was carried out in 1992 and repeated in 1994. An average life of 12.8 years for a Beovision LX 5500 is used on the basis of this study.

- English average: the above study with meters was also made for televisions abroad. The meters in England showed that English televisions are switched on for an average of 5.3 hours a day. The study covered 94 televisions from England. Bang & Olufsen's subsidiary in England receives discarded televisions from the English market. An average life of 8.5 years was assessed.

Table 27.6. The three use scenarios assessed for the Beovision LX 5500

Use scenario	Daily operating time (hours per day)	Lifetime of TV (year)	Total operating time (hours)	Total standby time (hours)
Dimensioning criteria from B&O	6.0	10.0	21,900	65,700
Danish average	4.0	12.8	18,688	93,440
English average	5.3	8.5	16,443	58,017

The two scenarios are shown in Table 27.6 together with the reference scenario. The operating time in the table is given as the number of hours for which the television is switched on in its entire life, classified into "operating time" and "standby time".

As in the reference scenario, an average electricity production in the EU is used in both scenarios. The consequence of a different electricity production is discussed below and illustrated in Figure 27.17.

It is clear from Figure 27.16 that the alternative use scenarios are of considerable significance for total resource consumption and environmental impact potentials. It is clear that a television used for 4 hours a day uses less electricity than a set used for 6 hours a day. The energy-related resource consumptions and environmental impact potentials are thereby reduced. A television which is only used for 4 hours a day will also have a longer life. This means that the consumer will replace his or her television less frequently, which means fewer total production and manufacturing processes. When exports of the Beovision LX 5500 are taken into consideration and compared with the fact that there are large

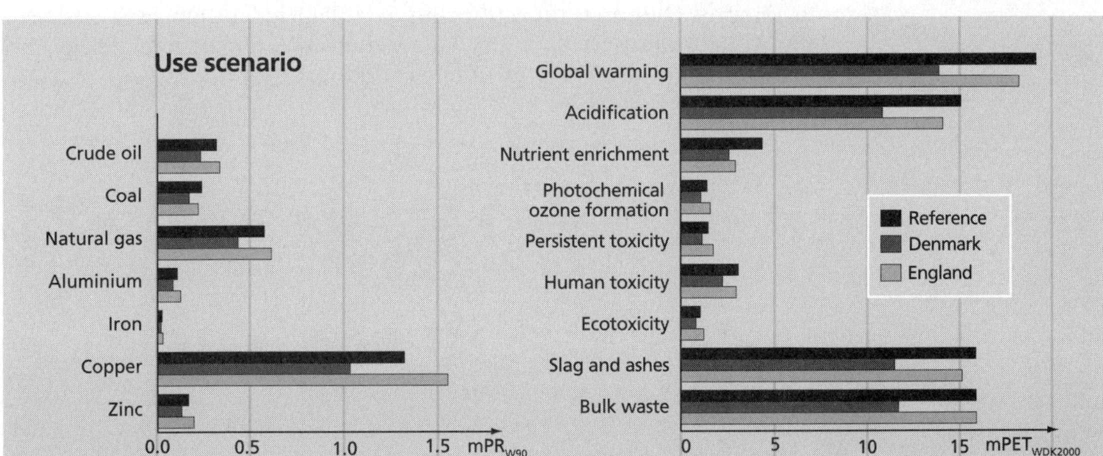

Figure 27.16. Three use scenarios for the Beovision LX 5500. The reference is Bang & Olufsen's dimensioning criteria shown here together with Danish and English averages. Weighted resource consumptions and environmental impact potentials

differences in TV viewing habits in different countries, the significance of the choice is considered to be large. Energy consumption in the use stage throughout the television's life is ca. 24% lower for the English scenario than for the reference. At the same time, the life is only 8.5 years, which gives a higher energy consumption for extraction of raw materials and production of the television. For televisions sold in England, the use stage therefore does not mean nearly as much for the total energy consumption. The significance of the use stage is probably somewhat overestimated with the operating time chosen.

The disposal stage

Disposal of the television is considered to be relatively well determined for Danish conditions, but for European conditions the scenario for disposal of household waste is used, as better information could not be obtained. There can be large differences in how disposal of household waste and electronic products occurs, and it was not possible to include local data on disposal in the countries to which Bang & Olufsen export. The disposal can therefore be significantly different from the way in which it is organized in the computations. Alternative disposal scenarios are given under the simulations in the next section, where the significance is discussed.

Like printed circuit boards from other electronic products, the printed circuit boards contain fire inhibitors which can be toxic on disposal. On incineration they can form compounds which are carcinogenic. This was not analysed further in the LCA and its significance cannot, therefore, be assessed.

During disposal, heavy metals in the television can have toxic impacts. Dumping in landfills can lead to percolation to the groundwater, and incineration can result in emissions to air and heavy metals in the slag and ashes. The significance of this was not assessed.

Transport

Transport of raw materials and processed raw materials up to production of materials is included as average transports for each material, which is considered to be better than specific information on transport for each individual delivery, as noted previously. Transport of materials to Bang & Olufsen's subcontractors is not included, but as it has been shown that transport is of very little significance for the total results, this omission is considered to be of no significance.

Production of electricity

The television's total resource and environmental profile depends to a high degree on where it is used, because of differences in electricity production. Figure 27.17 shows how resource consumption and potentials for environmental impacts appear for the television's entire product system in different countries with different electricity production:

- An EU average (the reference scenario)
- Denmark (mainly coal-fired co-generation power plants)
- France (a high level of nuclear power)
- Norway (almost exclusively hydroelectricity)

As is evident in Figure 27.17, the country's electricity production is of very great significance for the energy-related resources crude oil, coal and natural gas, while no change is visible for the resources entering into the television as materials. The same applies to the energy-related environmental impact potentials global warming, acidification, nutrient enrichment and the waste categories radioactive waste, slag and ashes and bulk waste, which are highly dependent on how the electricity is produced.

In Denmark, production of electricity in many places occurs together with production of heat in co-generation power plants. The emissions from co-generation of power and heat are divided between electricity and heat, but the division can be made in several different ways. The method is described in the sections on allocation in Parts II and IV. As the production of electricity is of decisive significance for the LCA of a Beovision LX 5500, the method of allocation is important for the result. The choice of allocation method is believed to lead to a variation of maximum 15% as a broad average for all impact categories.

Figure 27.17. The Beovision LX 5500 used in countries with different electricity production. Weighted resource consumptions and environmental impact potentials

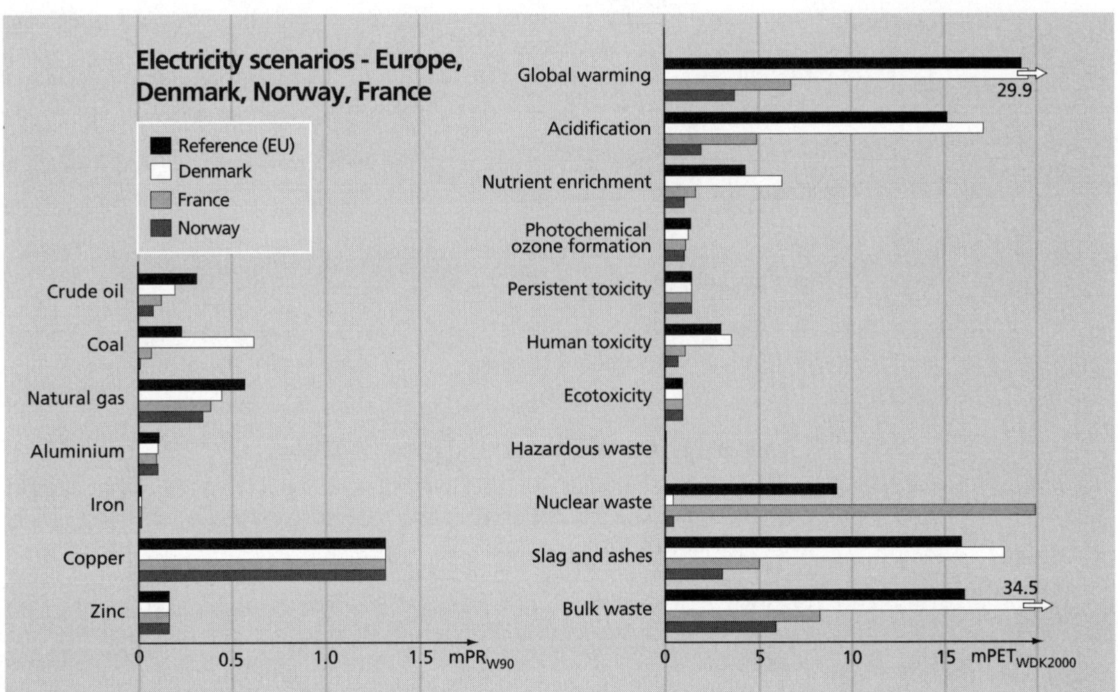

27.2. Environmental diagnosis for a television - the Beovision LX 5500

The environmental diagnosis includes simulation of various changes in the product or its product system. The result of the simulations shows the consequences of the changes. A review of the alternatives shows whether there are environmental potentials for improvement, and where they reside in the product.

Simulations of changes in the television or its product system

The simulations of the various environmental strategies show what can be gained in environmental terms, i.e. the potential for improvement if the various strategies are implemented. The simulations shown here answer questions such as:

- What are the environmental consequences if all metals are 100% recycled?
- What are the environmental consequences if the standby energy is halved?
- What are the environmental consequences if the casing is not painted?
- What are the environmental consequences if the life span of the television is increased?

The simulations are carried out by changing the various parts of the product and its product system. Some of the proposals for improvements have been implemented by Bang & Olufsen, and these are described in the last section of this chapter.

The simulations are made relative to a reference scenario, which is the system discussed in the previous section.

The potentials for improvement are to be found in one of the four areas under the MECO principle:

- Materials: consumption and disposal
- Energy: consumption
- Chemicals: consumption
- Others: especially life span

Consumption of materials and disposal

Many of the materials for the television can be replaced. For example, recycled materials can be used in place of new materials, or wood can be used in place of plastic or metal.

An example could be changing the material for the television's casing. Figure 27.18 shows a casing in wood in place of plastic. As is clear from the figure, the alternative is of lesser significance for the total resource consumption and the environmental impact potentials for the entire product system. About 5% less crude oil and natural gas is used, and the

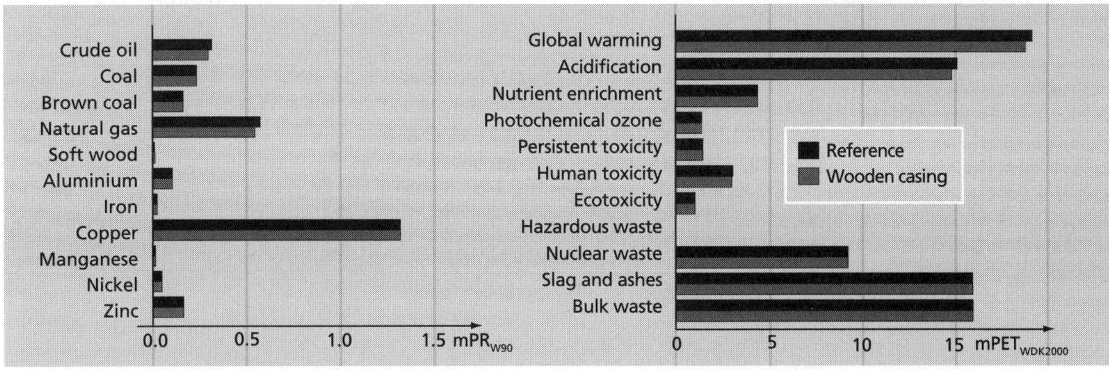

Figure 27.18. Simulation of various casings (wood and plastic). The figure shows weighted resource consumptions and environmental impact potentials

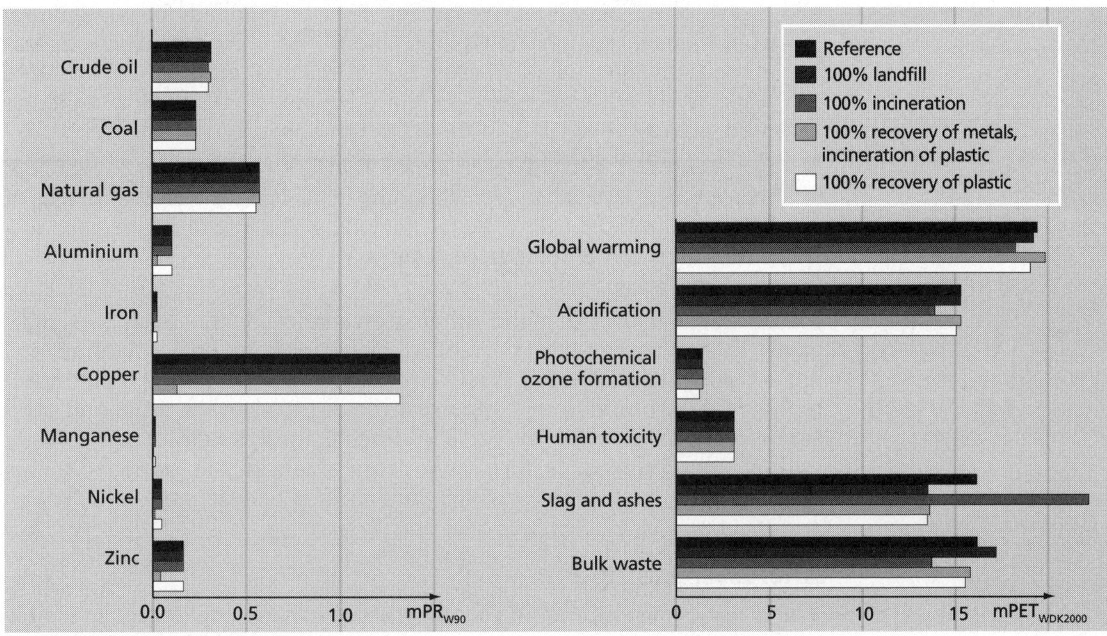

Figure 27.19. Simulation of various disposal routes. The figure shows weighted resource consumptions and environmental impact potentials

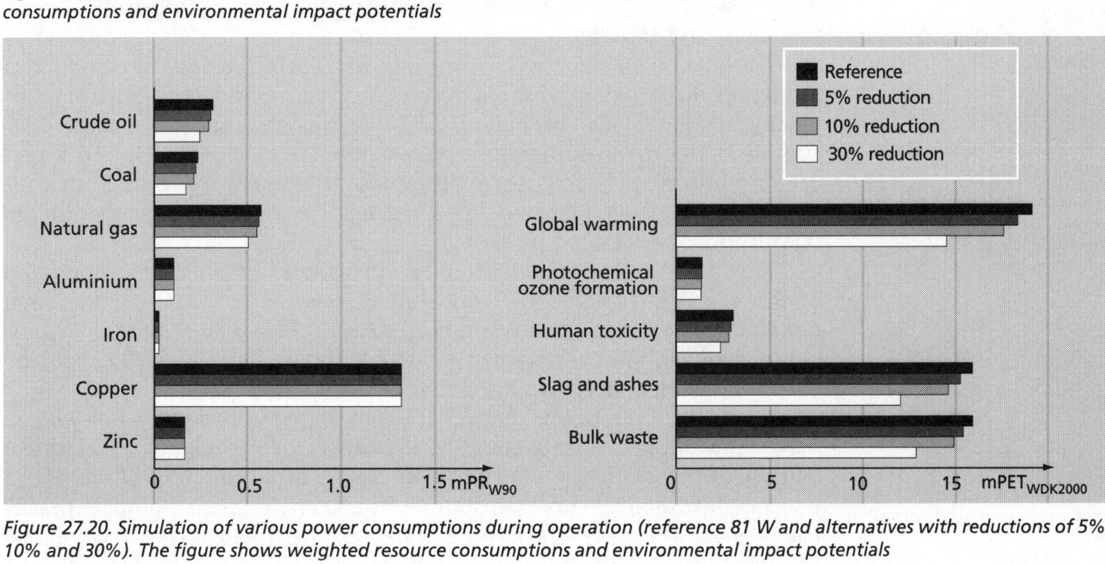

Figure 27.20. Simulation of various power consumptions during operation (reference 81 W and alternatives with reductions of 5%, 10% and 30%). The figure shows weighted resource consumptions and environmental impact potentials

contribution to global warming is ca. 2% less for a television with a wooden casing. However, when looking at the product system of the casing only, the improvements by a wooden casing are much larger, as can be seen from Figure 27.25 in Section 27.4.

Disposal of a Beovision LX 5500 can occur in several ways. In the reference LCA, 70% of the televisions are dumped, while the remaining 30% are incinerated in a waste incineration plant. Four alternative scenarios are set up here:

- 100% dumping.
- 100% incineration.
- Recycling of metals and 100% incineration of plastic. It is assumed that for the quantities recovered, 75% is credited for aluminium, iron, manganese, nickel and zinc, and 90% for copper. These figures relate to the allocation, where a loss of material grade of 25% or 10% respectively was assumed here. All other items, including the picture tube, are disposed of as in the reference scenario.
- Recycling of plastic. It is assumed that 80% of the plastic which is recovered is credited, corresponding to a loss of material grade of 20% per use. All other items, including the picture tube, are disposed of as in the reference scenario.

Figure 27.19 shows that the different scenarios do not change significantly in the energy-related resource consumptions and environmental impact potentials. There is, however, a significant saving to be found in recycling: consumption of copper can be reduced by ca. 90% and consumption of aluminium by ca. 75%. Incineration results in significantly greater contributions to slag and ashes than in the other scenarios because of the picture tube, which ends in the slag fraction.

Energy consumption

The energy consumption in the use stage is of very great significance for many environmental impact potentials and resource consumptions from a Beovision LX 5500. Electricity consumption can be reduced in both operation and standby.

A reduction in the energy consumption during operation results in a reduction in the energy-related resource consumptions and environmental impact potentials. In Figure 27.20, a reduction of the energy consumption during operation of a Beovision LX 5500 by 5%, 10% and 30% is simulated. Global warming represents the energy-related environmental impact potentials, and it is reduced by ca. 8% when the energy consumption is reduced by 10%.

Bang & Olufsen have achieved energy savings of 7% during operation of the television Beovision Avant, but it must be emphasized that large reductions in electricity consumption, such as 30%, will require a technological innovation beyond what is possible today. Figure 27.20 shows the effect of saving energy during operation of the television.

A realistic option for reducing electricity consumption would be to halve consumption during standby. It is evident from Figure 27.21 that a halving of the electricity consumption results in a reduction of ca. 4% in the energy-related resource consumptions and environmental impact potentials: consumptions of crude oil, natural gas, coal and brown coal; global warming, acidification, nutrient enrichment and all categories of waste.

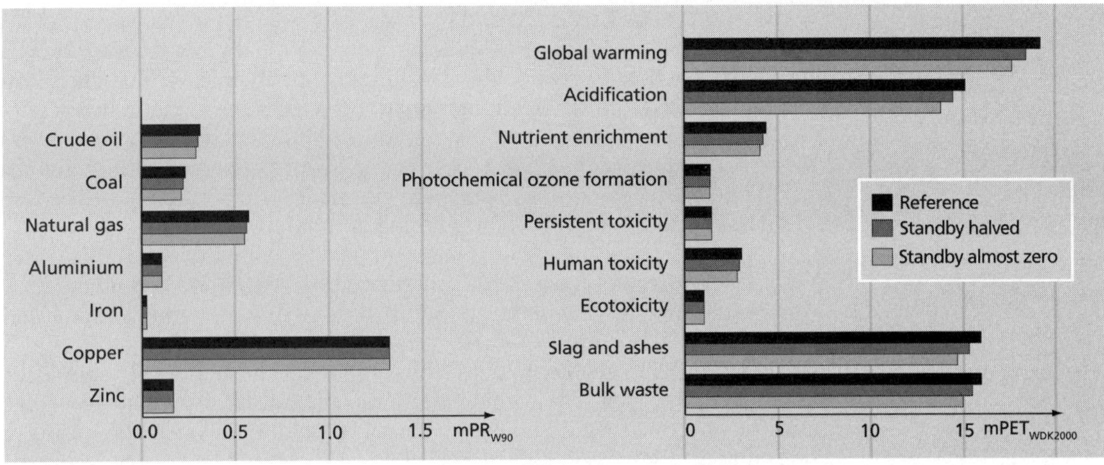

Figure 27.21. Simulation of different power consumptions in standby (the reference is 3 W with alternatives of 1.5 W and close to 0 W). The figure shows weighted resource consumptions and environmental impact potentials

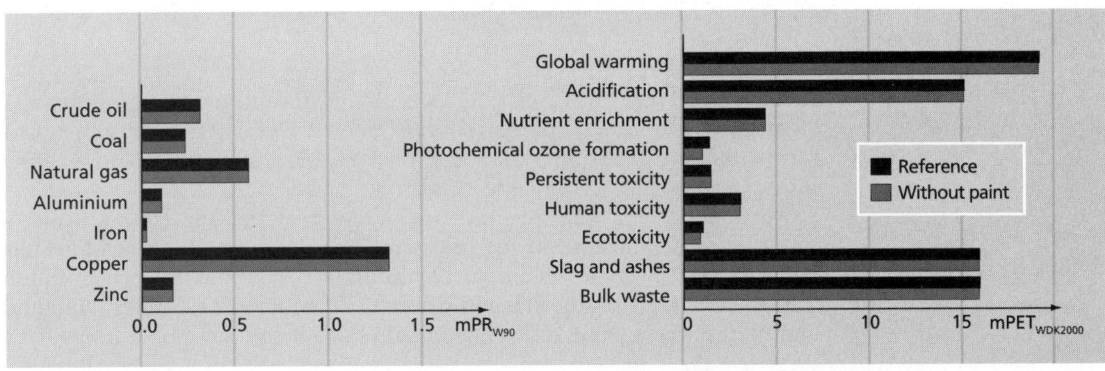

Figure 27.22. Simulation of casing, rear cover, front panel and fitting for front glass with and without painting. The figure shows weighted resource consumptions and environmental impact potentials

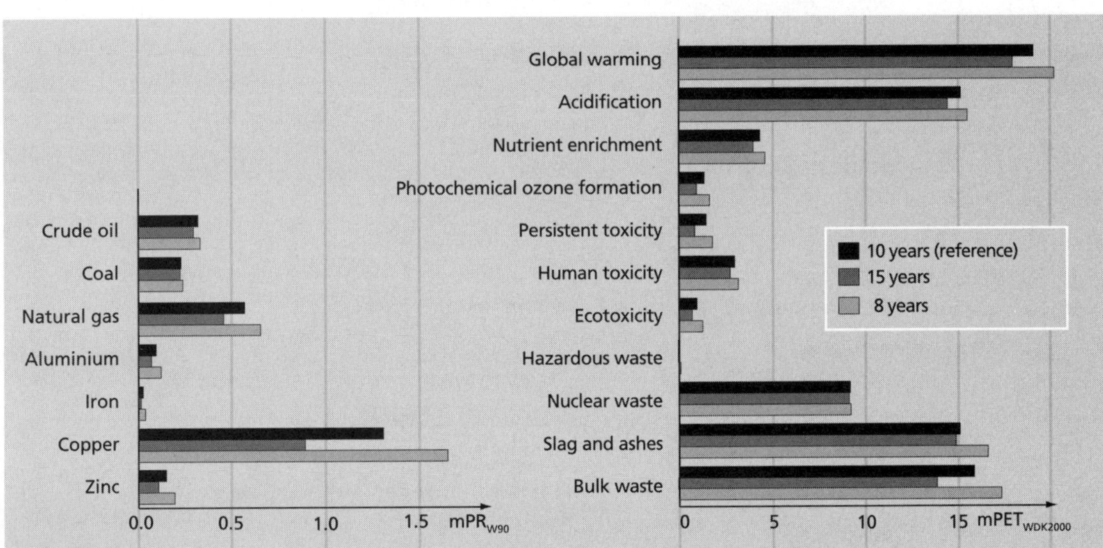

Figure 27.23. Simulation of different life spans (reference 10 years with alternatives of 8 and 15 years). The figure shows weighted resource consumptions and environmental impact potentials

Consumption of chemicals

It may be possible to reduce the consumption of chemicals. The chemicals used in surface treatment were examined. An alternative scenario was set up:

- Avoid painting the casing, rear cover, front panel and fittings for the front glass (the casing with paint is made of PS plastic, while ABS plastic must be used when painting is omitted).

As is evident from Figure 27.22, there are differences for individual environmental impacts in whether painting is done or not, namely about 72% reduction in the contribution to photochemical ozone formation and a smaller reduction for ecotoxicity for unpainted alternatives. It should be noted that the local environment around the painting process itself can also be improved significantly by avoiding painting. When the casing is not painted, the plastic can be recycled, and the impact of this is already shown in Figure 27.19. When looking at the product system of the rear cover only, the significance of the painting can be seen to be very large, cf. Figure 27.27 in Section 27.4.

Other - life span

A longer life span will reduce practically all resource consumptions and environmental impact potentials (cf. Figure 27.23). Longer life spans mean that the consumer does not buy as many new televisions during his or her life, and energy and resources for the extraction of raw materials, production of materials and manufacture of the television are therefore saved. An increase in the life span from 10 years to 15 years results in savings for all types of impacts, assuming that electricity consumption during operation is not changed in new televisions. For copper, zinc and aluminium, the savings will be around 33%.

Simulations – summary

The potentials for improvement from the simulations are reviewed in Table 27.7.

Table 27.7. Conclusions of the simulations

Source cf. MECO principle	Potentials for improvement
Materials	Choice of materials is significant for resource consumptions. For example, wood can be used in place of plastic for the casing, reducing consumption of crude oil and natural gas by ca. 5% and the contribution to global warming by ca. 2%. Recovery of the materials is significant for the resource consumptions of materials used in the television: the consumption of copper can be reduced by ca. 90% and consumption of aluminium by ca. 75%.
Energy	A reduction in energy consumption in the use stage is of great significance for both resource consumption and environmental impacts. The contribution to global warming will be reduced by ca. 8% when energy consumption is reduced by 10%.
Chemicals	It may be possible to substitute or avoid the chemicals used in surface treatment. The contribution to photochemical ozone formation and ecotoxicity can be reduced by omitting painting of the casing, the rear cover and the front panel. The contribution to photochemical ozone formation can be reduced by ca. 27% if they are not painted.
Others	The life span is significant for total resource consumptions and environmental impact potentials. An increase in life span from 10 to 15 years will result in savings up to ca. 33%.

Where do potentials for improvement reside in the television?

The results of the simulations and of the LCA were used to designate where individual parts of the television can be changed to achieve an environmental improvement. The alternatives, which are not intended to be exhaustive and which have not been weighed against other considerations, should be seen as proposals or examples. No accurate assessment of how a technical solution might appear was made at this stage. The alternatives are arranged at the levels of concept, structure and component in Tables 27.8-27.10.

The concrete initiatives applied by Bang & Olufsen in product development are reviewed in the last section, "Environmental design of new products".

Table 27.8. Alternative solutions - conceptual choices

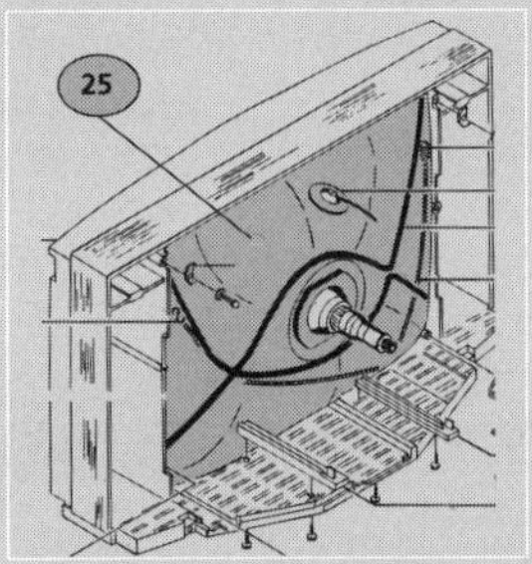

Picture tube: Most of the energy consumption in the use stage is ascribed to the picture tube. The picture tube is also the component which uses most energy in the material production stage (natural gas is used to melt the glass). The picture tube also contributes to 24 kg slag and ashes or bulk waste on disposal, according to whether it is incinerated or dumped.
Alternative: entirely new alternatives for the picture tube will appear if the LCD (Liquid Crystal Display) technology known from portable computers can be developed further for use in televisions. According to some estimates the LCD technology will use only half as much energy in operation, although this figure is disputed because the quality does not match that available with current picture tube technology. LCD television screens are currently made only for very small or very large screens; it is not known whether LCD technology will be used for the common TV formats. Contributions to toxicity may be associated with liquid crystals. Special provisions may therefore be required for manufacture and disposal. Several new technologies known as laser scanning and plasma display are under development, but their environmental impacts are unknown.

Energy consumption in the use stage: part of the energy consumption in the use stage is ascribed to the user forgetting to switch off the television or using it as background entertainment.
Alternative 1: it may be possible to install a sensor (similar to burglar alarm sensors) which will register whether somebody is or has been in the room during e.g. the preceding 15 minutes, so that the television can be switched off if nobody is watching it.
Alternative 2: activity can be registered from the remote control and a message given that the television will be turned off if there has been no activity for a certain time.

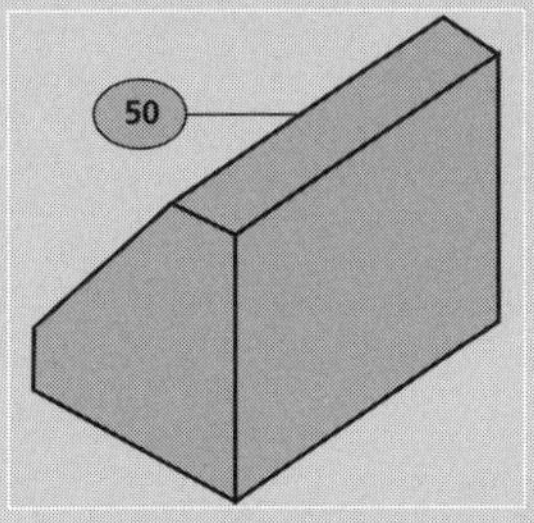

Electronics: the electronics contain numerous metals which are scarce resources, e.g. copper, tin, lead, silver and gold. Metals can have toxic impacts during production and on disposal. The printed circuit boards often contain fire inhibitors for safety reasons. Fire inhibitors can produce toxic impacts on disposal.
Alternative: a future alternative to the conventional printed circuit board is thick film technology, which is discussed further in the Danfoss example in Chapter 30. Use of thick film technology will reduce the amount of soldering and thus increase the average life span of the electronics. At the same time, consumption of scarce metals will be significantly reduced. In the Danfoss example, replacement of conventional printed circuits by thick film results in savings in resources of ca. 70% for tin, ca. 65% for lead and ca. 25% for copper. The exact savings depend very much on the construction of the printed circuit boards which are replaced and the construction of the new thick film.

Standby function: the standby function uses about 10% of the energy consumed in the television's use stage.
Alternative: standby consumption is reduced by, for example, constructing a television and video recorder together and having them use the same standby function. The standby consumption could be more than halved for the television and video recorder together.

Tabel 27.9. Alternative solutions - structural choices

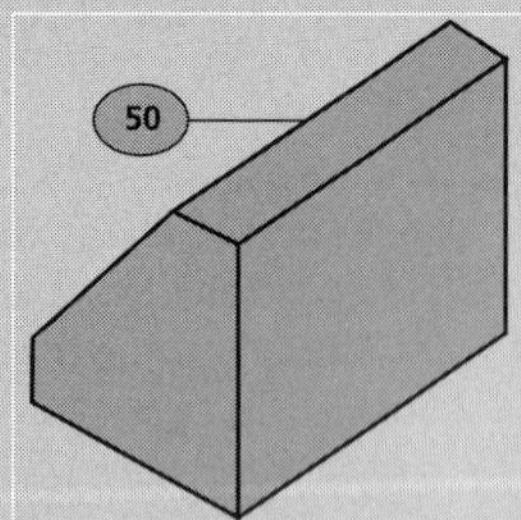

Soldering: temperature fluctuations can result in tension in the solder because of different coefficients of expansion. This can result in cracks which cause functional faults or which lead to the television not working.
Alternative 1: the construction of the electronics can be changed by providing better ventilation to soldered components, reducing heating of the solder during use and thus increasing its life span.
Alternative 2: changing the composition and application of the solder can result in stronger joints, but more tin is required.

Number of materials: recovery is made more difficult if there are numerous different types of materials in an apparatus, as it is more difficult to divide them into groups of materials for recycling. The casing and the rear cover are made of PS plastic, but the front panel is made of ABS plastic.
Recovery can potentially save up to 75% of the resource consumption of aluminium, iron and zinc and 90% of the copper consumption.
Alternative 1: attempts can be made to limit the number of materials, for example by using only very few types of plastic.
Alternative 2: consideration can be given to omitting some items altogether, e.g. decorative items.

Table 27.10. Alternative solutions - choice of component

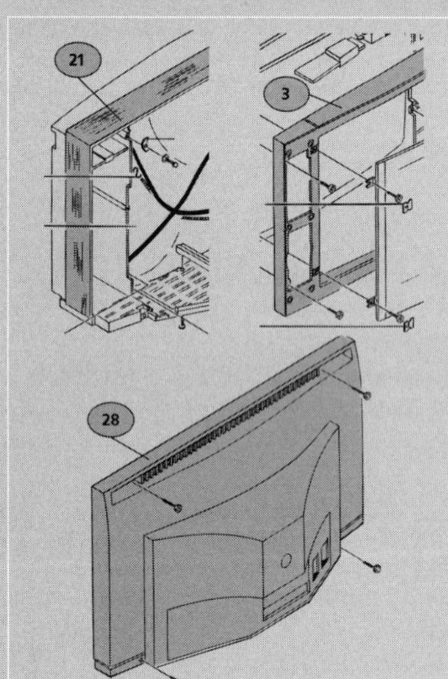

Casing, rear cover and front panel are made of painted plastic. The paint on the plastic reduces the qualities of the recycled plastic. In particular, the plastic's resistance to impact is reduced significantly. This significantly reduces the options for recycling the plastic.
Alternative 1: by avoiding painting the plastic and instead producing it directly with an acceptable surface structure (e.g. using an etched tool surface), the plastic will be able to be recycled. Resources and energy used for production of the plastic (crude oil and natural gas) will thus be saved. Emissions from the painting process will also be avoided. It will be easier to recycle the plastic if all parts are made of the same type. Bang & Olufsen have carried out LCAs on the rear covers with and without painting: see the section "Environmental design of new products".
Alternative 2: the casing can be made of entirely different materials, e.g. wood or metal. Bang & Olufsen have carried out LCAs on alternative materials for the casing: see Section 27.4, "Environmental design of new products".
Alternative 3: develop paints which do not reduce the recycled plastic's qualities, e.g. certain water-based paints.

Table 27.10. (continued)

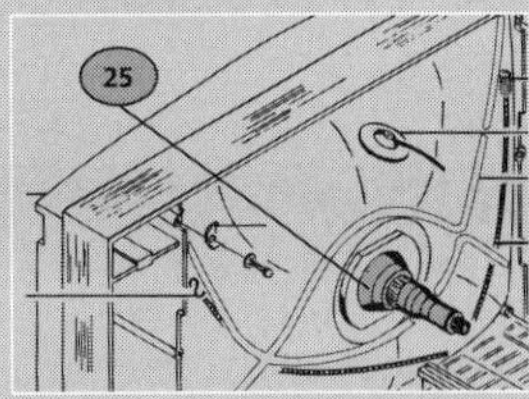

The deflection coil is the single component with the largest responsibility for the significant consumption of copper in the Beovision LX 5500.
Alternative: either choose a different material if something suitable can be found, or construct the television so that it is easier to separate, with the copper part being easy to remove and recycle.

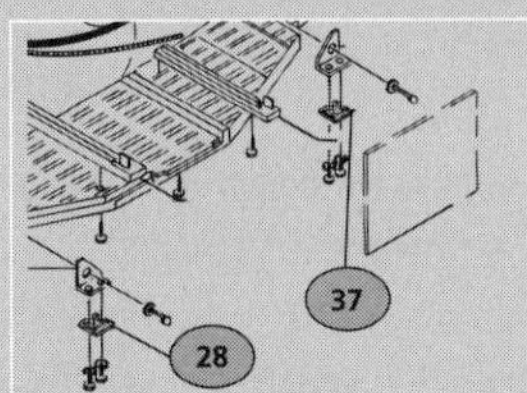

Spacers for the picture tube are made of zinc. The quantities are not large, but the supply horizon for zinc is only 20 years, and consideration can therefore be given to making the spacers in a different material.
Alternative: The spacers can possibly be made of steel, cast iron or recycled aluminium.

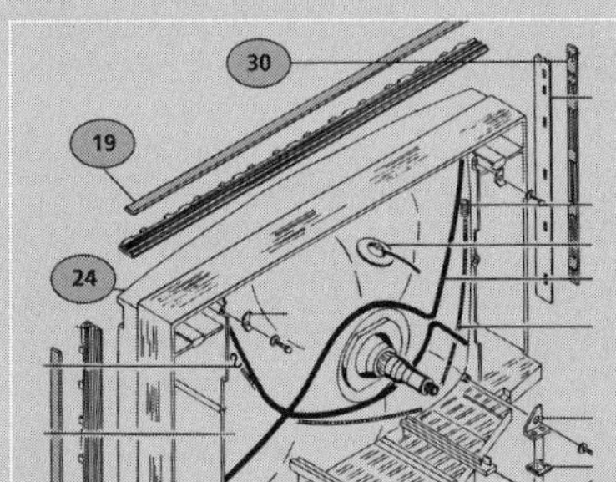

Profiles for the casing are made of primary aluminium. Production of primary aluminium requires large amounts of energy.
Alternative 1: consideration can be given to using a different material, e.g. painted recycled aluminium or painted steel, which could, however, cause problems with magnetic effects on the picture tube. Production of new aluminium uses much more energy than recycling of aluminium, so use of painted recycled aluminium will result in energy savings.
Alternative 2: it may be possible to construct the casing so that profiles are not necessary.
Alternative 3: again, it will be an improvement if the television can be taken apart easily so that the parts can be recovered.

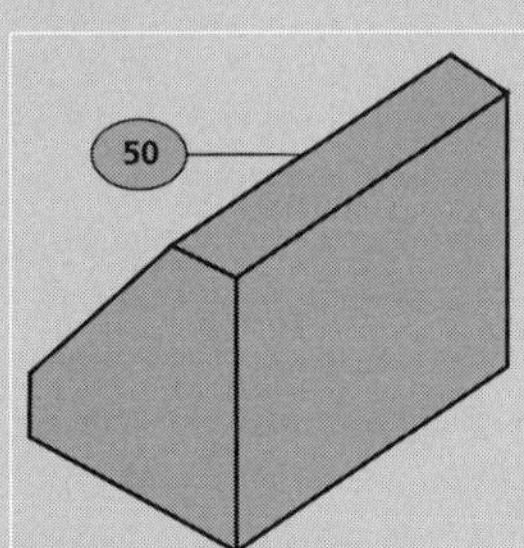

Electronics: as discussed under choice of concept, the electronics can contain fire inhibitors and other substances which can be toxic on disposal.
Alternative 1: several types of fire inhibitors are available on the market. The toxic characteristics of the various types can be assessed in the choice.
Alternative 2: ceramic printed circuit boards can be used. These can simply be heated up for removal of the electrical components on disposal.

Electronics: ca. 25% of the energy consumption in the use stage occurs in the electronics.
Alternative: this energy consumption can be reduced by choosing circuits which use less power by reducing the mains voltage and by changing to component types which use less power. Bang & Olufsen have implemented these in the Beovision Avant and achieved savings of 7% in energy consumption during operation (see last section). A reduction in power consumption (and therewith in heat given off by the components) will also increase the life span of the solder.

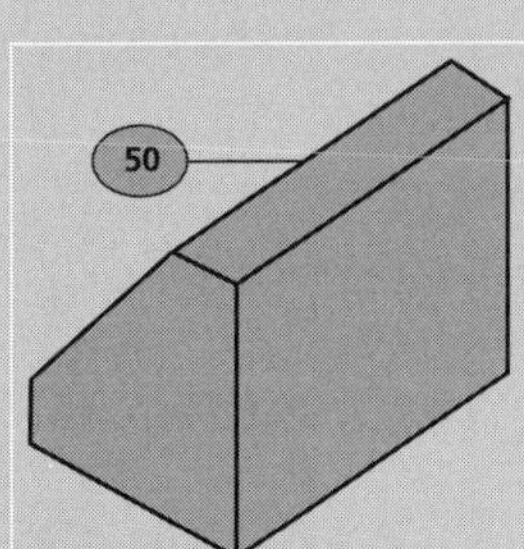

Standby function: the standby function uses 10% of the energy consumed in the television's use stage.
Alternative: the standby function's power consumption can be reduced under the same principle noted above, and consideration can also be given to using components which are specially designed for standby mode, e.g. a separate power supply (the power supply in a television will be designed to be able to handle the maximum power required in use).

Soldering: solder typically consists of 63% tin and 37% lead, both of which are scarce resources and potentially toxic on emission.
Alternative: solder can be replaced by glue (usually epoxy) in some cases. However, there are problems with poor mechanical strength, poor conducting ability, difficulties with repair, and possibly safety problems. According to the "Kamille Project" (Westphal *et al.*, 1995), it is better environmentally to glue than to solder, despite the fact that silver is often used as a conductor in glue with good conducting abilities, and that this can give problems on disposal because silver can be toxic. The epoxy itself can cause problems in the working environment.

Where can potentials for improvement be found in the surroundings?

Table 27.11. Alternative solutions - the surroundings

Disposal: local conditions are of major significance for how disposal occurs, whether it is a case of dumping or incineration or, in future, recovery. This is attributable to different technical possibilities, population density, traditions and statutory provisions. Low raw materials prices and high wages mean that there is no economic incentive today to recover the materials in a television.
Alternative: comprehensive national regulations could be promulgated for collection, recovery and disposal of electronic products, including televisions. A European regulation would be of even greater benefit. A change in the economic conditions for recovery will result in an increase. At the moment there are no satisfactory facilities for recovery of electronics. The degree of recovery is generally too low.

Energy consumption in the use stage
Alternative: the most significant reason why the use stage is of such significance for the television's contribution to energy-related resources and environmental impact potentials is society's choice of electricity generation without utilization of the waste heat. Future better energy utilization and more environmentally friendly generation of electricity will significantly reduce the impact of the use stage.

27.3. Environmental specification for televisions

When a manufacturer develops a product, the considerations include customer wishes, competitive parameters, and statutory requirements. These different parameters with results of the LCA and the environmental diagnosis have contributed to long-term environmental goals at Bang & Olufsen and to the environmental product specifications for a series of new products.

The competitive framework

Bang & Olufsen group the features of the product into three main categories: positioning, expectation and mandatory features.

For a Bang & Olufsen product, the positioning is the basis of the business and the product's competitiveness. It is the characteristics which especially distinguish and differentiate the product on the market and to the customer. No environmental features are included in this category.

The expectations are the qualities perceived by the target group and the market as natural features in a Bang & Olufsen product and which are not positioning characteristics. These include, for example, low energy consumption, absence of toxic substances in the product, and long product life.

The mandatory features must ensure the customer a safe and trouble-free use of the product throughout its life, and ensure market access for Bang & Olufsen's products. The mandatory features are defined in international standards and in internal Bang & Olufsen standards to which the product must conform, e.g. safety, possibility of recycling and standby energy consumption.

Bang & Olufsen's market analyses show that Bang & Olufsen's customers rank the environment fairly high in their choice of product, but the product's environmental qualities are not among the most significant. The environmental qualities belong under expectations and mandatory features.

Comparison with competitors for some important product properties

For some of the environmentally significant product properties noted above, it is possible to assess where Bang & Olufsen is ranked relative to its competitors. These concern:

- Energy consumption
- Possibility of recycling
- Product life span

An assessment was made of Bang & Olufsen's ranking relative to other televisions on the market for all three characteristics.

Energy consumption

In 1993, the National Consumer Agency of Denmark made a comparison of common 28″ televisions, including energy consumption. Power consumptions from 2 to 18 W were measured in standby and from 75 to 107 W in operation. However, neither the best nor the worst combinations of standby and operating power consumptions occurred. It was also noted that a single model with a standby consumption of 0.1 W exists. The Beovision LX 5500 has a power consumption of 81 W in operation and 3 W in standby, and the new Beovision Avant television uses 75 W in operation and 3 W in standby. Bang & Olufsen is therefore at the respectable end of the scale, and work is continuing on minimizing power consumption in both operation and standby.

Recycling

For the most critical resource, copper, consumption was judged to be of the same order for all televisions. The differences between televisions therefore lie in the dismantling options, i.e. how difficult it is to remove the rear cover. It was quickly found that there are no noteworthy differences among the individual products. Bang & Olufsen are working on improving their products' ease of dismantling.

Life span

When the customer decides to dispose of the television, it is most often because it is too expensive to repair relative to buying a new television. Repairs cost about the same, irrespective of whether the television is cheap or expensive, large or small. Cheap televisions will therefore be discarded earlier than expensive televisions, and they will thus have a shorter product life. Bang & Olufsen offer some spare parts and replacement modules which enable several repairs/replacements, increasing the television's total life.

Statutory requirements

A system of voluntary environmental regulations is currently in operation. The system includes provisions for take-back arrangements for electronics and maximum limits for standby power consumption, and Germany, Sweden and the Netherlands, for example, have gone a long way with respect to the take-back arrangements. Switzerland has already introduced voluntary maximum limits for standby power consumption in a large number of electronic products, and the EU is also working on their introduction.

Long-term objectives

The market's ranking of the television's environmental characteristics was weighted against experiences from the LCA and the environmental diagnosis. Based on a total evaluation, Bang & Olufsen have decided on the following ranking of priorities in their long-term work:

- Extend implementation of the conclusions from the life cycle assessments and environmental diagnoses in the development process.
- Carry out technology development projects with the object of reducing energy consumption during operation and standby.
- Carry out technology development projects with the object of increasing the possibilities of environmentally friendly disposal, e.g. by avoiding painted surfaces or by using a lacquer which permits recycling of the plastic, or by ensuring that the product can be dismantled with only one tool, and minimizing the time required for dismantling.

- Establish a formalized cooperation with suppliers on reduction of the environmental impact of their deliveries, with specifications for deliveries.
- Investigate the relevant environmental information for stakeholders in the product system after manufacture, e.g. guidelines for dismantling, degree of recycling possible, placing and handling of critical components and power consumption during operation and standby, and provide information on these aspects with the product.

Product specifications

Bang & Olufsen emphasize the following in their product specifications:

- Avoidance of toxic substances or substances which, for example, will give rise to toxins on disposal.
- Reduction of power consumption during operation.
- Reduction of power consumption during standby.
- Designing the product so that it is easy to dismantle (mechanically or manually).
- Choosing materials and processes which enable recycling, e.g. avoiding the painting of surfaces, or using a lacquer which permits recycling of the plastic.

These elements are part of the product specifications of the model as shown in Table 27.12.

Table 27.12. Model for environmental product specifications for a new product in 1996

Environmental specification for product NN

Manual dismantling done by persons without prior knowledge of the product must not take any longer than _____ minutes.

The product must be designed in such a way the it is possible to reuse
____ % copper,
____ % iron,
____ % aluminium and
____ % plastic.

Power consumption during standby must not exceed ____ W.

Power consumption during use must not exceed ____ W.
FR2 circuit boards are produced of laminate that does not contain: ___________
____
____

Condensers and other electrical components must not contain: ___________
____
____

Regarding the specifications in the first two items, it is recommended to use Bang & Olufsen's internal guidelines , "Design for recycling".

It is recommended that one or several reviews are made in order to see if product NN fulfils the environmental specifications.

27.4. Environmental design of new products

Product development at Bang & Olufsen

Consideration of the products' environmental characteristics must be integrated into the conventional product development routines and treated in the same way as other considerations.

Product development at Bang & Olufsen occurs as integrated product development, i.e. purchasing, service and production are included at an early stage of the project. The development plan and the methods and tools used by Bang & Olufsen are shown in Figure 27.24. Product development starts with the concept stage, where the general principles are fixed. This is followed by the structure stage, where the principles are made specific, and the detail stage, where solutions are found at the level of the components. The methods and tools currently used by Bang & Olufsen for environmental product development are described below.

- A simplified form of LCA and environmental diagnosis. On the basis of existing data from the inventory for the reference product in the EDIP project, the database is continually extended by assessments from new products. The LCA and the environmental diagnosis are carried out on this basis in a simplified form as the basis for environmental specifications for new products.
- Environmental specifications. These are developed in cooperation with an environmental specialist and the project group, and they include the project's goal and product specification as shown in the model in Table 27.12.
- DFR (Design for Recycling). Covers design guidelines for factors to be taken into consideration to enable the highest possible degree of recycling when the product is discarded. The guidelines are prepared with due account for the product's expected disposal route.

Figure 27.24. Plan for product development at Bang & Olufsen with integration of LCA methods and tools

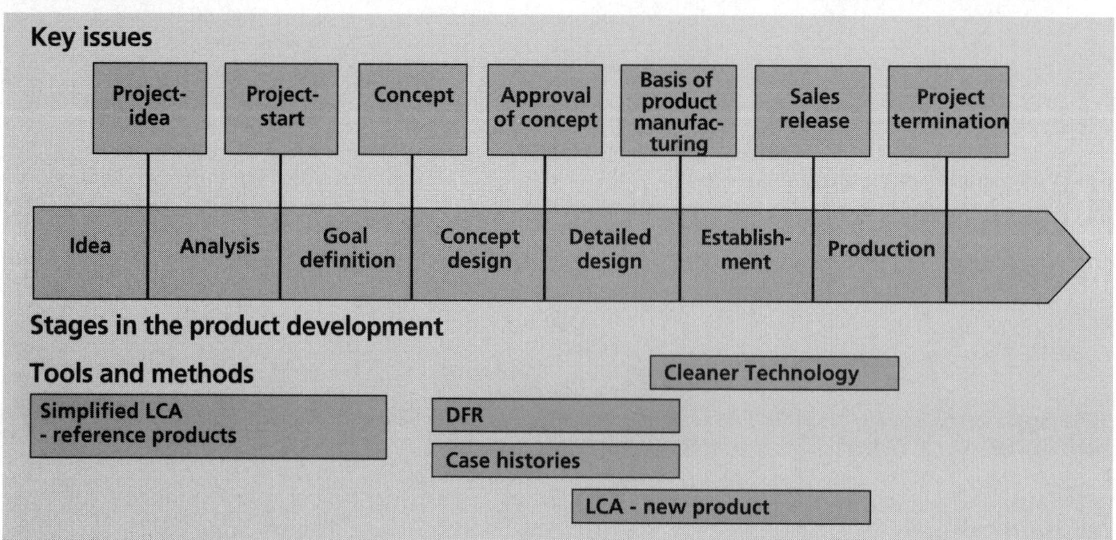

- A collection of case histories. This is continually extended and concerns various designs where the environment is considered during development. It is written by designers and intended primarily for designers. The following section describes the most significant parts of the case history collection's content in 1995.
- Cleaner technology. Covered at Bang & Olufsen by production technologists. Since work proceeds as integrated product development, cleaner technology can be used as early as the construction stage.
- LCA. Undertaken for new solutions on the basis of an assessment of materials which enter into it, particularly energy-heavy or polluting production processes, and Bang & Olufsen's dimensioning criteria with respect to use and disposal as they are expected to occur.

Examples of environmental design in new products

Bang & Olufsen have built environmental considerations into several new products. The most important examples are reviewed here.

Some of the examples are based on the fact that a more environmentally friendly design was desirable. In other cases, a more environmentally friendly design was obtained as an incidental gain deriving from other design requirements.

The examples can be structured after the MECO principle, which was also used in the simulations of potentials for improvements. These are shown in Table 27.13.

Source of the MECO principle	Examples from product development at Bang & Olufsen
Materials	• Main casing for the Beovision Avant
Energy	• Power consumption of Beovision Avant • Power supply in BeoSound Century • Class G amplifier
Chemicals	• Rear cover without paint for the Beovision Avant and the BeoSound Century
Others	• Increased lifetime of the Beovision Avant (is mentioned under power consumption in the Beovision Avant)

Table 27.13. Examples from Bang & Olufsen's product development

The main casing for the Beovision Avant

The Beovision Avant is a video system containing television, video and swivel base. The picture is in wide screen format. The picture is automatically adapted to the screen, but manual adjustment is also possible. The picture width can thus be "adjusted up and down". Twenty percent of all transmissions are currently in wide screen format, which is the generally accepted standard for the television of the future.

	M	E	C	O
Concept				
Structure				
Component	●			

During product development of the Beovision Avant, environmental considerations were included as early as the concept stage, when the television's main casing was to be chosen. The main casing is the support frame in a television, to which practically all other components, the rear, etc. are fixed.

Over the years, the main casing in Bang & Olufsen's televisions has changed in both form and materials. It was originally a regular box of veneered chipboard, later foamed plastic.

Several different options for the main casing for the new television were considered. The casing must fulfil the requirements concerning e.g. low cost, few changes relative to the original design, few items, fast and easy assembly and environmentally friendly design.

LCAs were carried out for main casings in the following materials:

- Moulded polystyrene. Consumption of materials: 3.30 kg PS, of which 0.05 kg is waste.
- Wood (MDF panelling). MDF (Medium Density Fibreboard) is made of fine grinded wood chips with glue and additives. The consumption of glue is simulated as plastic. Consumption of materials: 7.2 kg, or which 3.7 kg ends up as waste.
- Aluminium casing veneered with wood. Consumption of MDF: 3.1 kg, of which 1.0 kg waste. Consumption of aluminium: 2.9 kg, of which 0.3 kg is waste. (Assumption: 80% recycled aluminium and 20% new aluminium.)

The resource consumption and environmental impact potentials for the three proposed solutions are shown in Figure 27.25. Note that the profiles are shown only for the main casing itself. The rest of the television and use of the television are not included as they will be unaffected by the choice. No account was taken in the LCA of the paint, as the same quantity and type of paint would be used for all three solutions. It is assumed that the plastic and the wood casings are incinerated on disposal, while the aluminium casing is recycled.

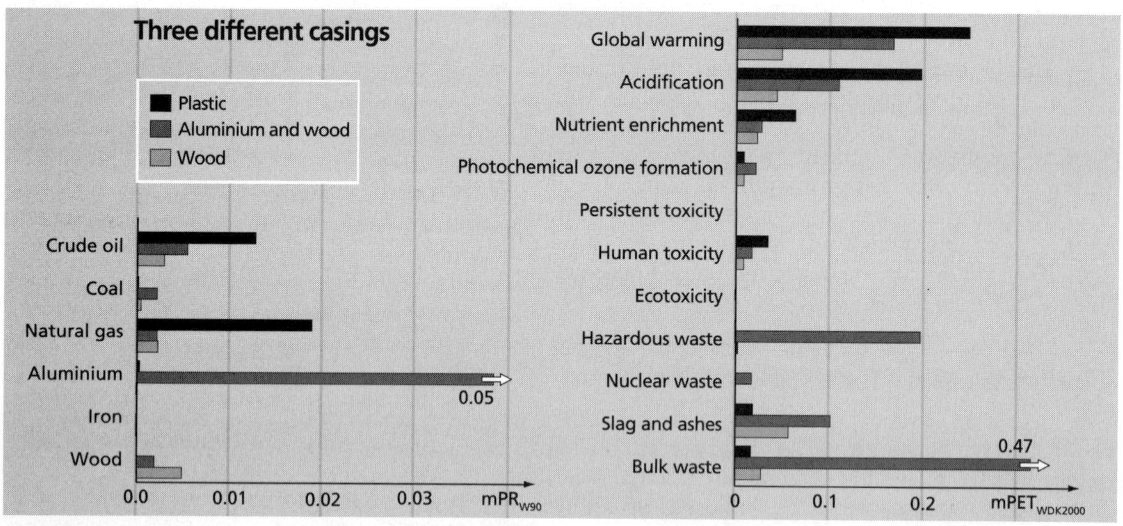

Figure 27.25. Resource consumptions and environmental impact potentials for three proposals for the main casing for the Beovision Avant: one casing of plastic, one of aluminium with wood veneer, and one of wood

Figure 27.25 shows that:

- Not unexpectedly, the aluminium casing consumes the most aluminium, and the plastic casing consumes the most oil and natural gas. These consumptions are significant relative to the other resource consumptions.
- The wooden casing's consumption of wood is not significant relative to the other casings' resource consumptions, and in a total weighting the wooden casing consumes fewest resources.
- The aluminium casing contributes more to radioactive waste, slag and ashes, and bulk waste than the other casings because of the production process for aluminium.
- The plastic casing contributes more than the other casings to global warming, acidification and toxicity to humans.
- The wooden casing contributes to some of the potentials for environmental impacts, but on the other hand contributes least to all categories except production of slag and ashes and bulk waste.

On this basis, Bang & Olufsen judged that a main casing of wood was the most environmentally friendly solution. The production of MDF wood results in emissions of formaldehyde and can result in problems in the working environment which were discussed in connection with the choice of MDF wood. It was decided that this situation does not alter the picture. The wooden casing option was investigated further for other requirements, including low cost and durability in various climatic conditions. As the wooden casing met these requirements, it was chosen by Bang & Olufsen.

Power consumption in operation in the Beovision Avant and increased product life

	M	E	C	O
Concept				●
Structure				
Component		●		

During development of the Beovision Avant, the placing of the electronics was considered at the concept stage. The objective was to increase the reliability of the television's electronics, and the solution meant that the total power consumption for the television in operation was reduced by ca. 7% (6 W). The energy savings meant significant environmental improvements.

Increased reliability was gained by:

- Constructing the electrical chassis with increased ventilation in the apparatus. This was decided as early as the project's concept stage.
- Selecting circuits which used less power.
- Reducing all supply voltages by ca. 6%, thus also reducing their electricity consumption by ca. 6%.
- Constructing a small SMPS (Switch Mode Power Supply) which can convert one voltage to another with very high efficiency and thus without losing power in the conversion.
- Saving 3 W in the power supply by changing to a different type of transistor.

In use, 3 W were saved in the power supply, 1 W in the adjustment and 1.25 W in the circuit, a total of 5.25 W. The efficiency in the power supply is ca. 90%, leading to savings of 5.8 W. The savings in power consumption in various circuits were achieved even though the television has several functions in the form of an additional SCART socket, an illuminated display, sockets for external speakers and camcorder, and the video controls. The lowered temperature also means that there is a greater range of electrical components to choose among.

The increased life of the electronics means an increase in the life of the television. The consequences of an increased life are shown in Figure 27.23 in Section 27.2.

The reduced power consumption during operation means savings in the energy-related resource consumptions and environmental impact potentials, as shown in Figure 27.20.

Figure 27.26. Sketch of traditional and new electrical chassis

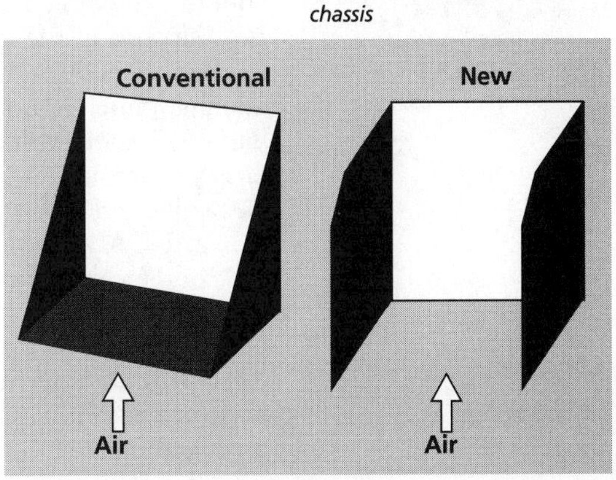

Power supply in the BeoSound Century

	M	E	C	O
Concept				
Structure		●		
Component		●		

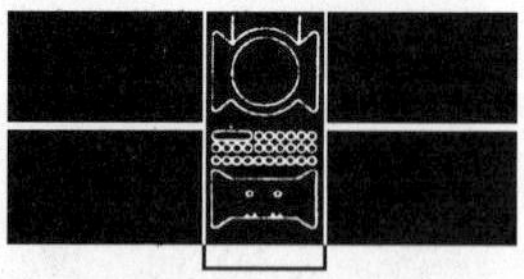

The BeoSound Century is a stereo system with FM radio, CD player, tape recorder and two active speakers. At only 11 cm in height the system is not large, but despite its size it produces good sound quality because of its active loudspeakers and Adaptive Bass Linearization.

For the past 10-15 years, standby power consumption in a stereo system has typically been 5-10 W. During development of the BeoSound Century, a requirement for a standby power consumption of 1 W or less was formulated for environmental reasons. The solution was found in the structure stage of product development.

The choice among possible solutions came down to an SMPS (switch mode power supply) or introduction of a separate standby transformer. The SMPS was rejected because of lack of experience with such circuits in stereo systems relative to the project's schedule. A separate standby transformer was introduced. Apart from the power supply, attention was directed to the necessary power consumers: the microprocessor, IR receiver (which receives infrared light from the remote control) and the display.

When using alternative components, optimization and changes in details of construction of the electronics for the power supply and for each of the necessary power consumers, it was possible to achieve a standby power consumption of 0.8 W.

Attention was also directed to energy consumption in operation. Only the circuits required for a given function (radio, tape, CD) are thus actually switched on. A thermal problem with respect to a cooling panel led to the implementation of class G exit amplifiers, described in the next section.

Energy savings during operation led to savings in energy-related resource consumptions and environmental impact potentials. An incidental gain in the reduced energy consumption was that the life of the Century was increased, as the temperature in the system was reduced as a consequence of the lower power consumption and the fact that electrical circuits are powered only when required. A longer life reduces both resource consumptions and environmental impact potentials, as described above.

Class G amplifier

	M	E	C	O
Concept		●		
Structure				
Component				

The power amplifiers are used by the loudspeakers and other components. Traditional loudspeaker power amplifiers normally have a power supply providing a constant 30 V. A significant proportion of this energy is dissipated in the loudspeaker as heat.

Bang & Olufsen considered several options for reducing the heat. A cooling panel is normally used, but in class G loudspeakers there is a variable supply voltage, such that only 15 V is used when the loudspeaker is playing at low volume. This means more components including two electrolyte condensers.

An LCA of the new class G loudspeaker relative to a traditional class AB loudspeaker showed that the new speakers use less electricity during their life, and thus have lower energy-related resource consumptions and environmental impact potentials, but because of substances from the electrolyte condensers they have a greater potential impact on the groundwater on disposal. This problem will be solved by a future return arrangement.

Rear covers without paint for the Beovision Avant and the BeoSound Century

	M	E	C	O
Concept				
Structure				
Component			●	

On the basis of the expected legislation on mandatory return of electronic products, including requirements with respect to recycling of materials, Bang & Olufsen have worked on a concept for increasing the options for the recycling of materials in their products. Bang & Olufsen have therefore conducted several experiments with the recycling of plastics. The results have shown that painted plastic does not provide a sufficiently good quality of recycled plastic, as the paint remains as flakes in the plastic and forms visible little whorls. The strategy employed in design of the rear covers for the Avant and the Century was therefore to avoid painting the items.

The rear covers must comply with a number of requirements with respect to price, quality and installation, and environmental considerations must also be included.

Obtaining an acceptable surface proved to be a problem. Several experiments were conducted in collaboration with the Danish Institute of Technology before production methods and tools were found which could fulfil the requirements for acceptable surfaces.

Figure 27.27 shows resource and environmental profiles for a rear cover for the Beovision Avant with and without paint. Note that the profiles are shown only for the rear cover. The rest of the television and use of the television are not affected by the choice and are therefore not included. Without paint, the rear cover can be recycled, and a recovery rate of 80% is expected as the raw materials suppliers estimate that the plastic can be used at least five times.

It is evident that the saving is mainly in consumption of crude oil and natural gas, as the recycled plastic replaces these raw materials. There is also a reduction in energy consumption for production of new plastic, which results in lower values for global warming, acidification, nutrient enrichment and the various categories of waste. Human toxicity and photochemical ozone formation are reduced, mainly because the paint solvents are not used for the unpainted rear cover.

For Bang & Olufsen, the changes led to savings in consumption of paint, but the tools were more expensive and the running in more difficult, and the running in time was therefore longer.

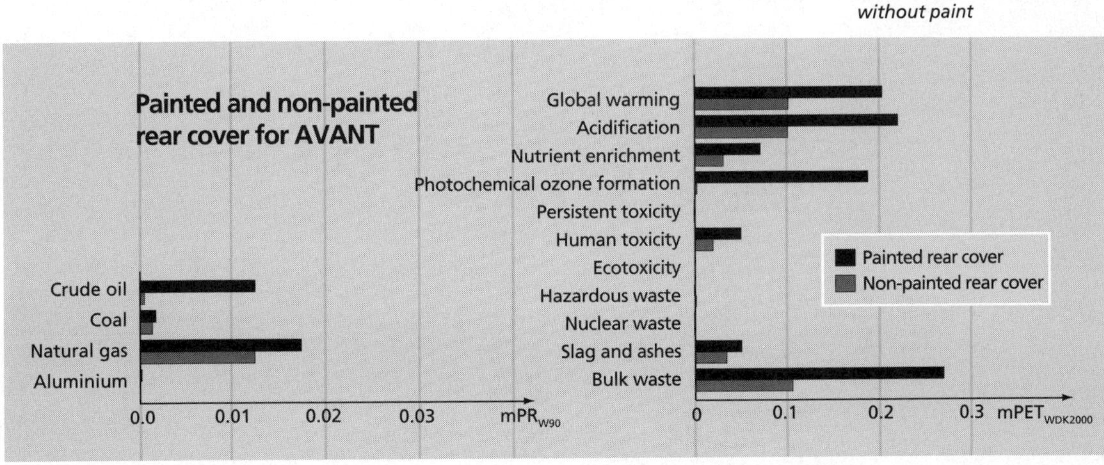

Figure 27.27. Rear cover for the Beovision Avant with and without paint

Bang & Olufsen's experience with environmental considerations in product development

Bang & Olufsen have concluded that it is generally possible to make more environmentally friendly products. The costs involved are primarily forethought, that is, more time spent during development.

It does not necessarily cost more (in variable costs) to make more environmentally friendly products, but it requires more consideration, among other reasons because there are more interrelationships to be considered. A greater insight into the processes which are selected is also required, together with a knowledge of disposal routes and of what can be recycled.

The nature of the work itself is not different from other product development procedures, and as Bang & Olufsen are used to working with subjective parameters, the environment is easier to handle than it might otherwise have been. Starting with broad general tasks has proved to be an advantage, and it is also an advantage that the environment can be a lever for accomplishing other tasks.

Designers do not necessarily require environmental knowledge for their tasks. Some of the tasks are purely technical, but there are often discussions of what is best, e.g. additional electrical components versus reduced power consumption. It can be difficult to find objective arguments which necessitate working with a formulation of the goal or obtaining help from an LCA such as those described above.

That Bang & Olufsen take the environment into consideration means that the environmentally aware among their employees do not have to suffer conflicts with their private attitudes. Environmental work makes their work more interesting, and provides greater challenges, satisfaction and professional pride.

28. KEW Industri: High pressure cleaners

Allan Sand, KEW Industri
Anton Sørensen, KEW Industri
Nina Caspersen, Institute for Product Development

KEW Industri A/S manufactures high pressure cleaners. The company is located in Denmark, has approx. 675 employees, 7 subsidiaries, and 50 importers all over the world

28.1. LCA of a high pressure cleaner - the Hobby 70

The Hobby 70, as illustrated in Figure 28.1, is a portable high pressure cleaner developed for the private consumer.

Choice of reference product

KEW selected the high pressure cleaner Hobby 70 because it is a new product. In contrast to earlier models it is portable. A new concept is therefore involved, but in its construction the cleaner resembles conventional high pressure cleaners.

Figure 28.1. The reference product, the Hobby 70

The functional unit

The high pressure cleaner is used for cleaning, where the combination of water and high pressure results in a strong cleaning effect. The cleaner's efficiency depends on quantity of water, water pressure, temperature and quantity of additives. The cleaner is typically used for washing cars and boats and for external cleaning of buildings and terraces.

In comparing products, a functional unit is used to which the inventory is related. This could be cleaning of a particular area. But cleaning can be difficult to define as a physically measurable quantity. For example, a pulsating stream can clean more efficiently than a non-pulsating one, even if pressure, quantity of water and temperature are the same.

The primary functional unit for the Hobby 70 is:

Cleaning of a surface for 125 hours distributed over a period of 5 years. The cleaning is further divided into seven different categories of objects cleaned.

Consumptions of electricity, water and chemicals are defined for each type of cleaning. A significant feature of the high pressure cleaner is also that it is portable.

Scope definition of the product system

The LCA is based on the parts list, which includes all components in the product with details of weight and material, and the individual manufacturing processes associated with the items. The parts list thus provides an overview of the product system for the high pressure cleaner from extraction of raw materials to and including production. The parts list contains well over 100 components, and for presentation purposes a simplified list was therefore prepared with the most significant components (see Table 28.1). An exploded view of the product is shown in Figure 28.2. The item numbers in the parts list refer to the numbers in the exploded drawing.

Pos.	Number	Item	Material	Weight total (g)	⏚	Manufacturing processes	* Incl. in model	Ancillary substances	Weight (g)	⏚
1	1	Motor	Steel	1620	✓	Pressure die-casting	÷			
			Copper		✓	Bending of wire	÷			
			Brass		✓	Bending of sheet	÷			
			ABS		✓	Injection moulding	÷			
2	1	Electric cable	PE	150	✓	Injection moulding	✓			
3	1	Motor casing	ABS	195	✓	Injection moulding	✓			
4	1	Noise condensator	Not known	11	÷	Injection moulding	✓			
5,6	2	Shield	PS	210	✓					
7	1	Packaging	Cardboard	540	✓					
8	0,5	Pallet	Wood	390	✓					
9,10	0,1	Pallet top and corner protection	Cardboard	183	✓					
11,12,13, 14,15,28, 29,31,33,40	7	Manual and miscellaneous	Paper	155	✓	Punching	÷			
14	1	Shoulder strap	PA	7	✓					
17	1	Gearbox	Aluminium	760	✓	Turning	✓	Lubrication oil, water		÷
						Pressure die-casting	✓	Lubricant	9	÷
								Aromatic amines	0.8	÷
								Water	150	✓
								Propylene glycol	6	÷
								Kerosene	4	÷
18	1	Disk	Aluminium	9	✓	Turning	✓	Lubrication oil, water		÷
						Pressure die-casting	✓	Lubricant	1	÷
								Aromatic amines	0.09	÷
								Water	18	✓
								Propylene glycol	0.7	÷
								Kerosene	0.5	÷
19	3	Sleeve for planet gear	Brass	0.3	✓	Pressure die-casting	÷			
20	1	Guidance	Aluminium	67	✓	Machine centre	÷			
						Pressure die-casting	÷			
21,25, 39	7	Miscellaneous	Steel	470	✓	Grinding	÷			
						Pressure die-casting	÷			
22	1	Cylinder casing	Aluminium	320	✓	Machine centre	÷	Lubricant	4	÷
						Pressure die-casting	✓	Aromatic amines	0.3	÷
								Water	64	✓
								Propylene glycol	3	÷
								Kerosene	2	÷
23		Grease BP	Oil	1.1	✓					
24	1	Cylinder head	Aluminium	150	✓	Turning	✓	Lubrication oil, water		÷
						Pressure die-casting	✓	Lubricant	2	÷
								Aromatic amines	0.15	÷
								Water	30	✓
								Propylene glycol	1.2	÷
								Kerosene	0.8	÷
						Impregnating	÷			
26,27		Oil	Oil	6	✓					
30,32	2	Cabinet shell left and right	ABS	540	✓	Injection moulding	✓			
34	1	Transverse handle	ABS	125	✓	Injection moulding	✓			
35,36	2	Nozzle casing End cover	ABS	5	✓	Injection moulding	✓			
37,38	2	Nozzle tube casing	ABS	220	✓	Injection moulding	✓			
Notes: See other product case histories										

Table 28.1. Simplified parts list for the Hobby 70

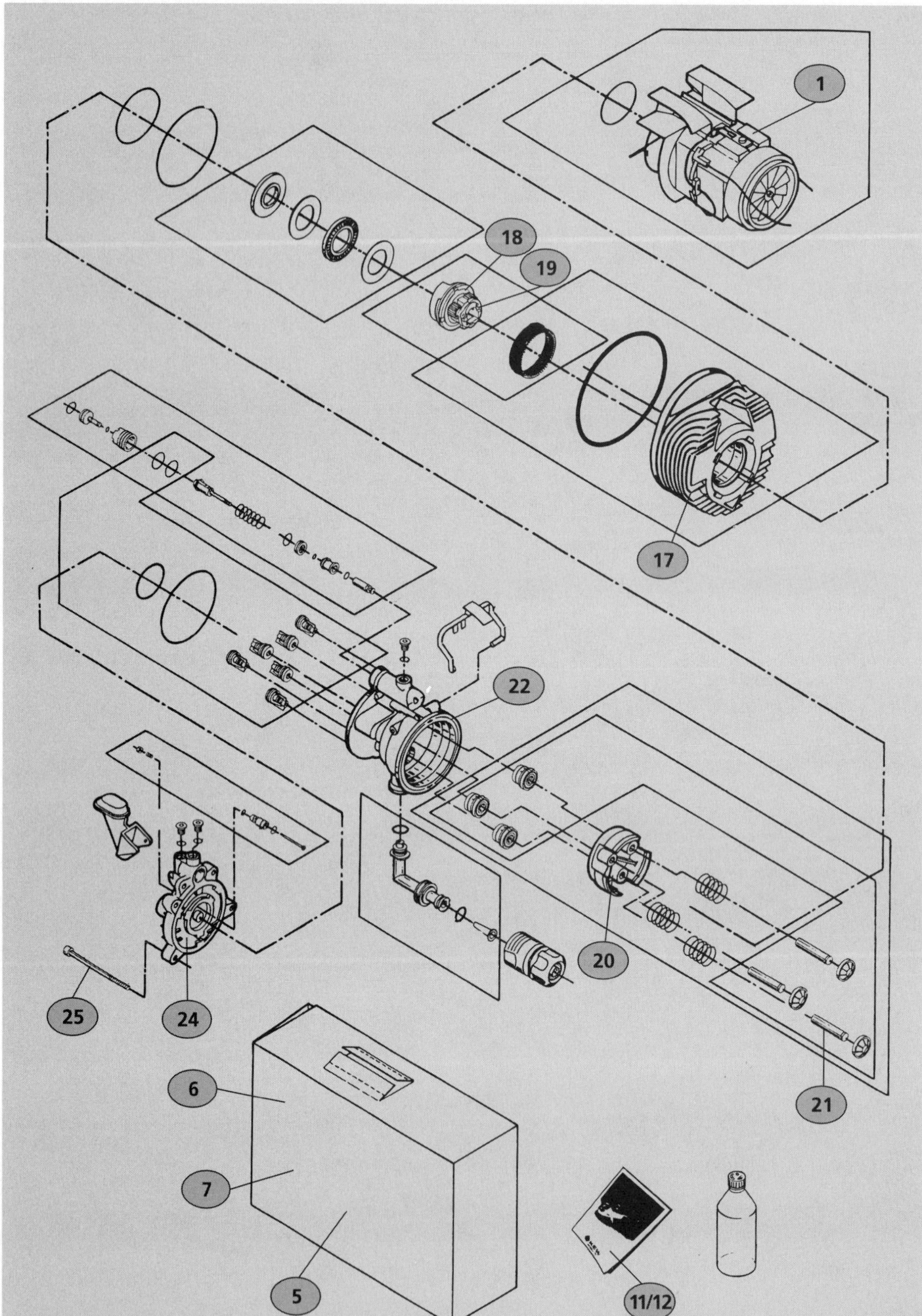

Figure 28.2. Exploded view of the Hobby 70

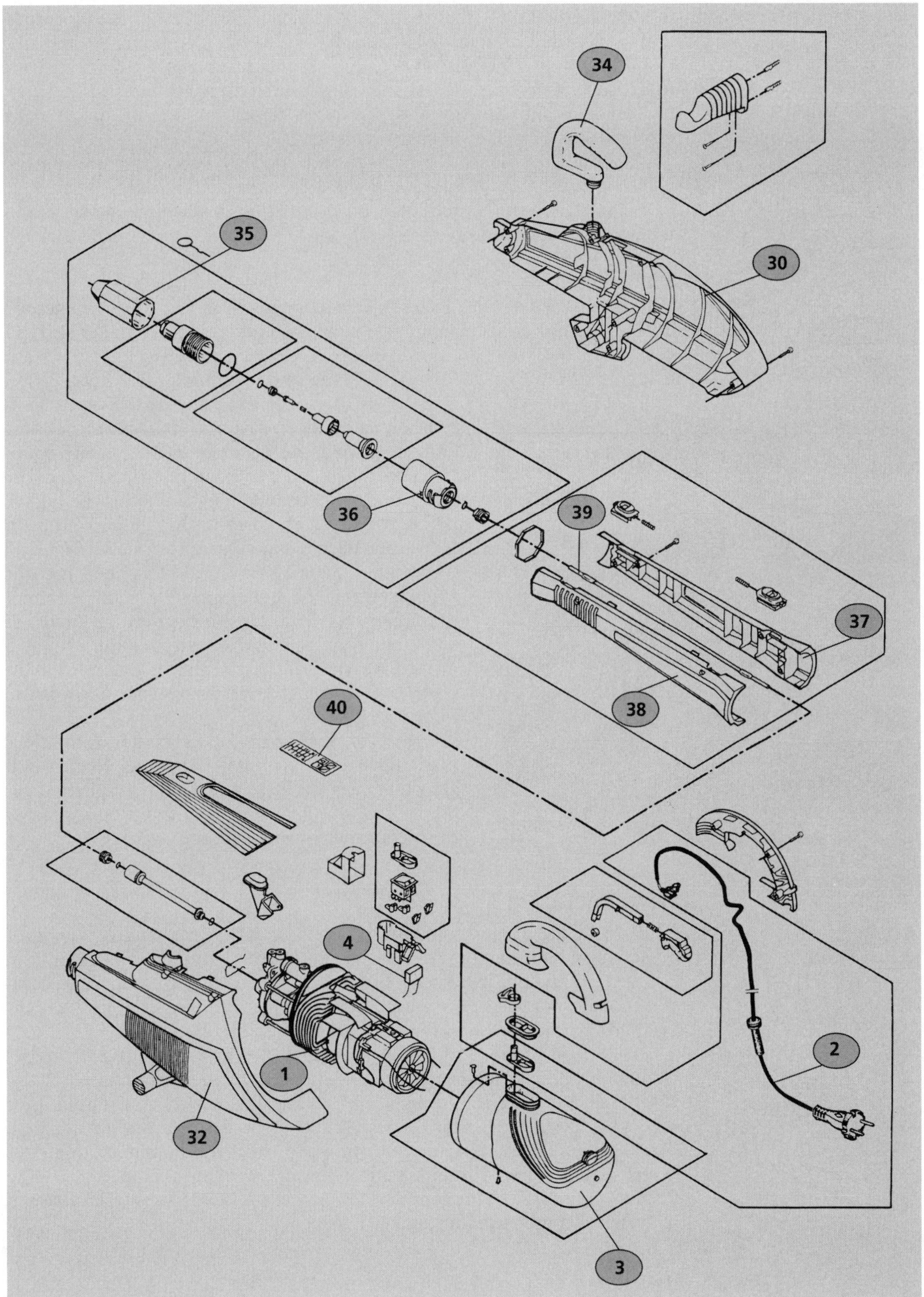
34
35
30
36
39
37
38
40
4
1
2
32
3

The parts list shows whether the component is included in the inventory, and which information is included. For some components only the resource for the material and energy for extraction and production are included. Other components include manufacturing processes, i.e. energy consumption, emissions, waste and potential impacts on the working environment as well as ancillary substances.

Materials, production processes and ancillary substances are marked with a ✓ if they are included and a ÷ if they are not included.

As is clear from Table 28.1, no production of ancillary substances is included in the inventory.

Materials

Production of materials includes extraction and refining of raw materials. The geographical distribution of extraction and refining of materials typically used is shown in detail in the LCA of the refrigerator. See the example from Gram in Chapter 26; the same picture applies here.

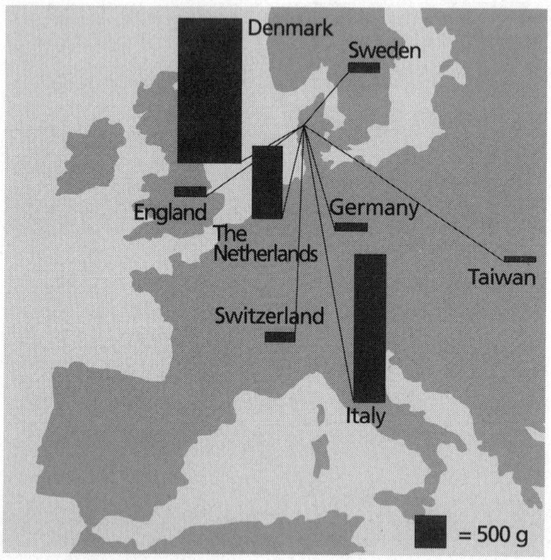

Figure 28.3. Geographic distribution for manufacturing of components

The subsequent processing of the materials is typically at subcontractors. Extraction, refining and processing can occur in each separate country, and the individual component can therefore have many thousands of kilometres behind it before it reaches the user in the finished product. It has not been possible to give precise details of where the materials are produced because the subcontractors have not provided this information. General data for production of materials were therefore used in the computation.

An overview of the country of origin for components in the high pressure cleaner is shown in Figure 28.3.

Most of the components come from Denmark. Companies in Italy and the Netherlands are otherwise the most significant suppliers. Less than 1% of the total weight of the product comes from countries outside Europe.

Manufacturing

Manufacturing at the company consists mainly of assembly. The processes at subcontractors are almost exclusively forming processes, e.g. pressure moulding, injection moulding or sheetmetal working. The components are not surface-treated. Very few chemicals are used in the manufacturing processes, and the associated emissions are small.

As illustrated in Table 28.1, some of the manufacturing processes occurring at subcontractors are not included in the inventory as it was not possible to obtain information on them. The energy consumption for the entire product system was estimated to be many times greater than for the manufacturing stage, and omission of some manufacturing processes is therefore of little significance for the total inventory of energy consumption. The emissions for the actual processes were also assessed as being of minor significance.

Figure 28.4. Assembly stand at KEW Industri

The inventory for the working environment derives from a report by Rytter (1994) prepared at KEW by the Danish Employees' Health Service. Data for electricity production in Denmark were also included. The working environment inventory does not include manufacturing at subcontractors. A large part of the manufacturing is at subcontractors, for which reason the complete inventory is not considered as covering the impact on the working environment throughout the product life. Data have, however, been included to make the area of the working environment visible.

The processes at KEW with the greatest impacts on the working environment are welding, grinding and assembly. Grinding and welding result in exposure to carcinogenic and allergenic substances and to substances which damage the reproductive system. The assembly work shown in Figure 28.4 involves exposure to allergenic substances and is also characterized by monotonous repetitive work.

Rejection rate

There will always be some items or products which have to be rejected because of faults in manufacture. More than one high pressure cleaner must therefore be produced for each unit sold. For the Hobby 70 the rejection rate is 3%, and the figures in the inventory for material and manufacturing stages were therefore multiplied by 1.03.

Use

The use stage for the high pressure cleaner covers the period from when the consumer buys the product until it is discarded. Electricity, water and cleaning agents are used in the use stage. The size of this consumption is based partly on estimates and partly on investigations.

The distribution of sales for the product is shown in Figure 28.5. It is evident that the product is sold predominantly in Europe.

Figure 28.5. Sales distribution for the Hobby 70

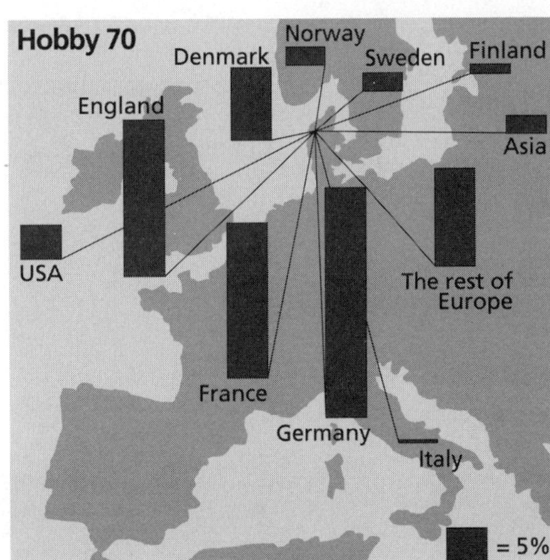

Use of cleaning agents

Most cleaning is exclusively with water, but for that part in which cleaning agents are used, a percentage distribution among the various types of use was estimated on the basis of sales figures for the various cleaning agents. The distribution is shown in Table 28.2.

Special cleaning agents are associated with all of these types of use. The principal ingredients in the cleaning agents are water and detergents, i.e. surface-active substances. Production of cleaning agents is not included in the inventory, but use of them enters into the assessment.

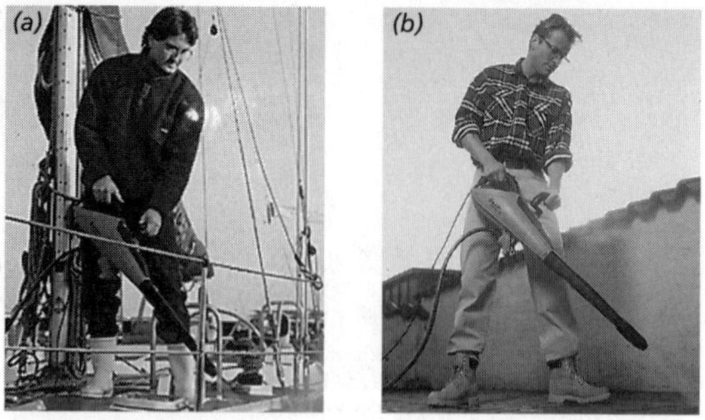

Figure 28.6. Cleaning of boat (a) and tiles (b) with a high pressure cleaner

Use	Distribution, %
Carwash	35
Carwash with wax treatment	10
Cleaning tiles covered with algae	20
Cleaning boat	10
Cleaning and wax treatment of boat	15
House cleaning, outdoors	8
Cleaning rims	2

Table 28.2. Distribution of types of use which employ cleaning agents

Fate of the cleaning agents after use

Depending on the use to which the high pressure cleaner is put, the water used will percolate down in the top layer of soil, run out into the surface water, i.e. stream, lake or sea, or be conveyed via the sewage system to a municipal waste water treatment plant.

Table 28.3 shows types of uses for the high pressure cleaner, where an average pattern of use is applied for an assessment of the quantities of the various products which typically finish in each of the final compartments.

The fraction which is conveyed to a municipal waste water treatment plant can be partly or entirely degraded by the microorganisms in the plant, reducing the quantity of the substances emitted to the aquatic compartment.

There can also be a redistribution of substances in the treatment plant among the various environmental compartments. Volatile substances are emitted to the air, and other substances which, for example, are water repellant, can bind to the surface of sludge particles and be spread with sludge on the soil (cf. Figure 28.7).

The extent of degradation and redistribution of substances in the waste water treatment plant was assessed with the aid of a computer model called FATE (Govind *et al.*, 1991).

Table 28.3. Drainage routes for washing water

Use	Discharge of water
Carwash with or without wax	Through sewers to waste water treatment plant
Cleaning tiles covered with algae	Directly to soil
Cleaning boat with or without wax	Directly to water
House cleaning, outdoors	Directly to soil

In some cases the water used in boat washing will be drained into the sewerage system, and a simulation is therefore shown in the environmental diagnosis in which the water from boat washing is ducted to the waste water treatment plant.

The effect of the emission on the treatment plant is not included. Examples of effects are inhibition of microbiological activity or the transport of heavy metals through the treatment plant. Some substances present in cleaning agents can bind heavy metals and prevent their precipitation in the treatment plant.

Figure 28.7. Redistribution at waste water treatment plant

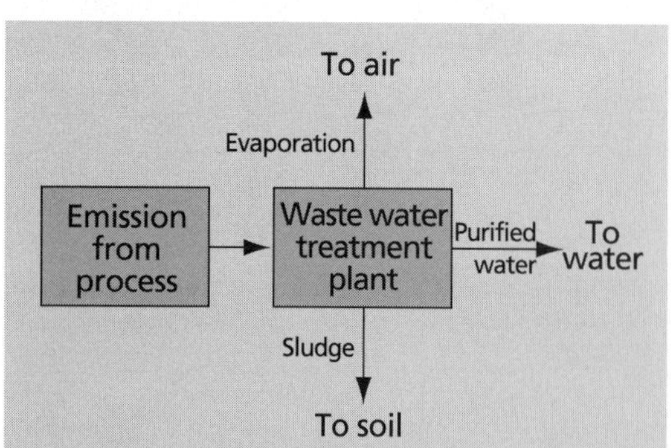

Other scope definition within the use stage

When chemicals are used in the high pressure cleaner's applications, the cleaning agents will probably form aerosols which can have an impact on the user. As it is difficult to assess the quantity of substances given off, the potential for human toxicity from the chemicals in the use stage is not included. One of the substances (nitrilotriacetic acid) is suspected of being carcinogenic. The substance is no longer included in KEW's cleaning agents.

Dirt released into the environment from the item washed is not included in the inventory.

The working environment is included only for electricity production and not for the use stage itself, as the working environment by definition covers only impacts on employees.

Life span and service

The total operating time is 125 hours. If the product is used for one hour a week for 25 weeks a year, a life span of 5 years is obtained, and this is used in the LCA. Repairs after the 5 years are possible, as the wear will be concentrated in only a few parts. But many people will choose to discard the product because they want a new one with a different design and other options. Service during the life of the product is not included as it will normally not occur during the fixed life span.

Disposal

A distribution among various disposal routes was prepared on the basis of questionnaires to the main markets. The distribution is shown in Table 28.4. All packaging is sent directly for incineration or dumping in landfills. Half of the products are sent to a scrap dealer. The rest go directly to the landfill. The scrap dealer sends the products through a shredder, and 30% of the metal content is then recovered. A total of 15% of the metal is therefore recovered. The materials remaining after the shredder are sent to the landfill.

Incineration results in larger or smaller emissions, depending on the material incinerated. Incineration of plastic has the greatest impact on the environment. Polyamide (e.g. nylon) in particular results in emissions of oxides of nitrogen. The Hobby 70, however, contains only small quantities of polyamide.

Source	Material	Landfill %	Incineration %	Reuse %
Packaging materials	Paper/cardboard	30	70	0
Product	Plastic	75	25	0
Product	Metal	85	0	15

Table 28.4. Disposal routes for the Hobby 70

Sources of data

Table 28.5 shows the sources of various data for the product system. Most of the data for materials are general, while data for production are mainly specific. The general data were chosen where no meaning would be provided by the specific data or where it was not possible to obtain specific data. Data for transport cover four aspects: distance, means of transport, and energy consumption and emissions. The distances were estimated on the basis of a knowledge of the location of the suppliers. A means of transport was then chosen. The means of transport is often known, e.g. a 16 tonne lorry or a tanker ship. Data for combustion of fuel for the individual transports are known at a general level.

General data, which are averages for the individual countries, were used for electricity production.

									Hobby 70
Data reference									
Product system & process type	Data specifity			Data source type					Comments
	Product specific	Site specific	General	1	2	3	4	5	
Extraction of raw materials									
Crude oil			x			x			Literature
Bauxite			x			x			Literature
Iron ore			x			x			Literature
Production of materials									
ABS plastic			x				x		
Other types of plastic			x			x	x		
Steel		x		x	x				
Aluminium		x		x					Lit. and sub-contr.
Product manufacturing									
Cutting (KEW Industri)		x		x					
Assembly and testing (only working env.)		x		x	x				
Cutting (sub-contractors)		x	x			x	x		
Injection moulding	x			x					
Use									
Energy consumption	x			x					
Consumption of chemicals	x				x				KEW's criteria for dimensioning
Lifetime			x			x			
Disposal									
Means of disposal			x					x	
Shredding		x		x					
Incineration			x			x			
Steel		x			x				Danish recycling centre
Aluminium		x		x					Danish recycling centre
Transport									
Distances and means of transportation	x		x		x	x			Estimated
Energy consumption and emissions			x			x			Based on measured emission factors
Energy systems									
Production of electricity			x			x			

Notes

1) Measurements
2) Computations (from mass balance considerations and input data for the process in question)
3) Extrapolation of data from similar process type or technology
4) Extrapolation of data from different process type or technology
5) Estimate

Product-specific data: concern processes especially dealing with the Hobby 70
Site-specific data: concern data from actual sites in the product system of the Hobby 70, but the process data are not specific for the Hobby 70
General data: all others

Table 28.5. Sources of data

Exchanges with the environment

The exchanges cover ingoing and outgoing substances in all stages of the product system as well as impacts on the working environment. These are shown in Table 28.6. Only the most significant contributions are included for resource consumptions and emissions.

The inventory was produced via a computerized modelling of the product at KEW. Data for the inventory were collected in a database (Frees, 1996).

In terms of quantity, the resource consumptions from the fossil fuels are the most significant. A negative resource consumption during disposal means that resources are recycled in a different product system and therefore credited.

The substances indicated under ancillary substances are some of the substances which enter into the cleaning agents. Some of these substances can be found in emissions to air, water or soil. The redistribution occurs in the treatment plant and by direct emission to e.g. water from the washing of boats.

The quantities of waste are considerable. The sources of waste are especially electricity production, waste from incineration plants, and dumping in landfills of the product which is disposed of.

Energy profiles

The energy profile in Figure 28.8 shows energy consumption converted to MJ (megajoules) for the various stages in the product life. The unit for the energy is primary energy, and the calculation of primary energy includes the efficiency of the power plants and the calorific value of the material.

Production of materials

Production of materials accounts for ca. 300 MJ. This includes the calorific value of the materials and energy for the mining of ore, extraction of oil and the further processing into e.g. steel plate and plastic.

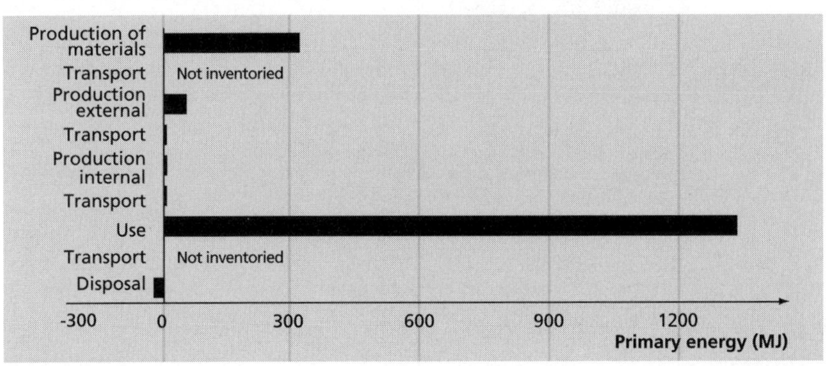

Figure 28.8. Energy profile. Primary energy consumption

Product manufacturing

Figure 28.8 clearly shows that the energy consumption at the company itself is minimal relative to the total energy consumption in the product system. Production at subcontractors is eight times greater than manufacturing at KEW with respect to energy consumption. Energy consumption for manufacturing corresponds to 20% of the consumption for production of materials, meaning that recovery of materials is of much greater significance for energy consumption than reduction in electricity consumption in the manufacturing processes.

			Production of materials	Product manufacturing	Use	Dis-posal	Transport	Total
Resource consumption								
Crude oil		g	2,670	90	4,920	-20	290	7,950
Natural gas		g	2,040	60	3,580	-1	20	5,700
Coal		g	1,860	1,800	17,700	20	1	21,400
Brown coal		g	290	-	13,200	-0.5	-	13,500
Uranium ore		g	0.05	-	1.80	-0.0001	-	1.9
Water for hydro power		liter	340	80	7,250	-0.10	0.0005	7,670
Aluminium	Al	g	580	0.01	0.9	-160	0.02	420
Iron	Fe	g	1,220	0.004	0.3	-230	0.02	990
Copper	Cu	g	460	-	-	-50	-	410
Manganese	Mn	g	9	-	-	-1	-	8
Nickel	Ni	g	7	-	-	-	-	7
Zinc	Zn	g	9	-	-	-	-	9
Calcium carbonate	$CaCO_3$	g	220	0.008	0.6	-0.005	0.04	220
Quartz	SiO_2	g	460	-	-	-30	-	430
Sodium chloride	NaCl	g	130	0.02	2	-0.003	0.04	130
Soft wood		g	1,670	6	840	-0.05	0	2,520
Groundwater		litre	-	5	-	0.3	-	5
Surface water		litre	-	0.1	-	0.006	0	0.1
Unspec.water		litre	50	0.01	41,300	-0.04	60	41,400
Ancillary substances for which resource consumption has not been included in the inventory								
Fatty alcohol (C12-18) ethoxylate		g	-	-	430	-	-	430
Secondary alkane sulphonate, Na salt		g	-	-	80	-	-	80
Diethylene glycol monobutyl ether		g	-	-	140	-	-	140
Nitrilotriacetic acid, 3Na salt		g	-	-	260	-	-	260
Quaternary coconut fat amine ethoxylate		g	-	-	110	-	-	110
Difatty acid isopropyl ester dimethyl ammonium methosulphate		g	-	-	20	-	-	20
Fatty alcohol ether sulphate, Na salt		g	-	-	80	-	-	80
Air emissions								
Carbon dioxide	CO_2	g	11,200	5,300	86,600	1,800	650	105,600
Carbon monoxide	CO	g	30	0.9	50	9	2	92
Nitrogen oxides	NO_x	g	60	20	350	0.7	8	440
Sulphur dioxide	SO_2	g	80	30	770	-0.05	3	880
Nitrous oxides	N_2O	g	0.3	0.3	7	0.003	0.01	8
Unspec. particles (dust)		g	60	3	40	-8	0.4	95
Hydrocarbons	HC	g	60	40	420	0.3	0.4	520
Volatile organic compounds	VOC	g	100	0.09	2	-0.09	1	100
Unspec. aldehyde		g	0.01	0.02	0.3	0.0002	0	0.3
Styrene		g	0.007	-	-	-	-	0.007
1-methyl-2-pyrrolidone		g	-	0.0007	-	-	-	0.0007
Unspec. methyl styrene		g	-	0.002	-	-	-	0.002
Dioxins		mg	0.0004	$2 \cdot 10^{-8}$	$1 \cdot 10^{-5}$	0	-	0.0004
Arsenic	As	mg	0.3	0.05	8	0	0	8
Cadmium	Cd	mg	0.07	0.005	1	$2 \cdot 10^{-5}$	0	1
Copper	Cu	mg	0.4	0.4	8	0.004	0	9
Mercury	Hg	mg	0.1	0.06	2	0	2	4
Manganese	Mn	mg	0.2	-	-	-	-	0.2
Vanadium	V	mg	30	4	230	0.02	0	264
Fatty acid (C14-18)-2-ethylhexyl ester		g	-	-	1	-	-	1
Terpine alcohol		g	-	-	0.08	-	-	0.08
Diethylene glycol monobutyl ether		g	-	-	2	-	-	2
1,2-benzoisothioazolin-3-one		g	-	-	0.01	-	-	0.01
Amino sulphonic acid		g	-	-	1	-	-	1

Table 28.6. Exchanges with the environment in the high pressure cleaner's product system

			Production of materials	Product manufacturing	Use	Dis-posal	Transport	Total
Air emissions								
Nitrilotriacetic acid, 3Na salt		g	-	-	80	-	-	80
Fatty acid (C8-10) Na salt		g	-	-	4	-	-	4
Emissions to water								
Chemical oxygen demand	COD	g	7	0.0006	0.04	-0.002	0.003	7
Total nitrogen	tot-N	g	0.1	0.0007	0.2	-0.00006	0.0008	0.3
Total phosphorus	tot-P	g	0.01	-	-	-	-	0.01
Fatty alcohol (C12-18) ethoxylate		g	-	-	20	-	-	20
Secondary alkane sulphonate, Na salt		g	-	-	70	-	-	70
Diethylene glycol monobutyl ether		g	-	-	150	-	-	150
Nitrilotriacetic acid, 3Na salt		g	-	-	70	-	-	70
Quarternary coconut fat amine ethoxylate		g	-	-	6	-	-	6
Fatty alcohol ether sulphate, Na salt		g	-	-	20	-	-	20
Difatty acid isopropyl ester dimethyl ammonium methosulphate		g	-	-	20	-	-	20
Hydrocarbons		g	0.60	0.001	0.1	-0.0004	0.006	0.7
Unspec. oil		g	0.3	0.002	0.2	-0.00001	0	0.5
Phenol		g	0.0003	0.00003	0.003	0	0	0.003
Arsenic	As	mg	10	-	-	-	-	10
Cadmium	Cd	mg	5	-	-	-	-	5
Chromium	Cr	mg	20	-	-	-	-	20
Copper	Cu	mg	30	-	-	-	-	30
Mercury	Hg	mg	0.0006	-	-	-	-	0.0006
Manganese	Mn	mg	0.05	-	-	-	-	0.05
Nickel	Ni	mg	20	-	-	-	-	20
Zinc	Zn	mg	30	-	-	-	-	30
Emissions to soil								
Nitrilotriacetic acid, 3Na salt		g	-	-	30	-	-	30
Fatty alcohol ether sulphate, Na salt		g	-	-	15	-	-	15
Diethylene glycol monobutyl ether		g	-	-	20	-	-	20
Secondary alkane sulphonate, Na salt		g	-	-	15	-	-	15
Fatty alcohol (C12-18) ethoxylate		g	-	-	50	-	-	50
Quarternary coconut fat amine ethoxylate		g	-	-	70	-	-	70
Difatty acid isopropyl ester dimethyl ammonium methosulphate		g	-	-	1	-	-	1
Waste								
Unspec. hazardous waste		g	0.07	-	-	-	-	0.07
Unspec. waste containing heavy metal		g	9	-	-	-	-	9
Unspec. nuclear waste		g	0.006	-	0.3	0	-	0.3
Unspec. chemical waste		g	30	-	-	-	-	30
Unspec. slag and ashes		g	100	140	3,400	1	0.7	3,600
Unspec. bulk waste		g	710	1,000	10,300	10	0.7	12,000
Working environment								
Monotonous repetitive work		hours	-	0.36	0.006	-	-	0.366
Hearing impairment inducing noise		hours	-	0.38	0.08	-	-	0.46
Substances causing damage to the nervous system		hours	-	-	0.003	-	-	0.003
Allergenic substances		hours	-	0.022	0.013	-	-	0.035
Substances causing damage to the reproductive system		hours	-	0.022	-	-	-	0.022
Carcinogenic substances		hours	-	0.022	0.009	-	-	0.031
Accidents		number	-	$14 \cdot 10^{-6}$	$3 \cdot 10^{-6}$	-	-	$17 \cdot 10^{-6}$

Notes

-	means that no information is available for the life cycle stage in question
0	means that the figure is very small compared to the other stages of the life cycle

Use
The greatest energy consumption is in the use stage. In the inventory the electricity consumption is converted to primary energy, taking the efficiency of the power plant into account.

Transport
The contribution from transport is small, but as will be evident, it is twice as much as the contribution from KEW's own manufacturing of the product.

Disposal
A negative energy contribution means that energy is credited. The contribution comes from heat generated in the incineration of plastic. The heat is used for e.g. district heating.

The energy content of the materials

The energy inventory is illustrated in Figure 28.9a. Both calorific value and process energy are given for each individual material. The process energy is the energy used for production of materials and the calorific value is the energy released on incineration. Metals are assumed to have no calorific value.

Figure 28.9a. Content of primary energy in the materials found in the high pressure cleaner

Figure 28.9b. Content of primary energy in the materials in the high pressure cleaner, by component

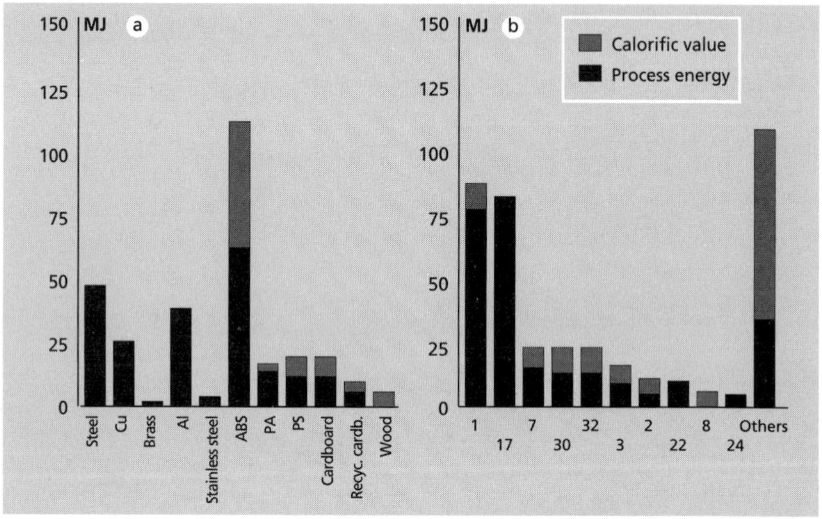

The greatest energy contribution is obtained from plastic of type ABS (acrylonitrile butadiene styrene), which is used in large quantities in the cleaner. A significant part of the energy is released on incineration. Apart from the contributions from the ABS, the most significant contributions come from aluminium, steel and copper. The aluminium used is recycled aluminium, which has a much lower energy consumption than primary aluminium.

The energy profile classified by components (Figure 28.9b) shows the energy consumption for the ten most energy-intensive components and for the remaining group. Item number 1 is the motor, which consists of several different materials. The other components contain only one material. The item numbers are found in the exploded view in Figure 28.2.

The motor's raw materials have a high energy content. This is attributed to the fact that it is the heaviest part of the cleaner, with a weight corresponding to ca. 20% of the product's total gross weight. The high pressure cleaner contains a few parts which together comprise ca. 60% of the total energy content, while the rest derive from over 100 smaller parts.

Impact assessment

The exchanges with the environment through the life of the product hold a potential for impacts, including resource consumption and impacts on the environment and the working environment.

Environmental impact potentials

The potential for impacts on the environment resides in the emissions to water and air and in waste. Conversion of the emissions to impact potentials is described in Parts II and IV. The result is shown in Table 28.7.

The high pressure cleaner distinguishes itself from the other product examples in this book by the fact that chemicals are used in the use stage. The ecotoxicity potential is therefore of special interest and will be reviewed here. The ecotoxicity potential of an emission is computed by multiplying the quantity of substance emitted by an ecotoxicity equivalency factor which reflects:

1) The substance's toxicity to ecosystems.
2) Its ability to undergo bioaccumulation, i.e. the more toxic the substance, the higher the factor.
3) Its ability to be degraded in the environment, i.e. the more easily it is broken down, the lower the factor.
4) Its distribution in the environment: even if it is emitted to the atmosphere, it may well end in the soil and water compartments.

Ecotoxicity potentials are computed for the water and soil compartments for the emissions occurring in the high pressure cleaner's product life. For the aquatic environment, the potential for both immediate impacts (acute toxicity) and for impacts on longer-term exposure (chronic ecotoxicity) are computed. For the soil environment, only the potential for chronic impacts is computed.

Table 28.8 presents computed ecotoxicity potentials in water for the most important of the individual substances in the cleaning agents used in the high pressure cleaner. The quantities emitted are shown at the far left. The potential for the environmental impact is obtained by multiplying the quantity emitted by the impact factor for the potential in question.

Depending on their intended use, the cleaning chemicals contain numerous different substances which fulfil the required function, e.g.

- detergents,
- complexing agents,
- preservatives,
- solvents,
- fragrances,
- wax and stabilizers.

Table 28.7. Conversion of the environmental exchanges to environmental impact potentials

Impact category	Unit	Potential impact Per product per lifetime
Global impacts		
Global warming	g CO_2-eq	1,008,200
Regional impacts		
Acidification	g SO_2-eq	1,200
Nutrient enrichment	g NO_3^--eq	600
Photochemical ozone	g C_2H_4-eq	4
Human toxicity (water)	m^3 water	19,700
Ecotoxicity (water, chronic)	m^3 water	25,874
Local impacts		
Human toxicity (soil)	m^3 soil	200
Ecotoxicity (water, acute)	m^3 water	6,020
Human toxicity (air)	m^3 air	68,000,000
Ecotoxicity (soil)	m^3 soil	1,441
Hazardous waste	g	90
Nuclear waste	g	0.3
Slag and ashes	g	3,700
Bulk waste	g	12,700

The environmental characteristics for many of these substances have not been investigated thoroughly, and it may in any case be difficult to obtain information on them from the supplier. The assessment of ecotoxicity potentials is thus often based on a rather fragile database.

It is clear from Table 28.8 that over 90% of the ecotoxicity potential for water is ascribed to three substances: a secondary alkane sulphonate, dialiphatic acid isopropyl ester dimethyl ammonium methosulphate and a fatty alcohol ethoxylate. All of the substances are detergents which comprise only a quarter of the total quantity of substances emitted from the high pressure cleaner's use stage. Data for the last two substances are particularly scarce. When the database is fragile, the most conservative figures are chosen. Very large equivalency factors are therefore obtained for these two substances.

As the only substance emitted to water, terpene alcohol, a fragrance, contributes to the ecotoxicity potential in soil. This is attributed to the fact that terpene alcohol is so volatile that it evaporates from the aquatic compartment, after which part of it is deposited in the soil compartment. The ability to evaporate and to be deposited again is included in the ecotoxicity factor for soil for substances emitted to water.

Apart from ecotoxicity, the most significant potential impacts are global warming, acidification and various types of waste.

The contribution to photochemical ozone formation is very small, and there is no contribution at all to ozone depletion.

Table 28.8. Computation of ecotoxicity potentials for chemicals emitted to water either directly or via a treatment plant. For each substance, the quantity emitted is shown in grams together with equivalency factors for acute and chronic impacts in water and for chronic impacts in soil. The ecotoxicity potentials are expressed for each emission by multiplication.

		Water, acute		Water, chronic		Soil, chronic	
Compound	**Quantity**	**Equivalency factor**	**Ecotoxicity potentials**	**Equivalency factor**	**Ecotoxicity potentials**	**Equivalency factor**	**Ecotoxicity potentials**
Unit	**g**	**m³/g**	**m³**	**m³/g**	**m³**	**m³/g**	**m³**
Fatty alcohol ether sulphate (C12-14), Na salt	24.5	8.8	210	17.5	430	0	0
Di-fatty acid isopropyl ester dimethyl ammonium methosulphate	20	100	2,000	500	10,000	0	0
Fatty acid (C14-18)-2-ethylhexyl ester	3.9	0.32	1.3	10.7	42	0	0
Laurylamine ethoxylate, 10 EO	16	3.6	58	18.2	290	0	0
Terpene alcohol	11.25	5.0	56	25.0	280	2.3	26
Diethylene glycol monobutyl ether	152.5	0.19	29	0.19	29	0	0
1,2-benzo-isothioazoline-3-one	0.375	10.0	3.8	100	38	0	0
Secondary alkane sulphonate (C13-17), Na salt	70.55	24	1,700	150	11,000	0	0
Nitrilotriacetic acid, 3Na salt	70.8	0.056	4.0	0.11	7.8	0	0
Fatty alcohol alcoxylate phosphoric acid ester	0.4	3.3	1.3	6.7	2.7	0	0
Aminosulphonic acid	1	7.0	7.0	14.1	14	0	0
Fatty alcohol (C12-18) ethoxylate, 8 EO	19.1	100	1,900	200	3,800	0	0
Quaternary coconut oil amine ethoxylate	5.6	7.7	43	38.5	220	0	0
Total			**6,000**		**26,000**		**26**

Resource consumption and potentials for impacts on the working environment

The figures in Table 28.6, expressed in grams and hours of impact on the working environment, are direct expressions of the potentials and are used in further impact assessment.

The magnitude of the product's contribution

The contribution to potential impacts from the high pressure cleaner is expressed relative to a person's average contribution. This is called normalization. Normalization makes it easier to compare the magnitudes of the various contributions. The magnitude of the high pressure cleaner's contribution is thus expressed in person-equivalents, or more correctly in milli-person-equivalents (mPE), which are the product's contribution in parts per thousand of an average person's contribution. For all contributions, 1990 is used as reference year.

The resource consumption in Table 28.6 was dominated by the fossil fuels. After normalization, copper and aluminium are the largest, as shown in Figure 28.10a. The copper and aluminium consumptions thus comprise a relatively greater proportion of the average person's consumption in 1990 than the fossil fuel.

An aggregation of several toxic impacts was made for the presentation in Figure 28.10a. Human toxicity covers toxicity to humans exposed via air. Ecotoxicity covers the acute ecotoxicity in water, while persistent toxicity is the average of chronic or long-term toxicity to both humans and ecosystems. For the environmental impacts, ecotoxicity has the greatest potential. When the persistent toxicity makes a significantly lower contribution than the ecotoxicity, this is attributed to the fact that the cleaning agents are relatively easily degraded.

In 1990, the total annual Danish consumption of cleaning agents for the washing of clothes and other cleaning was 51.6 kg per person (Damborg & Thygesen, 1991). By comparison, the high pressure cleaner uses an average of 1 kg of cleaning agent per annum. The figures are, however, difficult to compare directly because the content of detergents in the cleaning agents can be different.

According to the calculation of the total aquatic ecotoxicity potential emitted for an average Dane in 1990, the emission of detergents from treatment plants contributed about 6% (Hauschild *et al.*, 1997a). Figure 28.10b shows that the normalized contribution to ecotoxicity from cleaning agents for the high pressure cleaner amounts to over 2%. In other words, using a high pressure cleaner gives more than a third of the contribution to ecotoxicity from a person's average consumption of detergents, including that part which is due to industrial consumption. The high figure is attributed particularly to the fact that only a small portion of the cleaning agents is conveyed to a treatment plant in the scenario used for the high pressure cleaner.

For the working environment, normalization means that the actual contribution is expressed relative to the average impact per employee in Denmark per annum. For ease of comparison, all of the normalized fig-

Figure 28.10. Normalization of resource consumption and impact potentials for the high pressure cleaner

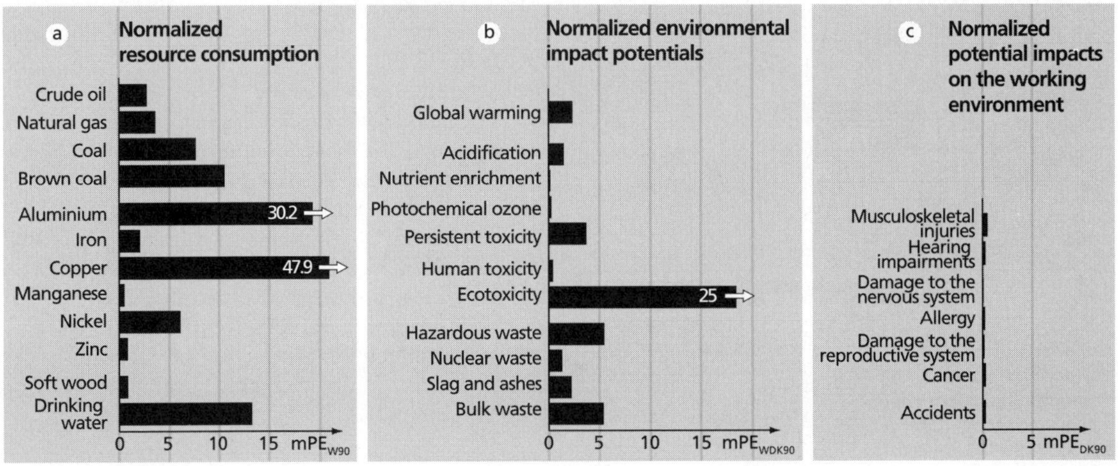

ures in Figure 28.10 have the same scaling. It is therefore obvious that the contribution to impacts on the working environment shown in Figure 28.10c is significantly less than environmental impacts and resource consumptions, which can probably be attributed primarily to the fact that only a small part of the product system is covered in the inventory for the working environment. The greatest contribution is potential injuries to the musculoskeletal system, which is attributed particularly to assembly processes.

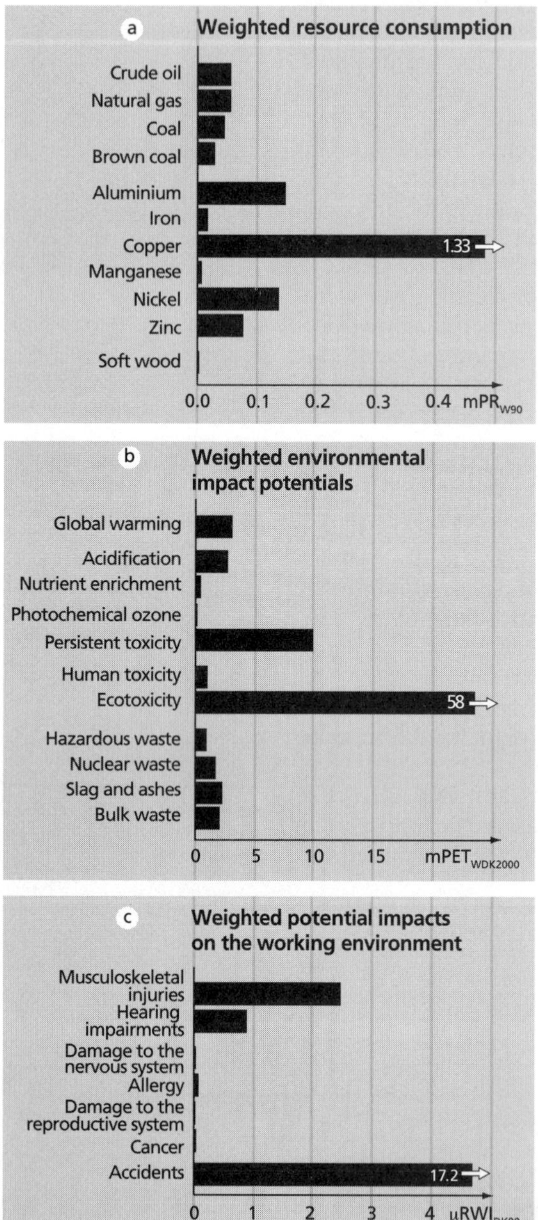

Figure 28.11. Weighted resource consumptions and impact potentials for the high pressure cleaner

The most significant impact potentials

The normalized values express the contribution to the environmental impacts relative to an average person's contribution. However, it does not show the degree to which the contribution is critical. For example, certain resources will be exhausted in a few decades at present rates of consumption. As it is not possible to determine the permissible level objectively, a weighting is made. The background to the different weighting factors is described in Part II.

As the weighting principles for the various impacts are not the same in the EDIP method, the weighted profiles cannot be compared among the main groups of resource consumptions, environmental impact potentials and potentials for impacts on the working environment, but only internally within each main group. The units used are described further in Chapter 10.

A weighting is used for resources relative to the size of the known reserves. The weighted resource profile in Figure 28.11a shows that the consumption of copper is now clearly dominant. This is attributed to the fact that known reserves of copper are small compared to the annual consumption. Aluminium and iron, which comprise the greatest contribution with respect to quantities, have lower weighting factors. Note that in spite of the small quantity involved, nickel is ranked highly in the weighted profile because of the small known reserves compared to the annual production.

The environmental impact potentials are weighted on the basis of international or Danish political reduction targets for the year 2000. For environmental impact potentials it is still the toxicity which is weighted most heavily (cf. Figure 28.11b). Ecotoxicity and persistent toxicity are very clearly the highest priorities.

The weighting factors for the working environment reflect the frequency of reported work injuries. When the normalized potentials are weighted, potential reported work injuries per product are obtained.

In the weighted profile for the working environment (Figure 28.11c), potential work accidents have the highest priority. Injuries to the musculoskeletal system and hearing impairment also have relatively high weight. Both categories derive principally from the manufacturing stage, but electricity production contributes ca. 20% of the expected hearing impairments.

The reason why the expected number of reported accidents is so large is that accidents are reported much more frequently than general work injuries. The reporting of accidents which lead to absence from work is mandatory, while other work-related injuries are reported only on the basis of an illness when the illness is judged to be work-related.

Materials	
Material	- consumption - disposal
Processes	
Energy	- thermal - electricity
Chemicals	- ancillary substances
Others	- overhead

Table 28.9. Sources of resource consumptions and impact potentials classified under the MECO principle

The most significant sources of impact potentials

The total environmental impacts for the entire product life are shown above. If the environmental impact is to be reduced, it is, however, important to know from where the contributions come. The sources of environmental impacts are grouped in Figures 28.12a and 28.12b into consumption of materials, disposal, ancillary substances, electricity consumption, thermal energy and overhead, i.e. administration, heating and lighting. These categories can be placed under the MECO principle which was introduced in Chapter 12 (see Table 28.9).

The MECO classification is used later to express the sources which are affected by changes in the product.

The classification is useful to the designer because his or her options are clearly reflected by the source classification. The figure shows which sources make the greatest contributions and where the first initiatives should be taken. The resources are used almost exclusively for materials and electricity. The electricity production uses fossil fuels, and both fossil fuels and metals are used for materials. Crude oil and natural gas are also used for production of plastic.

A more subtle picture is obtained for the environmental profile. Consumption of materials and electricity consumption contribute especially to global warming, acidification and various categories of waste. Thermal energy covers combustion of e.g. oil or coal. In this context it will typically be the transport, i.e. combustion of diesel oil, which is shown in Figure 28.12b. The most significant sources of environmental impact potentials for the high pressure cleaner are thus the electricity consump-

Figure 28.12. Weighted resource consumptions and environmental impact potentials classified by types of source

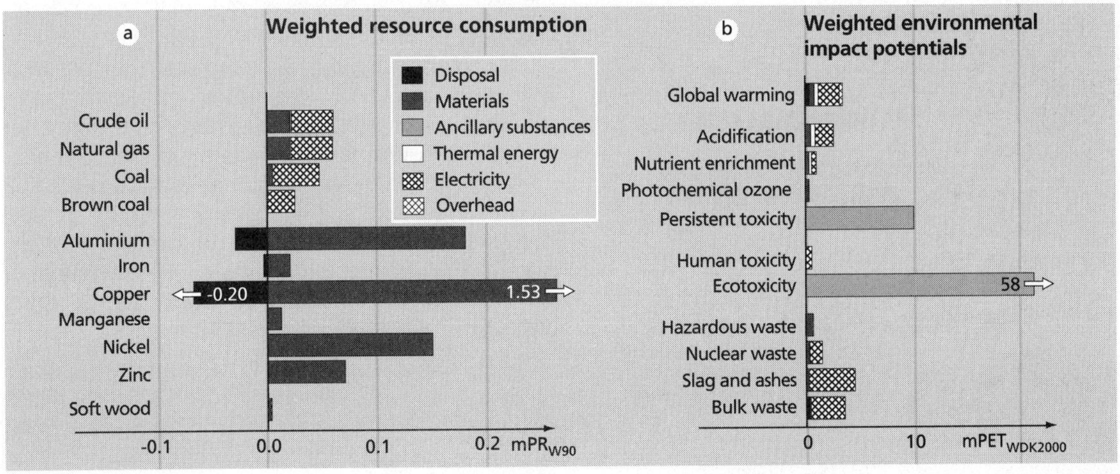

tion and the consumption of chemicals. For resource consumption, it is the source "materials" which contributes most, especially because of the consumption of copper.

The most significant stages of the product life

Like the classification by source, the classification of the environmental impacts by the stage in which they occur provides information on which parts of the product life have the greatest impact. For clarity of presentation, only a few resources and types of environmental impacts were selected. The impacts were selected on the basis of the classification of sources in Figures 28.12a and 28.12b.

For resources, natural gas, crude oil and coal were selected besides aluminium, copper and nickel. Several different fossil fuels were selected because some enter almost exclusively into electricity production (coal), while crude oil is also a raw material used in the production of plastic. Natural gas is used for electricity production and production of various materials.

The resource consumption is shown per life cycle stage in Figure 28.13a. Aluminium is used in the product in large quantities, but known reserves are substantial, and the weighted value is therefore not large. Copper and nickel are metals with small known supply horizons. The weighted values are therefore large in spite of the small quantities used in the product. Nickel is used as an alloying element in stainless steel.

The contributions from natural gas, crude oil and coal are small in comparison with the metals. The greatest consumption is in the use stage. Crude oil and natural gas are also used for production of materials, crude oil directly as a raw material, natural gas as a raw material and, for example, in the production of steel.

Naturally enough, the contribution for the metals lies in the material stage. Note the negative contribution on disposal, which expresses a credit for resource consumption on recovery of the material.

The selected environmental impact potentials seen in Figure 28.13b are global warming, photochemical ozone formation and human toxicity. Slag and ashes are also shown as representative of the categories of waste. The waste derives especially from the production of electricity.

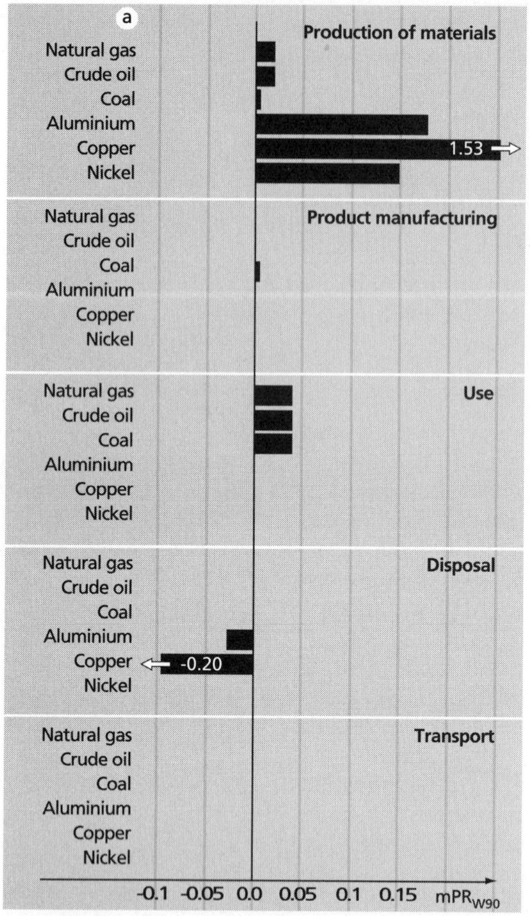

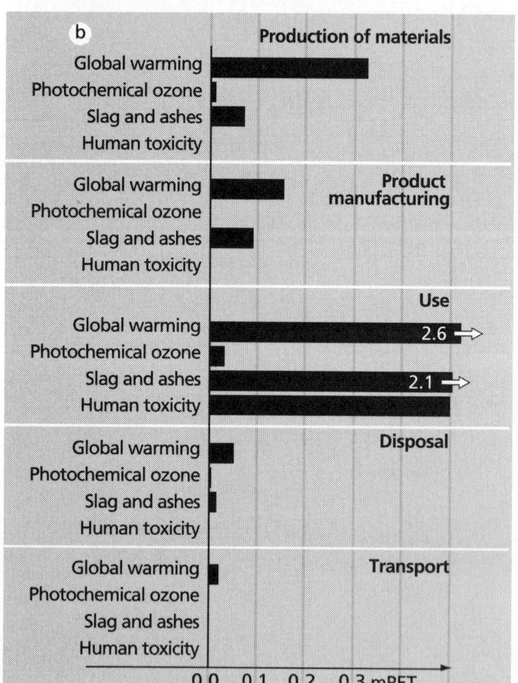

Figure 28.13a. Weighted resource consumptions divided by stages of the product life
Figure 28.13b. Weighted potentials for environmental impacts divided by stages of the product life

Human toxicity was selected as representative of the toxic impacts.

Apart from the toxic impacts, the global warming potential has the highest priority. The global warming derives from the electricity production. Photochemical ozone arises in connection with transport and other combustion where both VOCs and oxides of nitrogen are emitted.

The most significant contribution to global warming comes from the use stage with its production of electricity. The contributions from the other stages also derive from the electricity production. The photochemical ozone formation is somewhat evenly distributed across all stages, again with the use stage as the most significant. The toxicity to humans derives exclusively from the use stage.

The most significant components

The components with the greatest environmental impacts are shown as part of the basis for the ranking of priorities in product development. Unfortunately it was not possible to obtain information on the working environment at the level of components. This is made particularly difficult by the fact that most of the production is with subcontractors.

Only the stages of materials and manufacturing are included in this inventory for components. Thus, only the impact of the component before assembly is considered. The weighted values are shown in Figure

Figure 28.14. Top scorers among the high pressure cleaner's components

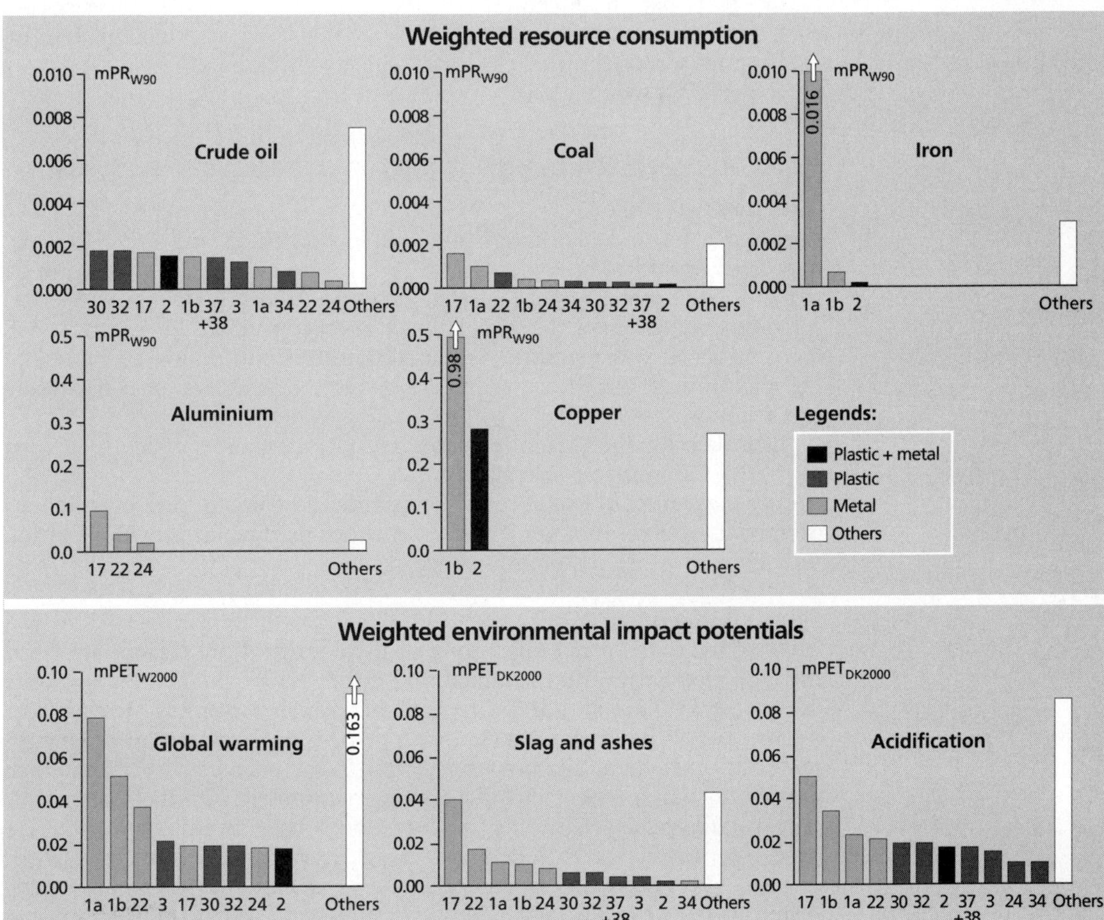

28.14. A list of top scorers was prepared for each of the selected resource consumptions and environmental impact potentials. The lists show which components make the greatest contributions.
The plastic ABS uses crude oil as a raw material. The large shells (nos. 30 and 32) which enclose the high pressure cleaner therefore have a significant consumption of oil.

Coal, which is generally used only for production of electricity, has the aluminium gearbox as top scorer. This is ascribed to the electricity consumption for the production of aluminium and the further associated manufacturing processes. The gearbox also has a high consumption of crude oil, which again derives from the production of electricity. Other top scorers for fossil fuels are the motor, which is divided into rotor (1a) and stator (1b), cylinder housing (22) and electric cable (2) made of polyethylene (cf. Figure 28.2). The motor is not divided into “1a” and “1b” in Figure 28.2.

The consumptions of aluminium, iron and copper are concentrated in few items only. Aluminium is found in the gearbox (17), the cylinder housing (22) and the cylinder head (24). Iron is found in the motor (1a and 1b), and copper is found in the motor (1a) and cables (2).

The environmental impact potentials shown derive mainly from the production of electricity. The top scorers are therefore the same as for consumption of coal. The differences in ranking are attributed *inter alia* to the fact that production of materials occurs in different countries, each with its own distribution of fuels for electricity production. Use of coal for the production of electricity yields, for example, more slag and ashes per kWh than electricity production from natural gas.

Lack of information, uncertainty and sensitivity analyses

The material stage

Data for steel and aluminium are well documented and are based on data from specific works (see Table 28.5). An assessment on the basis of working environment data for steel shows that the contribution is less by a factor of 1000 than the total impact on the working environment for the product. It was therefore estimated that the contribution to the working environment for the other materials will not be able to alter the result significantly.

The data for the plastic materials are average figures which may vary by ±50% for emissions, possibly more.

For extraction of metal ores, the quantities of waste were presumably estimated too low. In several cases it was not possible to provide data for the quantity of ore required for 1 kg of metal.

The manufacturing stage

Most of the data on manufacturing of the high pressure cleaner are from the final element in the manufacturing chain: KEW. Manufacturing processes carried out at KEW are well described. Assembly, testing and repair are not, however, included with respect to consumption of energy, materials and chemicals, but impacts on the working environment enter into the total assessment of the working environment. On the other hand, only certain processes at the subcontractors have been described. The processes which are included are assessed as being the most energy-intense. The uncertainty in the individual electricity measurements is estimated to be 10%. The uncertainty for emissions will be greater because the measurements are instantaneous readings, and the emissions vary as a function of time. The electricity measurements are converted to resource consumptions and impact potentials on the basis of assumptions about the electricity production. Data for electricity production are grouped for specific years, and changes in the distribution between various fuels as well as efficiency of the works influence the magnitude of the environmental impact.

The use stage

The consumption of cleaning agents was modelled on the basis of assumptions and information on the distribution of sales. The further fate of the cleaning agents after emission are based on assumptions of how much ends in the soil, water or a treatment plant. These assumptions are of major significance for the magnitude of the toxicity impact potentials. A worst case was selected for several of the user types, i.e. that the cleaning agents are emitted directly to the final recipient, e.g. soil or water. This does not apply to the washing of cars, where the water is assumed to enter the sewerage system and to be conveyed on to the treatment plant. This is a reasonable assumption in Denmark, where 90% of the water which enters the sewerage system passes through a treatment plant. In other countries this proportion may be much less.

The computation of toxicity potentials itself is quite uncertain, as the database for several of the substances is fragile.

As aerosols formed in cleaning with chemicals are not included, human toxicity includes only the contribution from the production of electricity.

The assumptions concerning how much electricity and water, and how many chemicals, are used in the use stage are based on estimates and experiments. Electricity consumption can be specified accurately, but consumption of chemicals and water is more difficult to assess. The consumption of chemicals is based on sales of KEW's own products. According to Damborg & Thygesen (1991), high pressure cleaning agents comprised 1.3% of the total quantity of cleaning agents in 1991. The estimated consumption of 5 litres of cleaning agents for the life of the high pressure cleaner, corresponding to 1 litre or 1 kg per annum, amounts to 1.9% of the total consumption per person per annum. The estimated quantity therefore sounds reasonable, but it may be too high.

Consumption of water is based on readings taken by KEW and is difficult to assess.

The operating time was assessed as 125 hours, with an uncertainty of ±20 hours.

The disposal stage

Assumptions concerning disposal of the high pressure cleaner are based on inquiries in the main markets, but they may be regarded as uncertain as the percentage of replies was low. The fraction recycled is generally set very low. This applies for example to the assumption that the packaging is not recovered, and that only 15% of the products are sent to scrap dealers. The significance of a recovery rate of 100% for the metals is shown in a simulation in Section 28.2.

28.2. Environmental diagnosis for a high pressure cleaner - the Hobby 70

The LCA in the previous section shows where the most significant environmental impacts are found. But it says nothing about what can be improved.

Improvement potentials are established in the diagnosis. The elements involved are simulation of changes and designation of design options.

Simulation of changes in the high pressure cleaner or its product system

The simulations were made within the four MECO categories:

- Materials
- Energy
- Chemicals
- Others

For materials, it was decided to examine increased recycling and substitution of materials. The category "energy" has two simulations: reduction of energy consumption in the manufacturing stage and reduction of electricity consumption in the use stage. Consumption of chemicals enters into the simulation of boat washing in which the waste water is lead to a treatment plant, and the life span represents the category "others". Figure 28.15 shows the types of impacts on which the actual changes have a significant effect. Some of the changes were introduced in new product types and are described in Section 28.4 on environmental design of new products.

Consumption of materials and disposal

In the present disposal scenario, only 15% of the metal is recycled. Realistically, this fraction should be increased. A 100% recovery of the metal was chosen for the simulation, as shown in Figure 28.15a. This alternative for improvement has an influence on the consumptions of aluminium, iron and copper. Resource consumption here is reduced to one-third.

Another option within materials is substitution of primary metals with secondary metals. This results in a saving in resources and for some metals also a reduction in energy consumption. Secondary aluminium requires only ca. 20% of the energy which goes into the production of primary aluminium. For steel, the energy reduction from primary to secondary is only a factor of two. The aluminium used in the high pressure cleaner is already secondary, but the steel is primary. Replacement of primary steel with secondary will result in a halving of the iron consumption. This is not shown in the figure.

Energy consumption

A 25% reduction in electricity consumption in the manufacturing stage and a 50% reduction in the use stage were simulated, as shown in Figure 28.15b.

A reduction in the manufacturing stage has a visible although small influence on the impact categories shown. The change is at the level of 1-2%. As the electricity consumption in the manufacturing stage comprises a relatively small part of the total consumption, reductions in this stage will have very little effect on the total contribution. Even the 25% selected here is at the limit of what is realistic.

The major effect is achieved by changes in the use stage. A reduction of 50% here will mean a reduction of ca. 40% in the total impacts. A reduction of 50% in the use stage may seem large, but a change in the design of the nozzle is expected to be able to yield reductions of this order of magnitude.

Consumption of chemicals

The chemicals are the reason for a significant portion of the impact potentials. The washing of boats makes the greatest contribution to the ecotoxicity potential. This is attributed partly to the content of cleaning agent and partly to the fact that the cleaning agent is assumed to be emitted directly to water without passing through a treatment plant. But it is not unrealistic to convey the water to a treatment plant, as could be done if the boat were washed on land, and the water would enter the sewerage system. Potentials for persistent toxicity and ecotoxicity are reduced significantly in the treatment plant. Conveying the water from the washing of boats through a treatment plant instead of directly into the water therefore gains considerable significance for the total picture for the high pressure cleaner, as shown in Figure 28.15c.

Both persistent toxicity and ecotoxicity are reduced by a third. The human toxicity potential is unaltered, and this is attributed to the fact that the contribution to this impact derives from the consumption of electricity.

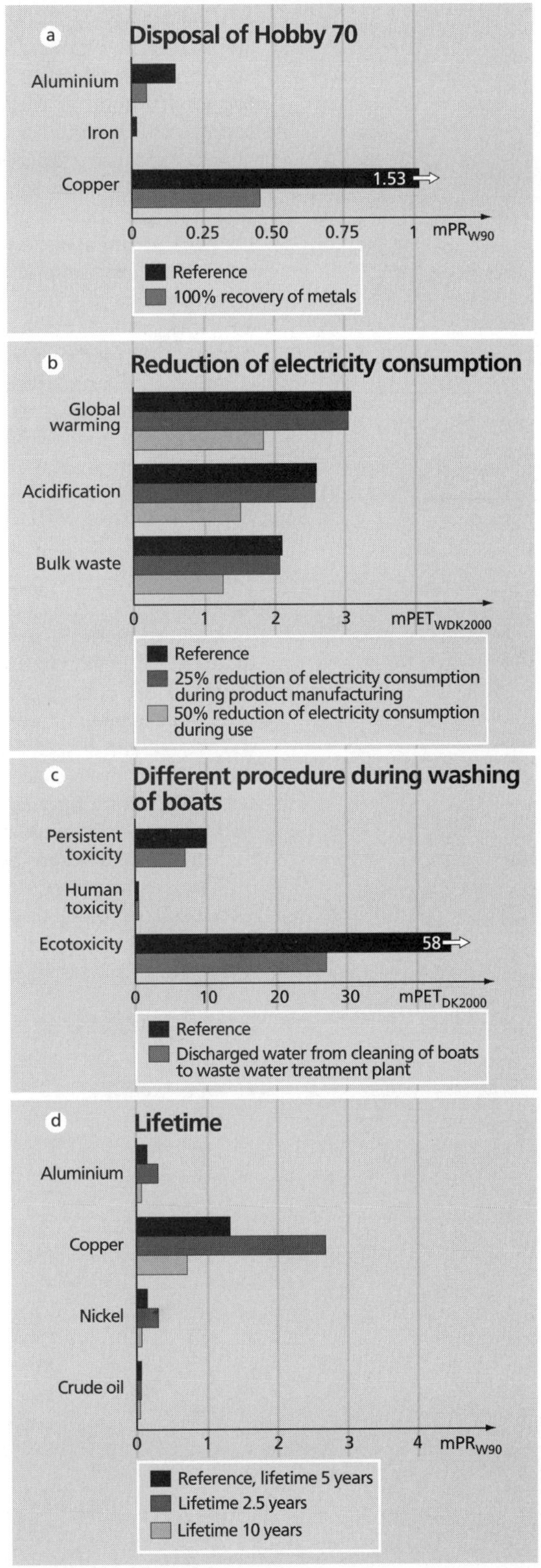

Figures 28.15. Simulation of changes in the product or its product system

Others - life span

As shown in Figure 28.15d, the operating time per annum is kept constant in the simulation of changes in the life span. As the weighted values for environmental impact potentials take account of the product's life span, the functional unit is thus the same, irrespective of life span. This also means that the resource consumption for electricity production is independent of the life span. On the other hand, the resource consumption for production of the high pressure cleaner will be affected by changes in the life span.

For the resource consumption for materials, there is a direct relationship between life span and consumption. A doubling of the life span to 10 years results in a halving of the annual resource consumptions of aluminium, copper and nickel. The same applies to that part of the consumption of crude oil which enters into the production of plastic. A halving of the life span to 2.5 years correspondingly means a doubling in the resource consumptions, as the weighted values are normalized with respect to the annual consumption.

Simulations - summary

The most significant environmental impacts affected by the simulations are the ecotoxicity potential and resource consumptions of metals and fossil fuels. A reduction in electricity consumption in the use stage results in large reductions in the consumption of fossil fuels, and thus also a reduction in the potentials for global warming and acidification. Optimization of the energy consumption therefore has the highest priority.

In general, an increased percentage discharge to the treatment plant will result in a lower ecotoxicity potential, as a large portion of the substances is degraded in the treatment plant. Cleaning agents, which contain complexing agents, can, however, result in a transport of heavy metals from the treatment plant. An assessment of the possibilities for reduction in the ecotoxicity potential should have a high priority.

The resource consumption of metals can be significantly reduced by increased recycling of the product. As the material already contains large quantities of secondary metals, a replacement of primary with secondary will have only a small effect.

Where do potentials for improvement reside in the product?

The LCA has shown which impact potentials are the most significant, and which components, stages and sources make the greatest contributions. Subsequently the potentials for improvement can now be sought at the levels of concept, structure and component. The areas in which efforts are already being made are described in Section 28.4, which deals with environmental design of new products.

Table 28.10. Examples of points of focus at the concept level

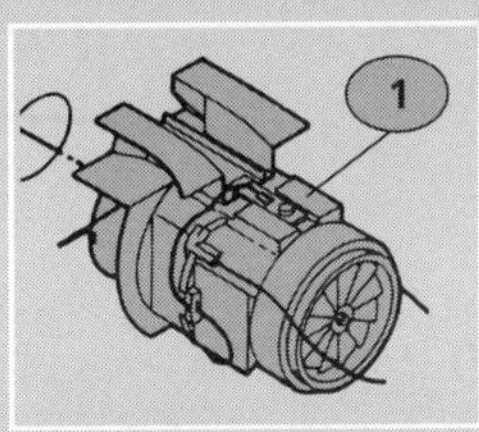

Idling: the motor in the high pressure cleaner idles even if the cleaner is not in active use.
Alternative: energy can be saved if the motor is stopped automatically when cleaning is not in progress. This can be achieved by an automatic start/stop mechanism of one form or another.

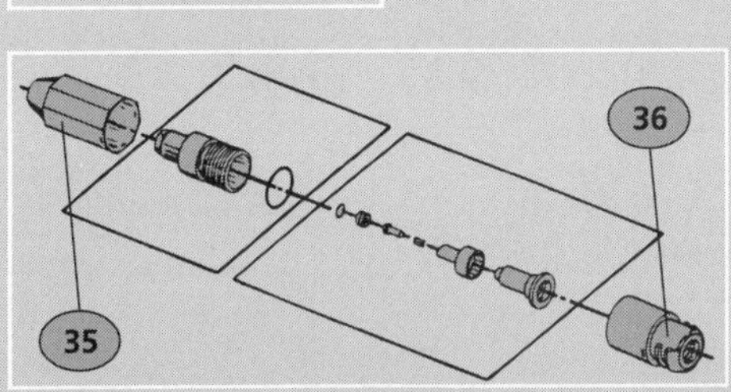

Nozzle: the design of the nozzle determines the form and pressure characteristics of the water jet, which can, for example, be pulsating or continuous. The design of the nozzle can thus be of significance for the time required for a particular cleaning task and for consumption of water and chemicals.
Alternative: an improved nozzle can save energy in the use stage and may mean that certain cleaning tasks can be done entirely without use of chemicals. This is significant for the environmental impact potentials, persistent toxicity and ecotoxicity.

Quantities of chemicals used: the largest impact potentials for the product derive from the use of chemicals. If the chemicals can be used more efficiently with less waste, a reduction in the contributions to toxicity will be obtained.
Alternative: the cleaning effect will be greater if the chemicals are foamed before they reach the surface to be cleaned. Foaming is achieved by adding the cleaning agent just before the nozzle, resulting in a 50% saving in chemicals.

Packaging for cleaning agents: the packaging used for the cleaning agents is typically polyethylene (PE), which is made from crude oil.
Alternative: use of more concentrated cleaning agents will mean savings in packaging. A solution which is five times more concentrated will result in a saving of 70% for PE.
Life span: resource consumption for the materials which enter into the product itself will be reduced if the life span is increased.
Alternative: introduction of an automatic start/stop function is expected to result in an increase of 5% in life span. Other alternatives are discussed under structural choices.

Table 28.11. Examples of points of focus at the structure level

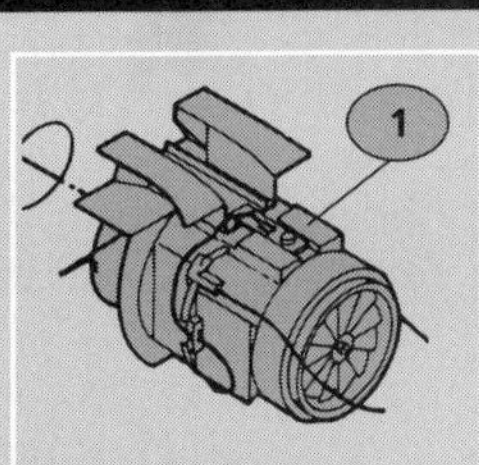

The motor's efficiency: an increase in the motor's efficiency will result in a lower energy consumption.
Alternative: one possibility is to use a different type of motor, e.g. an internal combustion engine, although this will not necessarily mean reduced energy consumption.
Quantity of materials: the product's environmental impact depends on its weight. The impact in terms of resources will also be affected by the rate of recycling and the life span of the product.
Alternative: a reduction in the quantity of material can occur via integration of functions and components.

Table 28.12. Examples of points of focus at the component level. Materials

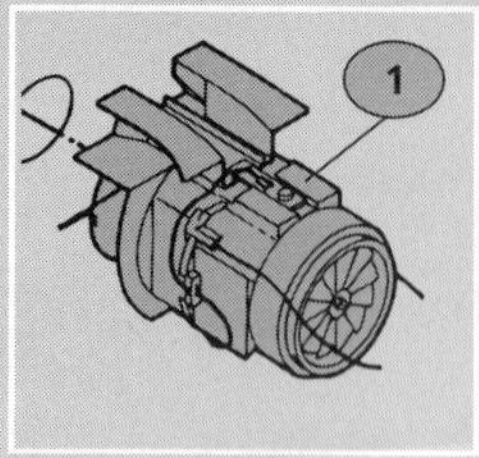

Copper: when resource consumption is weighted, copper makes the greatest contribution. Copper is found in the motor and in leads.
Alternative: motors without copper are available, but their efficiency is lower. Aluminium cables could be tried.
Stainless steel: nickel is used as an alloying element in stainless steel. As nickel is a scarce resource, the weighted contribution is large despite its small quantity. As the product contains such small quantities of stainless steel, it would be unrealistic to undertake a manual separation on disposal. When the product is sent through a shredder, the nickel content will be lost.
Alternative: stainless steel can be replaced in some cases by a plastic composite.

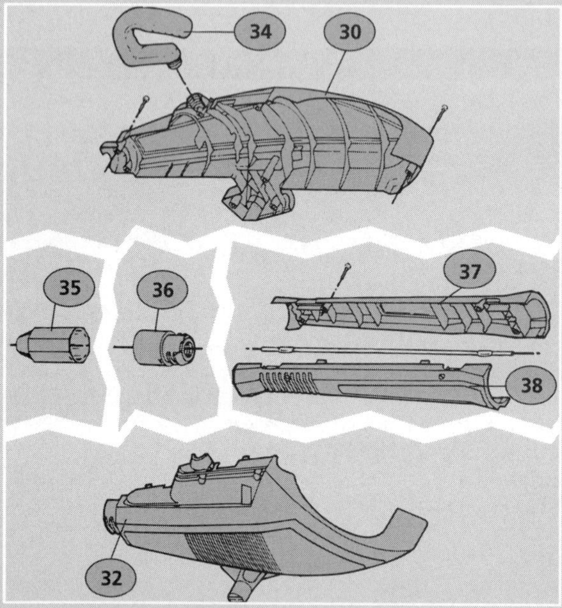

ABS: the product contains large quantities of ABS, which is shaped by injection moulding. Carcinogenic substances, e.g. styrene, are given off especially during injection moulding.
Alternative: a different material can be chosen, or the process can be managed so that it is kept within the temperature range in which carcinogens are not given off.
Recycling: an important precondition for recycling at the highest possible level is that it be possible to separate the various materials.
Alternative: labelling of all materials will make separation possible.

Table 28.13. Examples of points of focus at the component level. Others

Chemicals: the chemicals in the cleaning agents make a large contribution to ecotoxicity.
Alternative: apart from the options already noted, chemical substitutions can be made so that substances which are more easily degradable and less toxic are used.
Water consumption: the washing process consumes large quantities of water.
Alternative: use of stored rain water will eliminate the water consumption.

Where can potentials for improvement be found in the surroundings?

Some environmental problems or potentials for improvement lie beyond the product itself. These can be based on causes or traditions determined by the surrounding society.

The electricity system: the efficiency of electricity production is low (cf. the description in the example for the Gram refrigerator in Chapter 26).

Separate sewerage system: the large values for ecotoxicity from the cleaning agents are attributed particularly to the fact that part of the washing water goes directly to the recipient, e.g. the sea. In case of a separate sewerage system for pond areas, this part of the water will be conveyed directly to lakes and streams. In this case it is a disadvantage because the washing water from car washing, for example, will bypass the treatment plant.

28.3. Environmental specification for high pressure cleaners

The focal points for the product were designated in Sections 28.1 and 28.2 on the LCA and the environmental diagnosis. The experiences from these and a commercial assessment form the basis for the environmental specification for high pressure cleaners.

The competitive framework

The most important competitive parameters for KEW Industri's high pressure cleaner are quality and price. The quality parameters enter into all interfaces between customers and companies: in the product, in the service arrangements and in marketing.

This positioning is made through conscious choices of solutions for all circumstances which signal a particular profile for the products and for the company. Product quality is controlled via internal KEW standards and plans for testing.

KEW regards environmental considerations in the product, in manufacturing and among subcontractors as a part of the company's total quality profile, and activities in this area will supplement and strengthen the positioning in the market.

With increasing strength, feedback from the market is showing a consciousness about the product among the customer groups which prefer a KEW product. The feedback comes mainly from the still more dominant chain store dealers of products of the do-it-yourself type.

The dealer generally demands a greater product consciousness from the manufacturer, because a clear product profile is necessary in a market where the competition is becoming stronger and stronger.

KEW is one of two companies offering a total range of high pressure cleaners to the private consumer. The brand awareness of KEW is high in Scandinavia, and the company has a dominant role in the market, while for the rest of the world the contrary is the case. The lack of brand knowledge outside Scandinavia must therefore be weighed against other positioning aspects. Work is proceeding here with an active representation of environmental efforts and results to, in the first instance, selected chains.

The current competitive parameters in the area of the environment are consumptions of energy and chemicals and water consumption in the use stage. The market's attitudes are relatively influential for these aspects, and expectations of the product change rapidly when new products with improved qualities appear in the market.

It is therefore necessary to be at the leading edge with environment-related product features, and there are a number of possibilities here which can be used via goal-directed product development and marketing.

Long-term objectives

Environmental work at KEW is proceeding on a number of long-term objectives:

- Improved implementation of life cycle assessment during product development.
- Establishment of better cooperation with leading chain store dealers in the marketing of environmentally friendly products.
- Introduction of an environmental management system to ensure focus in management.
- Formalized assessment of suppliers and cooperation with suppliers with respect to environmental considerations. This applies especially to suppliers in the Far East and India.
- Preparation of a new service concept which in the longer term will enable the recycling of components in manufacturing.

Product specifications for new high pressure cleaners

Current focus on environmental efforts at KEW is on:

- Reduction in energy consumption in the use stage.
- Reduction in consumption of chemicals in the use stage.
- Reduction in consumption of water in the use stage.
- Elimination of all toxic substances in the product and in manufacturing.
- Design of the products such that they can be separated for recycling.
- Choice of pure materials and limitation in the quantity of materials (plastic) in the products.
- Development of packaging with low and uniform consumption of materials.

KEW is adopting these points on an on-going basis in environmental specifications for the company's new products.

28.4. Environmental design of new products

A large number of new development projects were introduced at KEW during work with the EDIP project. The interim results of the life cycle assessments have been used in these projects to the extent to which it was possible to do so.

Product development at KEW Industri

Product development at KEW proceeds in accordance with the model "integrated product development". A system has been developed which ensures that all of the company's functional areas are involved in the product development sequence. The system consists of a separate organizational unit in which project-oriented developmental activity occurs. The unit has its own decision-making structure from management level to project level. All projects are also carried out in accordance with the same general plan (cf. Figure 28.16).

The project sequence is divided into stages with specific activities in each stage. The activities in the various stages are pre-defined, and there is a specified requirement for documentation for decisions concerning the project's further course. The concrete division of responsibilities for the project is also defined in a description of the organizational and technical cooperation. The individual participants from different functional areas in the company thus have pre-defined tasks at particular times in the project. The system is integrated into the company's quality control system, thus ensuring that project activities are carried out and documented.

The actual content of the projects is not described here, but it is clear that environmental considerations in the products enter into all stages in the course of product development. The basic environmental concept or feature of the product is defined at the concept stage. Concept proposals are prepared in the design stage, and work proceeds to a given level, where the concept is usable and market conditions are in order. The stage finishes with preparation of actual specifications for the product, which are developed to include the product's environmental properties. The rest of the project is used to carry out and to document the specifications, and new product characteristics are thus developed in this part of the project.

Figure 28.16. Product development at KEW Industri

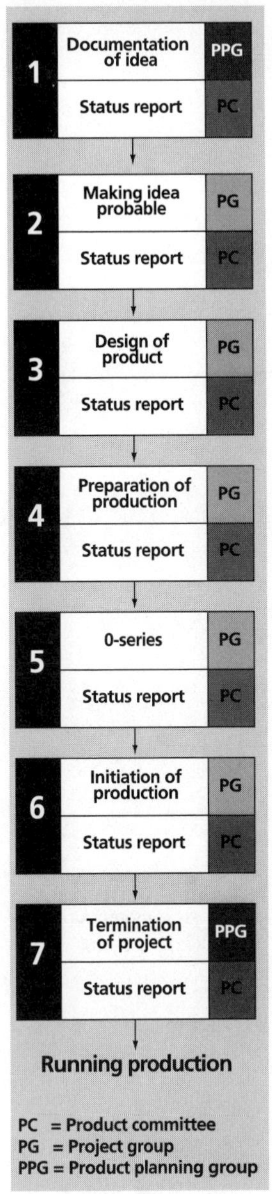

Examples of environmental design in new products

A catalogue is set up below for a number of projects and solutions which have established an improved environmental performance of the products. Some information is omitted for reasons of competitive secrecy.

A table is shown for each of the projects, giving the level at which the change has an influence, and the area in which the saving lies. The savings are divided into the MECO categories.

Project 1. Introduction of automatic start/stop

	M	E	C	O
Concept		●		●
Structure				
Component				

When using a high pressure cleaner, a flushing handle is applied by the user to turn the water supply to the nozzle on and off. When off, the pump normally cycles the water in an internal circuit in the machine at a relatively low pressure: a kind of idling.

To avoid this idling, an automatic start/stop system was developed. The system starts and stops the electric motor with operation of the flushing handle, i.e. a kind of hydraulic remote control is applied. This avoids mechanical wear on the motor and the pump during idling.

Energy: Ca. 10% saving of the energy consumption in the use stage.

Others: Increased life span for the product of ca. 5%.
Lower average noise level in use.

Project 2. Nozzle

	M	E	C	O
Concept		●	●	●
Structure				
Component				

There is a large potential for energy saving in the development of better energy transfer systems from the cleaner's nozzle to the item being cleaned. A large number of different nozzles were developed over time with differing effects. Most attempts to produce a better nozzle in terms of energy resulted in constructions with a large number of moving parts, which meant a short unit life and increased costs for the customer. A new type of nozzle has now been developed which uses a hydrodynamic principle to form the stream, where the stream was previously formed with the aid of the mechanical design of the nozzle opening.

Apart from a significant improvement in the nozzle's cleaning ability (sharper stream), the nozzle life was increased.

Energy: 0-30% saving of the energy consumption for a given cleaning task.

Chemicals: The increased cleaning power of the nozzle has to some extent replaced the use of chemicals.

Others: Increased nozzle life.

Figure 28.17. New nozzle lowers energy consumption in the use stage

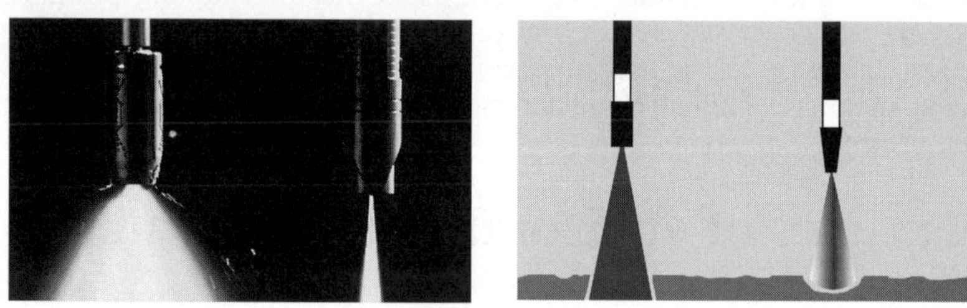

Project 3. Foaming system for chemicals

	M	E	C	O
Concept			●	
Structure				
Component				

A new chemical concept was developed, resulting in the mixture of chemicals and water foaming significantly more than in the previous concept. Mixing now occurs entirely at the nozzle, and air is added. When sprayed on to a car, the visual effect gives a clear feeling that cleaning agents are being used for the task, where previously it was difficult to see any purely visual effect with the use of cleaning agents. At the same time, the cleaning agent is more efficient when it is pre-foamed, as the foam adheres to the surface for a longer time (as is well known in the professional sector).

The concept has led to a halving in the applied concentration of cleaning agents.

Chemicals: Reduction of chemical consumption by 50%.

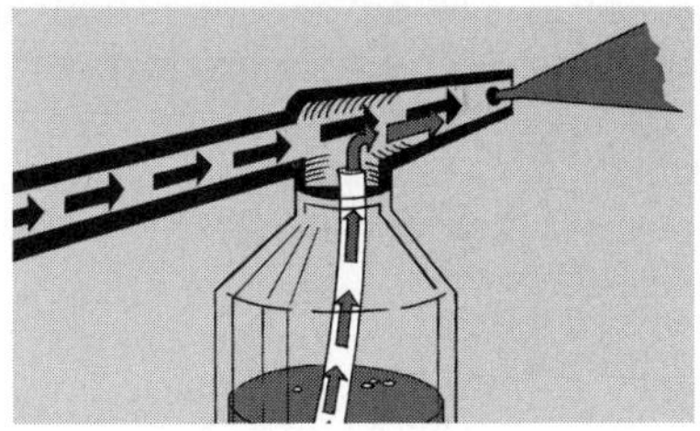

Figure 28.18. Washing a car with foamed cleaning agent

Project 4. Chemical concentrate

A new concept was developed where the chemicals are used in concentrated form. The chemicals are mixed to the desired strength by the consumer, saving packaging both for the actual bottle and also for its transport packaging.

Transport volume and weight are also saved, resulting in reduced transport.

	M	E	C	O
Concept	●	●		
Structure				
Component				

Materials: Packaging for chemicals was reduced by ca. 70%, with a saving in oil for the production of PE.

Energy: Transport volume and weight were correspondingly reduced, with a saving in oil for, and emissions from, transport.

Project 5. Integration of functions and components

Through the development of new constructions, it proved possible to integrate a large number of functions which were previously assigned to individual components or component assemblies into a few but more complex components. Increased use of injection and pressure moulding technology and investment in more complex tools made this development possible.

The project was carried out to achieve savings in the product's cost price. This is a good example of how economy and environmental considerations can be integrated at the same time.

Attention was also directed to achieving a uniform choice of materials in the products, so that separation and sorting are made as easy as possible in the disposal stage.

	M	E	C	O
Concept				
Structure	●			
Component				

Materials: Reduction in materials of ca. 25%.

Project 6. Design for disassembly and servicing

Designing for disassembly and designing for servicing are usually different aspects of the same task. It was therefore relatively easy to focus on this area.

Through cooperation with service representatives and via the EDIP project's knowledge of design rules for disassembly, a knowledge was gained of the general choice of product structures which was to be made and, at a lower level, of the methods of assembly which may and may not be used.

Among other innovations, snap locks in plastic constructions are no longer used in the product's service areas. They have been replaced by joined elements which can be separated with a simple tool.

	M	E	C	O
Concept				
Structure	●			
Component				

The development of new constructions with a high degree of component integration also offers new opportunities in the service area. Work is continuing today on a new service concept which in the first instance will simplify servicing and increase customer satisfaction, and which in the longer term can be developed into a basic recycling concept to supply the manufacture of new products.

Materials: A saving of resources can be achieved by increasing the degree of disassembly and recycling of materials in the product.

	M	E	C	O
Concept				
Structure				
Component	●			

Project 7. Labelling of plastic components

Since 1989, KEW has labelled plastic components. The labelling was not standardized at that time, and a labelling system used in the plastic bottle and packaging industry was chosen. A standardized labelling system has subsequently been developed, and all components over 50 g are labelled under this system.

Materials: Labelling of plastic components ensures a correct and fast sorting and is a precondition for the recycling of plastic materials.

	M	E	C	O
Concept				
Structure				
Component	●			

Project 8. Uniform packaging materials and reduction of the quantity

Through the joining of national recycling systems in countries such as Germany, France, England and, most recently, Sweden, the required packaging solutions were developed which as a starting point contain only a few types of materials and which are closed without use of clips or other metal elements.

KEW has gone the whole way, supplying its products today in packaging which consists only of cardboard. Both the carton and the stabilizing inserts are made of cardboard, and a new design, where consumption of materials has been optimized, has reduced the weight of the packaging by 25%.

Materials:
The uniformity means that the materials can be recycled without sorting, and 100% recycling is possible.

Reduction in the quantity of material, a saving of 25% was achieved in newly developed packaging systems.

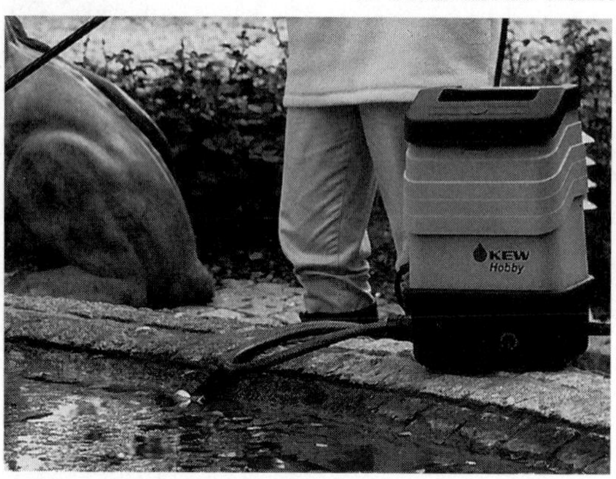

Figure 28.19. A new filter makes it possible to use collected rain water for cleaning

	M	E	C	O
Concept				
Structure				
Component			●	

Project 9. Substitution of water by use of a filter

A new suction filter was developed which in the first instance can be used with collected rain water for high pressure cleaning. Special conditions apply to this use, and KEW has solved the problems which are involved. The suction filter is required to ensure efficient operation of the machine.

Water: Substitution of drinking water with rain water/surface water. Reduction of 100% in consumption of drinking water.

KEW's experience with environmental considerations in product development

For KEW, the inclusion of environmental considerations in product development has been a process which has involved not only the development department, but which has been of significance for the whole company. Initial work was concerned especially with changing employee attitudes. As the work became more specific, the competitive aspect gained in significance.

The influence of the media, authorities and directly from the company's customers has meant that environmental problems have become topical and significant. The demands of the outside world have thus helped to enable the formulation of specific environmental projects.

The methodology of the EDIP project was integrated into the company's system of development, and its activities managed and carried out similarly to the other disciplines in the development process.

The increased focus on the environment has been received positively by KEW's organization and its partners. Internally at the company, employees are happy that the environment is part of KEW's profile, and many of the employees identify themselves personally with the company's attitude.

29. Grundfos: Pumps

Nils Thorup, Grundfos
Nina Caspersen, Institute for Product Development
Elisabeth Rasmussen, Institute for Technology and Society, Technical University of Denmark

Grundfos is a Danish-owned company with headquarters in Bjerringbro in the centre of Jutland, Denmark. Grundfos employs more than 9,000 people worldwide with a primary business area of supplying pumps for water transport

Page

29.1. LCA of a pump - the JetpaQ

The product selected by Grundfos for an LCA is a pump, or more correctly a compact household water works for a private water supply (cf. Figure 29.1). The JetpaQ pump is used for private water supply in houses, summer cottages or small farms. It is also suitable for increasing or stabilizing the pressure in situations where the pressure in the ordinary water supply is inadequate or unstable.

Figure 29.1. The reference product, the JetpaQ

Choice of reference product

The JetpaQ was chosen because it was at an early stage of product development when the LCA was started. It was therefore possible to introduce the findings made during the LCA into the product development.

The functional unit

The pump can supply a quantity of water at a constant pressure (cf. Figure 29.2). The higher the pressure, the higher the power consumption. If it is decided that the pump is to operate at a particular power level, a choice can therefore be made between high pressure and a small quantity of water or low pressure and a large quantity of water. The primary functional unit of the pump is:

Supply of 5 cubic metres of water per hour at an outlet pressure of 1.5 bar, or a correspondingly smaller quantity of water at a higher pressure of up to 5 bar, for 4,870 hours of operation.

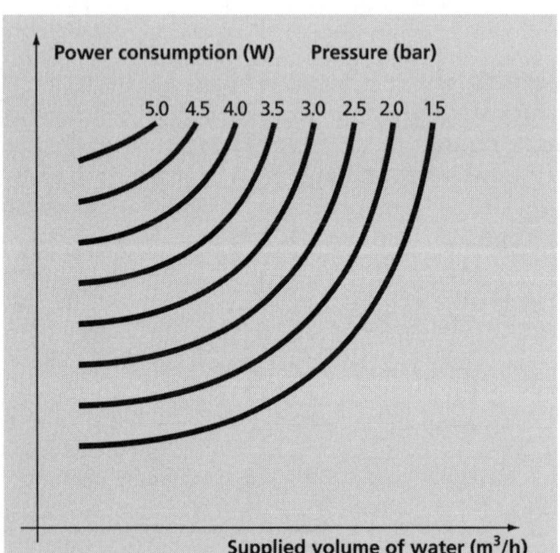

Figure 29.2. Relationship between quantity of water supplied, power consumption and pressure for the JetpaQ

The secondary qualities include protection against running dry, and the pump's ability to suck up water even if there is little water in it.

Scope definition of the product system

The exploded drawing shown in Figure 29.3 is a detailed breakdown of the product with item numbers. The JetpaQ is an integrated unit consisting of a pump, motor, membrane tank and electronic control.

The items in the drawing can be found in the parts list in Table 29.1, which also provides a description of the manufacturing processes through which the individual items pass at Grundfos. For both items and manufacturing processes, the parts list indicates with a ✓ if the item is included in the LCA or a ÷ if it is not included. The most significant omissions are the cooling oil and the electronic control. The LCA to and including manufacturing thus covers only the product's mechanical parts.

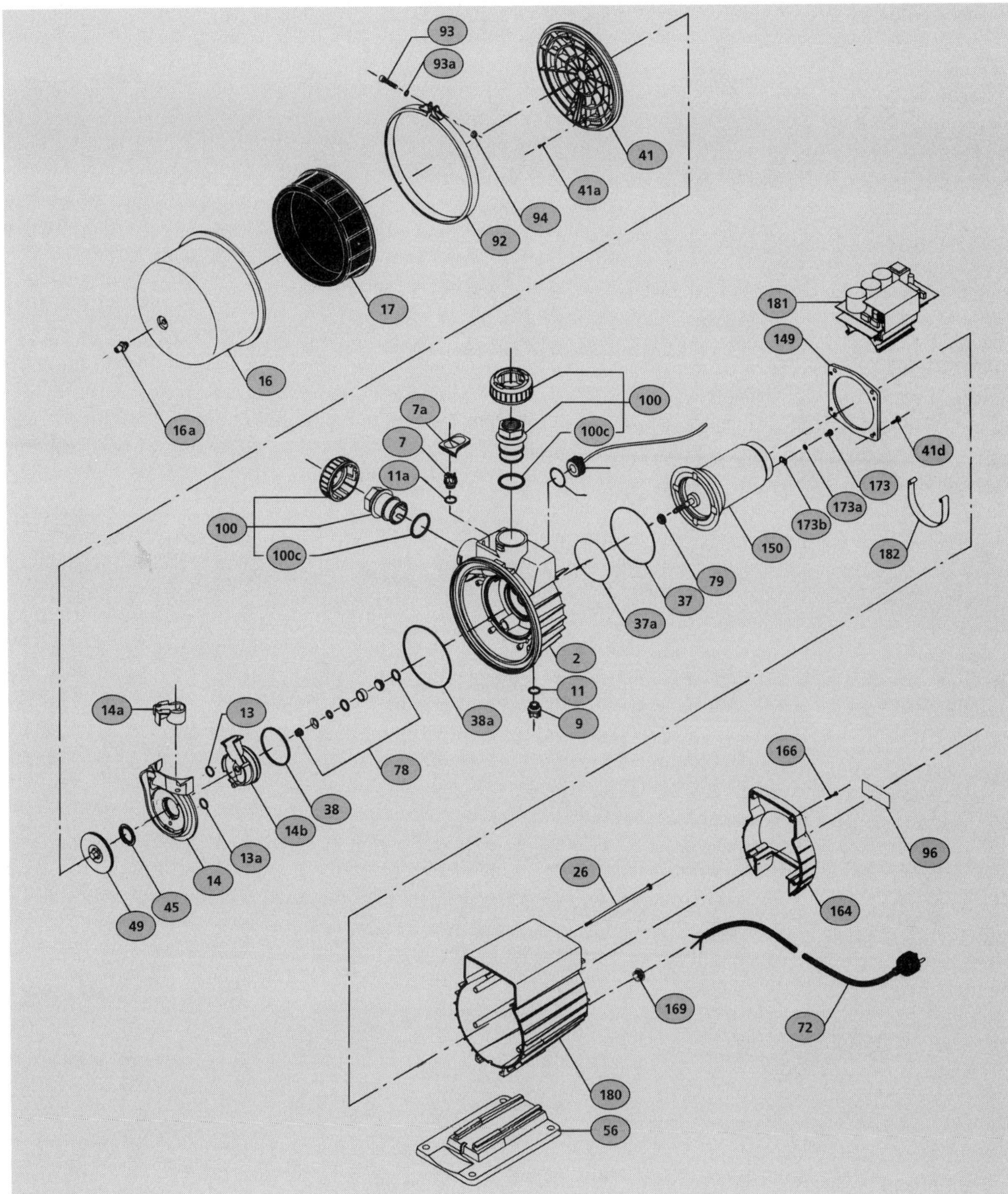

Figure 29.3. Exploded view of the JetpaQ

The materials which the individual items consist of, their weights and whether the LCA covers the environmental impacts associated with the extraction of raw materials and production of the materials, are also given in the parts list.

The parts list provides an overview of which materials and manufacturing processes are included. The scope definition of the product system is further reviewed in the next section.

Pos.	Number	Item	Material	Total weight (g)	⏚	Manufacturing processes	* Incl. in model	Ancillary substances	Weight (g)	⏚
Cabinet										
16	1	Tank	Stainless steel	1143	✓	Punching, pressing	✓✓			
						Cutting, cleaning	✓✓	Sodium metasilicate	0.07	÷
								Phosphate	0.07	÷
						Ball burnishing	✓			
92	1	Clip	Stainless steel	190	✓	Punching, cleaning	✓✓	Sodium metasilicate	0.007	
								Phosphate	0.007	÷
						Welding	✓			
2	1	Stator housing	Stainless steel	849	✓	Punching, pressing	✓✓			
						Cleaning, cutting	✓✓			
						Welding, assembling	✓✓			
16a	1	Air valve	Stainless steel	10	÷					
150		Rotor shaft	Stainless steel	190	✓	Shortening, cutting	✓✓			
150	1	Ball bearing	Stainless steel	37	÷					
149	1	Flange	Stainless steel	276	✓	Punching, pressing	✓✓			
						Cleaning, cutting	✓✓			
						Welding, electrodeposition	✓✓	Thinner	0.009	÷
								Acetic acid	0.001	÷
								Tetraethylene glycol	0.002	÷
								Sodium nitrite	0.0007	÷
								Iron sulphate	0.02	÷
								Various chemicals	0.26	÷
26,41a, 41d,93, 94,166, 173,182	8	Other steel items	Galvanized steel	136	÷					
150	1	Rotor	Recycled aluminium	132	✓	Collecting, moulding	✓✓			
			Sheet metal	474	✓	Pressing, cutting	✓✓			
						Grinding, painting	✓✓	Thinner	3.9	÷
								Paint	0.6	÷
150	1	Stator	Sheet metal	1067	✓	Assembling	✓			
			Copper wire	316	✓					
			Fibre reinforced PETP	35						
			Polyester foil	11	✓	Foil extrusion	✓			
181	1	Saddle for electrical unit	Recycled aluminium	300	✓	Moulding	✓	Coolant/lubricant	6.6	÷
								Propylene glycol	4.8	÷
								Paraffin	3	÷
								Aromatic amines	0.6	÷
41	1	Inserts in sheet	Brass	20	✓	Cutting	✓			
96	1	Label	Unknown metal	?	÷					
41	1	Duplicate plate	Fibre reinforced PPE	1070	✓	Moulding	✓			
7+9	2	Priming/ emptying plug	Fibre reinforced PPE	45	✓	Moulding	✓			
180	1	Motor shield	Fibre reinforced PPE	770	✓	Moulding	✓			
14	1	Spiral casing	Fibre reinforced PPE	155	✓	Moulding	✓			
49		Runner	Fibre reinforced PPE	20	✓	Moulding	✓			
						Ultra sound welding	✓	Unspecified xylene	3	÷
164	1	Lid	Fibre reinforced PPE	140	✓	Moulding, assembling	✓✓			
100	1	Connection piece with chaser	Fibre reinforced PPE	230	✓	Moulding	✓			
56	1	Sole plate	Fibre reinforced PPE	260	✓	Mouldinging	✓			
100	1	Check valve	PP	ca. 10	÷	Moulding	✓			
45	1	Gap sealer	PP	4.3	÷	Moulding	✓			
169	1	Cable coupling piece	PA (nylon)	7.5	÷	Turning	✓			

Table 29.1. Simplified parts list for the JetpaQ

Pos.	Number	Item	Material	Total weight (g)	⏚	Manufacturing processes	* Incl. in model	Ancillary substances	Weight (g)	⏚
Accessories										
78	1	Shaft sealer	Coal, ceramics and stainless steel	ca.20	÷					
72	1	Electric cable	Neoprene	200	÷					
181	1	Electric unit (electronics)	Various	ca.700	÷					
174a	1	Pressor sensor	Various	24.5	÷					
17	1	Membrane	Butyl rubber	320	✓	Moulding Vulcanization	✓ ✓			
11,13,	8	Gasket rings	Nitrile rubber	25	÷					
37,38,		Packing	Cardboard	300	✓	Punching, printing	✓✓			
78,79,		Manual	Paper	?	÷					
100c,		Silicone gel	Silicone gel	8	÷					
174b		Cooling grease	Cooling grease	120	÷					
		Grease for O-rings	Lubricant grease	?	÷					

Notes

⏚ The column indicates whether or not the product system of the material or the ancillary substance is included all the way back to the extraction of raw materials.

✓ Means that it is included.

÷ Means that it is not included.

* In the column "Incl. in model" is likewise indicated if the manufacturing processes itself is included or not. ✓ means that energy consumption and use of ancillary substances is measured, that emissions and waste is either measured or calculated and that impacts on the working environment are measured or assessed. The number of ✓'s indicates how many processes are included.

Tabel 29.1. Continued

Materials

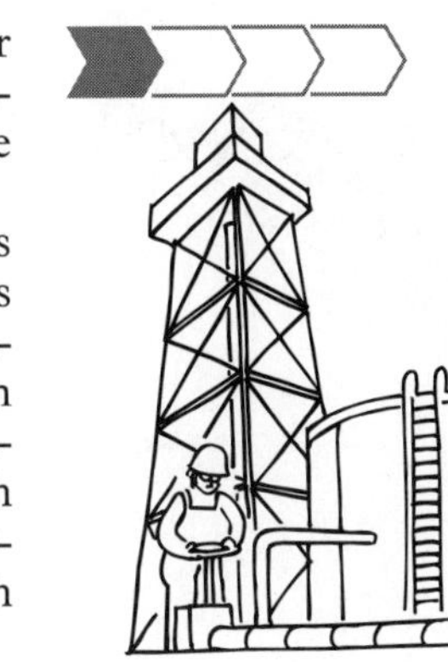

The materials commence as raw materials in the form of e.g. crude oil or ores, after which they are refined. The locations of extraction and refining in Europe are shown in the example for the refrigerator LER200 (see the Gram example, Chapter 26).

Smaller items are omitted for some materials. These amount to less than 10% of the weight of the respective materials from which the items are made. As the goal of the LCA is to find the most significant environmental impact potentials, these omissions are of minor significance. An attempt was made to include all items which could have specific environmental impact potentials. A significant exception is the electronics, which were not included in the inventory because of insufficient data. The electronic components typically contain metals which are toxic and for which the supply horizon is short.

The geographical locations of the subcontractors are given in the map in Figure 29.4. Suppliers of materials are distributed throughout Europe. The largest quantities, i.e. stainless steel, plastic and tinplate, come from southern Europe. Transport of materials to Grundfos is assumed to be by lorry on motorways.

Data on the working environment are included for the production of steel, which were the only data available. Production of stainless steel strongly resembles production of recycled steel, and data for the working environment for recycled steel were therefore also used for stainless steel. As steel and stainless steel comprise 40% of the total consumption of materials, the working environment was thus included for 40% of the consumption of materials.

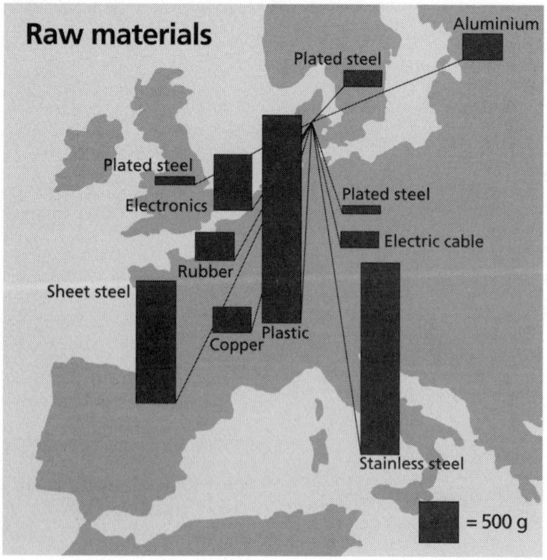

Figure 29.4. Geographic distribution of suppliers of raw materials.

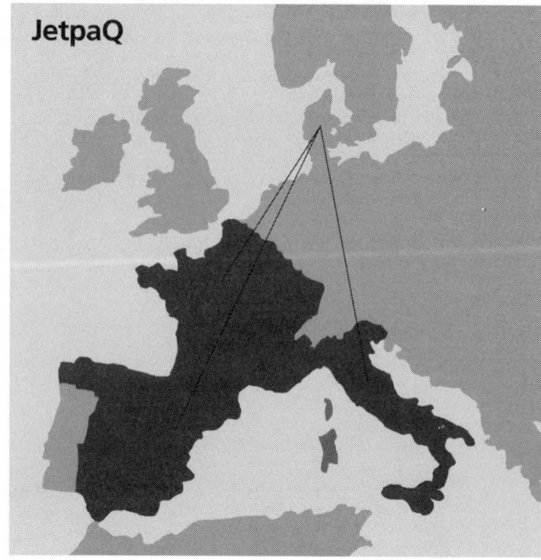

Figure 29.5. Sales area for the pump

Manufacture

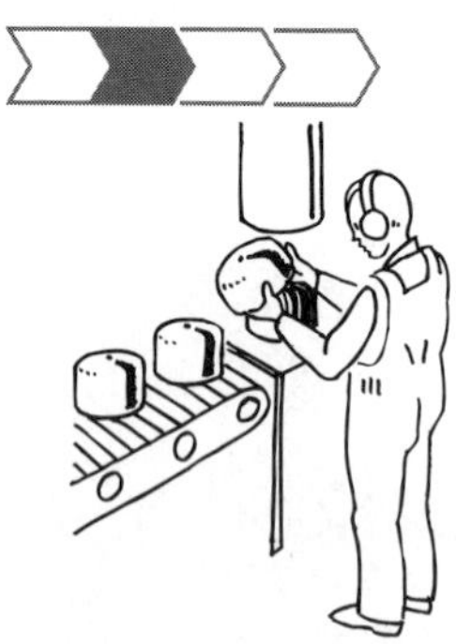

That part of the manufacture which takes place at Grundfos is included in the inventory. This is by far the greatest part.

With respect to the working environment, some specific remarks should be noted: as manufacturing occurs in halls with a high noise level, all of the manufacturing time was considered as having noise at a level which can cause hearing impairment. Separate measurements for monotonous repetitive work were not made either. All manufacturing was therefore included under monotonous repetitive work. The data used for the working environment were valid for 1993, and there have been changes since then.

Rejection rate

A rejection rate of 2% at Grundfos was assumed, i.e. for each JetpaQ delivered, 1.02 JetpaQs must be manufactured. Data for the material and manufacturing stages were therefore multiplied by 1.02.

Overhead

Heating of buildings, consumption of water and administration are called overhead. Grundfos have calculated all consumption at the company per kilogram of products. Consumptions of electricity, natural gas and water which enter into the inventory were computed accordingly.

Use

As shown in Figure 29.5, the main markets for the JetpaQ are Italy, Spain and France. Quantities have not been assigned to these markets as sales are assumed to be equally distributed among the three countries.

Transport of the product to the user is estimated as 300 km in a 3.5 tonne diesel van and 2,500 km in a 16 tonne lorry. The lorry transport is included in the inventory of the working environment.

The pump's electricity consumption in the use stage was computed for a family with two adults and two children. Grundfos has set up a use scenario which makes it possible to estimate annual power consumption. Assuming a water consumption of 960 litres and an operating time of 80 minutes per day, an annual consumption of 230 kWh is obtained. The idling consumption is 5 W or 41 kwh per annum. The impact on the working environment for the production of electricity, which covers the impact of the power plants, is included.

Product life span and service

The life span of the pump is set at 10 years. The product is designed such that at least 95% of pumps will last for at least 5 years with 10% operating time, i.e. time in active use. The operating time is typically much shorter. It is here set to 80 minutes per day. This corresponds to an operating time of 5%. An average life of 10 years was therefore chosen. No service is required during the life of the pump.

Disposal

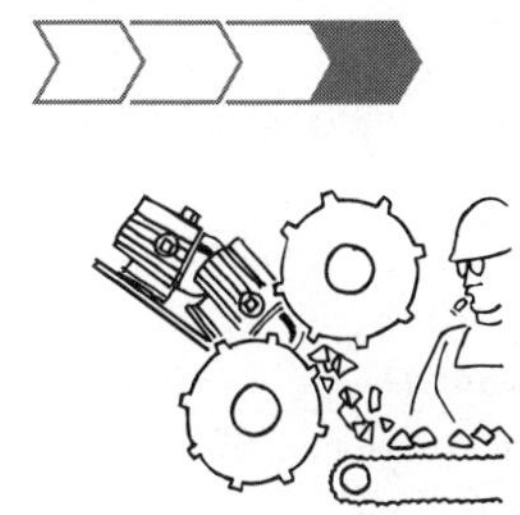

After use it is assumed that 70% of products are dumped, while 30% enter into material recycling. Figure 29.6 shows the disposal scenario. Material recycling occurs via the shredder, where the product is crushed into smaller pieces and most of the metal is separated. The combustible parts are burned in a municipal waste incineration plant with heat recovery. Transport after use is estimated as 15 km in a petrol-driven van for the 70% which are dumped, or 500 km of transport by lorry on highways for the 30% of the products which are recycled.

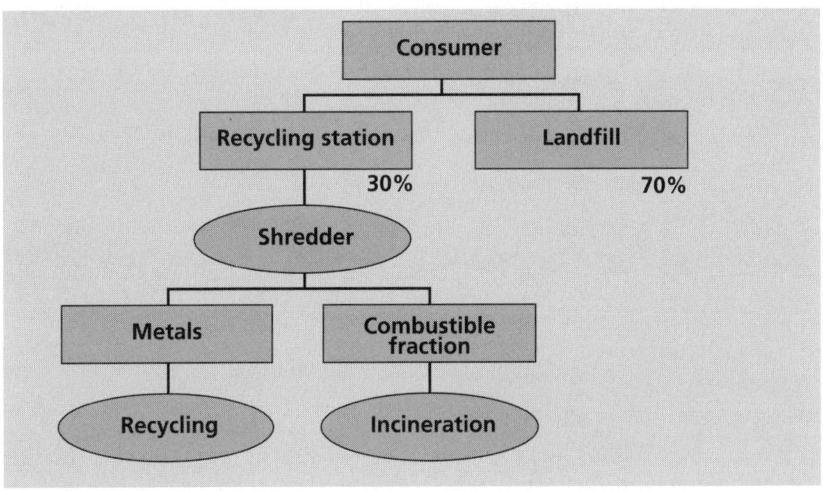

Figure 29.6. Disposal route for the JetpaQ

Data

Data were collected as site-specifically as possible. Table 29.2 provides a brief overview of the sources of data for the individual stages in the product life.

Site-specific information was not collected for the raw materials, but average data from the EDIP database were used (Frees, 1996). For sheet metal, which is used in the motor (item no. 150), there are no data in the EDIP database. The database's values for 90.5% recycled steel were therefore used. As sheet metal is thin steel plate and the plate in the pump is not surface-treated, this is a reasonable approximation. For the plastic composites, data for the individual ingredients were used, as there are no data for the production of composites.

									JetpaQ
Data reference									
	Data specifity			Data source type					Comments
Product system & process type	Product specific	Site specific	General	1	2	3	4	5	
Extraction of raw materials									
Crude oil			x			x			Literature
Iron ore			x			x			Literature
Copper ore			x			x			Literature
Production of materials									
Plastic			x			x	x		Literature and subcontractors
Copper			x			x			Literature
Stainless steel			x			x			Literature and subcontractors
Product manufacturing									
Grundfos	x	x		x					Grundfos
Sub-contractor, pressure moulding		x		x					Subcontractor
Use									
Energy consumption	x			x					Grundfos
Lifetime			x			x			Grundfos
Disposal									
Disposal routes			x				x		Grundfos
Shredding		x		x					Literature
Incineration			x			x			Literature
Recycling of steel			x			x			Literature
Transport									
Distances and means of transportation	x		x		x	x			Grundfos
Energy consumption and emissions			x			x			Based on measured emission factors
Energy systems									
Production of electricity			x			x			Literature

Notes

1) Measurements
2) Computations (from mass balance considerations and input data for the process in question)
3) Extrapolation of data from similar process type or technology
4) Extrapolation of data from different process type or technology
5) Estimate

Product-specific data: concerns processes specifically handling the JetpaQ
Site-specific data: concerns data from actual sites in the product system of the JetpaQ, but the process data are not specific for the JetpaQ
General data: all others

Table 29.2. Sources of data

Most of the manufacturing is at Grundfos, where direct measurements were made on the manufacturing equipment. Most data were, however, collected in the manufacture of products other than the JetpaQ.

The data for use were based on experiences with previous pumps, as the JetpaQ is a new product.

Transport distances were computed, and a means of transport was assumed for each transport type. Fuel consumption and emissions for all transport types were measured by Kilde (1995).

Exchanges with the environment

Resource consumptions, emissions to air and water, waste, and impacts on the working environment comprise the exchanges with the environment in the product system. All exchanges enter into the LCA, but for the sake of clarity, only the most significant exchanges are shown in Table 29.3. Further details and documentation of the mapping are available at Grundfos and in the EDIP unit process database.

Resource consumption

The resources which enter into the product itself are seen in the material stage. These are partly fossil fuels for the production of plastic and partly metals. The most significant metals are iron, chromium and nickel, which are used in stainless steel, and aluminium and copper. Large quantities of coal and brown coal, used for the production of electricity, are used in the use stage.

Negative values for both metals and fossil fuels appear in the disposal stage. The negative values express the crediting of resources on their recovery. When fossil fuels are credited, it is because energy produced in the incineration of waste is utilized and thereby substitutes for other energy production.

Emissions to air

The largest emissions derive from the production of electricity, i.e. carbon dioxide, sulphur dioxide and oxides of nitrogen. These emissions appear in all stages as they all use electricity. Emissions of arsenic, cadmium, copper, lead and vanadium derive from the combustion of fossil fuels for electricity production.

The greatest emissions in the material production stage are aluminium and manganese.

Emissions to water

The material production stage makes the greatest contribution to the emissions to water. The emissions are very small for the manufacturing itself.

Waste

The production of materials results in some waste in the form of slag and ashes from the production of metals. The chromium-containing slag derives from the production of stainless steel. The manganese-containing slag derives from steel production.

The extraction of coal used for the production of electricity makes a large contribution in the form of bulk waste. The waste derives from the cleaning of coal in the coal mines.

On disposal, part of the product ends in landfill and is designated as unspecified bulk waste in the table. Hazardous waste comes principally from the electronics in the product. As the worn out electronics are generally regarded as problem waste in environmental terms, this waste is included in the inventory as hazardous waste.

			Production of materials	Product manufacturing	Use	Dis-posal	Transport	Total
Resource consumption								
Crude oil		g	5,700	130	82,200	-80	750	88,700
Natural gas		g	5,200	380	59,700	-0.9	50	65,300
Coal		g	10,900	2,300	295,700	20	4	308,900
Brown coal		g	480	-	220,600	-2	221,090	442,200
Water for hydro power		litre	940	500	121,100	0.2	2	122,500
Aluminium	Al	g	190	20	0.1	-40	0.05	170
Chromium	Cr	g	280	-	-	-60	-	220
Iron	Fe	g	1,360	0.02	5	-310	0.05	1,060
Copper	Cu	g	280	-	-	-80	-	200
Manganese	Mn	g	50	-	-	-10	-	40
Nickel	Ni	g	120	-	-	-30	-	90
Zinc	Zn	g	8	-	-	0	-	8
Soft wood		g	290	40	14,100	-0.1	0	14,430
Groundwater		litre	50	20	-	-0.5	0	70
Surface water		litre	0.4	0.3	-	0.009	0	0.7
Unspec. water		litre	50	0.09	50	-0.03	0.2	100
Air emissions								
Carbon dioxide	CO_2	g	43,100	7,600	1,446,000	5,100	780	1,502,600
Nitrogen oxides	NO_x	g	220	30	5,800	1	20	6,100
Sulphur dioxide	SO_2	g	320	40	12,800	-0.06	3	13,160
Dinitrogen oxide	N_2O	g	2	1	120	0.004	0	120
Unspec. particles (dust)		g	30	20	690	-5	3	740
Hydrocarbons	HC	g	280	50	6,900	0.5	-	7,200
Volatile organic compounds	VOC	g	40	0.2	2	0	0.007	40
Unspec. aldehydes		g	0.02	0.1	5	0.0002	-	5
Styrene		g	0.02	-	-	-	-	0.02
Trichlorethylene		g	0.2	-	-	-	-	0.2
Unspec. xylene		g	3	-	-	-	-	3
Aluminium	Al	mg	4,000	-	-	-	-	4,000
Arsenic	As	mg	0.5	0.3	140	0.0009	-	140
Cadmium	Cd	mg	0.5	0.03	20	-0.00001	-	20
Copper	Cu	mg	0.6	3	140	0.006	-	140
Mercury	Hg	mg	0.5	0.4	2	0.0008	-	3
Manganese	Mn	mg	1,940	-	-	-	-	1,940
Nickel	Ni	mg	0.01	0	1	-0.02	-	1
Lead	Pb	mg	1	1	190	0.0009	0.5	190
Vanadium	V	mg	3	2	380	0.02	-	390
Emissions to water								
Chemical oxygen demand	COD	g	6	0.003	0.7	-0.001	0.007	6
Total nitrogen	tot-N	g	0.6	0	3	-0.00005	0	4
Hydrocarbons		g	2	0.01	2	-0.0002	0.1	4
Arsenic	As	mg	0.01	-	-	-	-	0.01
Cadmium	Cd	mg	0.005	-	-	-	-	0.005
Copper	Cu	mg	0.04	-	-	-	-	0.04
Nickel	Ni	mg	0.001	-	-	-	-	0.001
Zinc	Zn	mg	0.03	-	-	-	-	0.03
Unspec. oil		g	0.8	0	3	-0.00004	0	4

Table 29.3. Exchanges with the environment in the pump's product system

		Production of materials	Product manufacturing	Use	Dis-posal	Transport	Total
Waste							
Slag containing chromium	g	650	-	-	-	-	650
Slag containing manganese	g	20	-	-	-	-	20
Unspec. chemical waste	g	100	-	-	-	-	100
Soil or dust containing heavy metals	g	90	-	-	-	-	90
Unspec. slag and ashes	g	970	190	56,400	50	0	57,600
Unspec. bulk waste	g	5,950	1,340	171,700	5,560	2	184,600
Unspec. hazardous waste	g	0.07	-	-	500	-	500
Working environment							
Monotonous repetitive work	hours	0.0015	1.2	0.1	-	0.004	1.3
Hearing impairment inducing noise	hours	0.02	1.2	1.3	-	0.006	2.5
Substances causing damage to the nervous system	hours	0.001	0.0006	0.04	-	0	0.04
Allergenic substances	hours	0.003	0.02	0.22	-	0.0004	0.24
Substances causing damage to the reproductive system	hours	0.0003	0	0	-	0	0.0003
Carcinogenic substances	hours	0.002	0.002	0.16	-	0	0.16
Accidents	number	$1.4 \cdot 10^{-6}$	$37 \cdot 10^{-6}$	$54 \cdot 10^{-6}$	-	$0.16 \cdot 10^{-6}$	$92 \cdot 10^{-6}$

Table 29.3. (continued)

Working environment

For the working environment, the greatest contributions come from the manufacturing and use stages. The manufacturing stage contributes especially with monotonous repetitive work. The most significant contribution from the use stage is noise from the production of electricity.

Energy profiles

The energy profiles for the product system in Figure 29.7 show energy consumption expressed in megajoules (MJ) in the various stages. The energy is computed as primary energy, which includes the power plants' energy efficiency and the calorific values of the materials.

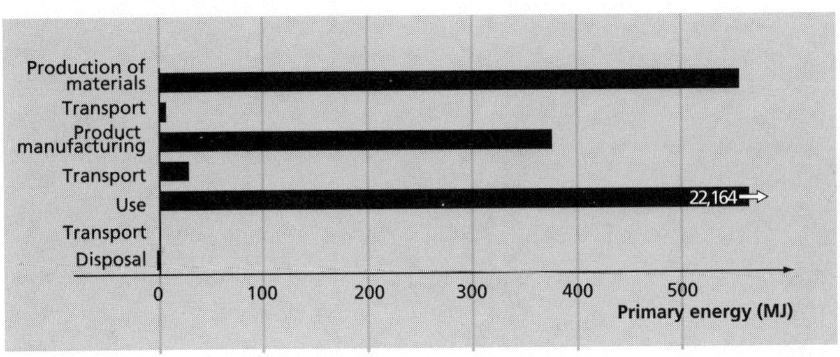

Figure 29.7. Energy profile. Primary energy consumption

Production of materials

The energy consumption for the production of materials is the second largest, but in any case it is ca. 50 times less than energy consumption in the use stage.

Manufacture

Most of the manufacturing is own manufacturing, i.e. at Grundfos, which is of the same order of magnitude as for the production of materials.

Use

By far the largest energy consumption is in the use stage. The use stage is therefore the most important area for initiatives if energy consumption is to be reduced.

Transport

The contribution from transport is not large, but it should be noted that transport from the company to the user is significantly greater than the other transport contributions. This is attributed to the fact that the pump must be transported from Denmark to southern Europe.

Disposal

The negative contribution is caused by the fact that energy is credited on material recycling.

The energy content of the materials

Figure 29.8a and 29.8b show the energy content in the raw materials and in the most energy-intensive components. Stainless steel accounts for the largest fraction, followed by the various types of plastic. That fraction of the energy content in the product which does not figure under a particular material is given in the last column under "others".

The component numbers in Figure 29.8b are those in the parts list in Table 29.1.

For stainless steel it is especially the motor (item no. 150), the motor housing (item no. 2) and the tank (item no. 16) which contribute.

The energy content of the plastic components is dominated by the separator plate (item no. 41) and the motor shield (item no. 180), which are made of a fibre glass-reinforced composite, and by the rubber membrane (item no. 17).

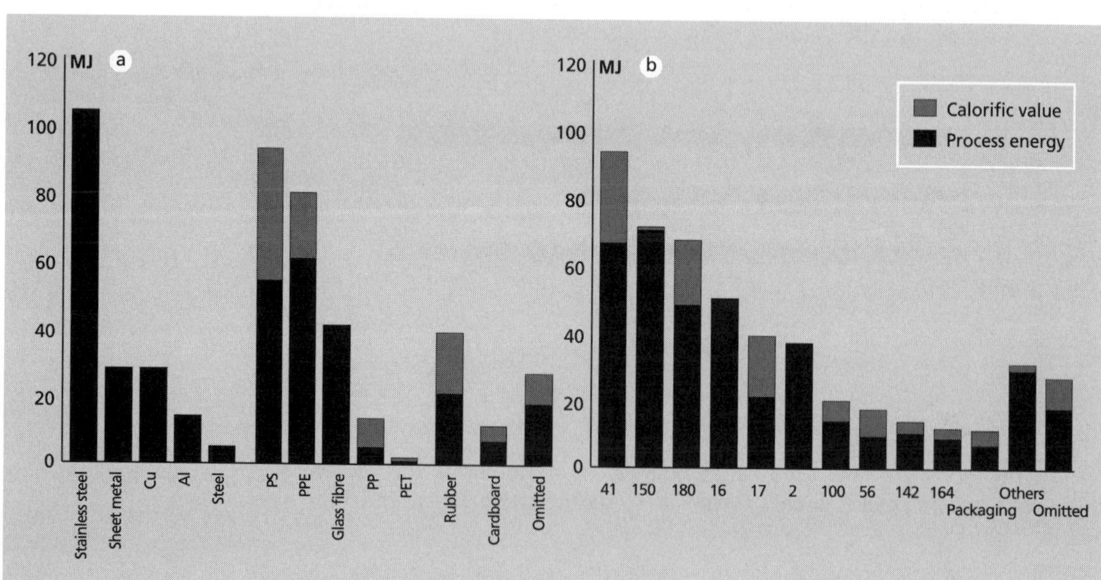

Figure 29.8a. The primary energy content in the materials found in the pump

Figure 29.8b. The primary energy content in the materials in the pump, by component

Impact assessment

Table 29.3 shows the exchanges with the environment expressed in grams for resources, emissions and waste, and in hours or numbers for impacts on the working environment. The exchanges are converted to potential impacts, which express the effect which can be triggered under the right circumstances.

Environmental impact potentials

Emissions and types of waste can be summarized into a number of environmental impact potentials as described in Parts II and IV. The result for the environmental impact potentials is given in Table 29.4.

The JetpaQ has a large potential for the environmental impacts associated with consumption of fossil fuels, i.e. global warming, acidification, and slag and ashes. On the other hand, it is evident that the JetpaQ has relatively small potentials for toxic impacts, with the exception of human toxicity deriving mainly from SO_2 and NO_x in emissions from power plants. In the category of hazardous waste, the electronic waste is also accompanied by various slag residues containing heavy metals from the production of materials.

Impact category	Potential impact	
	Unit	Per product per lifetime
Global impacts		
Global warming	g CO_2-eq	1,545,000
Regional impacts		
Acidification	g SO_2-eq	17,600
Nutrient enrichment	g NO_3^--eq	8,520
Photochemical ozone formation	g C_2H_4-eq	73
Human toxicity (water)	m^3 water	24
Ecotoxocity (water, acute)	m^3 water	116
Local impacts		
Human toxicity (air)	m^3 air	936,000,000
Ecotoxicity (water, acute)	m^3 water	6.4
Hazardous waste	g	1,270
Slag and ashes	g	57,500
Bulk waste	g	184,600

Table 29.4. Conversion of the environmental exchanges to environmental impact potentials

Resource consumptions and potentials for impacts on the working environment

The figures in Table 29.3 express the potentials directly in grams for resources and hours for the potentials for impacts on the working environment, and they should therefore not be converted.

The magnitude of the product's contribution

A normalized value which expresses the impact potential relative to the average contribution for a person can make the figures easier to compare. The size of the pump's contribution is thus expressed in person-equiva-

Figure 29.9. Normalization of resource consumptions and impact potentials from the pump

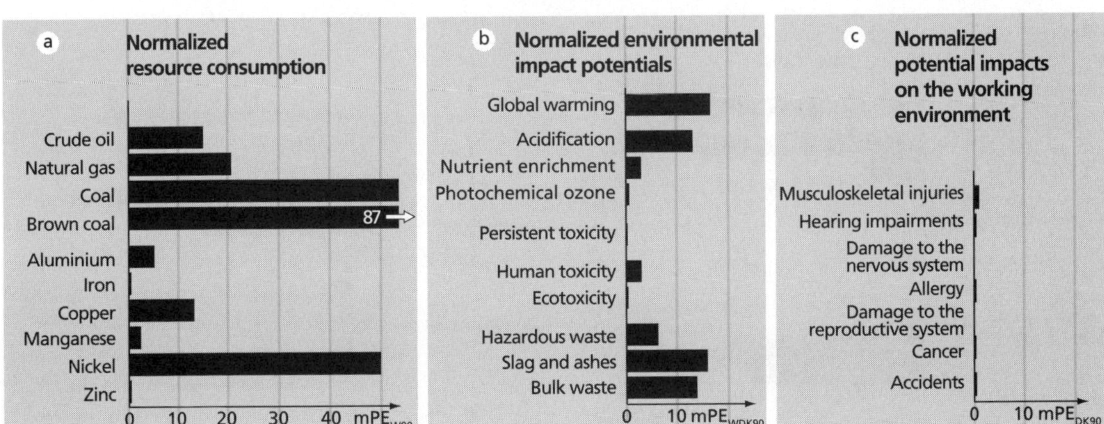

lents, or more correctly milli-person-equivalents, abbreviated mPE, which are the product's contribution in parts per thousand of an average person's contribution.

For resources shown in Figure 29.9a, fossil fuels and nickel make the greatest contribution. Nickel is used as an alloying element in stainless steel, and the consumption of nickel corresponds to 5% of an average person's annual consumption in 1990. The consumption of fossil fuels is of the same order of magnitude.

The environmental impact potentials are shown in Figure 29.9b. The JetpaQ's contribution to global warming, acidification and various categories of waste is of the order of 1-2% of an average person's contribution in 1990.

The impact on the working environment is normalized for a Danish employee's annual impact (cf. Figure 29.9c). The largest relative contributions to the impacts on the working environment are allergies, injuries to the musculoskeletal system, hearing impairment and work accidents. The large contribution to injuries to the musculoskeletal system comes from manufacture at Grundfos. The potential for allergic reactions derives mainly from electricity production. The electricity production contributes together with the manufacturing at Grundfos to potential hearing impairment. The definitions for impacts on the working environment are given in more detail in Part IV.

The most significant impact potentials

Although the normalized profiles in Figure 29.9 are expressed in the same unit (mPE), this does not necessarily mean that they can be compared directly. Equally large columns in the figure, for example, are not necessarily equally significant. Further information on the significance of the individual impact is therefore added in the form of a weighting factor.

Figure 29.10 shows the profiles, where the weighting factors are included. It is evident that the three profiles have different units, and they cannot, therefore, be compared directly. The Danfoss example explains all units in detail, and they are also included in the glossary.

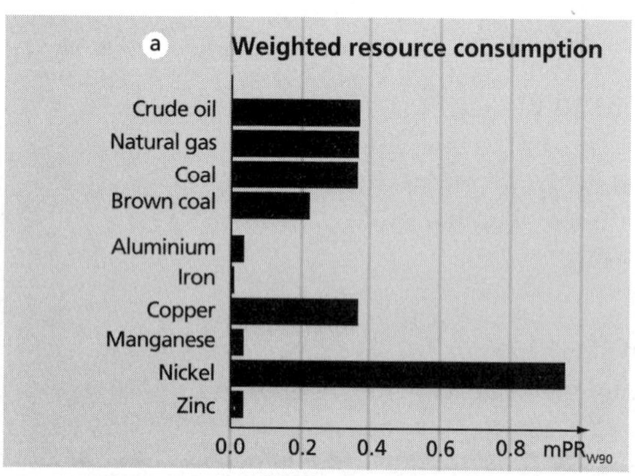

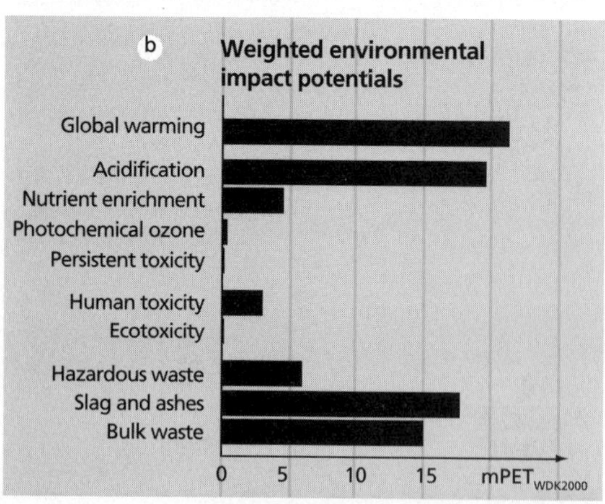

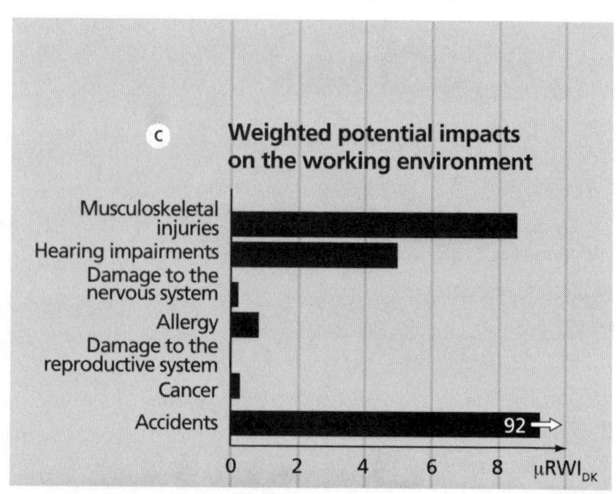

Figure 29.10. Weighted resource consumptions and impact potentials for the JetpaQ

The resource profile is obtained by weighting the normalized values relative to the size of known reserves. As the size of reserves is dependent on time, 1990 was chosen as reference year. The weighting of the resources expresses the proportion of known reserves remaining in 1990 for the individual person and his or her successors, i.e. the size of the reserves is set relative to the size of the population in 1990.

Nickel makes the greatest contribution in the resource profile, with a value of one milli-person-reserve (mPR). This corresponds to the pump using one thousandth of the nickel available to an individual and all of his or her successors. The reason for the large weighted contribution for nickel is nickel's short supply horizon. The consumptions of fossil fuels and copper are also significant. Values of 0.4 mPR are obtained for copper, crude oil, natural gas and coal.

The weighted environmental profile is obtained by multiplying the normalized values by a weighting factor based on actual Danish or international political reduction targets for the year 2000. The weighted environmental impact potentials are seen in Figure 29.10b. The picture is much the same with weighting and normalization. This is attributed to the fact that the weighting factors for global warming and the categories of waste which make the biggest contribution are close to 1. A weighting factor of 1 means that the contribution in 1990 corresponds to the reduction target set for the year 2000. Global warming and acidification are ca. 20 mPET, i.e. the pump's contribution corresponds to 2% of the maximum contribution which an average person should make in the year 2000. Slag and ashes and bulk waste are of the same order or magnitude.

The working environment profile is obtained by multiplying the normalized values by the annual number of work injuries which have been reported. This results in a direct expression of the number of work injuries which can be expected for the actual product.

For the working environment, a very large contribution is obtained for work accidents, of which ca. 60% derive from the production of electricity and 40% from manufacturing at Grundfos. When the figure for work accidents is so high, it is attributed to the fact that work accidents must be reported by law. It is therefore easier to class work accidents as injuries at work relative to other impacts on the working environment. The values for damage to the musculoskeletal system and hearing impairment are also significant: 92% of the potential injuries to the musculoskeletal system and 46% of potential hearing impairment are ascribed to manufacturing at Grundfos.

The most significant sources of the impact potentials

If the weighted profiles are to be used to improve the product, it is necessary for it to be possible to trace the sources of the individual environmental impacts. The environmental impacts in the following figures are divided into the six sources: consumption of materials, consumption of chemicals, consumption of thermal energy, consumption of electricity, overhead, and disposal. These sources can be grouped within Materials, Energy, Chemicals and Others as shown in Table 29.5. The grouping follows the MECO principle and it is described in more detail in Part II.

Materials	
Material	- consumption - disposal
Processes	
Energy	- thermal - electric
Chemicals	- ancillary substances
Others	- overhead

Table 29.5. Sources of resource consumptions and impact potentials classified after the MECO principle

Figure 29.11a shows that the consumption of fossil fuels is almost exclusively for electricity consumption, while the other resources enter into the product as materials.

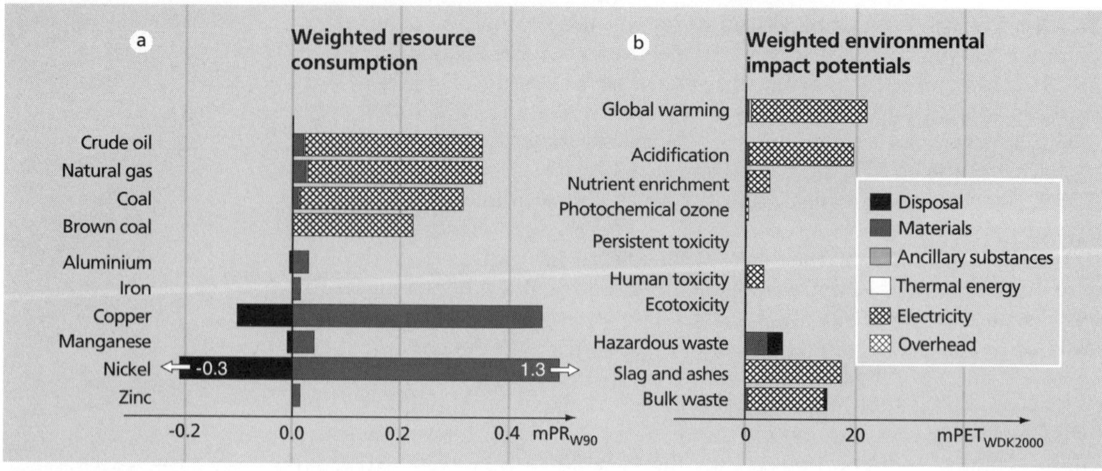

Figure 29.11. Weighted resource consumptions and environmental impact potentials classified by type of source

The environmental impact potentials in Figure 29.11b derive almost exclusively from the consumption of electricity. Hazardous waste and persistent toxicity come mainly from the consumption of materials.

Given these facts, it is possible to simplify the LCA for the significant sources. For the electricity consumption, consumptions of coal and crude oil, global warming, acidification and slag and ashes were selected as representing all impacts. For resource consumption in general, copper and nickel were selected as they give the most heavily weighted contributions. Hazardous waste is included as a category of waste which does not derive from electricity production.

Figure 29.12a. Weighted resource consumptions classified by stages in the product life

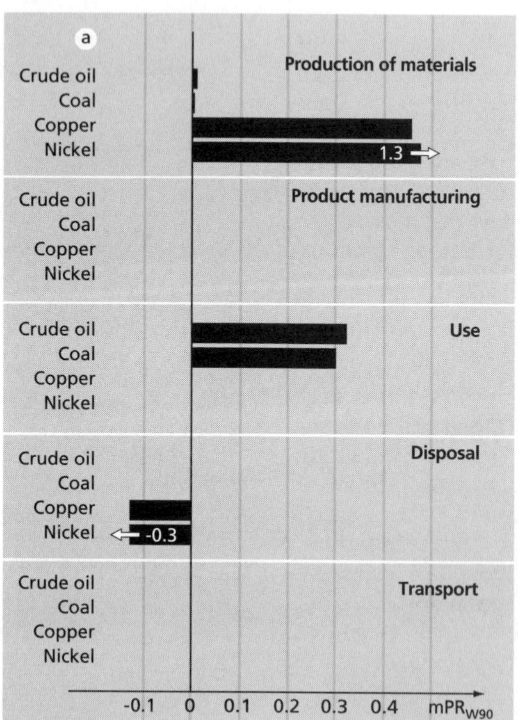

The most significant stages of the product life

The simplification with respect to impact potentials is made for the sake of clarity when the profiles are shown divided into stages. The resource consumption is shown in Figure 29.12a. The use stage is the most significant for crude oil and coal, while copper and nickel give the greatest contributions in the material production stage. Some of the crude oil and coal is, however, also used for the production of materials. The value for nickel means that ca. 1.3‰ of the person reserve in 1990 is used. As nickel is recycled after final use of the pump, part of it is credited during disposal, but a much higher degree of recycling and sorting of materials is required if it is to be possible to make pumps with stainless steel containing nickel in the future.

Figure 29.12b shows the environmental impact potentials. The significance of the use stage, which is dominated by the consumption of electricity, is stressed here. Almost all of the contribution to global warming, acidification and slag and ashes comes from the use stage. Hazardous waste comes from the production of materials and from disposal.

The impact potentials for the working environment are shown in Figure 29.12c. The distribution among the individual stages shows that the major contributions come from manufacturing and from the use stage, where it is the production of electricity which gives the potentials for impacts on the working environment. Production of electricity includes work at power plants, which are especially responsible for hearing impairment, allergenic substances and work accidents.
The contribution from the production of materials is ascribed to the production of steel, the main contributions of which are accidents. The contributions from manufacturing derive from manufacture at Grundfos and from production of electricity. Most of the contributions for the impact categories shown derive from manufacturing at Grundfos. Damage to the musculoskeletal system is ascribed to monotonous repetitive work.

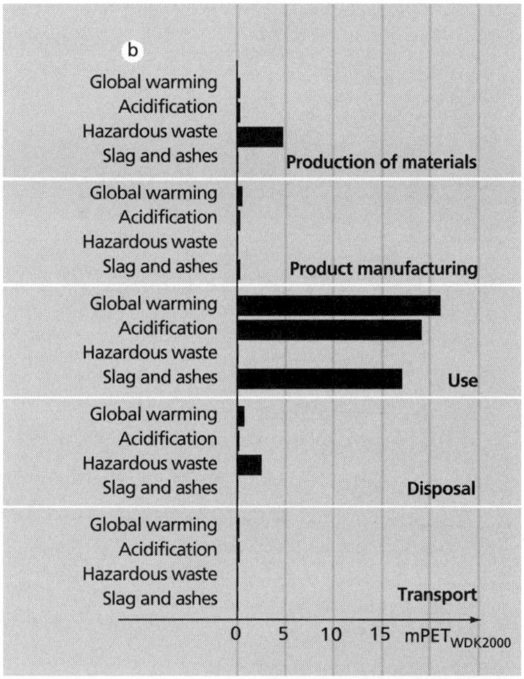

Figure 29.12b. Weighted potentials for environmental impacts, classified by stages of the product life

The most significant components

An LCA was made for each of the components, including production of materials and manufacture. The impact potentials from the LCA can be used to designate the components with the most significant environmental impacts. A top scorer list was prepared for each category of impact so that the components with the biggest contributions can be seen. A top scorer may well, however, be the best alternative for a particular function, and thus not necessarily represent a focal point, but the top scorer list is a good basis for discussion of alternatives among the components. Figure 29.13 shows the environmental impact potentials and resource consumptions calculated for the components.

For crude oil consumption it is the tank (no. 16) which gives the biggest contribution, followed by the separator plate (no. 41), the motor (no. 150) and the rubber membrane (no. 17). The separator plate is made of plastic composite, and crude oil may be used as a raw material as well as fuel in its manufacture.

The rubber membrane (no. 17), the motor (no. 150), the tank (no. 16) and the stator frame (no. 2) are the top scorers for coal.

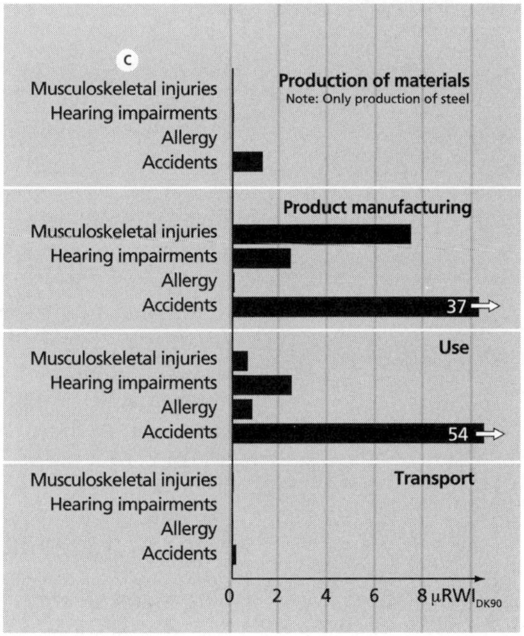

Figure 29.12c. Weighted potentials for impacts on the working environment, classified by stages of the product life

Copper is used especially in the motor (no. 150), where is it used in the windings.

Nickel is used as an alloying element in stainless steel. The top scorers are the tank (no. 16), the clamping collar (no. 149) and the motor

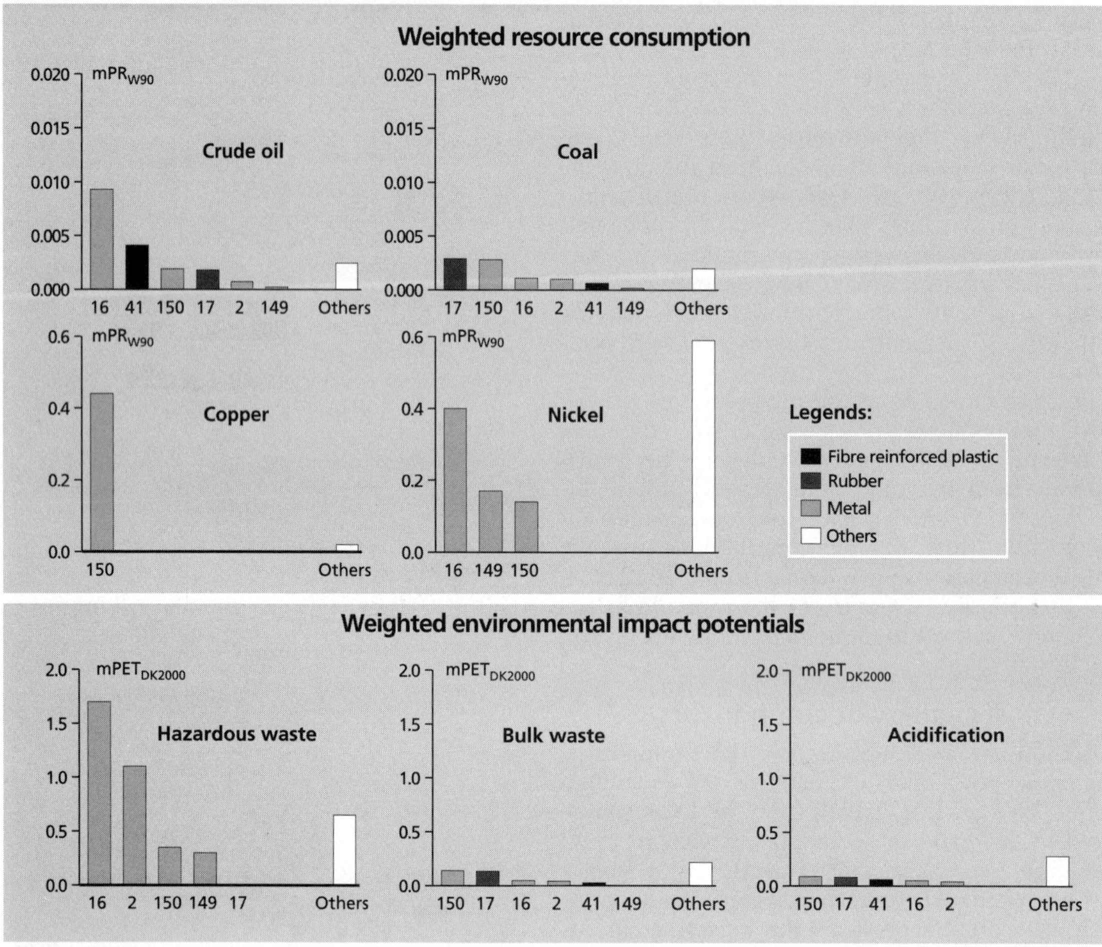

Figure 29.13. Top scorers among the pump's components

(no. 150). The rest, which are unspecified, contain the other stainless steel components.

The top scorers for the environmental impact potentials are the motor (no. 150) and the rubber membrane (no. 17). Hazardous waste is dominated by the tank (no. 16) and the stator frame (no. 2). Both parts are made of stainless steel. The hazardous waste derives from chromium-containing slag in the production of stainless steel.

It should be noted here that the most important source of environmental impacts, the consumption of electricity, cannot immediately be ascribed to a particular component, but is associated with the function of the whole product and its efficiency.

Lack of information, uncertainties and sensitivity analyses

The material stage

For the product itself, the most important omission is the electronics. Electronics typically contain scarce metals, and the weighted resource consumptions of these can therefore be significant, even if the quantities are not large. The toxicity of heavy metals in the electronic components is not included either. Data for production of the fibre glass-reinforced plastic ingredients of the composite are not available, but all ingredients of the composite are included, i.e. the resource consumption is correct.

On the resource side, chromium is lacking in the figures, although chromium is used in significant quantities in stainless steel: the chromium content of the alloy used is 18%. Chromium can therefore be found in Table 29.3 as resource consumption. As a normalization factor for chromium was not calculated, chromium does not appear in the normalized and weighted figures. According to Goeller & Zucker (1984), chromium does not, however, belong among the scarce resources, i.e. the supply horizon at present rates of consumption is long.

A general problem, which is attributed to lack of information, is the quantities of waste from the extraction of metals. The ore from which the metals are extracted typically contains a few percent of the desired metal. Large heaps of earth or rocks can therefore be seen around the mines. These quantities of waste are not included for the metals. The slags which comprise the waste from the further refining, such as chromium-containing slag from the production of stainless steel, are included.

More than 90% of the resources are included for the material stage.

The manufacturing stage

The data used for some manufacturing processes at Grundfos were measured specifically in the manufacture of another Grundfos pump. As all manufacturing processes take place at Grundfos, the uncertainty is assessed as being small. The uncertainty in electricity measurements is estimated to be 10%. The uncertainty is larger for measurements of emissions because the readings are instantaneous measurements and the emissions vary as a function of time.

The use stage

The life of the pump was estimated, and the figure can vary within ± 5 years. A simulation is made in the environmental diagnosis in Section 29.2 to show the significance of the life span.

The electricity consumption in the use stage depends especially on the time in operation and the outlet pressure of the water. The scenario which was set up for use was based on standard information for a family with two children. The more users of the pump, the longer is the time in operation. No estimates were made of the variation in time in operation. The outlet pressure is the second significant parameter, as each 0.5 bar additional pressure increases electricity consumption by 35 kWh per annum. Depending on the outlet pressure, power consumption can vary between 120 and 410 kWh per annum when the pump's time in operation is constant. Power consumption can thus actually vary by ca. ± 100 kWh per annum. A simulation in the LCA shows the consequences of changes in power consumption if the efficiency of the pump is changed. The simulation can be taken as a general expression of changes in power consumption.

The disposal stage

The distribution between recycling, incineration and dumping in landfills was made on the basis of estimates based on Danish conditions, and large deviations can occur when the pump is used in other countries. A simulation with 100% recycling of stainless steel and copper is shown in the environmental diagnosis.

29.2. Environmental diagnosis for a pump - the JetpaQ

The object of the environmental diagnosis is to designate the most significant potentials for improvement and to discover where they reside in the product.

Simulation of changes in the product or its product system

By simulating changes in the product, it is possible to see how great an effect the change will have. The effects are shown only for the potentials which are affected by the changes. The framework for the simulations consists of the four areas in the MECO principle discussed under the most significant sources of impact potentials. The simulations were made within the three areas Materials, Energy and Others. No simulation was made for Chemicals. The simulations are shown in Figure 29.14.

Greater recycling of metal

In the present scenario, only 30% of all metals are recycled. There is therefore a large potential for a saving of resources with increased recycling. If 100% of the copper and 100% of the stainless steel are recycled, the resource consumption will be significantly smaller. The example is shown in Figure 29.14a. Consumption of nickel is reduced by 60%, and for copper it is reduced by ca. 85%. When the consumption of copper is reduced by more than that of nickel, it is because the recycling system for copper is more efficient in terms of primary material savings than for stainless steel. For the scrap dealer, copper is easy to recognize because of its reddish colour, and it can be refined to 100% purity. The copper made from scrap thus has the same purity and qualities as primary copper. In the production of stainless steel from scrap, a quantity of primary metals is normally added to adjust the composition. Twenty percent of primary materials are typically added to stainless steel.

Reduction in electricity consumption

As the electricity consumption in the use stage makes the greatest contribution to the environmental impact potentials, a reduction in this consumption would be significant. The electricity consumption can be reduced by increasing the efficiency of the pump. The total efficiency is a combination of the efficiencies of the motor, the pump and the pump control. The second simulation is a reduction in idling from 5 to 2 W, i.e. a reduction of 60%. But idling comprises a minor part of the total power consumption, so the percentage reduction in total electricity consumption will only be 10%. The reduction in idling gives a reduction in fossil fuels of 3% (cf. Figure 29.14b), and the environmental impact potentials shown in Figure 29.14c are reduced by the order of 7%. An increase of 25% in the efficiency of the pump results in a reduction in the consumption of fossil fuels by ca. 6% and a reduction in environmental impact potentials of ca. 15%. All of the environmental impacts shown depend on the electricity consumption. The environmental impacts are therefore less when the electricity consumption is reduced.

Life span of the pump

In the simulation of changed life span shown in Figures 29.14d and 29.14e, it is assumed that the electricity consumption per annum is the same. As the normalized values take account of the life span, changes in the life span will be of significance only for impact potentials deriving from the production of materials or from manufacture. The normalized values are the same for the use stage. A replacement of the rubber membrane is simulated in the increased life span, i.e. the environmental impact of an extra rubber membrane is included.

For the resources shown in Figure 29.14d, the effect of the increased life span is large. All of the resources shown enter into the product, and an increased life span from 10 to 15 years therefore results in a reduction of 33% in resource consumption. Similarly, an increased resource con-

Figure 29.14. Simulation of alterations in the product or its product system

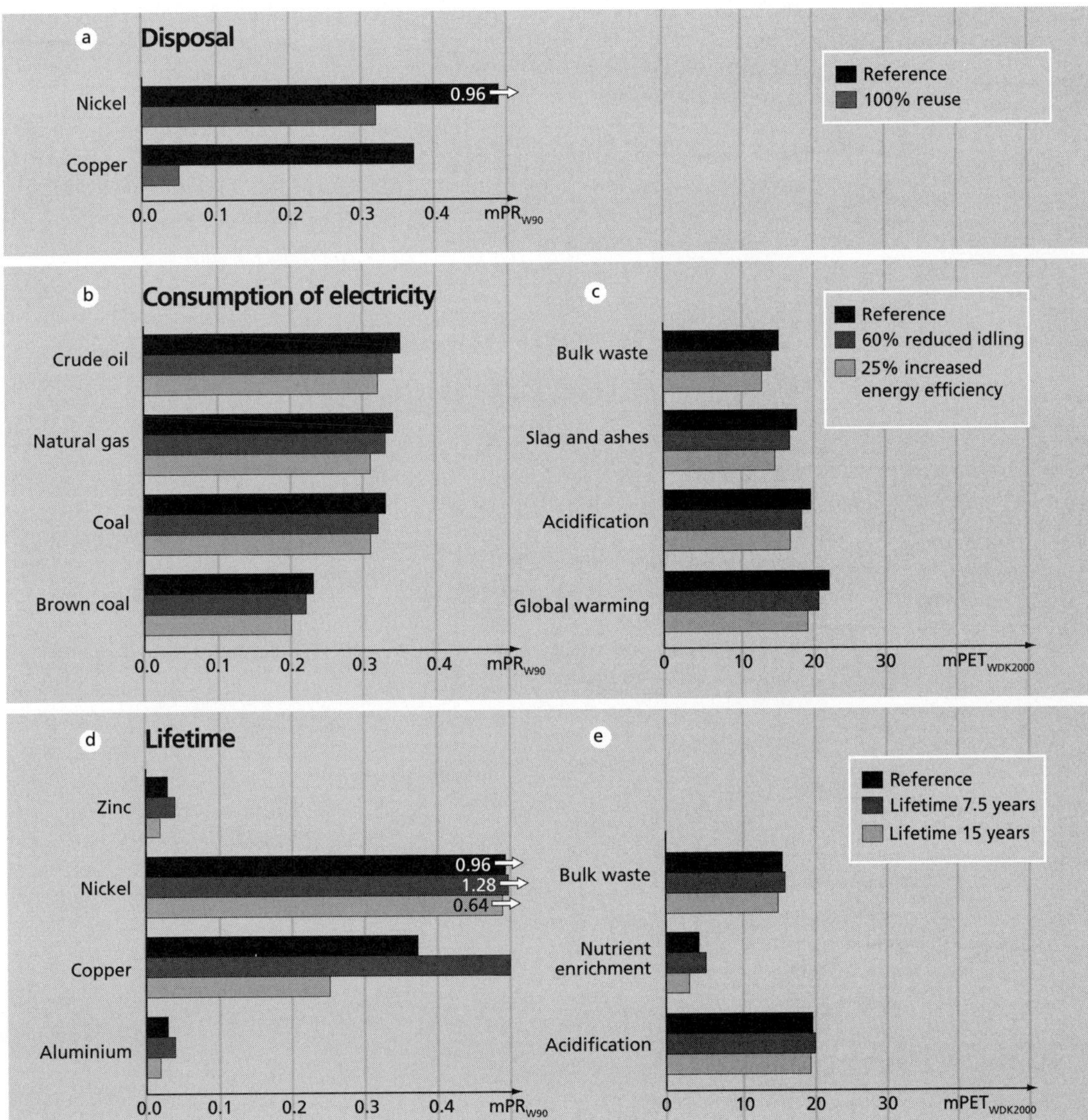

sumption per annum results if the life span is shorter. The consumption of crude oil for the rubber membrane is so small relative to the other consumptions that it is not shown here.

The significance of the life span is small for the environmental impact potentials (cf. Figure 29.14e). The potentials shown derive principally from the production of electricity, and only the consumption of electricity for production of materials or manufacturing is affected. The longer the life, the smaller the environmental impact potentials. The environmental impact potential for the extra rubber membrane is too small to have a visible effect.

Simulations - summary

The energy consumption has the greatest significance for the environmental impact potentials. A reduction in energy consumption is therefore the improvement with the highest priority. Increased recycling of metals follows. As the degree of recycling is not particularly large, there is a large potential here. The life span is also of significance for resource consumption, but the extent is uncertain as an increased life span will require more service. In the life span simulation shown, only a rubber membrane was assumed to be replaced. If an increased life span can be achieved with few replacements, there will be significant savings in resources. The more the servicing and transporting of the product, the smaller the advantage of an increased life span.

Where do potentials for improvement reside in the product?

This section makes use of the information collected in the LCA concerning the most significant environmental impact potentials for the product and the components which are the top scorers within the impact potentials. When compared with the simulations this information provides a good basis for designating focal points in the product. Grundfos have attempted to set up alternatives. Some alternatives are realistic, while others must be considered purely inspirational.

Potentials for improvements have been designated at the levels of concept, structure and component.

Table 29.6. Example of point of focus at the concept level

Idling consumption: the pump uses 5 W when idling. The energy is used by the control unit, which continuously measures the outlet pressure.
Alternative: reduction in idling consumption.

Table 29.7. Examples of points of focus at the level of structure

Efficiency: The electricity consumption in the use stage makes the most significant contribution to the environmental impact potentials. An increase in efficiency will therefore be of major significance.
Alternative: the efficiency can be increased in the motor or the pump.

Increased recycling: it is important to increase recycling so that resource consumption is reduced.
Alternative: easier disassembly can increase the percentage of recycling. This can be accomplished in several areas:

- Separation of components without use of special tools.
- Labelling of materials with details of type of material.
- Use of uniform materials and fewer types of materials.

Table 29.8. Examples of points of focus at the component level

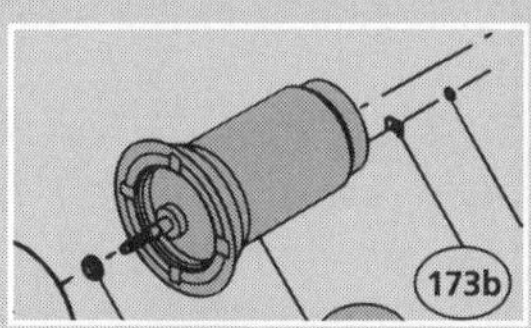

Copper: copper is a component of the motor. As copper is a metal with a short supply horizon, use of the metal should be limited and recycling should be possible.
Alternatives: copper consumption can be reduced by initiatives on three fronts:

- Increasing the degree of recycling will reduce consumption.
- Copper consumption can be reduced by a reduction in the number of windings, but this will result in lower efficiency.
- An entirely different motor concept can be used. One possibility is to replace copper with aluminium. The efficiency of these motors is currently less than in motors with copper.

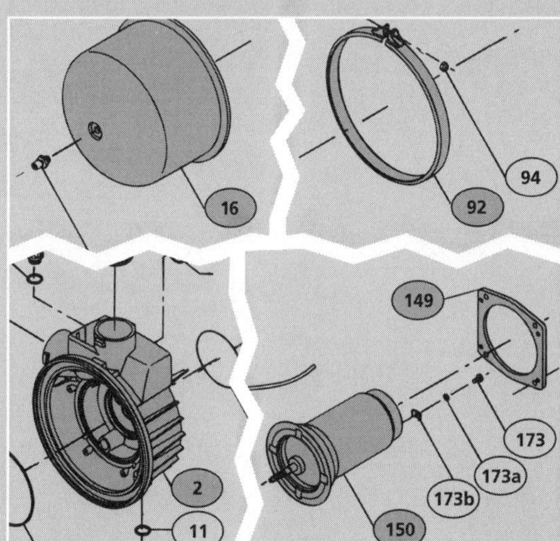

Stainless steel: the tank and several other parts are made of stainless steel. Nickel, which comprises ca. 9% of stainless steel, is a scarce resource. Its use should therefore be limited. As the tank is a major component, the possibility of recycling on manual disassembly is large, although increased recycling cannot entirely eliminate consumption of nickel.
Alternatives: it may be possible to use another material, e.g. plastic composite, or an alloying process with less nickel could be used.

Table 29.9. Example of point of focus at the component level. Working environment

Monotonous repetitive work: occurs especially in manufacturing (assembly).
Alternative: frequent changes between tasks with redeployment of staff will reduce the impact time associated with the same movement.

Where can potentials for improvement be found in the surroundings?

Some of the potentials for improvement are to be found beyond the product. Two examples are given here:

Shredding: when the pump is scrapped, it will typically be sent through a shredder together with other metal waste. If the product contains magnetic stainless steel it will go into the same fraction as iron, and finish as recycled steel at a steel works. Non-magnetic stainless steel will only be collected if there is a manual sorting.

Alternatives: manual disassembly gives the greatest degree of recovery and ensures that stainless steel continues in the stainless steel cycle so that the alloying elements are not lost.

The electrical system: the most significant environmental impact potentials derive from the production of electricity. This is ascribed especially to the large consumption of coal as fuel.

Alternatives: an increased efficiency of the power plants together with greater use of long-term energy sources will reduce the environmental impact.

29.3. Environmental specification for the product type

This section contains a comparison with competing products. The product specifications for new products are derived from general environmental objectives.

The competitive framework

The market for small decentralized water works is large, particularly in southern Europe, where the variation in pump types is large and the market is characterized by small local suppliers.

No special analysis of the market's expectations concerning the product's environmental qualities was made. This is mainly because the product is sold in southern Europe, where consumer interest in a product's environmental qualities is limited. The product is also a low price product, where price rather than environment or other qualities is the most important competitive parameter.

Marketing of the JetpaQ has been based on the product's reliability, ease of installation and the electronics, which provide some possibility of control which competing products cannot offer. At the same time, the JetpaQ is quieter than similar pumps.

The JetpaQ is a unique product within water supply systems to private households, primarily because of its compact size. Other products which have the same function as the JetpaQ are typically booster systems, where a pump is installed together with a pressure tank. Compared with a pump of the booster type, a large part of the metal content in the JetpaQ is replaced by plastic composite. Compared with a booster system from Grundfos called the JP5-Booster, the proportion of composite is increased from 8% to 30%. Production of the JP5-Booster commenced long before the JetpaQ, and it is still in production. The increased proportion of composites is not in conflict with the specifications for new products, which state that the number of composite types should be limited. The energy efficiency of the JetpaQ is not improved relative to the JP5-Booster.

Long-term objectives

Grundfos's environmental policy has the following objectives.

Within an economically responsible framework, impact on the environment and consumption of resources should be minimized in the development of new products and processes.

The environmental impact for the whole life of the product must be assessed and described. In practice this will mean that all forthcoming products must be subject to a life cycle assessment, and the environmental impact must be assessed relative to those of corresponding products at Grundfos. This also applies to the introduction of new technologies, which must be assessed relative to existing technology.

The service department at Grundfos must be able to accept and repair defective products or to indicate disposal routes for worn out products. Grundfos must attempt to recycle parts from worn out or defective pumps. This also means that account of the possibility of reuse or recycling must be taken in the design stage.

Grundfos has obtained environmental certification under EMAS, the European Management and Audit Scheme. This certification places requirements on the motivation of employees and suppliers towards increased environmental awareness. Suppliers must also be assessed on their environmental behaviour and rejected if their behaviour is in conflict with policy at Grundfos.

Product specifications

Specifically, the objectives imply the following demands on new products:

- The number of composite types should be limited as far as possible.
- The embedding of metal within plastic should be avoided.
- Optimization of design with respect to disassembly.
- Optimization of motor efficiency.
- Reduction of idling power consumption.
- Reduction in consumption of scarce resources, especially copper and nickel.

29.4. Environmental design of new products

When Grundfos introduces a new pump, the product can be an entirely new product which fulfils customer needs which were not covered previously. It can also be a product which replaces a previous product.

Product development at Grundfos

A typical development procedure for a Grundfos product is shown in Figure 29.15.

The project commences with a concept stage, where aspects such as customers, markets, price, technology and environment are assessed. Within the area of the environment, statutory requirements and customer wishes typically enter into the assessment. The concept stage concludes with proposals for a specific project.

If the project is approved by the product development committee, the project enters a new development stage, where the environmental specialist participates in considerations on construction of the product. Materials, principles of construction, patterns of use, transport and disposal of the product are assessed here as the most important environmental aspects.

When the product's function, composition and characteristics have been determined, a life cycle assessment of the product is made. If a product with the same function already exists at Grundfos, the LCA of the new product will be compared with the LCA of the old product.

The project then enters a stage where it is made ready for manufacture of the product.

Before the product enters final production, a series is commenced where manufacture of the product is run in and adjusted. If significant changes are made to the product, the environmental specialist will again be involved in the project. The work of the environmental specialist otherwise ends here.

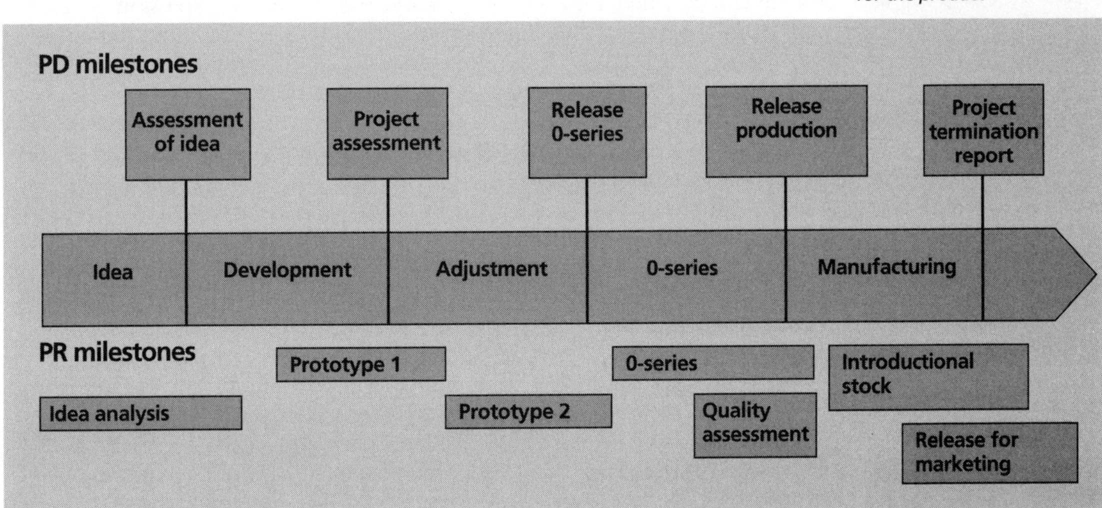

Figure 29.15. Product development at Grundfos. PD = product development committee and PR = persons responsible for the product

Examples of environmental design in new products

Product development at Grundfos has been able to draw on the experiences gained in connection with the LCA of the reference product. This has led to some changes at the concept, structure and component levels in a number of different product development projects. Within each project, the savings can be quantified and ascribed to one of the four groups in the MECO principle: Materials, Energy, Chemicals or Others, the latter being, for example, the working environment. Beside each project there is a table showing the level at which the project is carried out and where the savings lie.

	M	E	C	O
Concept				
Structure		●		
Component				

Project 1. Electronic control

Control of pump operation by using electronic metering and control systems can result in significant energy savings in the use stage. The electronic control was introduced into the JetpaQ.

Energy: Up to 75% of the energy consumption in the use stage of certain types of pumps is saved.

	M	E	C	O
Koncept				
Structure				●
Component				

Project 2. Installation of the electronic unit

To simplify removal of the electronics, the unit is clipped on so that it can be easily removed with the aid of a screwdriver.

Others: By making removal easier, it is easier to dispose of the electronics in an environmentally responsible way, and pollution of other material fractions is avoided.

	M	E	C	O
Concept				
Structure				
Component	●			●

Project 3. Self-tapping screws

Brass nuts are traditionally mounted into the plastic parts because of the requirement that the parts must be capable of being separated easily. However, this means that the materials are mixed together, which is inconvenient when they are to be recycled. A possible solution is use of self-tapping screws. This has been tested and successfully implemented.

Materials: There is a saving in brass for the nuts.

Others: Production of the units is simplified and recycling of the plastic is easier.

	M	E	C	O
Concept				
Structure				●
Component				

Project 4. Clip-in chassis

The clip-in chassis contains capacitors, which can emit toxic vapours on incineration. Clip-in chassis are now installed such that they can easily be disassembled after use.

Others: By removing the clip-in chassis before incineration, the emission of toxic vapours from the capacitors is avoided.

	M	E	C	O
Concept				
Structure				
Component	●			

Project 5. Packaging

Grundfos have prepared a positive list of the materials which may be used as packaging. The positive list prescribes use of corrugated cardboard and wood. Parts which are sensitive to moisture may, however, be packed in plastic. The cardboard may not be bleached or coloured, but there may be printing on up to 10% of the surface. Attempts are also being made to limit the quantity of packaging as much as possible.

Materials: Oil is saved by reducing the quantity of plastic.

Project 6. Labelling of composite items

All composite items over 10 g are labelled according to an internal Grundfos standard. Work on a standard for rubber is currently underway.

	M	E	C	O
Concept				
Structure				
Component	●			

Materials: Recycling of the materials is made easier by labelling them.

Project 7. Copper wire in the motor

The copper in the motor has a major impact on resources. It is therefore important that the copper can be recycled after final use. Almost all of Grundfos's motors are constructed such that the copper wire can be removed after scrapping. When the copper wire is removed before shredding, the copper is very pure. If it is sent to a copper works, it can be remelted directly without wastage of resources.

	M	E	C	O
Concept				
Structure				
Component	●			

Materials: If the copper wire is removed before shredding, 100% of the copper in the motor can be recycled.

Project 8. Monotonous repetitive work

To reduce the quantity of monotonous repetitive work, Grundfos is in the process of introducing production groups which can introduce internal job rotation.

	M	E	C	O
Concept				
Structure				●
Component				

Others: The quantity of monotonous repetitive work is reduced, but it is too early to place a figure on the change.

Project 9. Exchange pumps

Several parts in the pump do not wear during use. These parts can therefore be taken out and placed in another pump with advantage. The service department at Grundfos is offering exchange pumps to customers who return a damaged pump. The customer obtains money for the old pump, depending on its age and condition. The exchange pump contains both new and old components, but it carries the same guarantee as a new pump.

	M	E	C	O
Concept	●	●		
Structure				
Component				

Materials: Resources are saved when the components are recycled.

Energy: Energy for production of materials and of the component is saved by direct recycling of components.

Grundfos's experience with environmental design in product development

Participation in the EDIP project has provided Grundfos with the possibility of introducing environmental considerations into product development. Environmental considerations today are not merely statutory requirements. Customers are increasingly demanding environmental qualities in products and evidence of environmental initiatives at the company. Participation in the EDIP has made it easier for Grundfos to document the improvements made in the area of the environment.

One often finds designers who want to include environmental considerations in product development, but who cannot comprehend the task. The EDIP method has proved to be a tool which makes it possible to convert thoughts into actions.

A concrete example is the introduction of electronics into the product. The LCA indicated that electricity consumption in the use stage made the greatest contribution to the environmental impact potentials. The electronics regulate the pump's performance, so that the energy used in the use stage was reduced significantly. The LCA also showed that it is important to attend to recovery after use, so that resources are saved.

The EDIP method has helped to make the environment a parameter in product development at Grundfos, and the environment will in future be an integrated part of product development.

30. Danfoss: Electro-hydraulic activation units

Kirsten Stentoft, Danfoss
Hanne Erichsen, Institute for Product Development
Michael Hauschild, Institute for Product Development

Danfoss is a leading producer of precision mechanical and electronic components and of intelligent "mechatronic" devices. The company has over 17,000 employees worldwide. The headquarters is situated in the town of Nordborg in the southern part of Denmark, approx. 20 km north of the German border

30.1. LCA of an electro-hydraulic activation unit - the PVEH

The product selected by Danfoss as reference product is not a product which is visible in everyday life, such as a television set or a refrigerator. But who has not passed a construction site, where cranes are lifting and lowering building elements, or seen a lorry on some winter morning in the process of spraying salt on icy roads? In these situations, the PVEH, an electro-hydraulic activation unit for hydraulic proportional valves, has been in operation. The abbreviation stands for Proportional Valve Electric High performance. A PVEH is used in cranes, excavators, salt sprayers and other machines to control such functions as lifting, turning, gripping, etc.

Choice of product

Danfoss selected the PVEH as reference product because the company was in the process of developing a new generation of electro-hydraulic activation units. The company also wanted a life cycle assessment of a complex product, which in this case is a product comprised of mechanical, electronic and hydraulic parts.

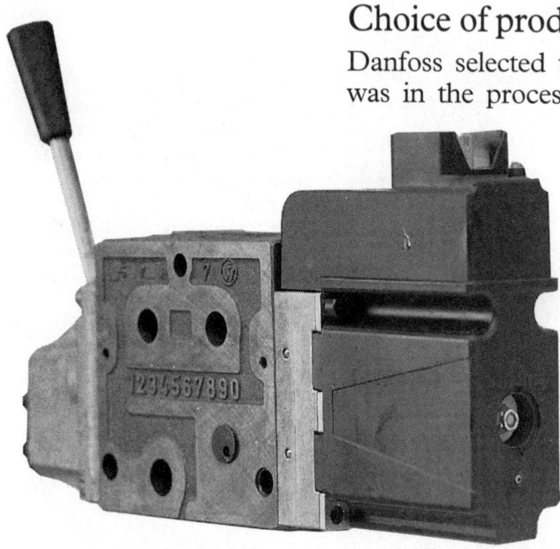

Figure 30.1. The reference product, the PVEH (right) installed on a hydraulic proportional valve (left)

The functional unit

The PVEH's function is to convert an electrical signal into a movement based on hydraulics in a machine. For each of the movements which the machine can make, there is one PVEH and one hydraulic valve. A machine with three functions thus has three PVEHs and three hydraulic valves.

The function of a PVEH is illustrated in the following example: the operator controls the crane with a handle which he places in lifting position. An electrical signal is thereby sent from the handle to the PVEH, where it is converted into a small oil flow in the PVEH's valve block. The small oil flow controls the main sliding valve in the hydraulic valve, which then admits an oil flow from a pump. The oil flow from the pump lifts the crane's load via a hydraulic cylinder.

The life of a PVEH depends on the type of machine in which it is installed and how often this machine is in use, but not on how often the PVEH itself is activated. A PVEH is exposed to temperature fluctuations as both the machine and the PVEH are heated up on starting and cooled down when the machine is stopped. When the excavator is started, the PVEH is heated up at the same time, irrespective of whether or not the excavation function is used. The temperature fluctuations in the PVEH can fracture solderings and components. The life span therefore depends on the number and extent of the temperature fluctuations to which the PVEH is exposed. The more numerous and the greater the temperature fluctuations, the shorter the life span. An average life of 5 years is assumed for the LCA.

The functional unit is therefore described as:

Regulation of one hydraulic proportional valve in a hydraulic system for 5 years.

The PVEH has a number of secondary qualities, including:

In-built error monitoring, fast reaction time and an inductive feedback transducer which monitors the main sliding valve's position in the hydraulic valve. The PVEH also has little hysteresis, that is, it regulates the hydraulic valve with a high level of precision.

Figure 30.2. The five main parts of a PVEH: valve block, top cover with printed circuit boards, magnetic valve set, assembled transducer and bottom cover

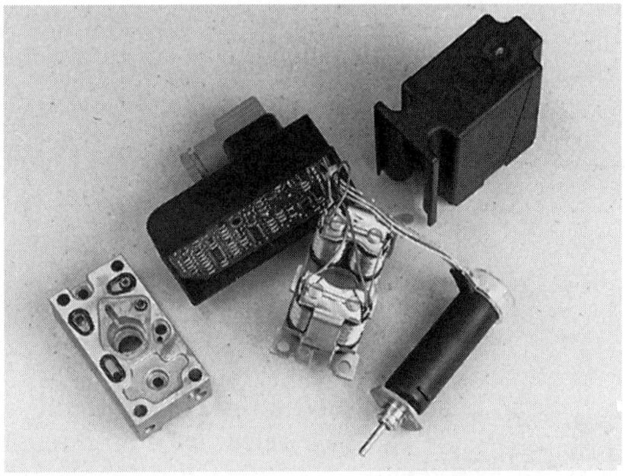

Scope definition of the product system

A PVEH consists of the five main components illustrated in Figure 30.2. Each part consists of numerous components; for example, a top cover with printed circuit boards contains over 100 parts. For the sake of clarity it has been necessary to group the components into assemblies. A component assembly is, for example, the valve block with the components which it contains. The component assemblies with item numbers are shown in the exploded view in Figure 30.3.

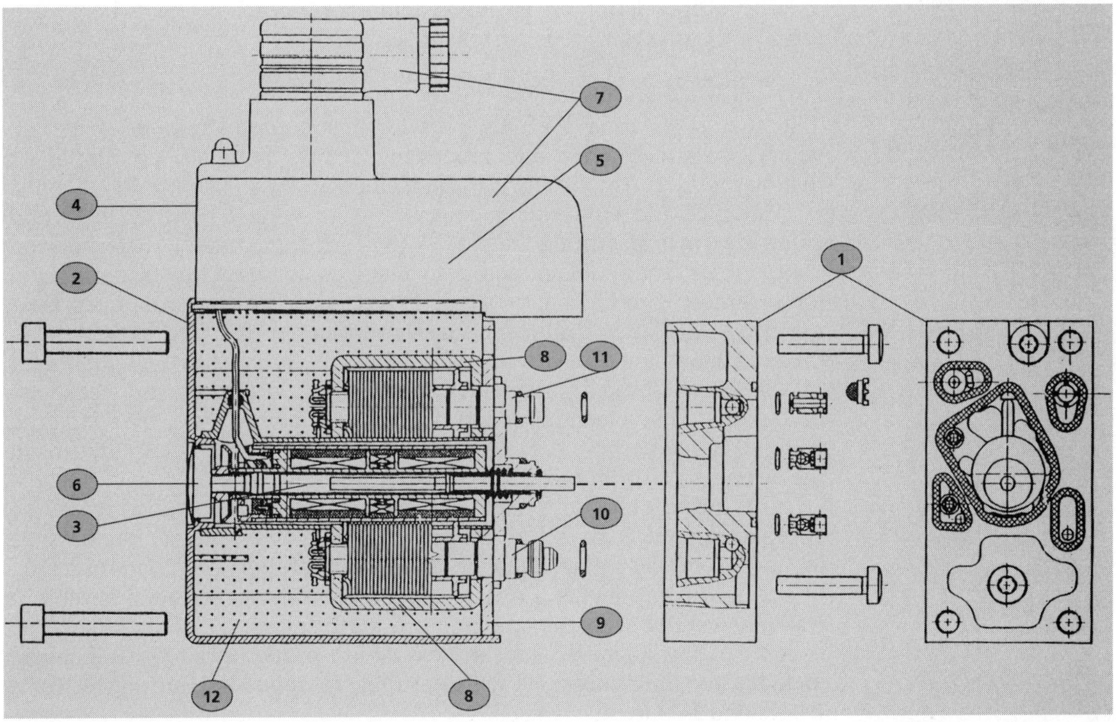

Figure 30.3. Exploded view of a PVEH

The parts list

The parts list provides an overview of what a PVEH consists of, and of how large a part of the product system from extraction of raw materials to and including manufacturing is included in the LCA. For each component, the symbols ✓ and ÷ are used to show whether the product systems of the materials and the ancillary substances are included back to extraction of raw materials. Whether the manufacturing processes are included is also shown.

The components' weights and material contents are described in the parts list (Table 30.1) for each item number. The weight is defined as the weight which the materials and the components have when they enter into manufacturing at Danfoss. The table also shows which manufacturing processes and ancillary substances are used to make a PVEH. The parts list is simplified relative to the product's original parts list in that:

- the printed circuit boards and their components are given with their total content of materials,
- small units of similar material are collected and specified under "miscellaneous".

For impacts on the working environment, all manufacturing processes at Danfoss, manufacturing of printed circuit boards at subcontractors, steel production in the material stage, oil extraction for the use stage and electricity production for the entire product system are included. A total of ca. 80% of the working environment was inventoried.

The scope definition is covered in more detail below, where each stage in the product life is described separately.

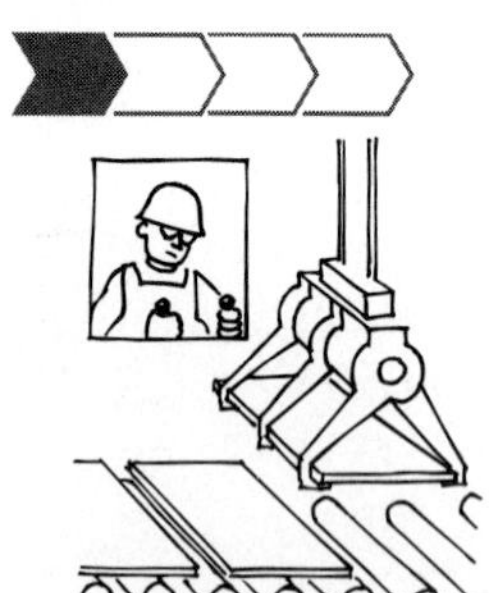

Production of materials

The first stage in a PVEH's product life is production of materials, which covers extraction and processing of raw materials, production of materials and transport between these steps.

Materials and components supplied from several parts of the world enter into manufacturing of the PVEH. All transport of materials and components from subcontractors to Danfoss in Nordborg is included in the inventory. Steel for the yoke and base plate is transported from Germany by lorry, train and ship. The valve block of recycled aluminium is made in Denmark and driven to Nordborg by lorry. The ingredients for polyurethane plastic are made in various places in Europe and transported by lorry. The electronic components are made in Asia, the USA and Europe. Some components are flown to Denmark and transported from the airport to Nordborg by lorry.

The entire product system for the materials is included with the exception of the ingredients for polyurethane, where it was not possible to find all of the information on energy for production. Manufacturing of standard components for the electronics is also not included because it was not possible to obtain the information from the subcontractors. The content of raw materials and the extraction proper of raw materials are included for the majority of the electronic components, but not for transistors and IC circuits.

The product systems of the ancillary substances from extraction of raw materials to processing and manufacturing are not included, as also

Pos.	Number	Item	Material	Total weight(g)	⏚	Manufacturing processes	* Incl. in model	Ancillary substances	Weight (g)	⏚
Valve block										
1	1	Valve block	Aluminium	236	✓	Die-casting[1]	✓			
1	2	Non-return valve	Steel	3.6	✓					
			Polyester	9.32	✓	Injection moulding	✓			
Bottom cover										
2	1	Cover	PPE	22	✓	Injection moulding	✓			
			Polyamide	22	✓					
12	2	Wire	Silver	0.01	÷	Subcontractor	÷			
			Copper	1.5	✓					
			Teflon	0.3	÷					
12	1	Embedment mass	Polyurethane	153	✓	Embedding	✓			
Transducer										
3	1	Transducer spring	Steel	0.9	✓	Winding	✓			
						Cleaning	✓			
						Tempering 460°C	✓			
3	1	Soldered armature pipe	Copper	0.1	✓	Copper soldering	✓			
6	1	Coil	Copper	27	✓	Subcontractor	÷			
			Polyamide	3.6	✓					
			Polyolefin	0.79	✓					
			Zinc	0.06	✓					
6	1	Crown	Polyester	2.6	✓	Injection moulding	✓			
6	1	Casing	Steel	18	✓	Assembly tube/crown	✓			
			Chromium	0.04	✓					
			Zinc	0.43	✓					
6	1	Embedment mass	Polyurethane	12	✓	Embedding	✓			
Top cover										
4	1	Cover	PPE	14	✓	Injection moulding	✓			
			Polyamide	14	✓					
5	3	Sheet	Epoxy laminate	0.6	÷	Drying	✓			
			board			Punching	✓			
						Cutting	✓			
5	1	SMD print	Copper	4.4	✓	Assembly SMD	✓	Isopropanol	0.003	÷
		with loose	Epoxy	7.4	÷			Loctite (glue)	0.2	÷
		components[2]	Fibre glass	12	✓	Spreading paste	✓			
			Lead	1.5	(✓)[3]	Printing copper conductors	✓			
			TBBPA	1.7	÷	Wave soldering	✓	Flux	0.3	÷
			Aluminium	2.6	✓			Nitrogen	0.3	÷
			Aluminium oxide	1.3	✓			Lead	0.6	(✓)[3]
			Fe	4.7	✓			Tin	1.1	(✓)[3]
			Tin	1.4	(✓)[3]	In Circuit Test	✓			
			$BaTiO_3$	2.1	÷	Manual soldering	✓			
			Gamma-			Laquering with silicone	✓			
			butyrolactone	1.3	÷					
			Silicone	5.6	÷					
7	1	Plug	Copper	1.9	✓	Subcontractor	÷			
			Polyamide	2.7	✓					
			Tin	0.08	(✓)[3]					
			Zinc	1.1	✓					
7	4	Wire	Copper	0.95	✓	Subcontractor	÷			
			PVC	0.38	✓					
			Lead	0.027	(✓)[3]					
			Tin	0.027	(✓)[3]					
7	1	Embedment mass	Polyurethane	140	✓	Embedding	✓			

Table 30.1. Simplified parts list for the PVEH

Pos.	Number	Item	Material	Total weight(g)	⏚	Manufacturing processes	* Incl. in model	Ancillary substances	Weight (g)	⏚
Magnetic valve set										
8	2	Yoke	Steel	240	✓	Washing	✓	4 detergents	1.02	÷÷÷÷
						Cutting	✓			
						Bending	✓			
						Screw cutting	✓			
						Electroplating with zinc	✓	Sodium metasilicate	0.1	÷
								Sodium hydroxide	0.2	÷
								Sodium hydrogensulphate	0.2	÷
								Phosphates	0.1	÷
								Zinc	0.5	✓
9	1	Base plate	Steel	148	✓	Washing	✓	4 detergents	0.8	÷÷÷÷
						Surface grinding	✓	3 lubricating oils	1.0	÷÷÷
						Punching steel sheet	✓			
						Electroplating with zinc	÷	Sodium metasilicate	0.1	÷
								Sodium cyanide	0.3	÷
								Sodium hydroxide	0.3	÷
								Sodium hydrogensulphate	0.2	÷
								Phosphates	0.1	÷
								Zinc	0.6	✓
10	2	Spring	Steel	0.4	✓	Winding	✓			
						Washing	✓			
						Tempering 350°C	✓			
10	2	Top	Steel	24	✓	Subcontractor	÷			
10	2	Armature pipe	Steel	16	✓	Copper soldering	✓			
			Copper	0.8	✓					
11	2	Armature pipe	Steel	28	✓	Copper soldering	✓			
			Copper	0.12	✓					
11	2	Stick	Steel	0.6	✓	Subcontractor	÷			
			Rubber	0.2	÷					
Miscellaneous										
13[4]	4	Coil	Copper	108	✓	Subcontractor	÷			
			Polyamide	13	✓					
			Tin	0.04	(✓)[3]					
13	1	Casing	Steel	8.9	✓	Subcontractor	÷			
13	1	Wire lead-in	Polyolefin	2.7	✓	Injection moulding	✓			
13	1	Conic spring	Steel	3.5	✓	Winding	✓			
						Washing	✓			
						Tempering 460°C	✓			
13	5	Plug	PVC	4.6	✓	Subcontractor	÷			
10,12,13	10	O-ring	Rubber	2.7	÷	Subcontractor	÷			
7,12	2	Tin solder	Lead	0.50	(✓)[3]	Subcontractor	÷			
			Tin	0.80	(✓)[3]					
			Copper	0.02	✓					
10,11	9	Tube core	Porcelain	1.8	✓	Subcontractor	÷			
3,10,13	5	Smaller steel items	Steel	4.5	✓	Subcontractor	÷			
13	3	Stainless steel items	Steel	2.3	✓	Subcontractor	÷			
			Polyamide	0.19	✓					
10,11	4	Valve seat	Steel	3.6	✓	Cutting	✓			
10,11	4	Ring	Steel	5.1	✓	Vibrating	✓			
			PE	1.5	✓	Cutting	✓			
			Rubber	1.5	÷					

Notes

1	Processed by subcontractor
2	Contains 81 resistors, 40 capacitors, 2 trimmers, 1 coil, 1 wirejumper and soldering paste and tin solder
3	Ressource consumption is back to the extraction of raw materials but not including emissions and resource consumption from the mining processes
4	13 is not an assembly of components, but the rest of the components in the product, i.e. those that do not fit into the other component assemblies
⏚	✓ ÷ * See other product case histories.

Table 30.1. (continued)

noted in Table 30.1. Note that lead, tin and zinc are also used as ancillary substances. For lead and tin, resource extraction is followed back to the extraction of raw materials, but the emissions and resource consumption for the extraction processes are not included. Zinc's product system is followed all the way back.

Manufacturing

Manufacturing of the PVEH is at Danfoss in Nordborg, and at subcontractors. All processes at Danfoss are summarized and included in the inventory. Energy consumption and resource consumption for processes are inventoried for subcontractors of printed circuit boards and valve blocks, but emissions from the manufacturing processes themselves are not. Most of the manufacturing processes are thus included in the inventory.

Processes

Apart from a few manual assembly processes, all processes use electricity. Building administration at Danfoss has measured the energy consumption for all processes on the individual machines while the items in question were being made. The various processes' consumptions of ancillary substances were also inventoried.

Emissions

Emissions to air from the processes were partly measured by an analytical laboratory in cooperation with the Employees' Health Service at Danfoss and partly determined on the basis of mass balances. Waste water is emitted both to Danfoss's own treatment plant and to a municipal treatment plant, and the emissions from the waste water treatment plants were inventoried. Those emissions from the waste water treatment plant at Danfoss, which derive from the manufacturing of PVEHs, were computed with the aid of mass balances, as were the emissions from incineration of waste oil from manufacturing. The waste oil is burned in an industrial boiler at Danfoss and used for production of hot water. Quantities of waste were also measured.

Figure 30.4. Manufacturing processes at Danfoss

Assembly of magnetic valve sets

Final test of PVEH

The working environment

Noise was measured directly at the processes, and all processes were reviewed for chemical impacts. Monotonous repetitive work was assessed by an expert on the working environment from the Employees' Health Service at Danfoss and by interviews with employees. Air pollution of the working environment was measured for embedment processes with polyurethane and for injection moulding machines in the plastics department. Work accidents were inventoried on the basis of Employees' Health Service statistics for Danfoss. The working time per product was inventoried for each process. The working environment for manufacturing of printed circuit boards at subcontractors was inventoried in the same manner.

Rejection rate

When faults are found in a product during manufacturing, the fault is rectified. The rejection rate is therefore insignificant, and it was decided to omit it from the LCA.

Overhead

Energy consumption for overhead, i.e. heating, lighting, ventilation and other joint facilities, was measured for the entire company. A fraction was then ascribed to the product on the basis of the sales value of a PVEH relative to the total turnover at Danfoss. Consumption of water for overhead was inventoried in the same manner.

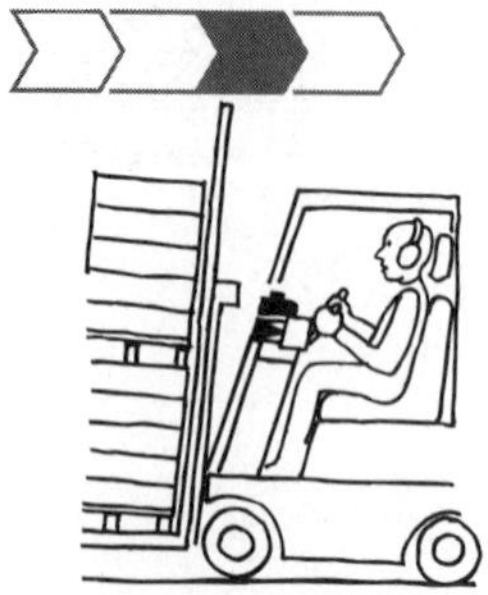

Use

PVEHs are sold almost exclusively in Europe, as also indicated in Figure 30.5. An average transport distance by lorry of ca. 800 km was assumed in the LCA.

As noted in the introduction, the PVEH is used in cranes, excavators and salt spreaders, but there are many more uses, a few of which include machinery used in forestry, the laying of asphalt, mining, and in lorry cranes and fork-lift trucks. An example is shown in Figure 30.6.

All of the machines noted operate for widely differing times, for example:

- Many of the machines used in mining run double shifts.
- Machines used in forestry also run in double shifts, but are only in operation from May to October.
- Machines used in the laying of asphalt run in double shifts except during very rainy weather or snow storms.
- Use in lorry cranes varies between an hour per day and an hour per week.
- Salt spreaders operate 24 hours a day when there is frost and snow.

A PVEH uses 8 W, supplied by a diesel generator on the machine in which the PVEH is installed. At a suggested average operating time of 5,000 hours, the maximum energy consumption is 40 kWh, which corresponds to combustion of 13 litres of diesel oil in the use stage.

Life span

The life span of a PVEH depends to a high degree on the frequency and magnitude of the temperature fluctuations to which it is exposed. Over time, the temperature fluctuations result in mechanical fractures in the materials. If an example with an excavator used for the clearing of snow is considered, the PVEH's temperature while in use will be +90°C. When the snow is cleared, the machines are turned off until the next snow fall. Depending on geography and how cold the winter is, the unit can be cooled down to as far as -30°C. There can thus be very large temperature fluctuations for some machines.

Figure 30.5. Distribution of sales for the PVEH

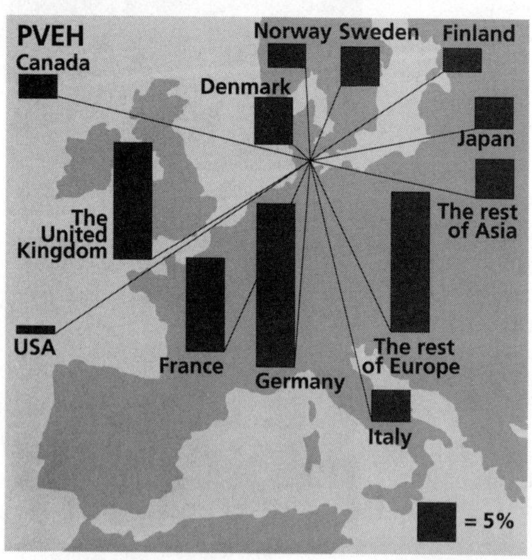

The life span of a PVEH depends on whether the machine is used 24 hours a day, in double shifts, or only occasionally. The PVEH's time in operation depends on how much the PVEH is used in the machine. The machine may well be used without activation of the PVEH. Using the example of snow clearing, the machine may have to travel large distances to where the snow is to be cleared,

Figure 30.6. A PVEH in use in an excavator

but during driving the PVEH will not be in use. An average life span of five years and time in operation of 5,000 hours is assumed for the PVEH.

The significance of different life spans and times in operation for the environmental impacts and resource consumption is simulated in the next chapter.

Service

A PVEH is not repaired. When it no longer functions it is replaced.

Disposal

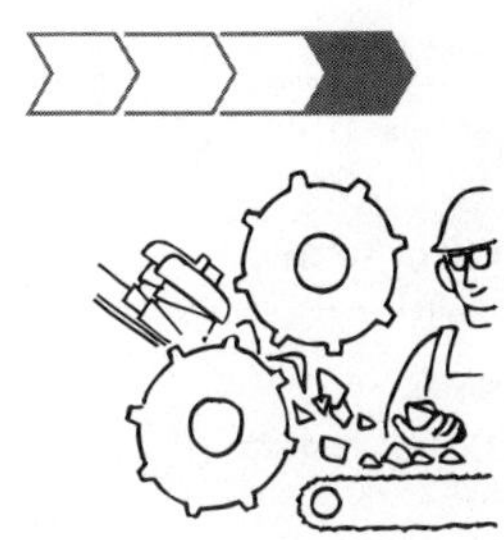

It is assumed that after final use, a PVEH is delivered to a recycling company. The valve block is removed here in order to recover the aluminium directly. The rest is disintegrated in a shredder, where iron and copper are separated and sent for recycling. The plastic is sent for incineration. The remaining fraction is deposited in landfills.

This manner of disposal is assumed on the basis of the method of disposal used for the PVEHs returned to Danfoss. Fewer than 1% of the products are returned to Danfoss. Most of the remaining products are replaced after final use by the workshop personnel responsible for servicing. Used units are sent to a scrap dealer, where the PVEH is disposed of in the same way as at Danfoss.

The significance of other disposal routes for environmental impacts is simulated in the next chapter. Both 100% consignment to landfills and 100% incineration of a PVEH are considered.

Data sources

Table 30.2 shows the sources of data for the inventory. Data for the material content of the electronic components were obtained partly from suppliers and partly from Danfoss.

The data cover a large part of the manufacturing processes both for new generations of electro-hydraulic activation units and for a large number of other products from Danfoss.

PVEH									
Data reference									
	Data specifity			Datasource type					Comments
Product system & Process type	Product specific	Site specific	General	1	2	3	4	5	
Extraction of raw materials									
Recycled aluminium		x		x					Subcontractor
Steel			x			x			Literature and subcontractors
Copper			x			x			Literature
PUR plastic			x			x			Literature
Plastic			x			x			Trade organization
Other materials			x			x			Literature
Product manufacturing by subcontractors									
Printed circuit boards	x			x	x				Printed circuit board manufacturer and the Danish Employees' Health Service
Aluminium valve block	x			x					Manufacturer: energy consumption and emissions
Product manufacturing at Danfoss									
Product manufacturing	x			x	x				Measurements of energy consumption and air performed by authorized laboratories. Assessment of the working environment performed by the Danish Employees' Health Service.
Use									
Lifetime	x			x		x			Empirical average
Energy consumption of PVEH	x			x					Dimensioning criteria
Diesel generator			x			x			Literature
Disposal									
Routes of disposal	x						x		Danfoss
Shredding			x			x			Literature
Recycling			x			x			Literature
Incineration			x			x			Literature
Transport									
Distances and means of transportation	x		x		x	x			Danfoss, literature
Emissions and energy			x			x			Literature/Common European transport database
Energy systems									
Energy production			x			x			Literature

Notes

1) Measurements
2) Computations (from mass balance considerations and input data for the process in question)
3) Extrapolation of data from similar process type or technology
4) Extrapolation of data from different process type or technology
5) Estimate

Product-specific data: concern processes specifically handling the PVEH
Site-specific data: concern data from actual sites in the product system of the PVEH, but the process data are not specific for PVEH
General data: all others

Table 30.2. Data sources

Exchanges with the environment

For each process, the exchanges with the environment were inventoried in the form of resource consumptions and emissions to air and water, and as waste. Impacts on the working environment were also determined as the time during which the exposure occurs per product. The most significant results from the inventory are presented separately in Table 30.3 for each stage of the product life.

			Production of materials	Product manufacturing	Use	Dis-posal	Transport	Total
Resource consumption								
Crude oil		g	470	190	11,400	-20	130	12,170
Natural gas		g	310	130	680	-2	8	1,126
Coal		g	740	3,790	50	2	1	4,583
Brown coal		g	50	-0.2	-	-0.3	-	49.5
Uranium ore		g	0.01	0	-	0	-	0.01
Water for hydro power		litre	140	180	-	-2	0	318
Aluminium	Al	g	200	-20	1	-110	0	71
Lead	Pb	g	0.2	2	-	-	-	2.2
Chromium	Cr	g	0.03	0.5	-	-	-	0.53
Iron	Fe	g	410	-140	1	-130	0	141
Copper	Cu	g	150	3	-	-140	-	13
Manganese	Mn	g	4	-1	-	-1	-	2
Nickel	Ni	g	0.4	0.2	-	-	-	0.6
Silver	Ag	g	0.07	-0.0002	-	-	-	0.0698
Tantalum	Ta	g	0.3	-	-	-	-	0.3
Tin	Sn	g	0.5	3	-	-	-	3.5
Zinc	Zn	g	3	0.2	-	-	-	3.2
Calcium carbonate	$CaCO_3$	g	50	-1	2	0	0	51
Quartz	SiO_2	g	100	-2	-	-30	-	68
Sodium chloride	NaCl	g	120	1	2	0	0	123
Soft wood		g	20	10	-	-0.2	0	29.8
Groundwater		litre	60	10	-	0	0	70
Surface water		litre	0	0.2	-	0.002	0	0.22
Unspec. water		litre	3	0.06	2	-0.05	0.03	5.04
Unspec. fuel (oil-eq.)		g	460	-20	-	-190	0	250
Emissions to air								
Carbon dioxide	CO_2	g	3,100	11,300	37,100	760	40	52,300
Carbon monoxide	CO	g	4	2	250	2	-	258
Nitrogen oxides	NO_x	g	20	40	430	-1	-	489
Sulphur dioxide	SO_2	g	20	60	70	0	-	150
Nitrous oxide	N_2O	g	0.2	0.7	-	0	-	0.9
Unspec. particles (dust)		g	10	5	40	-9	-	46
Hydrocarbons	HC	g	20	90	30	0	-	140
Volatile organic compounds	VOC	g	20	0	80	0	0	100
Unspec. aldehydes		g	0.006	0.04	-	0	-	0.046
Formaldehyde		g	0.3	-	-	-	-	0.3
2-propanol		g	-	0.003	-	-	-	0.003
Toluene		g	0.2	-	-	-	-	0.2
Dioxins		mg	0	0.000015	-	0.000004	-	0.000019
Epichlorohydrine		mg	130	-	-	-	-	130
Arsenic	As	mg	0.1	0.1	-	0	-	0.2
Cadmium	Cd	mg	0.02	0.01	-	0	-	0.03
Chromium (VI)	Cr	mg	-	0.0001	-	-	-	0.0001
Copper	Cu	mg	0.2	0.9	-	0	-	1.1
Mercury	Hg	mg	0.04	0.1	-	0	-	0.14
Manganese	Mn	mg	290	1	-	-	-	291
Vanadium	V	mg	7	8	-	-0.03	-	14.97
Emissions to water								
Chemical oxygen demand	COD	g	0.4	0.7	0.1	0	0	1.2
Total nitrogen	tot-N	g	0.4	0	0.03	0	0	0.43
Total phosphorus	tot-P	g	0.0005	0.01	-	-	-	0.0105

Table 30.3. Inventory of a PVEH's exchanges with the environment for the entire product system

			Production of materials	Product manufacturing	Use	Dis-posal	Transport	Total
Emissions to air								
Hydrocarbons	HC	g	0.04	0	0.2	0	0	0.24
Unspec. oil		g	0.03	0.004	-	0	0	0.034
Unspec. anionic detergents		g	-	0.01	-	-	-	0.01
Arsenic	As	mg	3	0.1	-	-	-	3.1
Cadmium	Cd	mg	1	0.03	-	-	-	1.03
Copper	Cu	mg	9	0.2	-	-	-	9.2
Mercury	Hg	mg	0.0002	0.00001	-	-	-	0.00021
Manganese	Mn	mg	0.01	0.001	-	-	-	0.011
Nickel	Ni	mg	0.5	0.01	-	-	-	0.51
Zinc	Zn	mg	7	1	-	-	-	8
Waste								
Unspec. hazardous waste		g	0.4	0.7	-	-	-	1.1
Unspec. radioactive waste		g	0.001	0	-	0	-	0.001
Unspec. slag and ashes		g	60	300	30	10	0	400
Unspec. bulk waste		g	420	2,210	30	1	0	2,661
Working environment								
Monotonous repetitive work		hours	0.0001	0.07	0.0003	-	0.0002	0.0706
Hearing impairment inducing noise		hours	0.001	0.009	0.001	-	0	0.011
Substances causing damage to the nervous system		hours	0.0001	0.006	0	-	0	0.0061
Allergenic substances		hours	0.0001	0.002	0.0003	-	0	0.0024
Substances causing damage to the reproductive system		hours	0	0.003	0	-	0	0.003
Carcinogenic substances		hours	0.00005	0.002	0	-	0	0.00205
Accidents		number	$0.1 \cdot 10^{-6}$	$25 \cdot 10^{-6}$	$0.03 \cdot 10^{-6}$	-	0	$25.1 \cdot 10^{-6}$

Notes

- means that no information is available for the life cycle stage in question
0 means that the figure is very small compared to the other stages of the life cycle

Table 30.3. (continued)

Resource consumption

The apparatus weighs about 1 kg. Resource consumption is shown in Table 30.3 as the pure metal content of the ores. Consumption of ore is not shown. The table shows that in the material stage, ca. 3 kg of resources are used to make the product, of which ca. 0.9 kg end as materials in the actual product. The resources used for materials which end in the product weigh less than the product. This is attributed to the fact that both the aluminium and most of the steel are recycled materials, and that recycled materials are not ascribed a resource consumption of 100%, but only a part of the resource consumption. This is described in more detail in Parts II and IV. Of the 2.1 kg which do not enter into the product, 81% is fuel for energy consumption in the material stage. The rest is lost.

The use stage consumes ca. 12 kg of fuel during the life of the apparatus, while the manufacturing stage consumes ca. 4 kg. On the other hand, the transport stage consumes only 0.14 kg. The transport stage covers transport from production of materials to manufacturing in Nordborg and transport on to the consumer. Transport between the extraction of raw materials and production of materials is included in data for the material stage. Relative to consumption in the manufacturing and use stages, the consumption of fuel in the transport stage is of very little significance.

Consumption of metals occurs in the material stage, but for tin, lead and chromium, the greatest consumption is in the manufacturing stage, as these metals, as noted above, are used as ancillary substances in the soldering and chrome plating processes. Negative resource consumptions appear in the manufacturing and disposal stages because of recycling. The explanation is that when recycling, a part of the resource consumption and the emissions from the extraction of raw materials accompanies the materials on to the next product system. In the total computation, consumption and emissions are therefore subtracted in the stage in which the recycling occurs. Some of the metals from manufacturing are collected as waste and sent for recycling. On disposal of the product after use, aluminium, copper and steel are sent for recycling.

Emissions to air

Most of the emissions derive from the production of electricity and thermal energy. Electrical energy is used mainly in manufacturing processes at Danfoss and subcontractors. Thermal energy is used, for example, in the production of plastics, for room heating, and in the use stage as diesel oil for the machine, for example the excavator. The use stage therefore produces emissions related to the combustion of diesel oil.

Formaldehyde and toluene are emitted during production of materials. Isopropanol is emitted from a wave soldering process during manufacturing and chromium is emitted from a chrome plating process in the surface treatment.

Emissions to water

The emission of organic substances from the entire product system is very small, as shown by the total chemical oxygen demand (COD), and so are the emissions of nitrogen and phosphorus. Oil-coated metals are washed during manufacturing, but the emission of oil and heavy metals from Danfoss's own waste water treatment plant is very small. Heavy metal emissions are significantly greater in the material stage than in the manufacturing stage.

Waste

The largest quantities of waste derive from the manufacturing stage in the form of unspecified bulk waste and unspecified slag and ashes. The bulk waste derives from the extraction of coal for electricity production. The slag and ashes derive from the production of electricity, especially from the combustion of coal. Aluminium, iron and copper are recovered on disposal of a PVEH, while the plastic is incinerated.

The working environment

The largest number of hours during which impacts on the working environment occur is in the manufacturing stage. The impacts come partly from manufacturing at the maker of printed circuit boards and partly from manufacturing at Danfoss, but also from production of the electricity used in the manufacturing stage.

Energy profiles

The energy consumption is a significant cause of resource consumption and impacts on the environment and the working environment throughout the entire product system. Figure 30.7 shows the primary energy consumption in all stages.

It is evident from the energy profile that:

- Energy consumption in the use stage is by far the greatest in the product system.
- The manufacturing stage has the second highest energy consumption, which is attributed to the fact that the energy consumption for overhead in manufacturing is large, amounting to ca. 86% of energy consumption in this stage.
- Energy consumption in the material stage is a little less than in the manufacturing stage.
- Energy consumption in the transport stage is very small relative to that in the other stages.
- The negative energy consumption for disposal derives from the incineration of plastic. It is assumed that the incineration of waste occurs as in Danish plants, where ca. 75% of the energy of combustion is utilized for district heating. As the metals are recycled on disposal, a minor fraction of the energy consumption is credited to extraction of raw materials, but relative to the energy utilization from the incineration of plastic, this is of little significance.

Figure 30.7. Energy profile

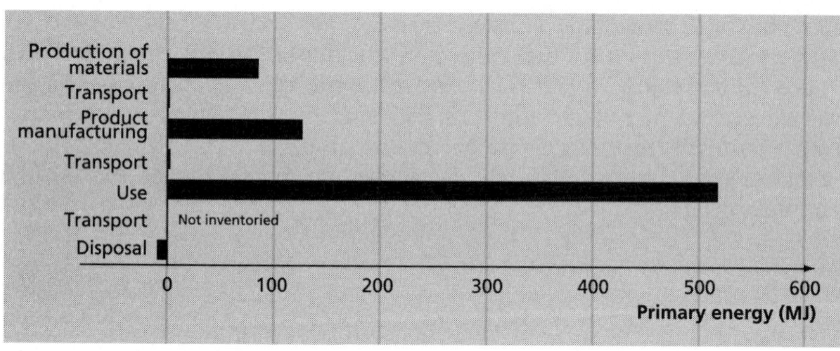

The energy content of the materials

In the finished product, the energy content of the materials is of interest in disposal. If the product is incinerated, the calorific value of the materials is in focus, but if the materials are recycled, the entire energy content is of interest. Figure 30.8a shows the energy which the various materials contain, and Figure 30.8b shows the energy content classified by the various component assemblies. The energy content is divided into:

- Process energy, which is the quantity of energy required for extraction of raw materials and production of the material.
- Calorific value, which is the energy content which can be utilized on incineration.

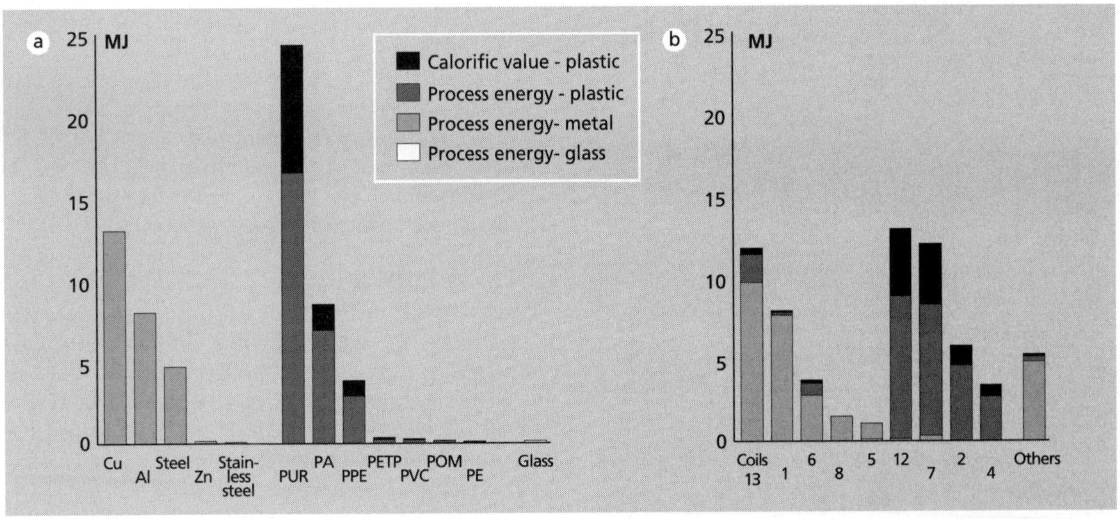

Figure 30.8a. The primary energy content in the materials found in the PVEH

Figure 30.8b. The primary energy content in the materials in the PVEH classified by component assemblies

As some of the component assemblies in Figure 30.8b contain different materials, the process energy is divided into glass, metal and plastic. In the lefthand component assemblies in Figure 30.8b, the metals' process energy comprises the main part of the energy content. For the rest, the main part of the energy content is from the process energy of the plastic and the calorific value.

It is evident from Figure 30.8a that the PUR plastic contains most of the materials' calorific value. The component assemblies (nos. 2, 4, 7 and 12) contain almost all of the product's calorific value, so that on incineration of the product, these four component assemblies will yield ca. 92% of the product's contribution to the energy for district heating. The base plate and the top cover (nos. 2 and 4) consist entirely of plastic. Embedded in PUR plastic, component assemblies nos. 7 and 12 also contain metals.

Apart from the energy content of the PUR plastic, Figure 30.8a is dominated by the energy content in the metals copper, aluminium and steel. More than 80% of the energy content in the valve block (no. 1) and in coils in the parts list's no. 13 is process energy from metals. For the transducer (no. 6), over 75% of the energy content derives from the metals' process energy, while the energy content in the yokes (no. 8) comes only from metals. The process energy of the metals for these four component assemblies amounts to a third of the energy content of the materials and over three-quarters of the metals' process energy. On recycling, it is therefore these component assemblies which are of interest in terms of energy.

Impact assessment

Table 30.4. The environmental exchanges converted to environmental impact potentials

Impact category	Unit	Potential impact Per product per lifetime
Global impacts		
Global warming	g CO_2-eq	53,100
Ozone depletion		-
Regional impacts		
Acidification	g SO_2-eq	500
Nutrient enrichment	g NO_3^--eq	660
Photochemical ozone formation	g C_2H_4-eq	52
Human toxicity (water)	m^3 water	6.8
Ecotoxicity (water, chronic)	m^3 water	41
Local impacts		
Human toxicity (air)	m^3 air	50,000,000
Ecotoxicity (water, acute)	m^3 water	2.2
Hazardous waste	g	9.5
Nuclear waste	g	0.0014
Slag and ashes	g	400
Bulk waste	g	2,800

Emissions to air and water and as waste from the various processes represent a potential for environmental impacts. The exposure in the working environment also comprise a potential for impacts. However, the potentials will not trigger actual impacts in all cases. Much depends on the circumstances where the process occurs.

Environmental impact potentials

The exchanges with the environment all represent a potential for environmental impacts and are converted to environmental impact potentials on the basis of the method described in Parts II and IV. All of the emissions which are known to have potential environmental impacts are included. The result of the conversion is shown in Table 30.4. The conversion was made on the basis of all of the inventoried emissions, and thus not only those shown in the simplified inventory in Table 30.3.

Resource consumption and potential impacts on the working environment

For resource consumption and potential impacts on the working environment, the figures in Table 30.3 are used directly for further calculations.

The magnitude of the product's contribution

To increase understanding of the magnitude of the contributions, the PVEH's contributions to the various impact categories are divided by the contributions which a person makes on average during a year. The product's contributions are expressed in milli-person-equivalents, abbreviated mPE. Milli-person-equivalents are defined as the product's contribution in parts per thousand of an average person's contribution. This conversion makes it possible to compare the various impact categories within resources, environment and the working environment. The conversion is called normalization.

Figure 30.9 shows the PVEH's profiles for resource consumption and potentials for impacts on the environment and the working environment. The profiles give the product's contribution relative to a person's average contribution, but they do not show which impact potentials are the most critical. This is, however, described in the section on the most significant impact potentials.

The resource profile in Figure 30.9a shows that the PVEH has a relatively high consumption of tin. The consumption amounts to 15 mPE_{W90} or 1.5% of a person's average consumption of tin in 1990, the reference year. The product uses aluminium and crude oil amounting to about 0.4% of a person's average consumption. For the remaining resources the unit uses 0.2% or less of a person's average consumption. The resource profile does not specify the normalized consumptions of silver and tantalum, as only the figure for reserves is known, not the figure for a person's average consumption; the latter figure is necessary for normalization.

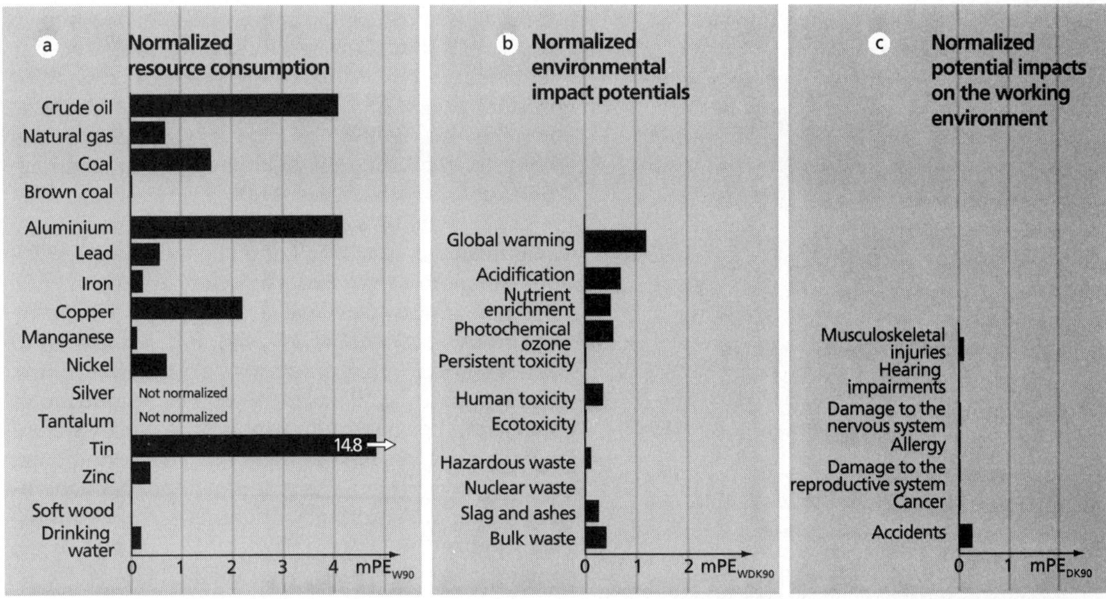

Figure 30.9. Normalization of resource consumptions and impact potentials for the PVEH

The environmental profile in Figure 30.9b shows that relative to a person's average consumption, a PVEH contributes especially to global warming. The contribution corresponds to 0.12% of a person's contribution to global warming. For acidification, the contribution is about 0.07% and for the other impact categories it is 0.05% or less.

It is seen from the working environment profile in Figure 30.9c that for work accidents and injuries to the musculoskeletal system, a PVEH contributes respectively 0.026% and 0.012% of the normal annual impact for workers in Denmark. The other impacts on the working environment are below 0.002% of the annual impact on an average Danish worker.

The most significant impact potentials

The various impact potentials are mutually weighted and compared in order to assess which are the most critical. The weighted impact potentials cannot be compared across the three categories. For example, resource consumption cannot be compared with environmental impact potentials.

It can be argued that resources which are burned to produce energy should be weighted more highly than resources used for materials, as the combustion is an irreversible conversion of the resource, while materials which are disposed of in landfills can be recovered if this is ultimately desired. However, this weighting criterion is not included in the present LCA and the resources used for the production of energy are not weighted more highly than the other resources. Resources are weighted on the basis of the sizes of known reserves, and the unit of weighted resource consumption is the milli-person-reserve (mPR_{W90}). The weighted resource profile in Figure 30.10a shows that a PVEH uses 15 mPR_{W90}

Figure 30.10a. Weighted resource consumptions for the PVEH

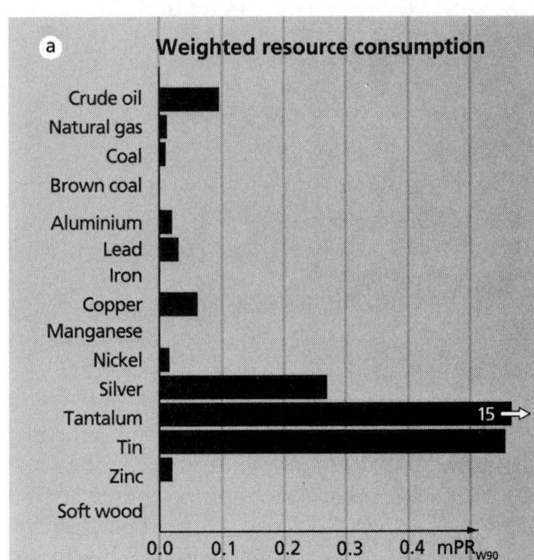

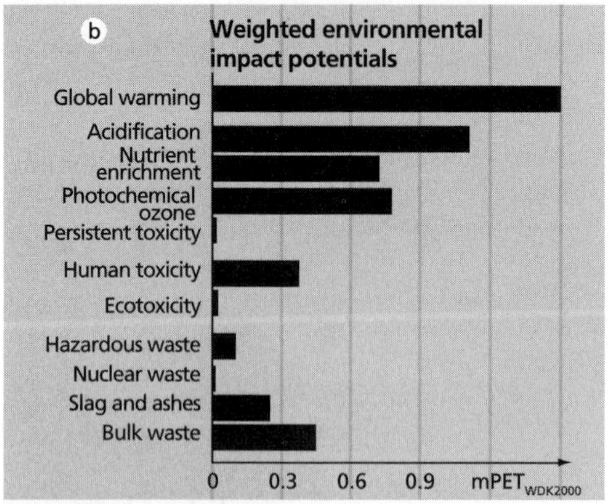

Figure 30.10b. Weighted environmental impact potentials for the PVEH

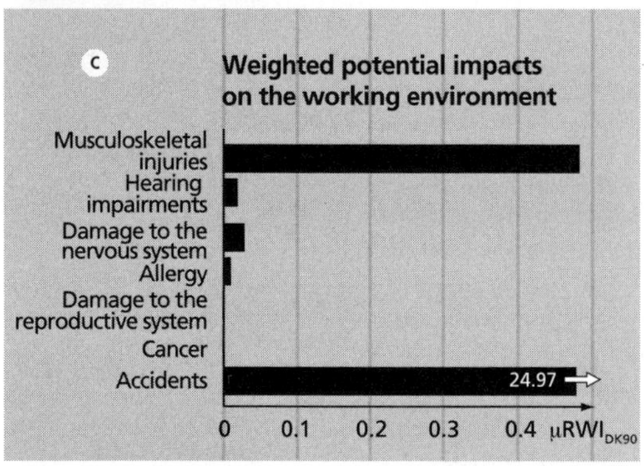

Figure 30.10c. Weighted potentials for impacts on the working environment for the PVEH

of tantalum, corresponding to 1.5% of the known reserves of tantalum remaining in 1990 for one person and all of his or her descendants. The product uses 0.055% of the tin reserve remaining per person, 0.027% of the silver reserve and 0.01% of the crude oil reserve. For the remaining resources, a PVEH uses 0.006% or less.

The environmental impact potentials are weighted with the international or Danish political reduction targets for the environmental impacts. The unit is called a targeted milli-person-equivalent ($\text{mPET}_{\text{WDK2000}}$). It is clear from the weighted environmental profile in Figure 30.10b that a PVEH potentially contributes 0.15% of the maximum amount of global warming which a person in the year 2000 should on average contribute. For acidification, photochemical ozone and nutrient enrichment, the contributions are in the range 0.07-0.11%. The rest of the environmental impact potentials lie below 0.05%.

Potential impacts on the working environment are weighted with the frequency of reported work injuries in Denmark, and the unit is therefore micro-reported work injuries ($\mu\text{RWI}_{\text{DK}}$). It is clear from Figure 30.10c that the PVEH is assessed as contributing 25 of each million reported work accidents in Denmark. For injuries to the musculoskeletal system, the product contributes a bare 0.5 of each million reported injuries. The contributions to the other impacts on the working environment are estimated to be 0.03 per million reported work injuries or less.

The most significant sources of impact potentials

Table 30.5 suggests a structure for classification of the various impact potentials to allow the designer to see how much the choices within "materials" and " processes" actually mean and to place the designer in a position to assess the consequences of different choices. Materials and processes are classified under the MECO principle described in Parts II and IV. The classification gives six sources of potential impacts. For example, the choice of materials also influences the disposal processes for the product after final use.

Materials	
Material	- consumption - disposal
Processes	
Energy	- thermal - electric
Chemicals	- ancillary substances
Others	- overhead

Table 30.5. Sources of resource consumptions and impact potentials after the MECO principle

It is clear from the resource profile classified by source in Figure 30.11a that it is mainly the use of materials in the PVEH which is of significance for resource consumption. For all metals apart from tin and lead, resource consumption is almost entirely due to use in materials. For both tin and lead, 85-90% of the resource consumption is ascribed to the use of metals as ancillary substances in the processes. The same applies to 2% of the copper consumption and 10% of the zinc consumption; 95% of the consumption of crude oil and 64% of natural gas consumption are ascribed to the consumption of thermal energy. The consumption of coal is attributable mainly to energy consumption for overhead.

The fact that total consumptions of aluminium, iron and copper are not greater than shown in Figure 30.10a is, as already noted, attributed

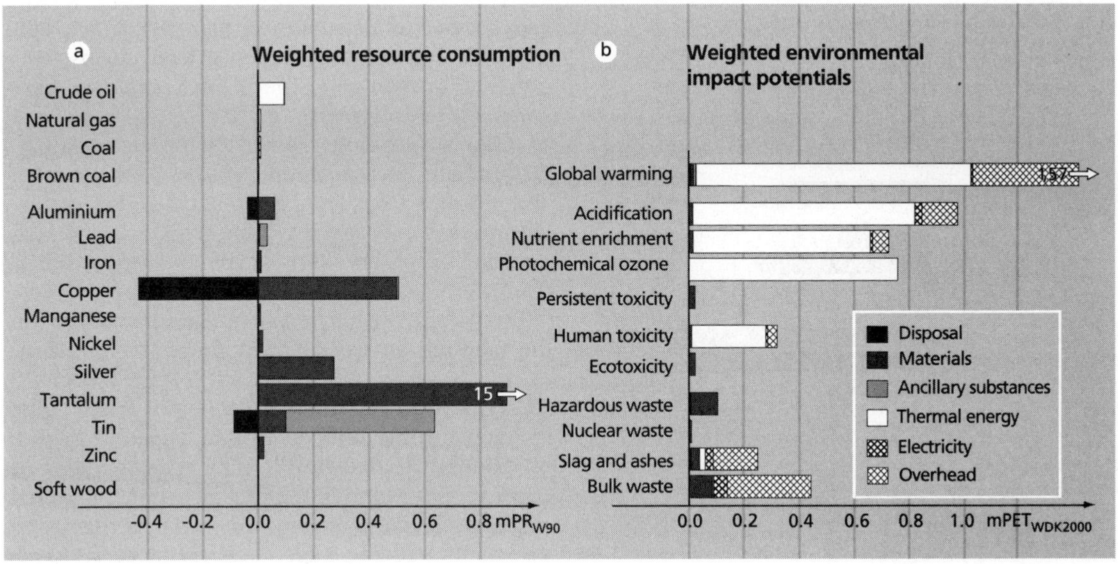

Figure 30.11. Weighted resource consumptions and environmental impact potentials classified by type of source

to the assumption that almost all metals are recycled on disposal. But even if the metal is 100% recycled, it is not 100% credited on disposal because there is a loss in material grade in the recycling processes. Resource consumption follows the recycled material only in proportion to the residual material grade, and the product is therefore always ascribed a certain resource consumption, even with 100% recycling.

The environmental profile classified by source in Figure 30.11b shows that by far the largest part of the contributions to global warming, acidification, nutrient enrichment, photochemical ozone and human toxicity derives from thermal energy for the processes. The materials dominate the contributions to persistent toxicity, ecotoxicity, hazardous waste and nuclear waste. Overhead accounts for over half of the contribution to slag and ashes and bulk waste. Whether disposal contributes positively or negatively to the environmental impacts varies.

The most significant stages of the product life

It is possible on the basis of the profiles classified by source to simplify the LCA. In Figures 30.12 and 30.13, the impacts which are either very small or which can be represented by other impacts are omitted. An impact can represent another impact if the contributions derive from the same sources. For resources in the PVEH product system, crude oil can represent natural

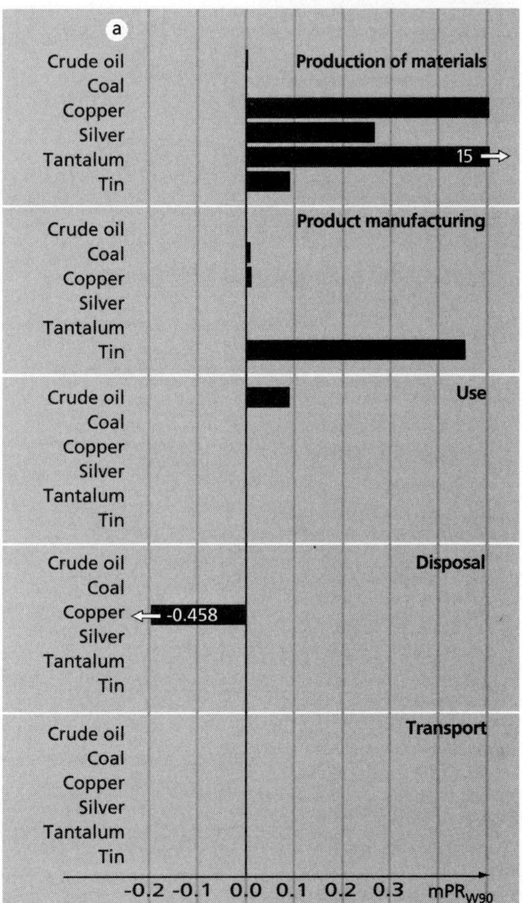

Figure 30.12a. Weighted resource consumptions divided by stages of the product life

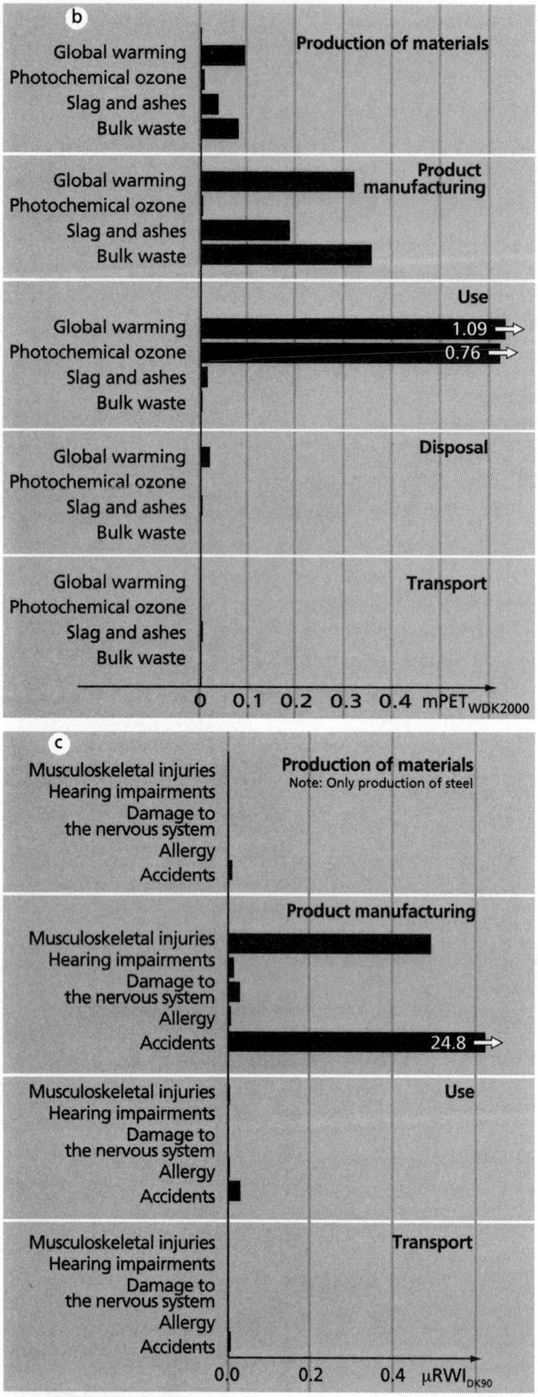

Figure 30.12b. Weighted potentials for environmental impacts divided by stages of the product life

Figure 30.12c. Weighted potentials for impacts on the working environment divided by stages of the product life

gas, copper represents iron, manganese and aluminium, while tin represents lead. Consumptions of brown coal, nickel and zinc are very small and are therefore omitted.

Global warming represents the environmental impacts acidification, nutrient enrichment and human toxicity. The potentials for persistent toxicity, ecotoxicity, hazardous waste and nuclear waste are very small and are omitted from the following presentations.

With respect to the potentials for impacts on the working environment, damage to the reproductive system and cancer are omitted because the contributions to these are very small. The resource profile divided by life stages in Figure 30.12a shows that the PVEH's consumptions of copper, silver and tantalum lie in the material stage, while the consumption of tin is mainly in the manufacturing stage. The use stage is dominated by the consumption of crude oil, as the power used by the PVEH is provided by a diesel generator. Disposal is dominated by the recycling of copper.

The environmental profile classified by stages in Figure 30.12b shows that a PVEH's contribution to global warming and photochemical ozone derives mainly from the use stage's combustion of diesel oil in the engine's generator. Bulk waste and global warming are dominant for the manufacturing stage, while the production of slag and ashes is about half as large. The same pattern applies to the material stage. In the disposal stage, incineration of plastic determines the contribution to global warming.

Only impacts on the working environment from the production of steel plates and the production of electricity for the manufacturing of these are included in the material stage in Figure 30.12c. Work accidents and, to a lesser degree, hearing impairment are expected to be the most frequent causes of the reported work injuries here. The manufacturing stage includes both manufacturing at Danfoss and the manufacturing of printed circuit boards as well as the production of electricity. For the manufacturing stage, work accidents and injuries to the musculoskeletal system dominate. The impacts on the working environment in the use stage come from the extraction of crude oil for use by the PVEH. For both use and transport, the greatest potentials are work accidents.

The most significant components

When environmental initiatives are to be ranked, it can be an advantage to show which components have the greatest environmental impact

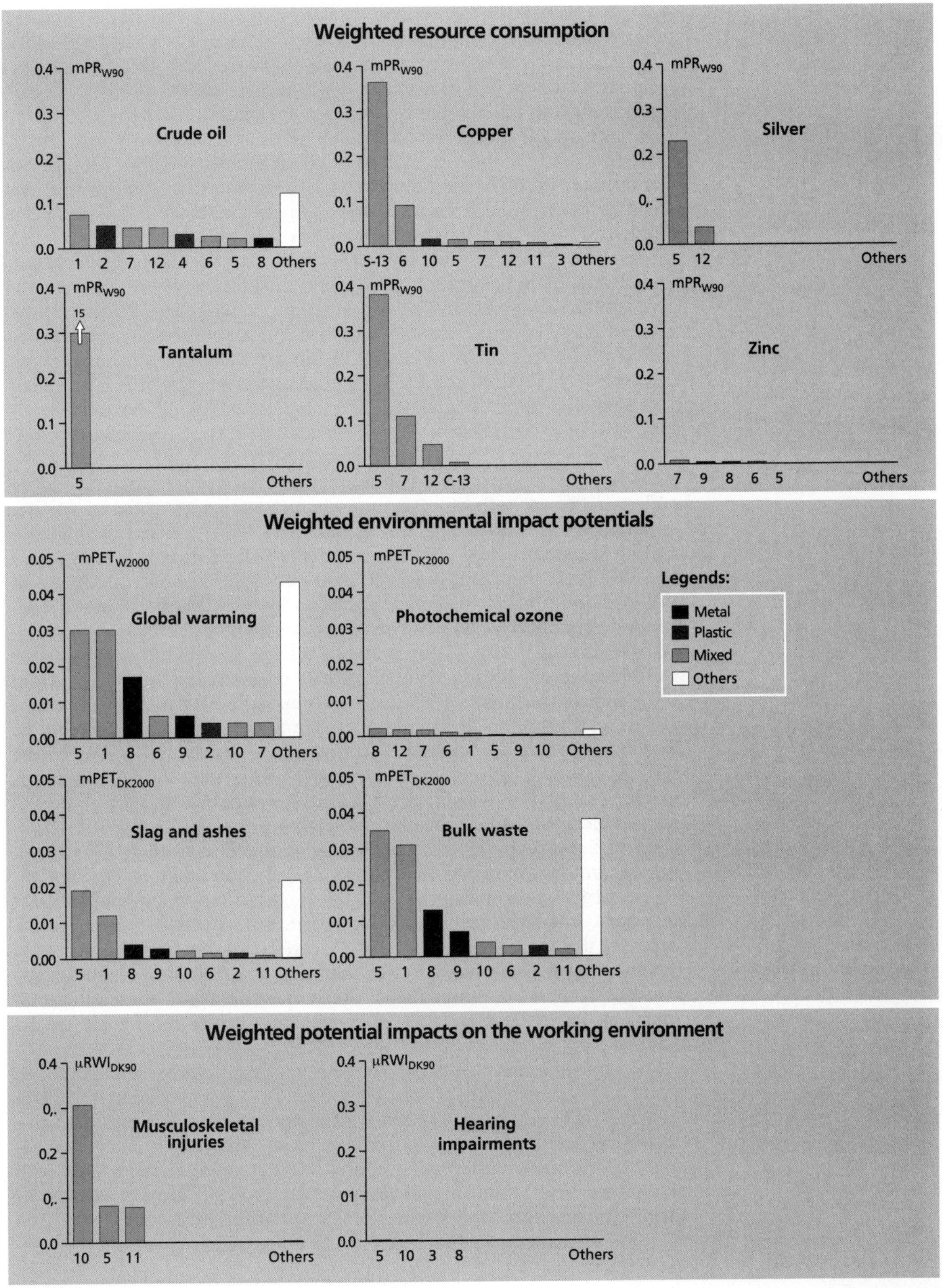

Figure 30.13. Top scorers among the PVEH's sub-assemblies

C-13 = coils in 13

potentials. The list of top scorers shown in Figure 30.13 is not a set of final answers, but it should be used in discussions of environmental rankings. It should be noted that a component may in any case be the best choice environmentally, even if the component is the top scorer. Everything depends on the alternatives which are available.

Because of the number of components in the PVEH, Figure 30.13 shows the component assemblies. The numbers for the component assemblies are given in the parts list (Table 30.1) as item numbers. Item no. 13 in the parts list comprises the components which it has not been possible to fit into the other component assemblies.

The presentation of top scorers in Figure 30.13 represents only the material and manufacturing stages plus transport between these stages. The manufacturing stage is given here without overhead. As noted, the entire product system is not included, and the resource consumption for some of the metals could be higher in the top scorer figures than in the inventories, where all of the PVEH's product system, including recycling in the disposal stage, is included. For impacts on the working environment, only manufacturing at Danfoss is shown in the presentation of top scorers.

Apart from the component assemblies which are top scorers, Figure 30.13 specifies an "others" category which contains all of the remaining components and the manufacturing processes which do not belong to a component assembly. In item no. 13 in the parts list, which is included under "others", there are, however, four coils which contain copper and tin in a quantity which has made it necessary to show the resource consumption separately, which was done by designating the coils as component assembly no. C-13 in the figure.

Printed circuit boards with electronic components (no. 5) are top scorers for consumption of scarce resources such as silver, tantalum and tin. The consumption of tantalum goes into three tantalum capacitors. Silver is used in certain electronic components. Tin is used partly in the components and partly in the soldering. Most of the copper consumption goes into the coils in item no. 13 in the parts list and into a coil in the transducer (no. 6). Zinc appears in a plug (no. 7), in the yokes (no. 8) and in the base plate (no. 9) because of the zinc plating of the steel. The consumption of crude oil is similar for all component assemblies.

The largest potentials for three of the four environmental impact categories come from printed circuit boards with components, the valve block and the yoke (nos. 5, 1 and 8). The contributions come mainly from consumption of electricity for manufacturing of component assemblies. The "others" category for all four environmental impact categories is high, as the consumption of electricity for testing the PVEH in the manufacturing stage is included here.

The contributions to impacts on the working environment come, as already noted, from the processes at Danfoss. It is evident in Figure 30.13 that printed circuit boards with components and magnetic valve sets are top scorers in the potential for the working environment impact "injury to the musculoskeletal system", while printed circuit boards with components (no. 5) and magnetic valve sets (no. 10) are top scorers for hearing impairment. Monotonous repetitive work is due to the insertion and removal of items in machines, while noise comes from the processing of metals.

Lack of information, uncertainties and sensitivity analyses

There are still areas in the LCA where more and better information could be desired, despite a large effort in data collection.

The material stage
The uncertainty in data for the production of aluminium is small, among other reasons because measurements were made at suppliers to Danfoss. Steel is provided by two different suppliers, but data for steel production are based only on information from one supplier, which of course results in a minor uncertainty. Data for the extraction of raw materials and the production of plastic, copper and other materials come from the EDIP unit process database (Frees, 1996). The information for plastic is average data from the European association of plastics producers, APME, and others. When average data are used there is some uncertainty involved, but as the choice of supplier in any case varies, data will also vary, and the picture presented is therefore realistic when average data are used. Data for copper and the other materials come mainly from the literature, with the uncertainties which this gives (Frees, 1996).

Information on the material content in the active electronic components - transistors, IC circuits and diodes - is lacking, but the content in the other electronic components is included in the assessment. The information on these is quite accurate, provided partly by suppliers and partly by investigations by Danfoss itself. As most of the non-active components contain scarce metals, it is probable that this is also the case for some of the active components. Consumption of some of the scarce metals is therefore probably greater than inventoried in the LCA.

In terms of energy, the manufacturing process for the electronic components is estimated as insignificant. The manufacturing process was estimated conservatively as consuming at most 2-3 times the energy content in the materials in the electronics themselves, and on this basis the rather light electronic unit means little. The emissions from the manufacturing processes of the electronic components may well be significant, but they cannot be estimated as very few data are available for these processes.

The ancillary substances comprise various detergents, tensides, cutting and lubricating oils and chemicals for the electroplating with zinc. During the manufacturing stage, emissions of the ancillary substances are included for the processes where they are used. There is no information on extraction and processing of raw materials and their production in the product systems of the ancillary substances. In general, the emissions of chemicals during their production are not significant relative to the emissions from their use, but there can be exceptions. For example, there may be special exposures in the working environment. The emissions from detergents in the product manufacturing stage are probably greater by a factor of at least ten than the emissions from their production as the percentage loss in the chemical industry during production is estimated to be significantly lower than 10%.

Impacts on the working environment in the material stage were assessed only for the production of steel. Steel comprises ca. 30% of the materials in the finished PVEH. Large quantities of steel are produced relative to the working time involved. The working time is probably greater per kilogram of plastic or chemical, but it is not known by how much. It is not possible to discuss impacts on the working environment from the manufacturing of electronic components as the processes are not known.

The manufacturing stage

There are no large gaps in the information on manufacturing at Danfoss, as all exchanges with the environment for all processes have either been measured or calculated on the basis of mass balances. Energy consumption and waste from manufacturing processes were inventoried at the subcontractors of printed circuit boards and aluminium blocks, but this does not apply for emissions. As the collected data are from sole suppliers and based on mass balances, the uncertainty is small.

The contributions to impacts on the working environment from manufacturing at Danfoss and from the manufacturer of printed circuit boards are included in the assessment. The uncertainty in the data is small, as they were measured or estimated at the respective sites by the Danish Employees' Health Service.

The use stage

Data for the use stage are based partly on Danfoss's dimensioning criteria for operating time and power consumption and partly on data from the literature for the combustion of diesel oil. There is some degree of uncertainty in using these data in place of measurements, but as the use scenario for a PVEH depends on the machine in which it is used, it seems appropriate to use average data for the use stage. The sensitivity to different life spans and patterns of use is shown in the next section with the aid of simulations.

The disposal stage

Danfoss does not know with certainty how most of the PVEHs are disposed of. The disposal scenario is documented only for the 1% of the product which ends at Danfoss. In the next section, the consequences of other disposal routes are described, and the significance of 100% incineration or 100% consignment to landfills of the product is simulated.

30.2. Environmental diagnosis for an electro-hydraulic activation unit - the PVEH

In the environmental diagnosis, the LCA is converted to a form which the designer can use in the development of new products. Where in the product the options for improvements in environmental terms reside is designated in the environmental diagnosis. This is done by first simulating the environmental consequences of various theoretical changes in the product or its product system. Alternative solutions for the product are then considered.

Simulation of changes in the product or its product system

The simulations are made in order to gain a picture of what the various changes mean for the impact categories. Will use of a new concept for the electronics result in improvements in e.g. resource consumption? Or will an increased life span for the PVEH result in lower environmental impact potentials? Such questions can be answered when the changes are simulated. The answers indicate which theoretical potentials for improvement are to be found.

The potentials for improvement are identified within the four areas of the MECO principle:

- Materials
- Energy
- Chemicals
- Others

Two simulations were made under the category Materials: use and disposal. No simulations were made under the categories Energy and Chemicals, but two simulations were made under Others: patterns of use and life span.

Figure 30.14 shows the result of the simulations. The first simulation shows the potentials for improvement if the concept for the electronics is changed by replacing conventional fibre glass-reinforced epoxy laminate with thick film in the printed circuit boards. The other three simulations show the significance of various changes in the product system for the PVEH.

Materials

Fibre glass-reinforced epoxy laminate vs. thick film technology in a PVEH

If conventional printed circuit boards are replaced by thick film technology, the concept for the electronics in the PVEH is changed. The material compositions in the printed circuit boards based on fibre glass-reinforced epoxy laminate and in the thick film boards are different. Some of the electronic components are also replaced by other types when the conventional technology is replaced by thick film technology. When the thick film technology is used, resistances are built into the thick film's circuitry, thus saving both materials and manufacturing processes. The number of solderings is also smaller if changes in the concept are made. The simulation takes account of both the different materials in the circuit boards themselves and the different materials from the change in some of the components.

The material content in the active components such as transistors and IC circuits is not included in the two inventories. The manufacturing of electronic components is also excluded. On the other hand, the energy consumption and the impacts on the working environment are included for the insertion of circuitry in the conventional printed circuit boards and the thick film. The simulation is shown in Figure 30.14a.

For all of the impact categories shown, the change to thick film technology means a reduction in the potentials, but for the environmental impact categories the fall is not particularly large. The reduction will be 4-18%, but the quantity of lead deposited in landfills when the PVEH is disposed of will fall significantly more because of the reduced number of solderings. Under resource consumption, consumptions of lead and tin especially will fall because the number of solderings is smaller. The reduction is 65-70% for lead and tin. The consumption of silver is also reduced by 70%. For the other metals, the reduction is between 5 and 23% - greatest for copper, which will not be used in such large quantities for circuitry in thick film boards. The consumption of coal falls by 5%, as electricity consumption for the manufacturing of thick film is a little less than in manufacturing of the traditional printed circuits.

In changing to thick film technology, the potential for damage to the nervous system in the working environment almost entirely disappears,

Figure 30.14. Simulation of changes in the product or its product system. Only the impact categories which are changed are shown

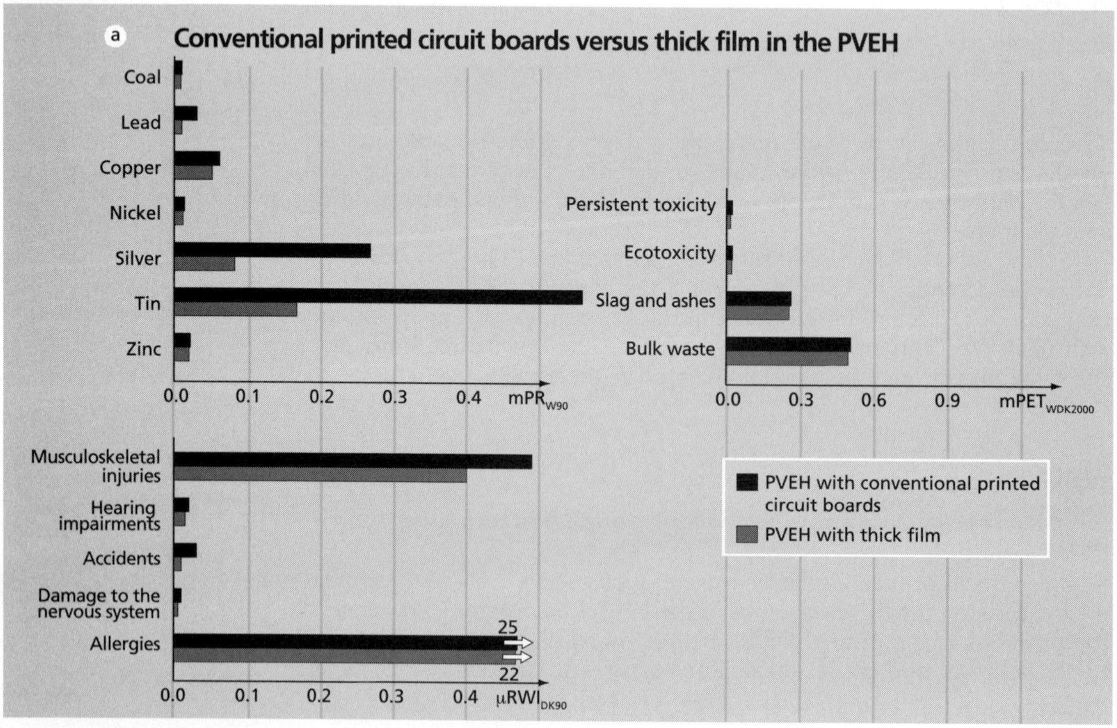

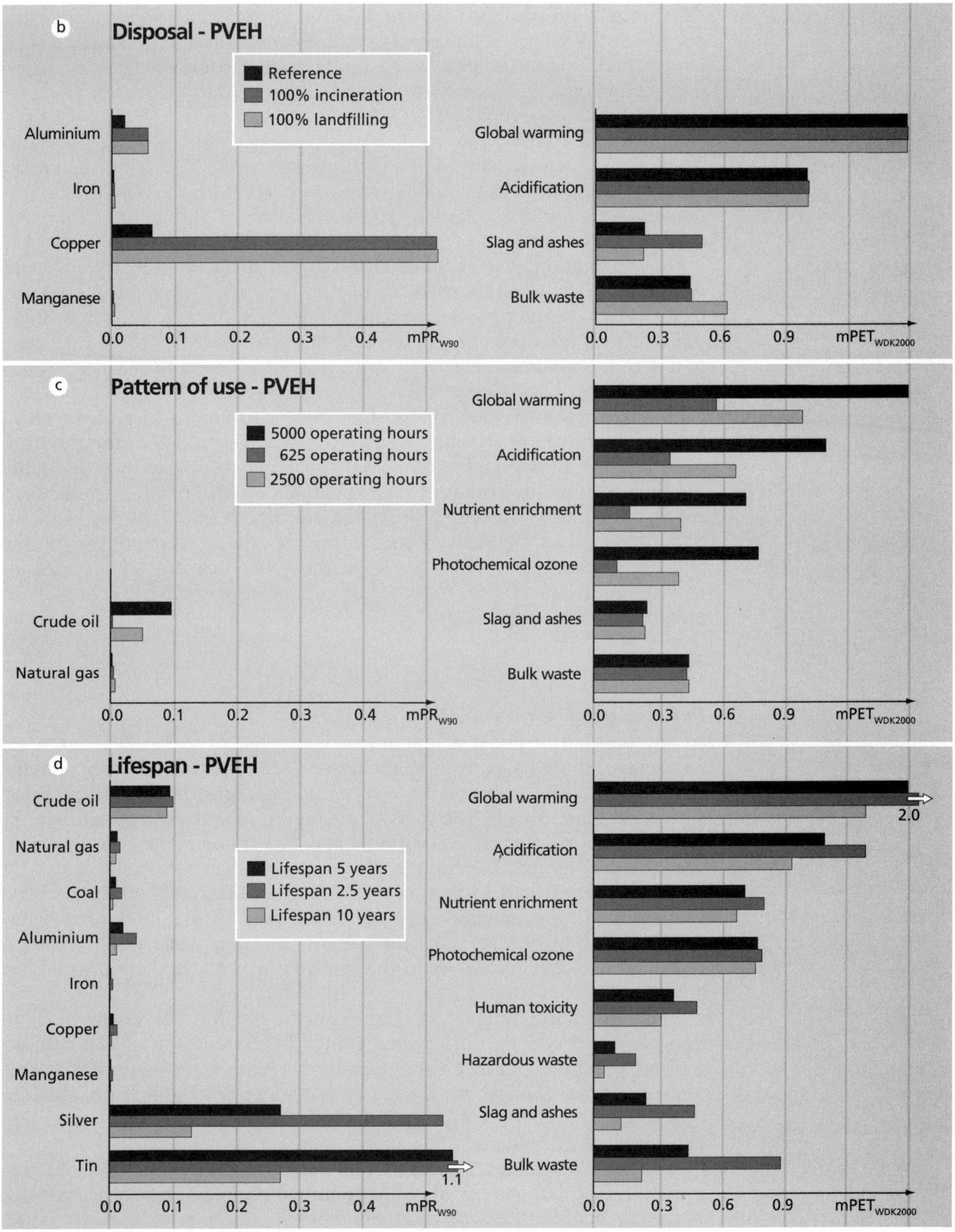

Figure 30.14. (continued)

as the use of lead and solvents is reduced. Impacts which can result in allergies are halved, as the use of formaldehyde is reduced. The remaining potentials for impacts on the working environment are reduced by 12-20% by the change.

The simulation shows that from an environmental perspective, the change to thick film technology is preferable to continued manufacturing with printed circuit boards based on fibre glass-reinforced epoxy laminate technology.

Disposal of a PVEH

Two scenarios were simulated for disposal of a PVEH: 100% incineration and 100% landfilling, which were compared with the assumption for the reference product, namely 100% recycling of aluminium, iron and copper and incineration of plastic.

In incineration, it is assumed that waste is incinerated in a plant with energy recovery which is used for district heating. This will replace heat from district heating plants, which are here assumed to use oil as fuel. The reduced quantity of oil is credited to the PVEH. On 100% incineration it is also assumed that 50% of the iron which ends in the slag is sent for recycling.

It is clear from Figure 30.14b that the contributions to global warming and acidification are almost equally large, irrespective of the disposal route used. For slag and ashes, 100% incineration results in a doubling of the contribution relative to the methods in the other two disposal routes, while 100% dumping in landfills results in an increase of 30% in bulk waste. The biggest difference for the three scenarios is in the resource consumption. It is clear from the figure that it is a good idea to recycle the metals in the product. For copper in particular, the reduction in consumption is significant.

Others

Patterns of use for a PVEH

As the number of hours in operation during use of a PVEH is taken as an average in the LCA, it is of interest to see how much the various patterns of use, i.e. variation in operating time, means for the product system. Figure 30.14c shows two examples of a reduced number of operating hours, namely 2,500 hours and 625 hours relative to the average of 5,000 hours. Assuming a lifespan of 5 years, the 5,000 hours of operation correspond to use of a PVEH when a double shift works all year or in 24 hour operation for part of the year. The 2,500 hours correspond to use in double shifts for part of the year, while the 625 hours represent use of a PVEH when the machine is used for 1 hour a day or a week.

The operating time is of significance only for the energy-related impacts, i.e. production of thermal energy by combustion of diesel oil in a diesel generator. The consumption of crude oil varies mainly with the operating time. Natural gas is used among other things for the extraction of crude oil, and the consumption of this thus varies in step with the consumption of crude oil.

As the changes correspond to the consumption of crude oil, the greatest contributions to the environmental impact categories global warming, acidification, nutrient enrichment and photochemical ozone formation derive from the combustion of diesel oil. The small changes in slag and ashes plus bulk waste come mainly from the consumption of electricity for the extraction of oil. Electricity consumption for the extraction of crude oil is much less than electricity consumption in the manufacturing stage. When a shorter operating time is simulated, less diesel oil is used, but as the simulation covers the entire product system, a lowered electricity consumption for extraction of crude oil is of little significance with respect to slag and ashes and bulk waste.

The life span of a PVEH

The actual life span of a PVEH depends, as noted, on the machine in which the apparatus is used. Figure 30.14b shows the impact potentials for life spans of 2½, 5 and 10 years. The number of hours in operation in the *use stage* is kept constant per annum for all three life spans. A life span of 10 years lies at the outer limit of the PVEH's potential. Under extreme conditions, a life span of 2½ years is realistic if, for example, the product is exposed to dramatic temperature fluctuations.

The silver in the resource profile of Figure 30.14d also represents nickel and tantalum. Tin represents lead and zinc. As the metals are generally only used in the material and manufacturing stages, their consumption is halved when the life span is doubled. The same also applies to much of the consumption of coal.

The consumption of crude oil and natural gas depends greatly on the use stage, and the consumptions of crude oil and natural gas do not, therefore, change much, even if the life span does.

The impact potentials for global warming are increased by 33% or reduced by 13% for life spans of 2.5 and 10 years respectively relative to a life span of 5 years. Acidification, nutrient enrichment, photochemical ozone formation and human toxicity follow the pattern for global warming, but are less pronounced. The other environmental impact potentials in Figure 30.14d rise by ca. 100% when the life span is doubled. The figure shows that an increased life span of a PVEH results in an environmental improvement, as fewer new PVEHs have to be made and there are therefore savings in energy and resources. This will, however, only be the case if the technology does not develop so rapidly that a PVEH in 5 years can be both made and used with a much lower environmental impact than today.

Summary of the simulations

The simulations show that the use of thick film technology in place of conventional printed circuit technology in the PVEH results in a reduction of 70% in the consumption of some of the scarce resources. The reduction for the other resources and for the environmental impact potentials is less than 23%. With an increased life span, consumption of the metals falls strongly, as is also the case for the environmental impact potentials for slag and ashes and for bulk waste. For resources and emissions from thermal energy, the impact potentials fall less markedly, but with a doubling of the life span they lie at or below 13%. With respect to the impact potentials from the energy-related resources and emissions, the influence of the life span is ambiguous, as there may be a technological development so that the PVEH's energy consumption in the use stage will be lower; in such a case it is not an advantage for the PVEH to have a greater life span.

If a ranking is to be made, it will be an advantage in the case of the metals both to replace conventional printed circuit technology with thick film technology and to increase the life span. For the energy-related resources, such replacement results in an improvement, as is also the case with an increased life span. For the environmental impact potentials, both the change in concept and the increase in life span are a gain, but this again depends on technological developments.

Where do potentials for improvement reside in the product?

The LCA and the interim environmental diagnosis from the simulation comprise some of the basis for designating the environmental points of focus in the product. The alternative given in Table 30.6 should be regarded as a proposal. It is not exhaustive and has not been weighed against other considerations. The changes which Danfoss has included in product development are reviewed in Section 30.4, "Environmental design of new products".

When the sites of potentials for improvement in the product are to be identified, choices of concept, structure and components are examined. Only the choices of concept and structure are described in the present case.

Table 30.6. Examples of points of focus at the concept level
Printed circuit boards: most of the consumption of scarce metals is attributable to the use of printed circuit boards in the product's electronics. A PVEH's life span depends to a large extent on how long the soldered joints between the printed circuit boards and the electronic components last, and on how long the components last: the more components and soldered joints on the printed circuit boards, the greater the chance of mechanical breakage. ***Alternative:*** replacement of the printed circuit boards by thick film, which does not require so many soldered joints, partly because some of the components are embedded in the film's circuitry. This increases the product's life span, as fewer soldered joints mean fewer possibilities of breakage between the thick film and the electronic components. Fewer components also mean fewer possibilities for breakage in the components themselves.

Table 30.7. Examples of points of focus at the level of structure
Embedding in PUR plastic: by embedding parts in PUR plastic, the product is made less suitable for dismantling, as the embedded parts are difficult to separate from the plastic. The embedding is done to protect components and soldered joints against moisture and vibrations. The product also gains greater mechanical stability. ***Alternative:*** no immediate alternative solutions are available.

Where can potentials for improvement be found in the surroundings?

The PVEH's surroundings are the systems into which it enters, both during and after use. The PVEH's surroundings carry some of the blame for the environmental problems in the PVEH's product system. The biggest problems posed by the surroundings are shown in Table 30.8.

Table 30.8. Examples of points of focus in the surroundings
Temperature fluctuations: the temperature fluctuations to which a PVEH is exposed derive from its surroundings. The machine in which the PVEH is used heats up during use, and if the machine is not operated 24 hours a day, machine and PVEH will cool down after use to the temperature of the surroundings. On renewed use, machine and PVEH will heat up again. ***Alternative:*** there are no immediate options for avoiding temperature fluctuations.
Recycling of electronic components: only a very small fraction of the electronic content in electronic products is currently recovered. When recovery is made it is usually of the noble metals plus copper and zinc; the rest are dumped. This is partly because shredding of the product does not separate the materials well enough. Only the large parts are separated, partly because it is not economic to extract the other materials. ***Alternative:*** removing the printed circuit boards from electronic products before the product is shredded will ensure better separation. It is also necessary to develop processes in which a greater fraction of the scarce resources can be recovered, and in which the loss of material grade for the individual resource is small.

30.3. Environmental specification for the product type

The competitive framework

The most important competitive parameters for the group of electro-hydraulic activation units from Danfoss are quality, performance, safety and product life. Safety, for example, is a very important parameter, as it can be a matter of life and death if safety fails.

The most important environmental priority based on competitive parameters derives from the customers' demand for increased life span for the PVEH. According to the simulations, an increased life span would result in an environmental improvement in the form of reduced consumption of scarce resources and reduced contributions to global warming, acidification and the other environmental impacts.

Relative to the competitors, Danfoss products are distinguished by integration of the electronics into the mechanical section, and by flexibility. As the electro-hydraulic activation unit is part of a larger hydraulic product range, there is considerable flexibility in constructing the hydraulic system according to the customer's wishes and requirements. With respect to integration of the electronics, the PVEH is distinguished from its competitors in that the electronics and the mechanical section are typically located in two separate parts in competing products. The PVEH is easy to install as the electronics are integrated into the mechanical section.

Long-term objectives

Work at Danfoss is proceeding on a number of long-term objectives. The following are of interest for this product type:

- Reduction in the amount of monotonous repetitive work.
- Reduction in energy consumption, also for overhead.
- Introduction of environmental management and environmental auditing.

Danfoss is in the process of qualifying under the EU's regulation on voluntary participation in environmental management and environmental auditing (EMAS). The first step is Danfoss's introduction of environmental management under British Standard BS 7750. Certification, which is for one Danfoss division at a time, has not yet reached the department which produces PVEHs. Under the plan, the entire company will be certified during the next 2-3 years.

Basic specifications for new products

Danfoss has prepared environmental specifications for the new generation of electro-hydraulic activation units, the PVE series 2. An extract is given in Table 30.9. The PVEH is one of the products in the PVE group, and series 2 is the generation after the reference product.

It is clear from the environmental specifications that Danfoss places weight on making the product easy to disassemble. This is done by labelling the individual parts and preparing guidelines for separating them. Disassembly versus non-disassembly is also assessed from an environ-

Environmental product specifications for the PVE series 2

Packaging

Suitable returnable packaging to be found

Choice of materials and design

When disposed of, the PVE is designed so that it can be dismantled in the following parts/materials:

- Hybrid (thick film)
- Magnetic valves in coils and valves
- Aluminium block in aluminium parts, pure plastic parts, O-rings and steel parts

The following alternative choices of materials and processes must be assessed environmentally, and environmental interests must be considered in the choice:

- Magnetic valves: sintergoods contra steel (rolled/processed)
- Paint: paint contra other kinds of surface treatment
- Casing: choice of plastic and embedment material

A total LCA regarding whether dismantling takes place or not, must be performed and the best result chosen

Prepare proposals regarding labelling of parts and instructions for disassembling

Prepare proposals for instructions regarding disposal

Table 30.9. Extract from environmental specifications for the PVE series 2

mental point of view. The choice of materials for the magnetic valves together with plastic types and embedment materials for enclosing must be considered. Surface treatment is also included in the considerations.

30.4. Environmental design of new products

Danfoss has been working on the PVEH's environmental characteristics on the basis of the life cycle concept since the EDIP project started in 1991. Some of the findings presented in the present case are entering into product development and have been included in forthcoming new products.

Product development at Danfoss

Product development at Danfoss proceeds as integrated product development. Purchasing, manufacturing and sales are thus involved at a very early stage in the development sequence. A development project requires close cooperation between various departments, and the project consists of a number of interdependent activities. The activities of all departments are linked together, in terms of time and economy, to a plan which is approved by the divisional manager. A person who will coordinate project activities in the area in question is designated within each of the areas of marketing, construction, manufacturing and quality control.

The tools used for environmentally correct product development include life cycle assessment, negative lists of prohibited or undesirable substances, and processes and checklists.

Examples of environmental considerations in new products

Some of the product development projects which have in-built environmental considerations in new generations of the product are presented in brief below. The projects have covered new solutions at both the concept and the component levels.

Project 1. Thick film

Replacement of conventional fibre glass-reinforced epoxy laminate technology by thick film technology. Apart from reducing the contributions to resource consumption and impacts on the environment and the working environment, as shown by simulation of the replacement, the average life of the new product is expected to be increased to 8 years. This will result in further reductions in resource consumption and environmental impact potentials. For example, the contributions to bulk waste will be reduced by ca. 40%, while the contributions to global warming will be reduced by ca. 10% and to ecotoxicity by ca. 45%.

	M	E	C	O
Concept		●	●	●
Structure				
Component				

Materials: The replacement and the improved life span result in reduced consumption of metals. Consumptions of lead, tin and silver fall by ca. 80%, of copper by 50%, and of aluminium, iron, manganese, nickel and zinc by 40%.

Energy: The fall in electricity consumption for manufacturing and for production of materials and the improved life span result in a fall of 40% in consumption of coal.

Others: The potential for damage to the nervous system falls by 97%, the potential for allergies is halved, and the rest of the potentials for impacts on the working environment are reduced by 12-20%.

Project 2. Tantalum capacitors

	M	E	C	O
Concept				
Structure				
Component	●			

Replacement of tantalum capacitors by ceramic capacitors. This reduces the resource consumption for tantalum from 15 mPR_{W90} to 0 mPR_{W90}, meaning that a significant resource consumption is avoided.

Materials: Units containing tantalum are eliminated. Consumption of tantalum is thereby avoided.

Project 3. Yokes

	M	E	C	O
Concept				
Structure				
Component	●			

Change in design of the yokes (component assembly no. 8) so that waste of steel from the punching process is significantly reduced. The waste of steel from the process was previously greater than the quantity of steel which entered into the finished yoke.

Materials: Reduction of ca. 30% in consumption of iron.

References

Ahbe, S., Braunschweig, A. and Müller-Venk, R. (1990) Methodik für Ökobilanzen auf der Basis ökologischer Optimierung. Bundesamt für Umwelt, Walt und Landschaft (BUWAL), *Schriftenreihe Umwelt* No. 133, Bern.

Albritton, D.L., Derwent, R.G., Isaksen, I.S.A., Lal, M. and Wuebbles, D.J. (1995a) Trace gas radiative forcing indices, in: *Climate change 1994, Radiative forcing of climate change and an evaluation of the IPCC IS92 emission scenarios* (eds J.T. Houghton, L.G. Meira Filho, J. Bruce, H. Lee, B.A. Callander, E. Haites, N. Harris and K. Maskell), Cambridge University Press, UK.

Albritton, D.L., Watson, R.T. and Aucamp, P.J. (Assessment Co-chairs) (1995b) Scientific Assessment of Ozone Depletion, 1994, *World Meteorological Organization, Global Ozone Research and Monitoring Project*, Report No. 37. World Meteorological Organization, Geneva.

Alting, L., Jørgensen, J. and Wenzel, H. (1996) *The environmental dimension in the product.* Danish Environmental Protection Agency and the Confederation of Danish Industries, Copenhagen. In Danish.

Andersson-Sköld, Y., Grennfelt, P. and Pleijel, K. (1992) Photochemical ozone creation potentials: A study of different concepts. *J. Air Waste Manage. Assoc.* **42**(9), 1152-1158.

AQUIRE (1992) *Aquatic Information Retrieval (AQUIRE) Toxicity Database.* Spectrum Research, Inc., Duluth, Minnesota, USA.

Association of Danish Electric Utilities (1992) *Danish electricity supply statistics 1991.* Rosenørns Alle 9, DK-1970 Frederiksberg C, Denmark. In Danish.

Association of Danish Electric Utilities (1995) *Danish electricity supply statistics 1994.* Rosenørns Alle 9, DK-1970 Frederiksberg C, Denmark. In Danish.

Blum, D.J.W. and Speece, E.E. (1991) A database of chemical toxicity to environmental bacteria and its use in interspecies comparisons and correlations. *J. Water Pollution Control Federation*, **63**, 198-207.

Boustead, I. (1993) Polyethylene and Polypropylene, *Eco-profiles of the European plastics industry*, Report 3, Association of Plastics Manufacturers in Europe (APME), Brussels.

British Petroleum Company (1992) *BP Statistical review of world energy*, British Petroleum Company plc, London.

British Geological Survey (1991) *World Mineral Statistics 1985-1989.* British Geological Survey, Derry & Sons Ltd., UK.

Bro-Rasmussen F., Boyd, H.B., Jørgensen, C.F., Knudsen, P. Lauridsen, E., Løkke, H. and Nielsen, K.M. (1996) *Non-assessed chemicals in EU*, The Danish Board of Technology, Copenhagen.

Bundesamt für Umweltschutz (BUS) (1984) *Ökobilanzen von Packstoffen*. Schriftenreihe Umweltschultz no. 24, Bern.

Caspersen, N. (1995) Personal communication. Institute for Product Development, Technical University of Danmark, DK-2800 Lyngby, Denmark.

Christiansen, K., Heijungs, R., Rydberg, T., Ryding, S.-O., Sund, L., Wijnen, H., Vold, M. and Hanssen, O.J. (1995) *Application of Life Cycle Assessments (LCA)*. Nordic Project on Environmentally Sound Product Development, Report 07/95, Oestfold Research Foundation, Norge.

Commission for the European Communities (1992) The Environmental Situation in the EC. **KOM (92)23** final edn, vol. III. 30 June 1992, Brussels, pp 36-39.

Consoli, F., Allen, D., Boustead, I., Fava, J., Franklin, W., Jensen, A.A., de Oude, N., Parrish, R., Perriman, R., Postlethwaite, D., Quay, B., Séguin, J. and Vigon, B. (1993) *Guidelines for Life-Cycle Assessment: A 'Code of Practice*', from the SETAC Workshop held in Sesimbra, Portugal 31 March-3 April, 1993, Society of Environmental Toxicology and Chemistry (SETAC), Brussels.

Damborg, A. and Thygesen, N.(1991) Surfactants - distribution and impacts on the environment. *Environmental project* no. 166, Danish Environmental Protection Agency, Copenhagen. In Danish.

Danish Energy Agency (1994) Proclamation on labelling and duty to provide information concerning electric refrigerators for household use. *Proclamation of 12 December 1994*, no. 1014, p 5816.

Danish Environmental Protection Agency (1994) *Note on quantities of waste in Denmark in 1993 compared with targets in the government's action plan for waste and recycling 1993-1997*. Danish Environmental Protection Agency, Office for Industrial Waste, **M356-0002**, 1994, Copenhagen. In Danish.

Danish Environmental Protection Agency (1995) *Overview of oxygen depletion in Danish coastal waters in 1995 to ca. mid-August, from mid-August to the end of September, and from the end of September to mid-October* (three separate summaries), Danish Environmental Protection Agency, Office for Waste Water and the Sea, Copenhagen. In Danish.

Danish Environmental Protection Agency (1997) *The EDIP PC Tool*, β-version, Copenhagen, (expected release June 1997).

Danish Ministry of Energy (1990) *Energy 2000. Action Plan for a sustainable Development, April 1990.*

Danish Ministry of Environment and Energy (1995) *Esbjerg Declaration*, 4th International Conference on the Protection of the North Sea, Esbjerg, Denmark, 8-9 June 1995.

Danish Ministry of Environment and Energy (1993) *Cleaner technology action plan 1993-1997*, Copenhagen. In Danish.

Danish Standards Association (1994) Specifications for Environmental Management Systems. *DS information DS/INF* **75**, 2. edition, Copenhagen, translation of BS7750.

Danish Working Environment Service (1990) Substances injurious to the nervous system in the working environment - a summary. *AT report* no. 13/1990, Copenhagen. In Danish.

Danish Working Environment Service (1994) *Circular* 5. In Danish.

Danish Working Environment Service (1994) Limit values for substances and materials. *Guideline* No. 3.1.0.2, Copenhagen.

Derwent, R.G. and Jenkin, M.E.(1990) Hydrocarbon involvement in photochemical ozone formation in Europe. **AERE R 13736**, AEA Environment and Energy, Harwell Laboratory, Oxfordshire OX11 0RA, UK.

EEC (1990) Commission communication pursuant to Article 13 of Council Directive 67/548/EEC of 27 June 1967 on the approximation of laws, regulations and administrative provisions relating to the classification, packaging and labelling of dangerous substances as amended by Directive 79/831/EEC - EINECS (European inventory of existing commercial chemical substances), Volumes I-II. *Annex to Official Journal of the European Communities, C146A,* vol. **33**, 15 June 1990, Brussels.

EEC (1993) Eco-Management and Audit Scheme (EMAS) *Official Journal of the European Communities,* **L168/1-18**, 10 July 1993, Brussels.

EEC (1994) *21. Announcement of Annex 1 to Council Directive 67/548/EEC on Classification, Packaging and Labelling of Dangerous substances*, December 1994, Brussels.

EEC (1994) Permission Regulation No. 1179/94 of 25 May 1994 concerning the first list of priority substances as forseen under Council Regulation (EEC) No. 793/93.

European Chemicals Bureau (1996) IUCLID, *International Uniform Chemical Information Database, Existing Chemicals*. ECB, Ispra, Italy.

European Commission (1994) *Technical guidance documents in support of: The Commission Directive 93/67/EEC on risk assessment for new notified substances and The Commission Regulation (EC) No. 1488/94 on risk assessment for existing substances.* Pre-print version, European Chemicals Bureau, Ispra, Italy.

European Commission (1996) *Technical guidance documents in support of: The Commission Directive 93/67/EEC on risk assessment for new notified substances and The Commission Regulation (EC) No. 1488/94 on risk assessment for existing substances*. Temporary version, European Chemicals Bureau, Ispra, Italy.

European Parliament and Council (1994) *Directive 94/62/EEC of 20 December 1994 on packaging and packaging waste.*

Fenger, J., Fenhann, J. and Kilde, N. (1990) Danish Budget for Greenhouse gases. *Nord,* **1990:97**, Nordic Council of Ministers, Copenhagen.

Fenhann, J. and Kilde, N. (1994) Inventory of emissions to the air from Danish sources 1972-1992. Risø National Laboratory, DK-3000 Roskilde, Denmark.

Fitzgerald, F. (1992) Energy use and management in British Steel plc. *Ironmaking and Steelmaking*, **19**(2), 98-106.

Frandsen, G. and Hansen, M. (1995) *Life cycle analysis - life cycle inventory for glucose*, Graduate thesis, Institute for Applied Chemistry, Technical University of Denmark, DK-2800 Lyngby, Denmark. In Danish.

Franke, M. (1984) *Umweltauswirkungen durch Getränkeverpackungen - Systematik zur Ermittlung der Umweltauswirkungen von komplexen Prozessen am Beispiel von Einweg- und Mehrweg-Getränkebehältern*, Technische Universität Berlin/Institut für Technischen Umweltschutz, Berlin.

Frees, N. (1996) *The EDIP* (*E*nvironmental *D*esign of *I*ndustrial *P*roducts) *unit process database*, Danish Environmental Protection Agency, Copenhagen. In Danish.

Goeller, H. and Zucker, A. (1984) Infinite Resources: The Ultimate Strategy. *Science*, vol. 223, 456-462.

Govind, R., Lei, L. and Dobbs, R. (1991) Integrated model for predicting the state of organics in waste water treatment plants. *Environmental Progress* **10,** 13-23.

Grau, P. (1995) Personal communication. Danish Environmental Protection Agency, Copenhagen.

Guldbrandsen, T. and Nørgård, J.S. (1986) Achieving Substantially Reduced Energy Consumption in European Type Refrigerators. Paper presented at the 37th Annual International Appliance Technical Conference, Purdue University, USA, May 6-7, 1986.

Guldbrandsen, T., Heebøll, J., Mehlsen, K. and Nørgård, J. (1986) Energy. Development of energy efficient electrical household appliances. Part one: refrigerators. *Report EUR* **10449 EN**, Commission of the European Communities, Brussels.

Gydesen, A., Maimann, D., Pedersen, P.B., Hansen, M.K., Bruhn, B. and Bidstrup, C. (1990) Cleaner technology in the area of energy. *Environmental project* no. 138, Danish Environmental Protection Agency, Copenhagen. In Danish.

Habersatter, K. (1991) Oekobilanz von Packstoffen, Stand 1990. *Schriftenreihe Umwelt* No. 132, Bundesamt für Wald und Landschaft (BUWAL), Bern.

Hansen, G. and Nielsen, A. (1992) *Statistics on waste for 1991, monitoring of material flow*, Project no. 205-31-2, 1992, Rendan, DK-2860 Søborg, Denmark.

Hass, U. Jakobsen, B.M., Brandorff, N.P., Kelnes, J.E. and Petersen, S.H. (1991) Chemicals in the working environment which damage the reproductive system. Report no. 35/1991, National Institute of Occupational Health, Copenhagen. In Danish.

Hauschild, M.Z. (1994) *Characterization of product specific toxicity in life-cycle assessment*, in: Integrating Impact Assessment into LCA. (eds H.A. Udo de Haes, A.A. Jensen, W. Klöpffer and L.-G. Lindfors), Proceedings of the LCA symposium held at the Fourth SETAC-Europe Congress, 11-14 April 1994, Free University, Brussels, Belgium, SETAC-Europe, Brussels.

Hauschild, M.Z. and Wenzel, H. (1997a) Photochemical ozone formation as assessment criterion in the environmental assessment of products, in: *Scientific Background for Environmental Assessment of Products*, 1997 (eds M. Hauschild and H. Wenzel), Chapman & Hall, London.

Hauschild, M.Z. and Wenzel, H. (1997b) Acidification as assessment criterion in the environmental assessment of products, in: *Scientific Background for Environmental Assessment of Products* (eds M. Hauschild and H. Wenzel), Chapman & Hall, London.

Hauschild, M.Z. and Wenzel, H. (1997c) Nutrient enrichment as assessment criterion in the environmental assessment of products, in: *Scientific Background for Environmental Assessment of Products* (eds M. Hauschild and H. Wenzel), Chapman & Hall, London.

Hauschild, M.Z. and Wenzel, H. (1997d) Global warming as assessment criterion in the environmental assessment of products, in: *Scientific Background for Environmental Assessment of Products* (eds M. Hauschild and H. Wenzel), Chapman & Hall, London.

Hauschild, M.Z. and Wenzel, H. (1997e) Stratospheric ozone depletion as assessment criterion in the environmental assessment of products, in: *Scientific Background for Environmental Assessment of Products* (eds M. Hauschild and H. Wenzel), Chapman & Hall, London.

Hauschild, M.Z. and Wenzel, H. (1997f) Normalization and weighting in the environmental assessment of products, in: *Scientific Background for Environmental Assessment of Products* (eds M. Hauschild and H. Wenzel), Chapman & Hall, London.

Hauschild, M. and Wenzel H. (eds) (1997g) *Scientific Background for Environmental Assessment of Products*, Chapman & Hall, London.

Hauschild, M., Wenzel, H., Damborg, A. and Tørsløv, J. (1993) Methods for screening and quantitative assessment of ecotoxicity in life cycle analysis of products. SETAC/SECOTOX workshop on Ecotoxicological assessments in LCA, 7-8 January 1993. Institute for Product Development, Technical University of Denmark, DK-2800 Lyngby, Denmark.

Hauschild, M.Z., Wenzel, H., Damborg, A. and Tørsløv, J. (1997a) Ecotoxicity as assessment criterion in the environmental assessment of products, in: *Scientific Background for Environmental Assessment of Products* (eds M. Hauschild and H. Wenzel), Chapman & Hall, London.

Hauschild, M.Z., Olsen, S.I. and Wenzel, H. (1997b) Toxicity to humans in the environment as assessment criterion in the environmental assessment of products, in: *Scientific Background for Environmental Assessment of Products* (eds M. Hauschild and H. Wenzel), Chapman & Hall, London.

Heijungs, R. (final edn) (1992) *Environmental life cycle assessment of products, Volume 1: Guide, Volume 2: Backgrounds*. Center for Environmental Science (CML) University of Leiden, The Netherlands.

Hemming, C. (ed.) (1995): Directory of Life Cycle Inventory Data Sources. Society for the Promotion of LCA Development, Brussels.

Hoffman, L., Weidema, B.P., Christiansen, K. and Ersbøll, A.K. (1995) Statistical analysis and uncertainties in relation to LCA. Special Reports No. 1, in: LCA-Nordic Technical Reports No 10 and Special Reports No. *1-2*. (eds L.-G. Lindfors, K. Christiansen, L. Hoffmann, Y. Virtanen, V. Juntilla, A. Leskinen, O.-J. Hansen, A. Rønning and T. Ekvall). *TemaNord,* **1995:503**. Nordic Council of Ministers, Copenhagen.

Houghton, J.T., Meira Filho, L.G., Bruce, J., Lee, H., Callander, B.A., Haites, E., Harris, N. and Maskell, K. (eds) (1995) *Climate change 1994, Radiative forcing of climate change and an evaluation of the IPCC IS92 emission scenarios*, Cambridge University Press, UK.

Houghton, J.T., Meira Filho, L.G., Callander, B.A., Harris, N. Kattenberg, A. and Maskell, K. (eds) (1996) *Climate change 1995, The science of climate change*, Cambridge University Press, UK.

Howard, P.H. (1990) *Handbook of Environmental Fate and Exposure Data for organic chemicals*. Lewis Publishers Inc., USA.

Howard, P.H., Boethling, R.S., Jarvis, W.F., Meylan, W.M. and Michalenko, E.M. (1991) *Handbook of Environmental Degradation Rates*. Lewis Publishers Inc., USA.

HSDB (1996) *Hazardous Substance Data Bank*. Published by National Library of Medicine, USA, available on CD-ROM, e.g. from Silver

Platter Information Inc., Norwood MA, USA.
Hunt, R.G., Franklin, W.E., Welch, R.O., Cross, J.A. and A.E. Woodall (1974) Resource and environmental profile analysis of nine beverage container alternatives, **EPA/530/SW-91c,** United States Environmental Protection Agency (US EPA), Office of Solid Waste Management Programs, Washington DC, USA.

IPCC (1995) *Technical summary forming part of the contribution of working group I to the IPCC Second Assessment Report (submitted by the Chairman of the IPCC for acceptance by the Panel). Eleventh session, Rome 11- 15 December 1995*, Intergovernmental Panel on Climate Change, Geneva.

IRIS (1995) *Integrated Risk Information System*, prepared by the US EPA, available on CD-ROM, e.g. from Silver Platter Information Inc., Norwood MA, USA.

ISWA (1991) *Energy from waste state-of-the-art*, Working Group on Waste Incineration, ISWA. Distributed by Swedish Association of Solid Waste Management, Malmö, Sweden.

Jakobsen, A. (1995a) *Energy Optimisation of Refrigeration Systems. The Domestic Refrigerator - a Case Study*. Ph D Thesis, report F-179-1. Technical University of Denmark, DK-2800 Lyngby, Denmark.

Jakobsen, A.(1995b) Personal communication, Technical University of Denmark, DK-2800 Lyngby, Denmark.

Jutland-Funen Power Pool/ Elkraft Power Company Ltd. Copenhagen (1994) *Inventory of waste resources in Denmark*, Management Committee 4 for Biomass, SK Power Company, Denmark, Project Division, Lautruphøj 5, DK-2750 Ballerup, May 1994. In Danish.

Kilde, N.,A.(1995) *Energy consumption and emissions in the transport of goods in 1990*. Risø National Laboratory, **Risø-I-906** (DA), DK-3000 Roskilde, Denmark.

Kindler, H. and Nickles, A. (1980) Energieaufwand zur Herstellung von Werkstoffen. *Kunstoffe*, **70**. Carl Hauser Verlag, Munich, Germany.

Lindfors, L.-G., Christiansen, K., Hoffman, L., Virtanen, Y., Juntilla, V., Hanssen, O.-J., Rønning, A., Ekvall, T. and Finnveden, G. (1995) Nordic Guidelines on Life-Cycle Assessment. *Nord,* **1995:20**, Nordic Council of Ministers, Copenhagen.

Lundholm, M.P. and G. Sundström (1985) *Resource and environmental impact of Tetra Brik carton and refillable and non-refillable glass bottles. Tetra Brik Aseptic environmental profile*, AB Tetra Pak, Malmö, Sweden.

Ministry of Housing, Spatial Planning and the Environment (1994): Policy Document on Products and the Environment. **VROM 94196/h/5-94**, The Hague, The Netherlands.

National Consumer Agency of Denmark (1993) The Journal. *Advice and Results*, no. 10, p 7-11, National Consumer Agency of Denmark.

Nedermark, R. (1995) Personal communication, Bang & Olufsen A/S, Struer, Denmark.

Nikunen, E., Leinonen, R. and Kultamaa, A. (1979) Environmental properties of chemicals. Research report no. 91, Finnish Ministry of the Environment, Environmental Protection Department, VAPK-Publishing, Helsinki.

Nørgård, J. S. (1979) Households and energy. Report no. 4 of the DEMO project. Physics Lab. III, Technical University of Denmark, DK-2800 Lyngby, Denmark.

OECD 301 (1992a) *OECD guideline for testing of chemicals* 301: Ready biodegradability. Organization for Economic Cooperation and Development, Paris.

OECD 302 (1992b) *OECD guideline for testing of chemicals* 302: Inherent biodegradability. Organization for Economic Cooperation and Development, Paris.

Olesen, J., Wenzel, H., Hein, L. and Andreasen, M.M. (1996) *Design for Environment.* Danish Environmental Protection Agency and confederation of Danish Industries, Copenhagen. In Danish.

Pedersen, J. (1995) Personal communication, Danish Energy Agency, Copenhagen.

Pedersen, P.H. and Lawætz, H. (1989) The LER refrigerator. Development, manufacture and testing of the low energy refrigerator LER200. Physics Lab. III, Technical University of Denmark, DK-2800 Lyngby. In Danish.

Pedersen, B. and Christiansen, K (1992) A meta-review on product life cycle assessment, in *Product Life Cycle Assessment - Principles and Methodology* (eds L. G. Lindfors). *Nord,* **1992:9**, Nordic Council of Ministers, Copenhagen.

Pedersen, F., Kristensen, P., Damborg, A. and Wenzel, H. (1992) Ecotoxicological Evaluation of Industrial Wastewater. *Environmental project* no. 254, Danish Environmental Protection Agency, Copenhagen.

Pedersen, F., Leinonen, R., Pettersson, I., Norrthon, S., Källquist, T., Hansen, H., Gjøs, N., Hægh, G.S., Lander, L. Petersen, G., Dahlbo, K., Kemiläinen, B. and Wenell, T. (1994) Environmental Hazard Classification - classification of selected substances as dangerous for the environment (I). *TemaNord,* **1994:643**, Nordic Council of Ministers, Copenhagen.

Rasmussen, E. (1997) The working environment as assessment criterion in the environmental assessment of products, in: *Scientific Background for Environmental Assessment of Products* (eds M. Hauschild and H. Wenzel), Chapman & Hall, London.

Ryding, S.-O. (1995) *Environmental oriented product development,*

Förlags AB Industrial literature, Stockholm. In Swedish.
Rytter, K. (1994) EDIP's (Environmental Design of Industrial Products) working environment method. T report no. 44. KEW Industri, BST Hadsund, DK-9560 Hadsund, Denmark. In Danish.

Schimel, D., Enting, I.G., Heimann, M., Wigley, T.M.L., Raynaud, D., Alves, D. and Siegenthaler, U. (1995) CO_2 and the Carbon Cycle, *in Climate change 1994, Radiative forcing of climate change and an evaluation of the IPCC IS92 emission scenarios* (eds J.T. Houghton, L.G. Meira Filho, J. Bruce, H. Lee, B.A. Callander, E. Haites, N. Harris and K. Maskell), Cambridge University Press, UK.
Skaarup, J. (1995) Personal communication, Danish Environmental Protection Agency, Copenhagen.

Simonsen, L., Midtgård, U., Lund, S.P. and Hass, U. (1995) Occupatonal neurotoxicity. Evaluation of neurotoxicity data for selected chemicals. Nordic Council of Ministers and National Institute of Occupational Health, Denmark, Copenhagen.

Solomon, S. and Albritton, D.L. (1992) Time-dependent ozone depletion potentials for short- and long-term forecasts. *Nature*, **357**, 33-37.

Solomon, S. and Wuebbles, D.J. (1995) Ozone depletion potentials, global warming potentials and future chlorine/bromine loading, *in Scientific Assessment of Ozone Depletion: 1994* (Assessment Co-chairs D.L. Albritton, R.T. Watson and P.J. Aucamp), World Meteorological Organization, *Global Ozone Research and Monitoring Project*, Report No. 37, World Meteorological Organization, Geneva.

Spangenberg, J.H. (ed.) (1995) *Towards sustainable Europe*. A study for the Friends of the Earth Europe, Luton, UK.

SRC(1994) *PCKOC*, Soil Adsorption Coefficient (K_{oc}) Program. Syracuse Research Corporation, New York, USA.

SRC (1995) *HENRY*, Henry's Law Constant Program. Syracuse Research Corporation, New York, USA.

SRC (1996) *AOPWIN*, Atmospheric Oxidation Program, Syracuse Research Corporation, New York, USA.

Statistics Denmark (1994) Oil and chemical waste delivered to the municipal hazardous waste treatment centre. *Statistical News, Environment* **1994:2**, Copenhagen. In Danish.

Statistics Denmark (1995) State of the ozone layer 1995. *Statistical News, Environment,* **1995:13**, Copenhagen. In Danish.

Sustainability (1993) *The LCA Sourcebook. A European business guide to life cycle assessment*, Sustainability Ltd., London.

Swedish Nuclear Fuel and Waste Management Company (1994), Information department, Stockholm.

Teoh, L. (1989) Electric arc furnace technology: Recent developments and future trends, *Ironmaking and Steelmaking*, **16**(5), 303-313.

Thomsen, K.G. (1991) Allergens and substances causing hypersensitivity in the working environment, AMI report no. 33/1990, National Institute of Occupational Health, Copenhagen. In Danish.

Thorsen, M. (1993) Collection of refrigerators and freezers, *Environmental project* **no. 217**, Danish Environmental Protection Agency, Energy, Copenhagen, In Danish.

Toppari, J., Larsen, J.C., Christiansen, P., Giwercman, A., Grandjean, P., Guilette, L.J., Jégou, B., Jensen, T.K., Jouannet, P., Keiding, N., Leffers, H., McLachlan, J.A., Meyer, O., Müller, J., Rajpert-De Meyts, E., Scheike, T., Sharpe, R., Sumpter, J. and Skakkebæk, N.E. (1995) Male Reproductive Health and Environmental Chemicals with Estrogenic Effects. *Environmental project* no. 290, Danish Environmental Protection Agency, Copenhagen.

US Bureau of Mines (1990) *Minerals Yearbook*. US Department of the Interior, Bureau of Mines, US Government Printing Office, Washington DC.

Udo de Haes, H.A. (ed) (1996) *Towards a methodology for life cycle impact assessment*. Society of Environmental Toxicology and Chemistry - Europe, Brussels, Belgium.

Ullman Encyclopedia of Industrial Chemistry (1993), 5th edn, volume A22, VCH Verlagsgesellschaft, Weinheim, Germany.

UNECE (1991) *Protocol to the 1979 convention on long-range transboundary air pollution concerning the control of emissions of volatile organic compounds or their transboundary fluxes*, **ECE/EB.AIR/30**, United Nations Economic Commission for Europe, Geneva.

UNECE (1994) *1994 Major review on strategies and policies for air pollution abatement, tables and figures*, **EB.AIR/R.87/Add.1, 21. September 1994**, United Nations Economic Commission for Europe, Geneva.

UNEP (1993) *Handbook for the Montreal Protocol on substances that deplete the ozone layer*, 3rd edn, Ozone Secretariate, United Nations Environmental Programme, August 1993.

United Nations (1992) *Long term range population projecting: Two centuries of population growth 1950-2150*, New York, ISBN 92-1-151241-9.

United Nations (1993) *The Global Partnership for Environment and Development, A Guide to Agenda 21, Post Rio Edition*, United Nations, New York.

US EPA (1992) *Product Life-Cycle Assessment: Inventory Guidelines and Principles*, **EPA/600/R-92/036**, United States Environmental Protection Agency, Cincinnati, Ohio.

Verschueren, K. (1996) *Handbook of environmental Data on Organic Chemicals*. 3rd ed, Van Nostrand Reinhold, New York.

Wenzel, H. (1997) Background for the EDIP method's allocation model, in: *Scientific background for environmental assessment of products* (ed. Hauschild, M. and Wenzel, H.), Chapman and Hall, London.

Wenzel, H. and Hauschild, M. (1993) Key issues for toxicity assessment of products, SETAC/SECOTOX workshop on Ecotoxicological assessments in LCA, Lyngby, 7-8 January 1993, Institute for Product Development, Technical University of Denmark, DK-2800 Lyngby, Denmark.

Wenzel, H., Damborg, A. and Jacobsen, B.N. (1990) Danish emissions of industrial waste water, *Environmental project* no. 153, Danish Environmental Protection Agency, Copenhagen. In Danish.

Westphal, H., Nielsen, I.R. and Hvims, H.L. (1995) *Kamille - can the environment be made cleaner by the use of conducting glues in the manufacture of electronics?* Report **SPM-127**, DELTA Danish Electronics, Light & Acoustics, DK-2800 Lyngby, Denmark.

World Commission on Environment and Development (1987) *Our Common Future*, Oxford University Press, United Kingdom.

World Meteorological Organization (1991) The global climate system in 1990. *WMO Bulletin,* **40**, 217-221.

World Resources Institute (1990) *World Resources 1990-91. A report by the World Resources Institute in collaboration with The United Nations Environmental Programme and The United Nations Development Programme*, Oxford University Press, UK.

World Resources Institute (1992) *World Resources 1992-93. A report by the World Resources Institute in collaboration with The United Nations Environmental Programme and The United Nations Development Programme*, Oxford University Press, UK.

Glossary

Allocation: finding a total and uniform measure for all the services to which the joint processes contribute, and on the basis of this partition the joint exchanges to each of the services to which they contribute

Background load: the contribution to a resource consumption or impact category from the reference area, expressed per person in the area. Used for normalization

Bioconcentration: tendency of a substance to accumulate in living organisms

Biodegradability: degradability by microorganisms in the environment. Assessed from standardized tests

CAS number: Chemical Abstracts Service Registry Number is a numeric designation assigned by the American Chemical Society's Chemical Abstract Service uniquely identifying a specific chemical compound regardless of name or naming system

Co-products: two or more products leading to a service for a user and deriving from or entering into a range of processes

Data format: a template for the three categories of information: a description of the process, an inventory of the process's exchanges with the environment, and a characterization of the information

Elements: defined tasks or "steps" in the LCA method. Each of the method's phases consists of one or more elements

Environmental latitude: the environmental impact acceptable on average for each person in a sustainable society

Environmental diagnosis: designating environmental points of focus in the product: designating which impact potentials are problematic, identifying improvement potentials and localizing where in the product they reside

Environmental exchanges: an exchange with the environment is defined as an input to a process, an output from the process or an internal interaction with an operator of the process

Environmental impacts: impacts on the external environment including human health

Environmental specification: retainment of the weighting of environmental considerations in the product's total commercial basis and determination of the specific requirements for the product's properties imposed by environmental considerations

Environmental assessment of a product: to define and quantify the service provided by the product, to identify and to quantify the environmental exchanges caused by the way in which the service is provided, and to ascribe these exchanges and their potential impacts to the service

Exposure: the degree to which a substance reaches parts of the environment where the impact can be exerted, e.g. the degree to which a toxic substance reaches an organism

Functional unit: the defined and quantified service provided by the product. The functional unit is the object of the environmental assessment

Goal definition: clarifies what the LCA shall and shall not be used for, including the decisions which it must support and the environmental consequences to which these decisions can lead. A phase of the LCA method

Inventory: collection of environmentally relevant information from processes identified during scope definition and included in the product system. A phase of the LCA method

Joint environmental exchanges: environmental exchanges from a joint process which can be ascribed to more than one service

Joint processes: processes which contribute to more than one service for a final user

Key figures: the environmentally most important data of the LCA, including both data on environmental exchanges from the processes and on the composition of the product system, i.e. data for the product life time, use patterns, disposal routes and other product system parameters

Life cycle assessment: see environmental assessment of products

Manufacturing: the processing of materials into the final product. Designates a stage in the product's life

Material grade: the utility value residing in the material expressed as a fraction of the initial utility value of the material. The material grade thus lies between 0 and 1

MECO principle: a method to structure and systematize the work of the LCA. M, E, C and O designate Materials, Energy, Chemicals and Others

Non-renewable resources: resources which cannot be regenerated or regenerated only within a time horizon beyond current interest, e.g. oil, coal, metals

Normalization: Relative scaling of the product's impact potentials and resource consumptions through comparison of each of them with a reference, which is the background contribution to the same impact potentials and resource consumption. In the EDIP method the average contribution from one person in 1990 is used as reference

NO_x: Nitrogen oxides, i.e. NO and NO_2

Parts list: (~ list of components) a hierarchic list of the product's composition from top level, which is the product itself, to bottom level, which is materials and components purchased externally

PCBs: Polychlorinated biphenyls - a group of highly persistent organic compounds

Phases: here used to subdivide the LCA-method into four: the goal definition phase, the scope definition phase, the inventory phase and the impact assessment phase

Product service: the service provided by the product; a definition of a product is that it is of benefit to the user: it provides a service

Product system: all processes into which the product enters are collectively termed its product system

Projection: assessing the future magnitude of key figures for the LCA

Reference product: the product selected to represent provision of the service

Renewable resources: resources which can be regenerated, e.g. wood, water, grain

Resource consumption: consumption of both renewable and non-renewable resources

Resources: the primary raw materials from which the materials, the ancillary substances and the energy in the product system derive

Scope definition: designates a phase of the LCA method and comprises the following elements: definition of the object of the study including specification of the functional unit, identification of the assessment criteria, definition of the scope of the product system, definition of the time boundaries of the study, definition of the technological level of the processes in the product system, and definition of allocation procedures

Scope definition of the product system: identifying and including the most significant processes from an environmental perspective, thus excluding the insignificant ones

Screening: an introductory rough environmental assessment used to plan the data collection for the actual environmental assessment. See also scope definition of the product system

Sensitivity analysis: idenfication of the key figures of the LCA and analysis of the overall variability of results and conclusions as a consequence of uncertainties in these figures

Service: see product service

Sustainable development: development that meets the needs of the present generation without compromising the ability of future generations to meet their needs with focus on material resources, biological diversity and the health of the environment

Technological scope: the technological level which must be met by the processes which comprise the product system

Uncertainty analysis: assessment of the uncertainty of the total results and conclusions due to uncertainties in the data used (in the scope definition, the inventory and the impact assessment)

Unit process: data for a process expressed per unit of process in a generalized form; for materials a unit will typically be weight, for a painting process typically surface area

Working environment, impacts on: negative influence on safety and health to which the operator is exposed during work. All impacts on the working environment are expressed as duration of exposure in hours or millihours except for accidents, which are expressed in numbers

Use stage: a stage of the product life comprising all processes related to use and maintenance of the product

Utility value: a measure of the total potential for use of the material, substance or component in question

Weighting: a quantitative comparison of the seriousness of the different resource consumptions or impact potentials of the product

Variables and parameters

B_b	Meat transfer factor
BCF	Bioconcentration factor
BIO	Biodegradability factor
B_m	Milk transfer factor
EF(ac)	Equivalency factor for acidification potential
EF(et)	Equivalency factor for potential for ecotoxicity
EF(etp)	Equivalency factor for potential for ecotoxicity in waste water treatment plant
EF(etsc)	Equivalency factor for potential for chronic ecotoxicity in soil
EF(etwa)	Equivalency factor for potential for acute ecotoxicity in water
EF(etwc)	Equivalency factor for potential for chronic ecotoxicity in water
EF(gw)	Equivalency factor for global warming potential
EF(ht)	Equivalency factor for potential for human toxicity
EF(hta)	Equivalency factor for potential for human toxicity on exposure via air
EF(htg)	Equivalency factor for potential for human toxicity on exposure via groundwater
EF(hts)	Equivalency factor for potential for human toxicity on exposure via soil
EF(htw)	Equivalency factor for potential for human toxicity on exposure via surface water
$EF(j)_i$	Equivalency factor for potential contribution from substance (i) to environmental impact category (j)
EF(N)	Equivalency factor for potential for nutrient enrichment (N only)
EF(ne)	Equivalency factor for potential for nutrient enrichment (N and P together)
EF(od)	Equivalency factor for ozone depletion potential
EF(P)	Equivalency factor for potential for nutrient enrichment (P only)
EF(po)	Equivalency factor for photochemical ozone formation potential
EP(ac)	Environmental impact potential for acidification
EP(et)	Environmental impact potential for ecotoxicity
EP(etp)	Environmental impact potential for ecotoxicity in waste water treatment plant
EP(etsc)	Environmental impact potential for chronic ecotoxicity in soil
EP(etwa)	Environmental impact potential for acute ecotoxicity in water
EP(etwc)	Environmental impact potential for chronic ecotoxicity in water
EP(gw)	Environmental impact potential for global warming
EP(ht)	Environmental impact potential for human toxicity
EP(hta)	Environmental impact potential for human toxicity on exposure via air
EP(htg)	Environmental impact potential for human toxicity on exposure via groundwater

EP(hts)	Environmental impact potential for human toxicity on exposure via soil
EP(htw)	Environmental impact potential for human toxicity on exposure via water
EP(j)	Environmental impact potential for impact category (j)
$EP(j)_i$	Environmental impact potential for impact category (j) from substance (i)
EP(N)	Environmental impact potential for nutrient enrichment (N only)
EP(ne)	Environmental impact potential for nutrient enrichment (N and P together)
EP(od)	Environmental impact potential for ozone depletion
EP(P)	Environmental impact potential for nutrient enrichment (P only)
EP(po)	Environmental impact potential for photochemical ozone formation
EP(pt)	Environmental impact potential for persistent toxicity
$ER(j)_{90}$	Normalization reference for environmental impact category (j) based on emissions in 1990
$ER(j)_{T2000}$	Society's target contribution in the year 2000 for environmental impact category (j) based on political reduction targets
ETF_p	Ecotoxicity factor for effects in waste water treatment plant
ETF_{sc}	Ecotoxicity factor for chronic effects in soil
ETF_{wa}	Ecotoxicity factor for acute effects in water
ETF_{wc}	Ecotoxicity factor for chronic effects in water
f(loss)	Loss of material grade
f(new)	Material grade of new material
f(scrap)	Material grade of scrap
f_a	That portion of the emission which contributes to toxicity via air
f_g	That portion of the emission which contributes to toxicity via groundwater
f_s	That portion of the emission which contributes to toxicity via soil
f_{sc}	That portion of the emission which contributes to chronic toxicity in soil
f_w	That portion of the emission which contributes to toxicity via surface water
f_{wa}	That portion of the emission which contributes to acute ecotoxicity in water
f_{wc}	That portion of the emission which contributes to chronic ecotoxicity in water
H	Henry's law constant
HRC	Human reference concentration
HRD	Human reference dose
HTF_a	Human toxicity factor for exposure via air
HTF_g	Human toxicity factor for exposure via groundwater
HTF_s	Human toxicity factor for exposure via soil
HTF_w	Human toxicity factor for exposure via surface water
I_a	Intake factor for air
I_g	Intake factor for groundwater
I_s	Intake factor for soil
$I_{s,b}$	Intake factor for soil via meat

$I_{s,d}$	Intake factor for soil on direct ingestion
$I_{s,m}$	Intake factor for soil via milk
$I_{s,p}$	Intake factor for soil via plants
I_w	Intake factor for water
K_d	Coefficient of adsorption in soil
L	Life span
NEP(j)	Normalized potential for environmental impact category (j)
NP(j)	Normalized impact potential or resource consumption (j)
NR(j)	Normalized resource consumption (j)
NWP(j)	Normalized potential for working environment impact category (j)
P(j)	Impact potential or resource consumption (j)
PNEC	Predicted no-effect concentration, the highest concentration assessed not to produce ecotoxic effects in the environment
$PNEC_{sc}$	The highest concentration assessed not to produce chronic ecotoxic effects in the soil compartment
$PNEC_{wa}$	The highest concentration assessed not to produce acute ecotoxic effects in the water compartment
$PNEC_{wc}$	The highest concentration assessed not to produce chronic ecotoxic effects in the water compartment
P_{ow}	Octanol-water partitioning coefficient
Product $\overline{NN}$	Average of products which draw on the material pool
Q_i	Environmental exchange of substance (i)
$Q_{i,p}$	Exchange of substance (i) from process (p)
$Q_{i,up}$	Environmental exchange of substance (i) from the use process and its non-terminal exchanges
Q_p	Quantity emitted to waste water treatment plant
R(j)	Normalization reference for impact potential or resource consumption for one year
R(j)	Consumption of resource (j)
$RR(j)_{90}$	Normalization reference for resource consumption (j) based on consumption in 1990
SCF	Stem concentration factor
$SF(j)_p$	Site factor for process (p) for use in computation of environmental impact potential for environmental impact category (j)
T	Duration of service as defined in the functional unit
T_a	Transfer factor from air
T_g	Transfer factor from groundwater
T_s	Transfer factor from soil
$T_{s,b}$	Transfer factor from soil via meat
$T_{s,d}$	Transfer factor from soil on direct intake
$T_{s,m}$	Transfer factor from soil via milk
$T_{s,p}$	Transfer factor from soil via plants
T_w	Transfer factor from water
WEP(j)	Weighted potential for environmental impact category (j)
WF(j)	Weighting factor for impact potential or resource consumption (j)
WP(j)	Weighted impact potential or resource consumption (j)
WP(j)	Potential for contribution to impact (j) on the working environment

WR(j)	Weighted consumption of resource (j)
$WR(j)_{90}$	Normalization reference for working environment impact category (j) for the years circa 1990
WWP(j)	Weighted potential for working environment impact category (j)
μ	Mean
σ	Standard deviation
σ/μ	Coefficient of variation (standard deviation divided by mean)

Abbreviations

ABS	Acrylonitrile-butadiene-styrene copolymer
CEN	Comité Europeén de Normalisation
CFC	Chlorofluorocarbon
D	Statistical distribution
EAA	European Aluminium Association
EDIP	Environmental Design of Industrial Products, acronym for the LCA method presented in this book
GWP	Global warming potential
HCFC	Hydrochlorofluorocarbon
HDPE	High density polyethylene
HFC	Hydrofluorocarbon
HIPS	High impact polystyrene
IPCC	Intergovernmental Panel on Climate Change
ISO	International Organization for Standardization
LC_{50}	Lethal concentration 50%. Concentration lethal to 50% of test organisms
LCA	Life cycle assessment
LC_{Lo}	Lethal concentration low. The lowest concentration which causes death among test organisms
LD_{50}	Lethal dose 50%. Dose lethal to 50% of test organisms
LDPE	Low density polyethylene
LLDPE	Linear low density polyethylene
LOAEC	Lowest observed adverse effect concentration. The lowest concentration observed to result in injurious effects in test organisms
LOAEL	Lowest observed adverse effect level. The lowest dose observed to result in injurious effects in test organisms
LOEC	Lowest observed effect concentration. The lowest concentration observed to result in effects in test organisms
logN	Log(arithmic) normal distribution
NMHC	Non-methane hydrocarbons. Hydrocarbons except methane
NOAEC	No observed adverse effect concentration. The highest concentration observed to result in no injurious effects in test organisms
NOAEL	No observed adverse effect level. The highest dose observed to result in no injurious effects in test organisms
NOEC	No observed effect concentration. The highest concentration observed to result in no effects in test organisms
ODP	Ozone depletion potential
OECD	Organization for Economic Co-operation and Development
PCBs	Polychlorinated biphenyls
POCP	Photochemical ozone creation potential
PP	Polypropylene
PS	Polystyrene
PU	Polyurethane
PVC	Polyvinyl chloride
REPA	Resource and environmental profile analyses
SETAC	Society of Environmental Toxicology and Chemistry

TD_{Lo}	Toxic dose low. The lowest dose which results in injurious effects in test organisms
UNCED	United Nations Conference on Environment and Development
UNECE	United Nations Economic Commission for Europe
VOC	Volatile organic compound

Units

C_2H_4-eq	Ethylene equivalents. Unit for potential contribution to photochemical ozone formation
CFC11-eq	Unit for potential contribution to stratospheric ozone depletion
CO_2-eq	Carbon dioxide equivalents. Unit for potential contribution to global warming
dB(A)	Decibels: unit of sound pressure ('A' weighting)
m^3 air	Unit for potential contribution to human toxicity via inhalation
m^3 soil	Unit for potential contribution to chronic ecotoxicity in soil or human toxicity via soil
m^3 water	Unit for potential contribution to acute ecotoxicity in water or waste water treatment plant, chronic ecotoxicity in water or human toxicity via water
N-eq	Nitrogen equivalents. Unit for potential contribution to nutrient enrichment
NO_3^--eq	Nitrate equivalents. Unit for potential contribution to nutrient enrichment
P-eq	Phosphorus equivalents. Unit for potential contribution to nutrient enrichment
$PET_{WDK2000}$	Target person-equivalents. Unit for weighted environmental impact potentials expressed via weighting relative to society's target impacts for the year 2000. Given as PET_{W2000} or PET_{DK2000} depending on whether the weighting factors are based on global (W) or Danish (DK) emission targets in the year 2000
PE_{WDK90}	Person-equivalent computed as the average resource consumption or impact potential per person in the world (W) or in Denmark (DK) in 1990. Unit for normalized resource consumption or normalized potentials for environmental impacts or impacts on the working environment
ppbv	Parts per billion by volume. Unit for concentration in air
PR_{W90}	Person-reserve. Unit for weighted consumption of resources which is expressed via weighting as the fractions of the person-reserve computed in 1990. Given as PR_{W90} as the reserve is computed for the world (W) in 1990
RWI_{DK90}	Expected number of reported work injuries based on figures for Denmark in the years circa 1990. Unit for weighted impact potentials on the working environment
SO_2-eq	Sulphur dioxide equivalents. Unit for potential contribution to acidification

Index

A

B

C

D

E

F

G

H

I

J

K

L

M

N

T

U

V

W